Bioreactors
Analysis and Design

Bioreactors
Analysis and Design

TAPOBRATA PANDA

Professor of Biochemical Engineering,
Indian Institute of Technology Madras,
Chennai

Tata McGraw Hill Education Private Limited
NEW DELHI

McGraw-Hill Offices

New Delhi New York St Louis San Francisco Auckland Bogotá Caracas
Kuala Lumpur Lisbon London Madrid Mexico City Milan Montreal
San Juan Santiago Singapore Sydney Tokyo Toronto

Tata McGraw Hill

Published by Tata McGraw Hill Education Private Limited,
7 West Patel Nagar, New Delhi 110 008.

ISBN (13 digits): **978-1-25-900572-5**
ISBN (10 digits): **1-25-900572-0**

Vice President & Managing Director—Asia Pacific: *Ajay Shukla*
Assistant Sponsoring Editor—Science, Technology and Computing: *Simanta Borah*
Manager—Production: *Sohan Gaur*
Asst. General Manager—Sales and Business Development: *S Girish*
Deputy Marketing Manager—Science, Technology and Computing: *Rekha Dhyani*
General Manager—Production: *Rajender P Ghansela*
Asst. General Manager—Production: *B L Dogra*

Typeset at Tej Composers, WZ 391, Madipur, New Delhi 110 063 and printed at Gopsons Papers Ltd., A1 & 2, Sector 60, Noida, U.P. 201 301

Cover Printer: Gopsons

Cover Designer: Kapil Gupta

RXYYCRQGLLRZC

The McGraw·Hill Companies

To

My Parents

Preface

With the recent developments in all spheres of biology, a good knowledge about the bioreactor and its operation has become inevitable for all professionals and students working in this area. Bioreactor operation, analysis, and design are complex phenomena of mass, momentum, and heat transfer aspects in biological system. A comprehensive book with the scope of solving important practical problems in biological processes is necessary. This book is aimed at addressing such difficulties faced by the readers.

The book begins with the introduction of preliminary concepts required for a bioreactor. An overview of biological reactions, elements of bioreactor design, and fundamentals of mass and energy balances in biological reactions are explained here. Chapter 2 considers definition of bioreactor, its development in submerged liquid reactions and in solid-state reactions, essential components of bioreactors, systematic classification of bioreactors with emphasis on advantages and disadvantages. Chapter 3 deals with the practical operations of bioreactors. Generalized concepts in the areas of reactions involving microbial cells, free enzyme systems, animal cells, plant cells, and waste treatment are discussed here. Exhaustive knowledge in the operations is required to carry out experiments in bioreactors. Homogeneous and heterogeneous reactions are discussed here in depth. In this chapter, the reader can find basic information on practical handling of bioreactors, their accessories, and the best reactor one can look for the problem at hand. In Chapter 4, basic reactors are analyzed with proper design equations. To get a better understanding of these equations, relevant and fundamental problems are solved. Thus, one can find important working theory for batch, continuous-flow, and semi-continuous bioreactors. The deviation from ideality is emphasized for all three basic bioreactors. In Chapter 5, non-ideal behavior of bioreactors is dealt with in such a way that the reader can glide on the basic concepts of bioreaction engineering. Stability analysis, idea of phase-plane behavior, and the application of bifurcation analysis are explained for continuous flow bioreactor. Various models, given in Chapter 6, have been illustrated so that the reader can visualize the behavior of real reactors. Classical phenomena of inhibition, wall growth behavior, multiple culture interaction, etc. are discussed in terms of mathematical models with physical interpretation.

Reactions in bioreactors are influenced by energy and mass transfer operations. Ultimately, the performance of a reactor is associated with these phenomena. Chapter 7 discusses those concepts. Important parameters are highlighted with appropriate numerical problems. The transfer operations need to be controlled through measured variables, which are discussed with basic control theory in Chapter 8. As there is diversity in biological reactions, bioreactor design, operations and problems are noted specifically. Chapter 9 discusses a few important reactors, viz., airlift reactor, reactors for animal cell and for plant cell processes. Mathematical models are presented in this circumstance. CFD in bioreactors are considered in Chapter 10 with typical numerical problems. For general theory, practice problems of bioreactors are discussed with reference to basic bioreactors.

For small scale operations, one can find a large number of successful configurations. The scenario changes if the large scale process is considered with the reactor configuration used in small scale. A few scientific approaches that are necessary to comprehend in this context are discussed in Chapter 11. As mentioned in Chapter 1, bioreactor design is a combination of biochemical and mechanical aspects. This necessitates the devotion of Chapter 12 for the mechanical design consideration for the bioreactors.

The book is tagged with related "Appendices" which will take care of extra discussions that are not directly relevant to all chapters. However, it will help the readers to conceive new ideas. Tools for easy calculations are provided to curious readers as further development and expansion.

I am tempted to mention here that this book result from my 22 years of teaching and research with my students at the IIT Madras and useful interaction with the students in other institutions, viz., Madurai Kamaraj University, India; Anna University, Chennai, India; Asian Institute of Technology, Bangkok; and the Department of Biotechnology, IIT Kharagpur, India.

Finally, suggestions, comments, constructive criticisms about the book are always welcome. I feel that I could not mention other relevant literature in the references due to space limitation.

Tapobrata Panda

Acknowledgements

First, I would like to thank my mentors and teachers Prof. Tarun K Ghose, former Professor of IIT Delhi; Prof. Saroj K. Majumdar, former Professor of Jadavpur University, Kolkata; Professors S.N. Mukhopadhyay, V.S. Bisaria, Vikram Sahai, Barun K Guha, Subhash Chand of IIT Delhi; Prof. K. B. Ramachandran, former Professor of IIT Delhi and currently at IIT Madras, India; Professor Purnendu Ghosh, former Professor of IIT Delhi and currently Director, Birla Institute of Science and Technology, Jaipur; Prof. Christian P. Kubicek of Techinical University of Vienna, Austria; and Professor Peter J. Reilly of Iowa State University, Ames, Iowa, USA.

At the outset, this book was prepared by me and Dr. Sanjoy K. Ghosh, formerly with Anna University, Chennai, India, and currently with the Department of Biotechnology, IIT Roorkee, India. However, Dr. Ghosh was unable to contribute to this book due to various other important assignments.

My students, Mrs. Praseetha P. Nair, Lecturer, Department of Chemical Engineering, Government Engineering College, Thrissur, Kerala; Miss. Naga Pallavi Chakka and Miss. K. Deepa went through the manuscript in detail and came up with critical suggestions and necessary alterations. I acknowledge the contributions of my former students, Drs. P. Arthur Felse, Thomas Théodore, Kandula Jagannadha Rao, Veeranki Venakata Dasu (IIT Guwahati), Prof. B. S. Gowrishankar, Head of the Department of Biotechnology, Siddaganga Institute of Technology, Tumkur, and Dr. G. N. Rameshaiah, Assistant Professor at BMS College of Engineering, Bangalore (for photographs), Mr. Abhisekh Neemuri (currently doing MBA at IIM Kozhikode); Mr. K. Sarathbabu for line drawing; and my present students, Mr. A. Seenivasan (for line drawing and collection of references), Miss. G. Saraswathi and Mr. Subin Poulose, Lecturer, Department of Chemical Engineering, Government Engineering College, Trichur, Kerala, for manuscript preparation and Mr. Siddharatha Singha for critical observation. Valuable comments and suggestions were also received from Mr. M. Sudhakar.

May heartfelt thanks to Mr. A. Pandian for excellent line drawing of all figures. Masterly help from Mr. S. Venkatesan and Mr. S. Ravikumar for manuscript formatting are sincerely acknowledged here.

Thanks to Prof. K. Krishnaiah, Professor of Chemical Engineering and Dean (Academic Research) IIT Madras, for permitting me to reproduce a figure, which is part of his lecture notes.

Thanks to M/s. Bioengineering AG, Switzerland, for giving me the permission to reproduce two important figures in this book.

Finally, thanks are due to the Tata McGraw Hill family, especially to Mr. Ramachandruni Chandra Sekhar, Ms. Sindhu Ullas, Mr. Simanta Borah, Ms. Nimisha Goswami and Mr. Vibhor Kataria for their patience and guidance since 2005.

My inspiration for this work is my wife Meeta and daughters Dr. Smriti and Shradha. Overall development for the success comes from my parents, brothers, and sisters.

Tapobrata Panda

Contents

CHAPTER 1

Introduction

OBJECTIVE

- Preliminary concept required for a bioreactor and its design are highlighted to review the knowledge gained earlier which are essential for the understanding of the following chapters in the book.

1.1 OVERVIEW OF BIOLOGICAL REACTIONS

Bioreactors are designed with the aim to nurture the knowledge in bioprocesses. Design of a bioreactor, understandably involves various critical parameters. The size and shape of a bioreactor differ to a great extent depending on the various applications in bioprocesses. While at its infancy, bioreactor design was based on "manufacturing art" and experiences, it has undergone many changes ever since. The processes demands critical design, development, and analysis of bioreactors.

Although to understand its concept, some analogy with chemical reaction system may be helpful, one should be extremely careful about the fact that a bioreactor is not simply the extension of a chemical reactor. A bioreactor is also sometimes loosely called a fermenter, irrespective of the type of biological reaction taking place in it. In this book, "bioreactor" or "reactor" has been interchangeably used throughout.

The general definition of bioreactors covers bio-reactions involving cells, enzymes, and specific components of cell as catalyst, in submerged (i.e., components and catalysts present in liquid phase) as well as in solid state (i.e., components and catalysts present in solid phase predominantly) culture processes. A glimpse of classification of bio-reactions is given in Figure 1.1.

1.1.1 Submerged Liquid Fermentation (SLF)

Submerged liquid fermentations include reactions involving microorganisms, animal cells, and plant cells. Brief information about cultivations of animal cells and plant cells is described here.

(a) Cultivation of Animal Cells

The cultivation of mammalian cells, cells of cold-blooded animals, and insect cells is important due to increased demand for industrial production of high-value products, viz., vaccines, interferons, hormones, immunological agents, mAb from hybridoma, clotting factors such as plasminogen and plasminogen activator.

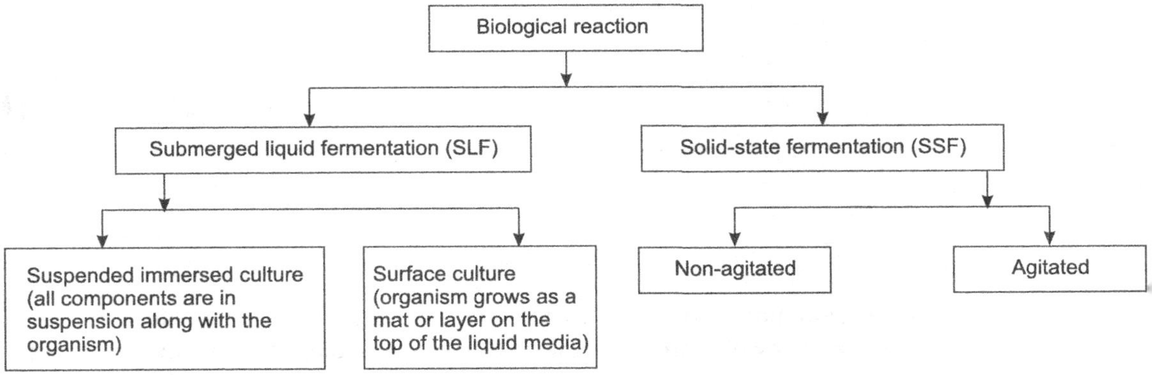

Figure 1.1 Brief classification of biological processes.

Improvement in animal cell culture is done by:

- cell fusion (hybridoma)
- chromosome transfer
- engineered plasmid infections

Animal cells or its components are used in reactions (Fig. 1.2).

(b) Cultivation of Plant Cells

Culture of plant cells produces steroids, cardiac glycosides, anti-tumor alkaloids, insecticides, food flavoring and coloring agents, perfumes, and emulsifiers. Plant cell cultivation is less expensive than the methods adopted for mammalian cell culture. Plant cells are not anchorage dependent. In many cases, suspension cultures have been found to give similar or even higher yields of products, compared to the parent plant.

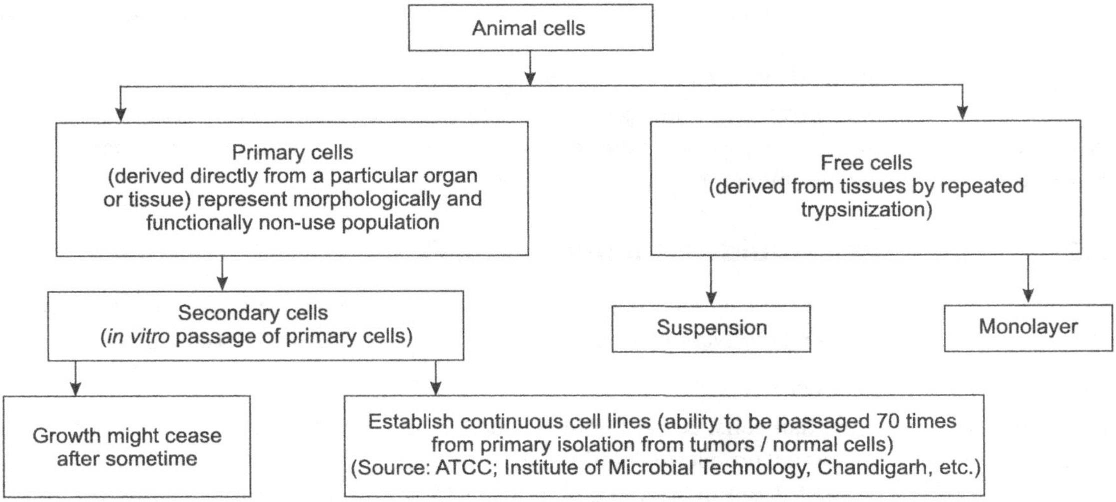

Figure 1.2 Type of animal cells used in bioreactor.

1.1.2 Solid-State Fermentation (SSF)

We have discussed about the use of animal cells and plant cells in SLF. We need another cultivation process particularly for microorganisms and plant cells in SSF. It is important to note here that a great variety of products are obtained from solid-state fermentation (Table 1.1).

Table 1.1 *Range of products from SSF*

Product	*Organism*	*Reference*
Antibiotics (e.g., Tetracycline, peptide antibiotics)	*Streptomyces iridifaciensis* *Bacillus subtilis*	Mial, 1975; Ohno *et al.*, 1992
Aroma compounds	*Bjerkandera adusta*	Lapadatescu and Bonnarme, 1999
Bio-insecticide	*Coniothyrium minitans*	Weber *et al.*, 1999
Bio-filter	Number of organisms	Wu *et al.*, 1998
Bio-pulp	*Pleurotus* sp.	Camareroe *et al.*, 1996
Enzymes (eg., lipase, glucoamylase, pectinase, lignolytic enzymes, etc.)	*Candida rugosa* *Trichoderma reesei* *Aspergillus niger* *Phanerochaete chrysosporium*	Benjamin & Pandey, 1997; Berovic & Ostroversenik, 1997; Gutierrez-Correa&Tengardy, 1999; Rodriguz *et al.*, 1999
Organic acids (eg. Citric acid, amino acids, kojic acid, etc.)	*Aspergillus niger* *Aspergillus oryzae* *Rhizopus oligosporus*	Mial,1975; Lonsane *et al.*, 1992
Polymers (eg. Succinoglycans, xanthan gums)	*Rhizobium hedysarcis* *Xanthomonas campestris*	Streadansky & Conti, 1999a & b
Small organic molecules (e.g., ethanol, methane, cephamycin c, etc.)	*Saccharomyces cerevisiae* Organisms from natural source *Streptomyces clavuligerus*	Kota and Sridhar, 1999; Sree *et al.*, 1999
Pigments	*Monaseus* sp.	Carvalho *et al.*, 2001
Mushrooms	*L. edodes*	Leifa *et al.*, 2000
Gibberellic acid	*Gibberella fujikuroi*	Soccol and Vanderberghe, 2003
Plant tissue culture		Soccol and Vanderberghe, 2003

A few important reports on SSF indicate the future applications, advantages, and reactor concept. Important factors influencing SSF processes are indicated in the literature (Banerjee and Bhattacharyya, 2003; Durand, 2003; Manpreet *et al.*, 2005; Mitchell *et al.*, 2003; Pandey, 2003; Raghavendrarao *et al.*, 2003; Shrikumar, 2003; Soccol and Vanderberghe, 2003; Viniegra-González *et al.*, 2003).

(a) Parameters that Influence SSF

SSF processes are influenced by the following parameters, some of which are different from that of SLF:

- **Water activity:** 0.95 – 0.98 (Mitchell *et al.*, 2000).

- **Organisms:** Generally, fungal systems find better application.
- **Aseptic practice:** Rotary cooker is used for sterilization of solid materials.
- **Inoculum preparation and use:** Spores or mycelia are developed from liquid culture.
- **Degree of mixing:** Low shear causing device is preferred in SSF.
- **Heat:** Control of heat inside the solid reactant is a typical problem in SSF as it is a poor conductor of heat.

(b) Advantages and Disadvantages of SSF

Advantages

- Low water activity provides less chances of contamination.
- High reactant concentration is used and hence, higher product concentration is expected in SSF.
- Supply of air requires less power.
- Liquid waste is less in this case.

Disadvantages

- Range of products is limited.
- Mixing within particles is difficult in SSF.
- Control of metabolic heat removal is difficult in SSF.
- Samples to measure kinetic parameters are not representative. Hence, growth kinetics and transport phenomena are poorly characterized in SSF.
- Ideal mode of operation of reactors is impossible.

SLF and SSF reactions are carried out in a vessel called bioreactor. In any biochemical process, economics of any process strongly depend on the bioreactor's design.

1.2 ELEMENTS IN BIOREACTOR DESIGN

The components in the design of a bioreactor can be categorized in a very simple fashion (Fig. 1.3).

From Figure 1.3 one can notice that the two important information for the design of a bioreactor are:

1. rate expression (with transport process effects included), and
2. the type of reactor.

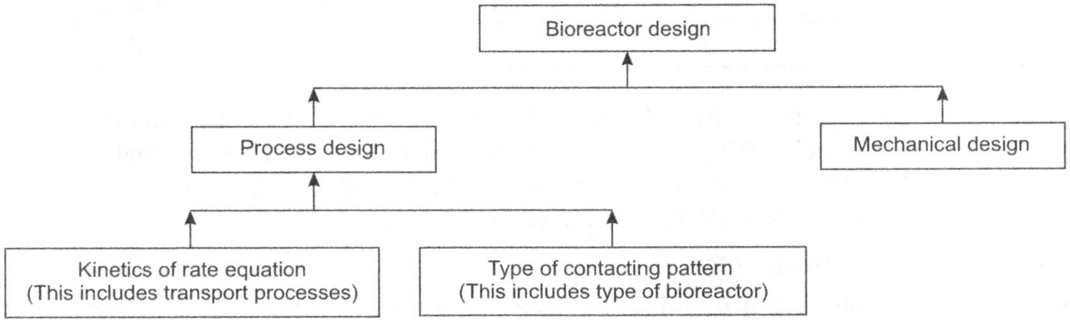

Figure 1.3 Bioreactor design chart.

Rate expression (with transport effects) will be discussed in detail from Chapter 4 onwards. However, for the convenience of smooth transition from chemical processes to biochemical processes, basic information is dealt in this chapter. The type of reactor is discussed in detail in Chapter 2.

1.3 RATE EXPRESSION IN BIOLOGICAL SYSTEMS

The rate equations of chemical reactions cannot be simply extended to biological reactions. In this context, let us look into the differences between chemical and biological reactions (Table 1.2).

Table 1.2 *Differences between chemical reactions and biological reactions*

S. No.	*Chemical reactions*	*Biological reactions*
1.	They are classified into homogeneous and heterogeneous reactions.	Except enzymatic reactions using soluble reactants, other reactions are heterogeneous.
2.	They may be catalytic or non-catalytic.	All of them are catalytic reactions.
3.	In catalytic reactions, catalysts are not so specific as in biological reactions.	All catalysts are highly specific.
4.	Broader ranges of pressure and temperature can be applied.	Very narrow range of pressure and temperature are employed (viz., in general, maximum temperature of reaction is 70°C).
5.	Except multiple reactions, reactions are elementary, i.e., rate equations correspond to a stoichiometric equation.	It is difficult to get a clear stoichiometric analysis in biological reactions.
6.	Catalysts, if they are inactivated, can be recovered by physical treatments.	If catalysts are inactivated (cells, enzymes or antibodies) they cannot be recovered in their original form.
7.	Reactions are, in general, simple.	Reactions are complex.
8.	Products can be defined and easily separated in chemical reactions.	Products (except for a few enzymatic reactions) are produced in a complex mixture. Separation and purification costs are higher.
9.	We can predict rates of reactions very easily.	So far it is unpredictable.
10.	A few autocatalytic processes are available. However, a catalyst does not multiply during reactions.	They are autocatalytic reactions and behavior of catalyst changes during reaction, i.e., activity and growth of catalyst change during reactions except in enzymatic reactions.

Biological reactions are mediated by:

- Enzymes,
- Cells, and
- Antibodies.

Enzymes and antibodies are derived from cells. Cells can be considered as large source of these catalysts. So, the reaction kinetics for enzymes and cells will not be the same. There is some analogy in both of these catalysts. One can find first order, pseudo order, and zero order, reactions in both enzymes and cellular reactions.

Typical example is considered here. Enzymatic reactions are generally described by Henri–Michaelis–Menten equation. Of course, this is an oversimplified scheme.

$$\underset{\text{(Enzyme) + (reactant)}}{E+S} \underset{k_2}{\overset{k_1}{\rightleftarrows}} \underset{\text{(Enzyme-reactant complex)}}{ES} \xrightarrow{k_p} \underset{\text{(enzyme) + (product)}}{E+P} \quad (1.1)$$

for which the rate is

$$v = \frac{v_m C_s}{K_m + C_s} \quad (1.2)$$

where

v = instantaneous rate of reaction or initial velocity, (wt) / ((vol)(time))
v_m = maximum rate of reaction, (wt) / ((vol)(time))
C_s = Reactant concentration, (wt/vol)
K_m = Henri–Michaelis–Menten constant, (wt/vol)

On the other hand, cellular reactions are considered in a similar fashion by Monod's equation.

$$\underset{\text{(Cell) + (reactant)}}{X+S} \longrightarrow \underset{\text{(more cell) + (products)}}{nX+mP} \quad (1.3)$$

$$\mu = \frac{\mu_m C_s}{K_s + C_s} \quad (1.4)$$

where

μ = specific growth rate = $(1/C_x)(dC_x/dt)$, (time^{-1})
μ_m = maximum specific growth rate, (time^{-1})
C_s = reactant concentration (wt/vol)
C_x = cell concentration (wt/vol)
K_S = Monod's constant (wt/vol)

Rate of enzymatic reactions can be determined from basic calculation. However, rate of cellular reactions cannot be determined easily. We shall treat them here separately. Even in the design of reactors for different applications (using cells or enzymes), we need to stress on different kinetic expressions.

This will be emphasized from Chapter 4 onwards. Here, we try to give some of the kinetic expressions which will be dealt in the later part of this book.

1.3.1 Enzymatic Reactions

Enzymes are efficient catalysts which are often superior to chemical catalysts. They have a number of distinct advantages over conventional chemical catalysts, viz.,

- Specificity and selectivity not only for particular reactions but in the discrimination between similar parts of molecules (region-specificity) or between optical isomers (region-specificity).

- Catalyze only the reactions of very narrow range of reactants, i.e., the chosen reaction can be catalyzed to the exclusion of side-reactions, eliminating undesirable products.
- Product is generated in an uncontaminated state.
- Often, a lesser number of steps may be required to produce the desired end-product.
- Nearly mild processing conditions of temperature, pH, and pressure.
- High reaction velocities and straight forward catalytic regulations allow an increase in productivity.

Some disadvantages of enzymes are:

- High cost of enzyme isolation and purification
- Unstable nature (mostly for intracellular enzymes).

It is important that one generates needful experimental data to get quantitative idea of enzymatic reactions. This is necessary to get an idea about the importance of the process. Generally, one calculates initial velocity at various concentrations of reactants. Initial velocity data is generated by:

- *Continuous method:* Either reactant or product of the reaction possesses unique physical property which can be measured in real time under the conditions of reaction.
- *Discontinuous method:* Neither the reactant nor the product can be selectively measured under the conditions of the reactions. The reaction in this case is stopped and measurement is carried out under different conditions.
- *Coupled methods*: To avoid discontinuous assay method, a product of the reaction of interest can be the reactant of a second enzyme whose reaction rate can be measured continuously.

In all of these methods, reaction conditions which affect reaction rates (temperature, pH, ionic strength) must be maintained constant.

To calculate initial velocity from data, one can follow the different procedures, viz.,

(a) In continuous measurement methods, the initial velocity can be determined from the graph by the construction of the tangent and calculation of the slope.

or

From meaningful data one can fit 2nd or 3rd order polynomial. The coefficient of the first-order term is the initial velocity.

(b) In discontinuous methods, a single point measurement looks up for a proper approximation.

Initial velocity data should be verified by testing it for linearity by using a fit of a polynomial equation or by the integrated Michaelis-Menten equation.

In all of these cases, the determination of goodness of fit to mathematical models, including the Henri-Michaelis-Menten model may be done in the following way:

$$\text{The least square error function} = \Sigma\,(V_{\text{experimental}} - V_{\text{calculated}})^2$$

One needs to derive mathematical models from the chemical models for the mechanism of the reaction. The derivation of Henri-Michaelis-Menten equation is discussed in all works in enzyme kinetics and biochemistry.

1.3.2 Cellular Reactions

Microbial kinetics is concerned with the mathematical description of metabolic processes of microorganisms which are contained in a controlled reaction vessel. Three broad classes of mathematical models are used to describe microbial growth and product formation kinetics, viz., unstructured models, structured models, and cybernetic models. In unstructured models, all cellular components are pooled into a single cellular component represented by the total cell concentration. The unstructured model indicates the most fundamental observations concerning microbial growth process (Nielsen and Villadsen, 1994): (i) the rate of cell production is proportional to the biomass present (ii) there is a saturation limit for microbial growth on each reactant, and (iii) cells need reactant and synthesize the product even when they do not grow. In major part of batch fermentation, the reactant concentration is usually high compared to saturation constants and hence a "balanced growth" situation prevails. In such cases, unstructured models are adequate. Control and optimization of fermentation process become easy with unstructured models as they have little mathematical complexity.

The Monod's model is the most simple and fundamental model proposed to explain microbial growth Equation (1.4). The Monod's model is analogous to the Michaelis-Menten kinetics (Equation 1.2) used to explain enzymatic reactions. This model follows a black box approach.

The Monod's equation has been shown to describe fermentation data of various organisms. The major drawback of the Monod's model probably is that it predicts a constant growth rate at high reactant concentration, while this is not observed in many fermentation. The Monod's equation assumes one limiting reactant and hence cannot explain growth on multiple reactants and diauxic growth. Hence, many empirical modifications have been proposed to the Monod's model.

(a) Modification of the Monod's Equation

To account for situation of reactant, product, and cell inhibition, many improvements of the Monod's model have been attempted by various researchers (Luong, 1987; Edwards, 1970). The improvement is not properly explained in the literature and proposed to fit a limited set of experimental data. A list of unstructured models reported in the literature based on Monod's kinetics is given in Table 1.3 (Felse, 1999).

(b) Product Formation Kinetics

Product formation kinetics becomes important as it will be involved in process design and further scale-up of the process. The type of kinetic expression used to describe product formation is similar to those

Table 1.3 *A few models for cell growth kinetics*

Name	*Kinetic model*	*Used to explain*
Haldane	$\mu = \dfrac{\mu_m C_s}{K_s + C_s + \dfrac{C_s^2}{K_I}}$	Substrate inhibition (Edwards, 1970)
Andrews	$\mu = \dfrac{\mu_m C_s}{(K_s + C_s)\left(1 + \dfrac{C_s}{K_I}\right)}$	Substrate inhibition (Andrews, 1968)

Contd.

Table 1.3 *Contd.*

Name	*Kinetic model*	*Used to explain*
Wayman - Tseng	$\mu = \frac{\mu_m C_s}{K_s + C_s}$, when $C_s < C_s^*$ $\mu = \frac{\mu_m C_s}{K_s + C_s} - i(C_s - C_s^*)$, when $C_s > C_s^*$	Substrate inhibition (Wayman and Tseng, 1968)
Moser	$\mu = \frac{\mu_m C_s^n}{K_s + C_s^n}$	Substrate inhibition (Moser, 1988)
Tessier	$\mu = \mu_m(1 - e^{(-C_s/K_I)})$	Substrate inhibition (Nielsen and Villadsen, 1994)
Contois	$\mu = \frac{\mu_m C_s}{C_x K_s + C_s}$	Substrate inhibition (Nielsen and Villadsen, 1994)
Blackmann	$\mu = \frac{\mu_m C_s}{2K_s}$, when $C_s \le 2K_s$ $\mu = \mu_m$, when $C_s \ge 2K_s$	Substrate inhibition (Nielsen and Villadsen, 1994)
Logistic law	$\mu = \mu_m\left(1 - \frac{C_x}{K_I}\right)$	Cell inhibition (Nielsen and Villadsen, 1994)
Aiba	$\mu = \frac{\mu_m C_s}{K_s + C_s} e^{(-C_s/K_I)}$	Substrate inhibition (Aiba *et al.*, 1968)
Mason and Millis	$\mu = \frac{\mu_{asym} C_s}{K_s + C_s} + r_s C_s^n$	Substrate inhibition (Mason and Millis, 1976)
Powell	$\mu = \frac{\mu_m C_s}{(K_s + K_D) + C_s}$	Substrate inhibition (Powell, 1967)
Luong	$\mu = \frac{\mu_m C_s}{K_s + C_s}\left[1 - \frac{C_s}{C_s^*}\right]^n$	Substrate inhibition (Luong, 1985)
Han and Levenspiel	$r_c = k\left[\prod_{i=1}^{i}\left(1 - \frac{C_{1i}}{C_{1i}^*}\right)^{n_i}\right] \frac{C_c C_s}{C_s + C_M\left[\prod_{i=1}^{i}\left(1 - \frac{C_{1i}}{C_{1i}^*}\right)^{m_i}\right]}$	Substrate, cell and product inhibition (Han and Levenspiel, 1988)
Ghose and Tyagi	$\mu = \mu_m\left(1 - \frac{C_P}{C_{Pm}}\right)$	Product inhibition (Ghose and Tyagi, 1979)
Levenspiel	$\mu = \mu_m\left(1 - \frac{C_P}{C_{Pm}}\right)^n$	Product inhibition (Levenspiel, 1980)
Nielsen and Villadsen	$\mu = \frac{\mu_m C_s}{K_s + C_s} \cdot \frac{K_P}{K_P + (C_P/K_P)}$	Product inhibition (Nielsen and Villadsen, 1994)

used to describe cell growth. In most cases product formation is given as a function of cell growth. The number of published works on product formation is not as exhaustive as the works on cell growth, but a few models that explain most of the fermentation data are available. The most classical and widely used kinetic expression for product formation is the Luedeking and Piret (2000) model which has been proposed to explain the production of lactic acid by *Lactobacillus delbrueckii* (Table 1.4).

Table 1.4 *Models for microbial product formation*

Name	*Kinetic model*	*Used to explain*
Luedeking and Piret	$q_p = \alpha\mu + \lambda$	Lactic acid production (Luedeking and Piret, 2000)
Giona	$q_p = K_1 \cdot O_L + K_2$	Penicillin production (Giona *et al.*, 1976)
Terui	$q_p = q_{p\,\max} \cdot e^{-K2(t-t_{\max})} + \omega\,[e^{-K1(t-t_{\max})} - e^{-K2(t-t_{\max})}]$	Enzyme production (Terui, 1972)
Ryu and Humphry	$q_p = q_{p\,\max} \cdot e \dfrac{\varepsilon\mu/\mu_{\max}}{1+(\varepsilon-1)\mu/\mu_{\max}}$	Enzyme production (Ryu and Humphrey, 1972)

From the review on microbial kinetics, it is clear that a wide variety of models are available to explain cell growth as well as product formation. The most appropriate models that are applicable to the present system will be chosen and an attempt will be made to fit the experimental data. Apart from using the models available to fit the experimental data, new model(s) will be proposed for the system under consideration.

1.4 BASIC CONCEPT OF ENERGY TRANSFER

For bioreactor design, energy transfer, like mass transfer, in the reaction influences the design criteria. Roles of combined energy and mass transfer are useful in the design. Let us consider the basic aspect of energy transfer involved in bio-reaction. In biological reactions, heat generation is inevitable. Usually this is produced over some specific volume. Since it is diffusive, the heat production is considered to be uniform.

In general, there are two types of heat generation in biological processes.

- **Anabolism:** Specialized components are formed from a number of stored precursors in the cell.
- **Catabolism:** Utilizing the reactants, various components are synthesized in the cell.

The basic source is adenosine tri phosphate (ATP). When ATP is hydrolyzed to phosphate and adenosine di-phosphate (ADP), energy is released by the reaction. This depends on the efficiency of the specific biochemical reactions. The rate of heat generation is temperature dependent. The following equation is generally used to describe this effect mathematically,

$$q_T = q'_T Q^{(T-T_O)/10} \quad (1.5)$$

where Q = Vant' Hoff's coefficient

q_T = heat generated at temperature, T

However, at very high temperature, biological reactions are stopped due to irreversible change in biological catalysts.

1.4.1 Metabolic Energy

Many biochemical reactions release large amount of energy, viz., oxidation of glucose to CO_2 and H_2O or oxidation of H_2 to produce H_2O. If this energy is released at once, a little of the energy of reaction could be captured. Cells have mechanisms to release the energy of such reactions slowly. To capture the large amount of energy created by the reaction, cells first split H_2 into H^+ and e^-. Electrons are passed to the electron transport chain comprising of cytochromes. As the electrons pass from one cytochrome to the other, the reduction in energy level occurs. The energy lost in each step is captured in chemical form which is used later. ATP is such a carrier component of metabolic energy.

1.4.2 Factors Affecting Performance in Bioreactors

Following are the major factors which affect the performance in bioreactors:

- Agitation
- Shear effect
- Modes of heat transfer

1.4.3 Effect of Agitation

The stirrer or stirring device facilitates energy transfer to the aerobic biological reactions in order to achieve

- Mixing
- Better oxygen transfer.

Hydrodynamics of the fluid is described by Re(Reynolds number), Fr(Froude number), Np(power number), etc.

1.4.4 Effect of Shear

Fluids are either Newtonian or non-Newtonian in bioreactors. It is observed that energy transfer is better in turbulent condition than in laminar region. Brodkey and Hershey (1988) suggested a minimum fluid velocity of 3 m/sec.

1.4.5 Effect of Modes of Heat Transfer

Heat transfer can be achieved by

- External device: double jacket

  or
- Internal coil.

Rate of heat transfer is defined by the classical equation

$$q = UA(T_2 - T_1) \tag{1.6}$$

To calculate q, it is necessary to incorporate the roles of biological reaction, role of mixing, power of agitation, etc.

One can express

$$q_{\text{total}} = q_{\text{agitation}} + q_{\text{growth}} \tag{1.7}$$

where, $q_{\text{agitation}} = q$ due to agitation
$q_{\text{growth}} = q$ due to growth.

Since temperature difference is usually small, UA should be large. The heat transfer through baffles can also be considered in the bioreactor.

Bioreactor design is a complex engineering problem. Under optimal conditions, microorganisms or cells are able to perform their functions with better efficiency. The environmental conditions of a bioreactor, like gas (air, oxygen, nitrogen, carbon dioxide) flow rates, temperature, pH, dissolved oxygen level, agitation rate/ circulation rate of fluid need to be monitored and controlled accurately. Fouling can harm the overall sterility and efficiency of a bioreactor, especially the heat exchanger. To avoid fouling, bioreactor should be clean and smooth. The other factor is to maintain constant temperature in biological reaction. This is achieved by proper refrigeration and using cooling jacket or coil devices. Lydersen *et al.*, (1994) discussed this topic in detail.

1.5 BASIC CONCEPT OF MASS BALANCE

Mass balance is an application of conservation of mass for the analysis of a physical system. For the design of bioreactor, mass balance has basic importance in analysis which is influenced by the reaction kinetics, operation conditions, etc. The general form of this equation is

$$\frac{d(C_i V)}{dt} = V \cdot r(C_i, C_j) \tag{1.8}$$

where V = reactor volume
C_i = concentration of component i
C_j = concentration of component j which influences/controls component i.
$r(C_i, C_j)$ = volumetric rate of consumption of component i.

Detailed application is discussed in Chapter 4.

EXERCISES

1.1 Following are the reaction schemes:

(a) $\underset{\text{(reactant)}}{A} \longrightarrow \underset{\text{(Product)}}{B}$ (homogeneous elementary chemical reactions)

(b) $\underset{\text{(reactant)}}{A} \xrightarrow{\text{Enzyme}} \underset{\text{(Product)}}{B + \text{Enzyme}\ (E)}$ (Enzymatic reaction)

where A and B are soluble in aqueous phase.

(c) $\underset{\text{(reactant)}}{A} \xrightarrow{\text{cell}} \underset{\text{(Product)}}{B + \text{Cell}\ (Cx)}$ (cellular reaction)

Write the rate equation for the above cases.

1.2 In the above problem (1.1), which reaction scheme satisfies the conditions for autocatalytic reactions?

1.3 For enzymatic reactions do you need to protect the catalyst? How will you do that?

1.4 Give some industrial examples of enzymatic reactions and cellular reactions.

1.5 In the above problem (1.4), do you believe that similar process design considerations are required?

REFERENCES

Aiba S, Shoda M, Nagatani M (1968) Kinetics of product inhibition in alcohol fermentation, *Biotechnology and Bioengineering*, 10, 845-864.

Andrew, JF (1968) A mathematical model for the continuous culture of microorganism utilizing inhibitory substrate, *Biotechnology and Bioengineering*, 10, 707-723.

Banerjee R, Bhattacharyya BC (2003) Evolutionary operation as a tool of optimization for solid state fermentation, *Biochemical Engineering Journal*, **13**, 149-155.

Benjamin S, Pandey A (1997) Coconut cake – A potent substrate for the production of lipase by *Candida rogosa* in solid state fermentation, *Acta Biotechnologica*, **17**, 241-251.

Berovic M, Ostroversinik H (1997) Production of *Aspergillus niger* pectolytic enzymes by solid state bioprocessing of apple pomace, *Journal of Biotechnology*, **53**, 47-53.

Camareroe S, Bockle B, Martinez MJ, Martinez AT (1996) Managanse – mediated lignin degradation by *Pleurotus pulmonarius*, *Applied and Environmental Microbiology*, **62,** 1070-1072.

Carvalho JC, Soccol CR, Miyaoka MF (2001), "Producao de pigrmentos de *Monascus emmeios* a base de bagacao de mandioca," in Proc. VIIth Encontro Regional Sul de Ciencia e Technologia de Alimentos ABM 2-15, Regional Parana, SBCTA-PR.

Durand A(2003) Bioreactor designs for solid state fermentation, *Biochemical Engineering Journal*, **13**, 113-125.

Edwards VH (1970) The influence of high substrate concentrations on microbial kinetics, *Biotechnology and Bioengineering*, **12,** Vol. 2, Issue 2, 679-712.

Felse PA (1999) "Process development for extracellular chitinase production by *Trichoderma harzianum*." Ph.D. Thesis, IITMadras, India.

Ghose TK, Tyagi RD (1979) Rapid ethanol fermentation of cellulose hydrolysate I. Batch versus continuous systems, *Biotechnology and Bioengineering*, 22, 1387-1395.

Giona AR, Marrelli L, Toro L, De Santis R (1976) Kinetic analysis of penicillin production by semicontinuous fermenters, *Biotechnology and Bioengineering*, 18, 473-492.

Gutierrez-Correa M, Tengardy RP(1999) Cellulolytic enzyme production by fungal mixed culture solid substrate fermentation, *Agro. Food – Ind. Hi-Technol*, **10,** 6-8.

Han K, Levenspiel O (1988) Extended Monod Kinetics for substrate, product and cell inhibition, *Biotechnology and Bioengineering*, 32, 430-437.

Kota KP, Sridhar P (1999) Solid state cultivation of *Streptomyces clavuligerus* for cepharomycin C production, *Process Biochemistry*, **34**, 325-328.

Lapadatescu C, Bonnarme P (1999) Production of aryl metabolities in solid state fermentation of white rot fungus *Bjerkandera adusta*, *Biotechnology Letters*, **21**, 763-769.

Leifa F, Pandey A, Raiunbault M, Soccol CR, Mohan R (2000) "Production of edible mushroom *Lentinus edodes* on the coffee spent ground," in: Proceedings of 3nd Int. Sem. on Biotechnol. in the coffee Agroindustry, Iapar / IRD, Londrina-PR, Brazil, pp. 377-380.

Levenspiel O (1980) Kinetics of product inhibition in alcoholic fermentation, *Biotechnology and Bioengineering*, 22, 803-809.

Lonsane BK, Saucedo-Castaneda G, Raimbault M, Roussos S, Viniegra-Gonzalez G, Ghildyal NP, Ramakrishna M, Krishaiah MM (1992) Scale-up strategies for solid state fermentation systems, *Process Biochemistry*, **27**, 259-271.

Ludeking R, Piret EL (2000) A kinetic study of the lactic acid fermentation: Batch process at controlled pH, *Biotechnology and Bioengineering*, **67**, 393-401.

Luong JHT(1985) Generalization of Monod kinetics for analysis of growth data with substrate inhibition. *Biotechnology and Bioengineering*, **29**, 242-248.

Lydersen BJ, D'Elia NA, Nelson KL (Eds) (1994). *Bioprocess Engineering: Systems, Equipment and Facilities*, John Wiley & Sons, Inc., New York.

Manpreet S, Swaraj S, Sachin D, Pankaj S, Baneerjee UC (2005) Influence of process parameters on the production of metabolites in solid-state fermentation, *Malaysian Journal of Microbiology*, **12**, 1-9.

Mason TJ, Millis NF (1976) Growth Kinetics of yeast grown on glucose or hexadecane, *Biotechnology and Bioengineering*, **18,** 1337-1339.

Mial LM (1975) "Historical developments of the fungal fermentation industry": In Smith JE, Berry DR, and Kristiansen B (Eds) *The Filamentous Fungi*, vol.1, Edward Arnold, London, p 104.

Mitchell DA, Kriegar N, Stuart DM, Pandey A (2000) New developments in solid-state fermentation. II Rational approaches for bioreactor design and operation. *Process Biochemistry*, **35**, 1211-1225.

Mitchell DA, von Meien OF, Krieger N (2003) Recent developments in modeling of solid-state fermentation heat and mass transfer in bioreactors, *Biochemical Engineering Journal*. **13,** 137-147.

Moser A (1988) "Bioprocess Kinetics," In: *Bioprocess Technology*, Springer-Verlag, New York, USA, pp 197-217.

Moser A (1985) Continuous cultivation, In: *Biotechnology*, Rehm H-J and Reed G (Eds), vol. 2, Verlag VCH, p. 303.

Nielsen J, Villadsen J (Eds) *Bioreaction Engineering, Principles*, Plenum Press, New York and London, 1994.

Ohno A, Ano T, Shoda M (1992) Production of the antifungal peptide antibiotic, iturin, by *Bacillus subtilis* NB 22., using wheat bran as a substrate, *Biotechnology Letters*, **14**, 817-822.

Pandey A (2003) Solid state fermentation, *Biochemical Engineering Journal*, **13**, 81-84.

Powell, EO (1967) The growth of microorganism as a function of substrate concentration, In: *3rd Int. Symp. Physiol. Cont. Culture*. Her Majesty's Press, London, UK, pp 23-33.

Raghavararao KSMS, Ranganathan TV, Karanth NG (2003) Some engineering aspects of solid-state fermentation, *Biochemical Engineering Journal*, **13**, 127-135.

Rodriguez VR, Cruz CT, Fernendiz SJM, Roldan CT, Mendoza CA, Saucedo CG, Tomasini CA (1999) Use of sugarcane bagasse pith as solid substrate for *P.chrysogenum* growth. *Folia Microbiology*, **44**, 213-218.

Ryu DY, Humphrey AE (1972) Unstructured modelling of microbial product formation, *Journal of Fermentation Technology*, **50**, 424-429.

Shrikumar S (2003) Current industrial practice in solid state fermentations for secondary metabolite production: the Biocon India experience, *Biochemical Engineering Journal*, **13**, 189-195.

Soccol CR, Vanderberghe LPS (2003) Overview of applied solid-state fermentation in Brazil, *Biochemical Engineering Journal*, **13**, 205-218.

Sree NK, Sridhar M, Suresh K, Rao LV (1999) High alcohol production by solid substrate fermentation from starchy substrates using thermotolerant *Saccharomyces cerevisiae. Bioprocess and Biosystems Engineering*, **20**, 561-563.

Streadansky M, Conti E (1999a) Succinoglycan production by solid-state fermentation with *Agrobacterium tumefaciens*, *Applied Microbiology and Biotechnology*, **52**, 332-337.

Streadansky M, Conti E (1999b) Xanthan production by solid state fermentation, *Process Biochemistry*, **34**, 581-587.

Terui G (1972) "Kinetics of product formation," In: *Microbial Engineering*, Sterbackeiz (Ed.), Butterworth, London, UK, pp. 377-380.

Viniegra–González G, Favela-Torres E, Aguilar CN, Rómero–Gomez SdeJ, Diáz–Godinez G, Angur C (2003) Advantages of fungal enzymes production in solid state over liquid fermentation systems, *Biochemical Engineering Journal*, **13**, 157-167.

Wayman M, Tseng MC (1976) Inhibition-threshold substrate concentrations, *Biotechnology and Bioengineering*, **18**, 383-388.

Weber JT, Tramper J, Rinzema A (1999) A simplified material and energy balance approach for process development and scale up of *Coniothyrium minitans* conidia production by solid-state cultivation in a packed-bed reactor, *Biotechnology and Bioengineering*, **65**, 447-458.

Wu G, Chabot JC, Caron JJ, Heitz M (1998) Biological elimination of volatile organic compounds in a biofilter, *Water Air Soil Pollution*, **101**, 69-78.

APPENDIX 1

Standard growth medium for microorganisms

(a) ***For aerobic bacteria*** (Babu, 1991)

Component	*Concentration(kg/ m³)*
Peptone	10
Beef extract	10
NaCl	5

Agar agar -20 kg/m^3 for slant growth and plating experiments. Components are dissolved in distilled water. The pH of the medium is adjusted to a suitable value by 1 M HCl or by1M NaOH.

(b) ***For anaerobic bacteria***

(i) CM3 Medium (Kundu, 1983; Weiner and Zeikus, 1977)

Component	*Composition(kg/m³)*
Cellulose	10
KH_2PO_4	1.2
K_2HPO_4	2.9
$(NH_4)_2SO_4$	1.3
$MgCl_2$	1.0
$CaCl_2$	0.15
Yeast extract	2.0
$FeSO_4$(5% solution)	25 µl
Cysteine HCl	0.5
Resazurin (0.2 % solution)	1 ml

The components are mixed in a fashion to avoid precipitation. The pH is adjusted to 7. The solution is sparged with nitrogen to scavenge oxygen. The medium is poured in a serum bottle followed by closure and autoclaving.

(c) ***For yeast*** (Anjani Kumari, 1992)

Component	*Composition (kg/m³)*
Malt extract	3
Glucose	10
Yeast extract	3
Peptone	5

Components are dissolved in distilled water and pH is adjusted by standard alkali or acid.

(d) ***For fungus Czapek-Dox medium*** (Kundu *et al.*, 1984)

Component	*Composition (kg/m³)*
Glucose	50
$NaNO_3$	2

KH_2PO_4	1
$MgSO_4.7H_2O$	0.5
$FeSO_4.7H_2O$	0.001

Components are dissolved in water.

(e) ***Potato – dextrose agar (PDA) for fungus***

Component	*Concentration (kg/m³)*
Peeled potato	200
Dextrose	25
Agar	20

Components are suspended/dissolved in distilled water.

(f) ***Cultivation medium for Clostridium sp.*** (Kundu, 1983)

Component	*Concentration (kg/m³)*
Xylose	15
Yeast extract	5
KH_2PO_4	1.5
Na_2HPO_4	3.0
Urea	1.0
$MgCl_2.6H_2O$	1.0
$CaCl_2$	0.15
$FeSO_4.7H_2O$	0.00125
Na-thioglycolate	0.5
Resazurin(0.2% solution)	1 ml.

The pH is adjusted to 7. Other conditions are stated above.

References to Appendix 1

Anjani Kumari J (1992) Studies on protoplast generation and reversion to cells of *Trichoderma reesei* and *Saccharomyces cerevisiae* and their protoplast fusion for direct conversion of cellulose to ethanol, Ph.D. Thesis, IIT Madras, India.

Babu PSR (1991) Studies on biosynthesis of penicillin amidase in *E.coli*, stabilization and immobilization of the enzyme associated with whole cells, Ph.D. Thesis, IIT Madras, India.

Kundu S (1983) Microbial conversion of cellulose to ethanol, Ph.D.Thesis, IIT Delhi, India.

Kundu S, Panda T, Majumdar SK, Guha B, Bandyopadhyay KK (1984) Pretreatment of Indian cane molasses for increased production of citric acid, *Biotechnology and Bioengineering*, **26**, 1114–1121.

Weiner PJ, Zeikus JG (1977) Fermentation of cellulose and cellobiose by *Clostridium thermocellum* in the absence and presence of *Methanobacterium thermoautotrophicum. Applied and Environmental Microbiology*, **33**, 289-297.

Chapter 2

Understanding of Bioreactors

OBJECTIVES

- Definition of bioreactor
- Components of bioreactor
- Classification of bioreactors
- Idea of contacting pattern

2.1 WHAT IS A BIOREACTOR?

It is a vessel in which bio-reactions are carried out under controlled conditions. The system maintains aseptic conditions during entire period of reactions (e.g., Fig. 2.1). The schematic representation of a typical bioreactor is given in Figure 2.2.

Besides these, there are a number of other components which will be discussed later in this book.

It has provisions for continuous monitoring and control of gases (air for aerobic reactions, N_2 for anaerobic reactions, and CO_2 as a metabolic product), agitation, pH, temperature, redox, etc. Also, a bioreactor has features such as *in situ* sterilization, vessel geometry, surface furnish, clean-in-place, etc.

The design criteria, performance of the reactor, and accessories are different for bioreactors applied in Submerged Liquid Fermentation (SLF) and Solid-State Fermentation (SSF).

2.2 WHY SHOULD WE STUDY BIOREACTORS?

Bioreactor analysis gives an insight into the following aspects of bioprocesses:

- Kinetics of biological reactions carried out in the reactor
- Typical mass transfer problems involving gas (air/CO_2), liquid (reaction medium) and solid (some reactant and cells or immobilized catalysts) phase or gas–liquid–solid phase or multiphase reaction
- Mass balances involving reactant and catalysts (particularly for free cells)
- New product development demands modification of accessories / components in bioreactor. For example, simultaneous separation of products from the reaction phase requires specific / selective membrane or the assistance of vacuum (Kundu *et al.*, 1984). Further importance of study of bioreactor is highlighted in the Section 2.3.

Figure 2.1 Typical photograph of a bioreactor from M/s Bioengineering AG, Switzerland (Published with kind permission from BIOENGINEERING, copyright Bioengineering AG).

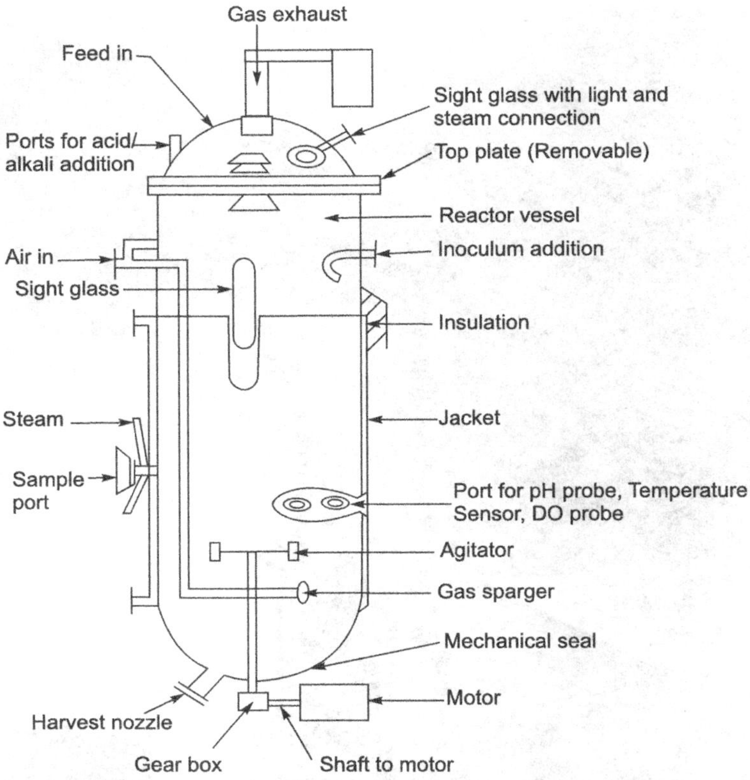

Figure 2.2 A typical bioreactor for submerged reaction (immersed / suspended) process.

2.3 DEVELOPMENT OF BIOREACTORS

The different stages of bioreactor development are given here to stress the emphasis for the full-fledged and properly controlled bioreactor.

(a) Slant Culture

The growth of a particular organism is usually maintained in agar medium. It facilitates limited production of bio-molecules. The cotton plugging on the test tube serves as the filter to maintain sterility in the tube through which gases (air/CO_2) are exchanged (Fig. 2.3).

Advantage

Complete sterility can be maintained during the desired period of incubation.

Disadvantages

- The process is laborious as a large number of tubes are required for desired production.
- Large scale process cannot be carried out using test-tube culture.
- Reliability is a question since sampling is difficult as a time function.

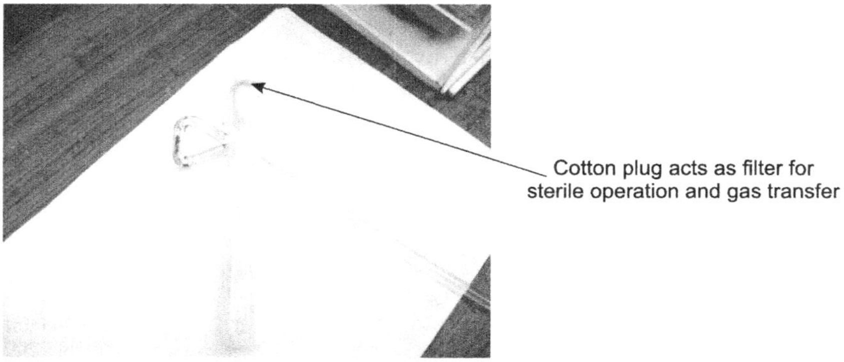

Figure 2.3 A slant culture.

- Isolation and purification processes from solid medium are additional expenses.
- Control is difficult throughout the reaction phase.
- Evaporation loss may be encountered in this case.

So, it was thought to use higher scale. In this aspect, the choice of Petri-plate is another avenue.

(b) Petri-plate Culture

The fermentation may be carried out in a sterile Petri-dish either on solid or liquid media. The volume is higher than the test-tube culture. The age-old story of penicillin synthesis in Petri-culture may be recalled in this context. The gap between the two matched-dishes of a Petri-plate may be considered as unique way to filter the air required for reaction or exchange of CO_2 evolved during reaction (Fig. 2.4). The management of aseptic condition is fulfilled in this strategy.

Advantage

Maintenance of aseptic condition is ensured in this case. However, the *disadvantages* stated in test tube culture are not rectified in this strategy.

In the third stage, Erlenmeyer flasks, containing the reaction medium, kept in stationary condition, are used.

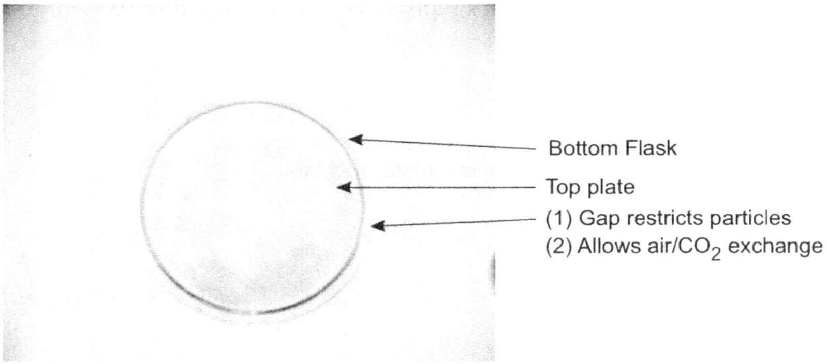

Figure 2.4 Growth of organism on Petri-plate.

(c) Stationary Flasks Culture

The reaction volume is higher than provided by either test-tube or Petri-plate culture. Sterility is maintained through the cotton plug in the flask. The diffusion of sterile air required for the reaction is possible through the top liquid layer (Fig. 2.5). This is again not an effective procedure of oxygen mass transfer in the system.

Figure 2.5 Stationary flask culture.

Other problems associated with the first two culture techniques are also equally applicable in this case. For stationary culture fermentation, horizontal rotary reactor is a further development (Fig. 2.6). The concept of reactor development based on stationary culture technique is elaborately mentioned in bioreactors used for SSF in this chapter.

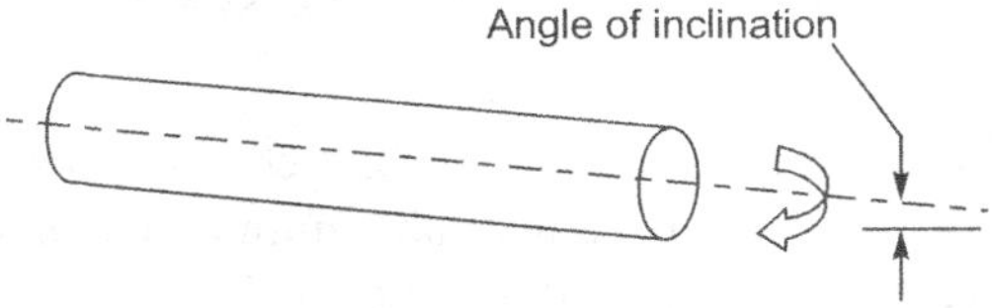

Figure 2.6 Horizontal rotary reactors.

The next generation of improvement in bioreactor concept for submerged reaction system is considered by the use of shake flask in baffled (Fig. 2.7b) and un-baffled condition (Fig. 2.7a).

(d) Unbaffled Shake Flask

Volume of a reaction may go up to one liter. Bioreaction in this case is also limited for oxygen supply. Shake flasks are usually placed on shaking machines (called shakers). Air is purified by sterile cotton plug placed in the mouth of the flask. Due to shaking effect, oxygen is forced into the reaction medium. Availability of oxygen is limited in the reaction phase. The distribution of oxygen in the reaction phase is not necessarily uniform.

Figure 2.7 (a) Unbaffled.

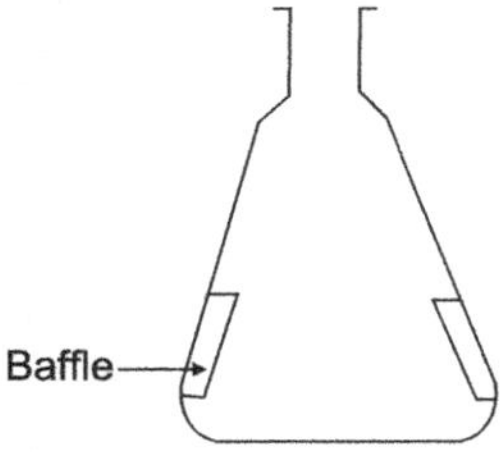

2.7 (b) Baffled.

(e) Baffled Shake Flask

Problems in an unbaffled shake flask can be partially solved using designed glass baffles or by placing stainless steel baffles (Fig. 2.7b). This is an improvement over unbaffled shake flask. Distribution of oxygen (from air) is much better. It also avoids vortex formation.

Problems

Common problems encountered in flask cultures are listed below.

- Small volume
- Limitation of gases utilized for the process
- Non-uniformity in production system
- Non-representative samples
- Difficult to control fermentation parameters
- Measurement/on-line estimation of parameters is difficult, i.e., difficult to implement automation.
- Difficult to minimize evaporation loss
- Require more space per unit production
- Batch type of reaction mode

For these obvious reasons, the studies of bioreactors are gathering momentum. Initial development of reactor has considered classical bubble column principle (Fig. 2.8). This configuration appears to be less effective.

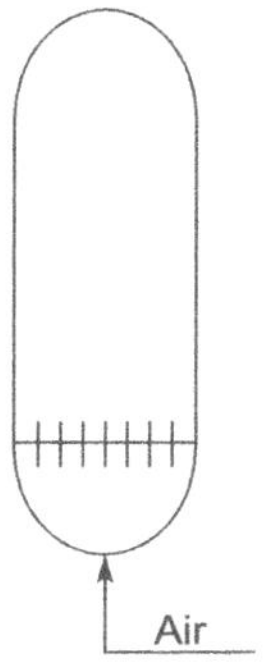

Figure 2.8 Sketch of a bubble column reactor.

Reason

Heat and mass transfer problems are prominent.

Question 1 ***How can one improve heat and mass transfer problem?***
Answer: Mechanical mixing is a solution to this problem.

2.4 PURPOSE AND IMPORTANCE OF BIOREACTORS

2.4.1 Necessary Functions of Bioreactors

It is appropriate to consider the functions of bioreactors here. They include

- To provide environment for the biocatalysts (e.g., growth of organisms or cells) under controlled aseptic conditions
- To facilitate better distribution / contact of medium (reactants) and gases (O_2) with the biocatalysts
- Easy monitoring and/or control of reaction (fermentation) parameters (viz., dissolved oxygen concentration, pH, temperature, agitation rate, redox value, pCO_2, etc.)
- Easy on-line analysis
- Ability to carry out constant and variable volumes of reaction

For special reactions, it may be necessary that additional components or features are included in a bioreactor. For example, to retain cells in the reactor, it is necessary to use a filter or a membrane (Lee, 1992). To measure on-line respiratory quotient in fed-batch culture, on-line CO_2 measurement device needs to be attached to the exhaust gas outlet of the reactor (Suga *et al.*, 1982).

2.4.2 Requirements for a Bioreactor

Additionally, one needs to know the requirements of a bioreactor.

Reliability of Operation in Days

For example, citric acid biosynthesis by *Aspergillus niger* requires about 7 days (Panda *et al.*, 1984) whereas typical animal cell culture will require a few weeks (Venkataraman *et al.*, 1991). So the bioreactor should maintain cells without any modifications or changes.

To Maintain Aseptic Conditions during the Period of Operation

With reference to the above examples, a bioreactor maintains aseptic conditions properly without any contamination.

Flexibility in Operation

Conditions controlled by the design and operation of the individual components of the bioreactor should be easily alterable. For example, a reactor used for batch operation can definitely be used for continuous mode of operation. This can be done by fixing one peristaltic pump each to the feed and to the out flow side of the existing reactor used in batch mode. The outflow is made by inserting suitable connections as overflow pipe.

Another example could be the use of different organisms separately for the particular reaction in mind. The reactor used for reactions involving microbial cells can be extended for plant and animal cells with specific modifications. Some of the accessories, viz., aeration device and agitation mode may be different. This will allow shear-sensitive biocatalysts to function in the reactor.

2.4.3 Major Components and its Purposes

Functions of the components of a bioreactor are given in brief to understand the complex requirement of a biological reaction compared to a chemical reaction.

Temperature control: Jacket around the reactor or a suitable heat exchanger *in situ* may be used with proper temperature sensor, monitoring and control device.

pH control: This consists of acid and alkali feed to the reactor by two separate peristaltic pumps and a pH probe.

Agitation device: Power-driven shaft with a proper controller is used in the bioreactor.

Impeller design: This depends on the use of catalysts. A few classical impellers are given in this chapter.

Aeration rate: This is controlled by agitation rate and gas flow rate. Gas flow rate is measured by a rotameter.

Sparger (baffled/unbaffled): This has been already discussed in this chapter.

Means for Addition of Inoculum, Antifoam Agents, and Necessary Reagents

Such required arrangements are sketched in Figure 2.2. For example, aseptic sampling device (Fig. 2.9) and harvesting system (Fig. 2.10) are mentioned here.

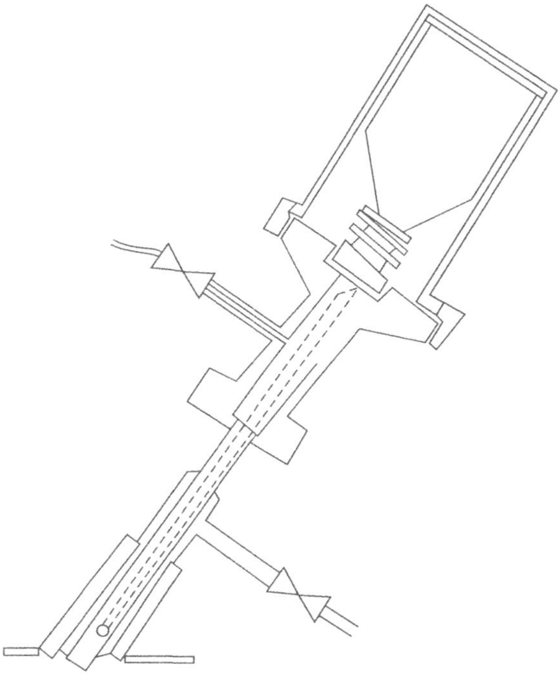

Figure 2.9 Schematic of sampling system.

Functions of some of the important components of bioreactor with their materials of construction are given in Table 2.1. Related concepts are described below and in **Chapter 12**.

2.4.4 Additional Information on Important Components

(1) Reaction Vessel

Vessel shapes are determined by the application. Basically two different categories of vessels are reported, viz.,

Pressure vessels: For reaction / fermentation/sterilization

Non-pressure vessels: For storage purposes

Pressure vessels used for microbial fermentation are generally tall, narrow, and thick-walled. This is for better mass and heat transfer. However, for mammalian and plant cell cultures, wider vessels are used to provide gentle treatment.

Table 2.1 *Components of a typical bioreactor with functions and materials of construction*

S. No.	*Component*	*Functions*	*Feature*	*Materials*
1.	Reaction vessel	To carry out biological reactions	(i) Top end open (ii) Both ends open with supported cover plate	(a) Small vessels (up to 20 L) Glass (Pyrex) (b) Large vessels of stainless steel (SS) (BS No. En. 58 J (Solomon, 1968)
2.	Jacket	For heat exchange (coolant / steam)	(i) For small reactor – Sometimes glass / stainless steel Usually 3/4th of the length of the vessel	(a) For glass vessel – it is usually glass (b) For stainless steel vessel – jacket of stainless steel (1/8 to 3/16 inch thick) (Solomon, 1968)
3.	Sight glass	To view reactions / interior of vessel	~ 15 cm window (Lydersen *et al.*, 1994) with sanitary clamps mounting probably used for large scale SS vessel	Good quality glass
4.	Baffles	Prevent liquid swirl so that power supplied by impeller to liquid can be improved	Vertical, approximately 1/10th of vessel height, SS; sometimes fixed to the wall of the vessel for SS reactor; for glass reactor they are loosely placed inside on hoops. Generally 4-baffles, each of 10% of vessel diameter, placed at equidistant around reactor periphery is also a choice (Solomon, 1968)	Usually SS of reactor vessel grade
5.	Sparger	Gas distribution	(i) Ring sparger (ii) Single orifice sparger	SS
6.	Impeller	• Mixing : biocatalysts, reactant and gas • Gas distribution • Effect on gas-trails	(i) Disc turbine – disc on which even number of vertical blades extending equally on both sides (ii) Vaned disc (iii) Open turbine (iii) Marine propeller	SS

Contd.

Table 2.1 *Contd.*

S. No.	*Component*	*Functions*	*Feature*	*Materials*
7.	Agitator Assembly	Supply power to impeller		
	Drive Assembly		(i) Top drive (ii) Bottom drive From a single phase motor through V-belts or pulleys, power is transmitted to agitator shaft.	
	Seals	• Transmission of power, proper alignment between drive shaft and agitator shaft, prevent contamination through the entry of agitator shaft	Best choice is mechanical seal (Lydersen *et al.*, 1994, Solomon, 1968) Other seals-oil seals, packed gland.	
	Shaft	• Transmit power from drive motor through seal to impeller • house impeller on its vertical axis	Vertical shaft for microbial culture; For mammalian cells shaft mounted at an angle of 15° to the vertical (Lydersen *et al.*, 1994)	
8.	Head plate	• Cover the ends of reactor body • provisions for valves, probes • holding agitator assembly	(i) Both dished- ends (ii) Both flat-ends (iii) Top flat end and bottom dished end, thickness decided by vessel and cover design	SS
9.	Sampling arrangements	To collect representative sample from the on-going reaction	Pipe/silicone tube connection through a top end port with the aid of peristaltic pump	
10.	Inoculation device	For aseptic transfer of inoculum	Similar to S. No. 9 Sometimes assisted by steam for local sterilization	

Contd.

Table 2.1 *Contd.*

S. No.	*Component*	*Functions*	*Feature*	*Materials*
11.	Inlet and outlet air filters	**Inlet filter supplies sterile air/ gas required** for bioreaction **Outlet filter restricts** organism performing reaction, to vent gases	Generally they are attachments though the ports on the top end cover through aseptic seals.	(i) **Membrane filter** (Fig. 2.10) 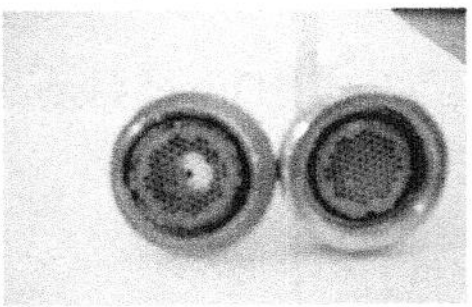 **Figure 2.10** (ii) Packed glass-wool filter (Fig. 2.11) 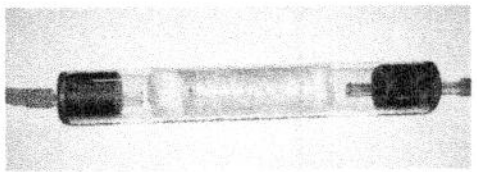 **Figure 2.11**
12.	Addition ports for • feed in and product out for continuous flow reactor system • Antifoam addition system • Probes Temperature pH pO_2 pCO_2 **Redox** Other speciality probes • Gas analyzer	For operation of continuous flow reactor Controls foaming during reaction		
13.	Steam locks	Provide aseptic operation during entire reaction cycle	Diaphragm valve welded to the reactor vessel (Lydersen *et al.*, 1994)	
14.	Static seals	Provide aseptic attachment of various components pipelines, etc.	O-ring	Teflon, SS, ethylene propylene diene **monomer (EPDM),** Viton (Lydersen *et al.*, 1994)

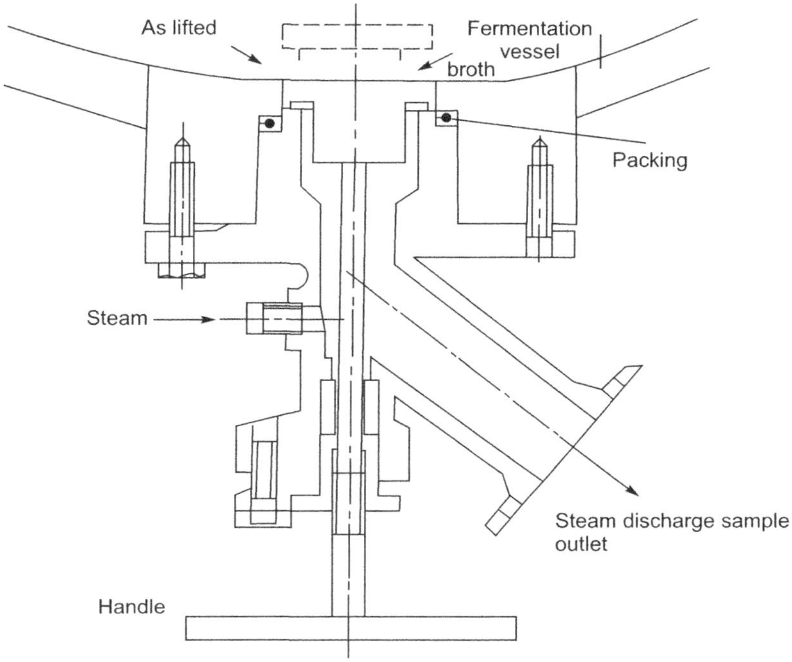

Figure 2.12 Schematic of a harvesting system.

There are certain different configurations of reactor vessels used for special processes.

(a) Small Bioreactor (Capacity < 20 L)

Generally, they are made of special quality glass with following configurations (Figs. 2.2 and 2.13). They are round bottom and flat-bottom vessels.

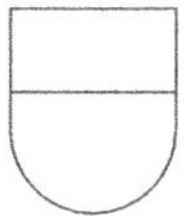

Figure 2.13 Round bottomed bioreactor.

Flat bottomed reactor with top flange carrying a metal head plate with all ports for necessary connections is shown in Figures 2.1 and 2.14. They can have either top- or bottom-driven system.

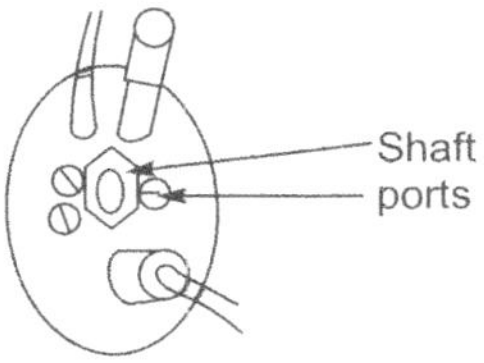

Figure 2.14 Metal top plate of a bioreactor with all ports.

Advantage of a Glass Bioreactor

- Glass vessel appears to be non-toxic to biological catalysts.
- Reaction can be easily followed in the system.

Bottom of the reactor is designed in various shapes for following objectives:

- To avoid non-idealities, viz., mixing, dead zones, interconnected fluid layers, and recirculation.
- To provide better mass transfer of oxygen.
- For better distribution of operating pressure.

For fermentation industries, the dished end bottom reactor is used. In laboratory scale reactor, the bottom has flat end. However, there are a few bottom configurations available for special types of reactors (Fig. 2.15), viz., spherical, tapered, and flat-bottom.

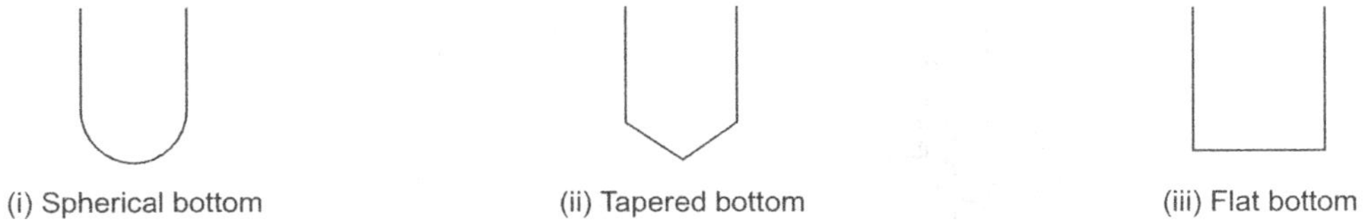

Figure 2.15 Bottom shape of reactor.

These reactor bottoms are suitable for non-pressurized processes. Reactors with such reactor bottoms are sterilized in an autoclave for special study purposes in the laboratory.

(b) Large Bioreactor (Capacity > 20 L)

The body of a large reactor is made up of stainless steel (SS 304, SS 304 L, SS 316 or SS 316 L, etc.) (Lydersen *et al.*,1994; Dillon *et al.*, 1992 a and b).

Factors to be considered

- Compatibility with biocatalysts
- Characteristics of fermentation broth
- Lifetime
- Nature of failures
- Safety
- Cost

It is preferable to fabricate larger vessel at the site of use (Lydersen *et al.*, 1994).

This reduces following risks/costs:

- Transportation
- Specific fabrication/alteration at site as per use in the application
- Installation requirements, viz., special pipeline, electrical connections, etc.
- Clean-in-place device
- Special steam connections.

Configuration

They are cylindrical with dished end covers. Drive is usually at the bottom of the vessel.

Advantages

- Easy production scale operation is possible.
- Varieties of reactions can be carried out.
- Reactions under various pressure and temperature conditions are possible.

(2) Jacket

Glass jacket with glass vessel for small laboratory fermentor is available commercially (Fig. 2.16). In some commercial fermentor with glass vessel, jackets are not provided. Instead, temperature control is done with immersed coil in the reactor (Fig. 2.1).

For large scale reactor, jacket is of SS 304 (Lydersen *et al.*, 1994). The configurations of jackets vary depending on the preference and flexibility of fabrication.

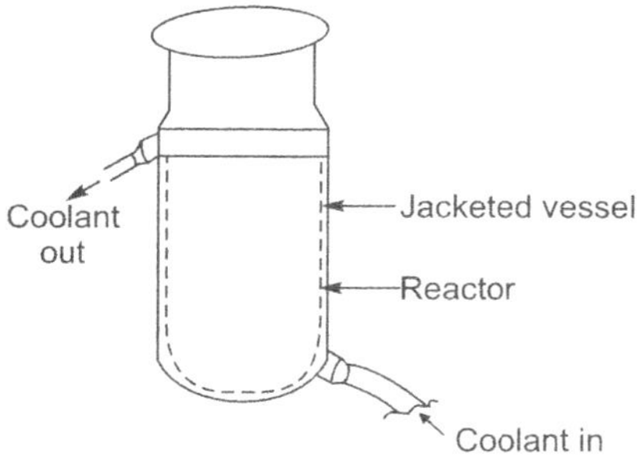

Figure 2.16 Glass reactors with glass jacket.

(3) Baffles

This is generally made of stainless steel. The mode of attachment to the reactor vessel depends on the material of the vessel.

- For glass vessel, baffles are arranged loosely on a separate peripheral structure (Fig. 2.17). Sometimes they are fixed to the flat end bottom plate (Fig. 2.1).
- For SS reactor vessel, baffles are welded to the reactor wall.

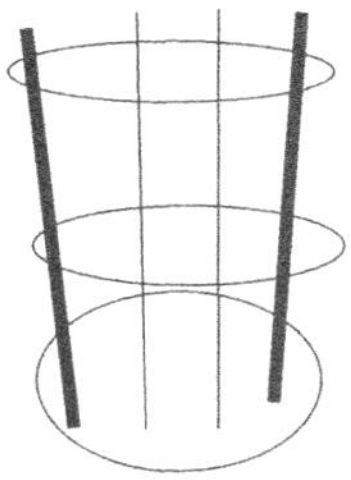
Figure 2.17 Baffles.

(4) Spargers

There are a few concepts in sparger configurations. Those are indicated here.

(a) Ring sparger with minute holes on the entire ring (Fig. 2.18).

(b) Circular and open nozzle (Fig. 2.19) configurations are other choices.

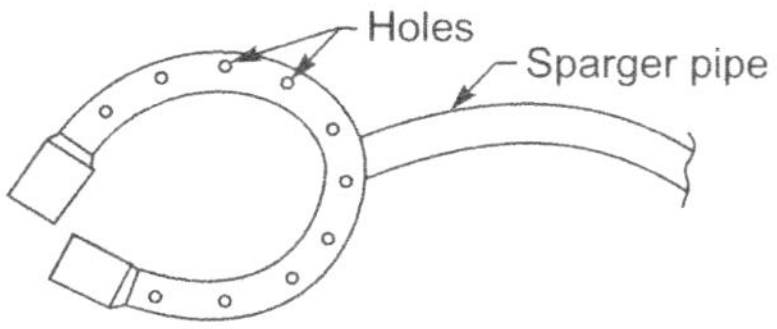

Figure 2.18 Ring sparger with minute holes.

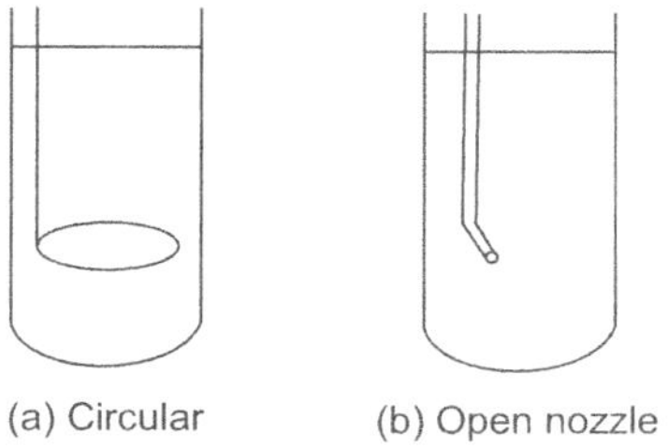

Figure 2.19 Circular and open nozzle sparger.

(5) Agitation Devices

This consists of the following components (Fig. 2.20).

- Motor
- Belt or gear device: Power transmission device to shaft.
- Shaft : Simple supported beam or cantilever beam system (Fig. 2.2)
- Mechanical seal for entry of shaft (conforming total aseptic condition during reaction)
- Impellers

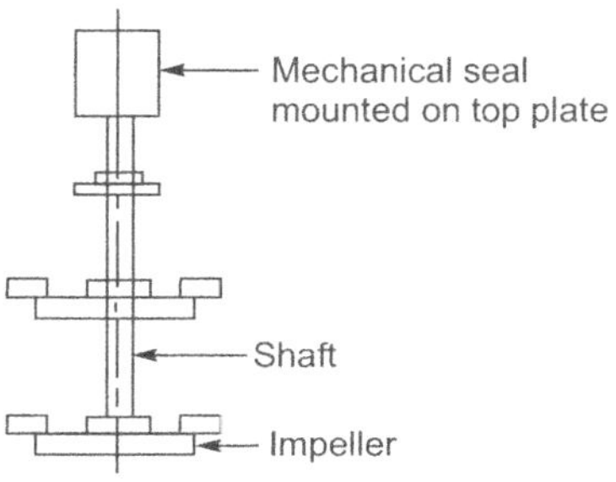

Figure 2.20 Top entry shaft.

There are two different configurations for entry of shaft to the inside of reactor vessel, viz., top entry and bottom entry.

(a) **Top entry**: Figure 2.20 gives some of the details of top entry shaft.

Advantage

Entry through top-end cover of the reactor plate does not cause leakage of materials, as the reactor is always under positive pressure during operation.

Disadvantages

- This requires longer shaft length to reach almost to the bottom of the reactor and hence material requirement is more.
- Loss of power transmitted to the shaft is more.
- It does not allow sufficient space on the top-end cover for the ports through which inoculum, feed, antifoam agent, other devices to be fixed and/or probes are inserted.

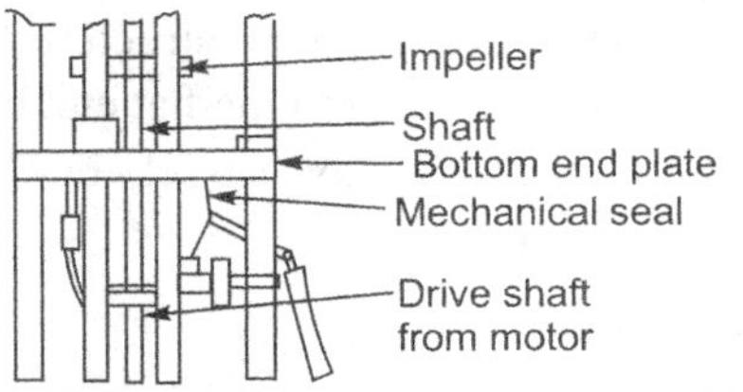

Figure 2.21 Bottom entry of agitator shaft.

(b) **Bottom entry**: Figure 2.21 shows the concept of bottom entry shaft.

Advantage

- Top end cover of the reactor can be freely used for necessary addition of reactants or insertion of probe (Fig. 2.22).
- Shaft length is shorter than the top-entry device which is an advantage over top entry system.

Disadvantage

At the shaft entry point to the bottom-end cover of the reactor, one has to make sure that no leak exists. This is a common failure for this type of entry device.

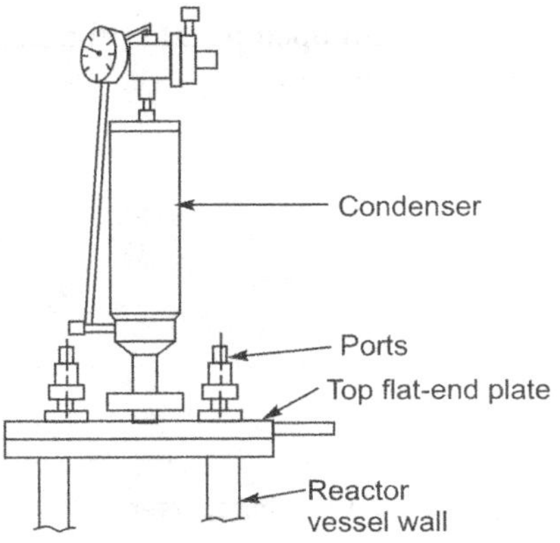

Figure 2.22 Top end plate of bioreactor having bottom entry shaft.

Impellers

Power transmitted through the shaft is received by the impellers which ultimately delivers the energy to the reacting fluid. Schematic diagram of impellers are given in Figure 2.23.

They influence mixing in the reaction phase. This phenomenon can be analyzed based on three physical processes:

(a) distribution – known as macro-mixing
(b) diffusion – called micro-mixing
(c) dispersion – is a combination of distribution and diffusion depending on the degree of fluid motion.

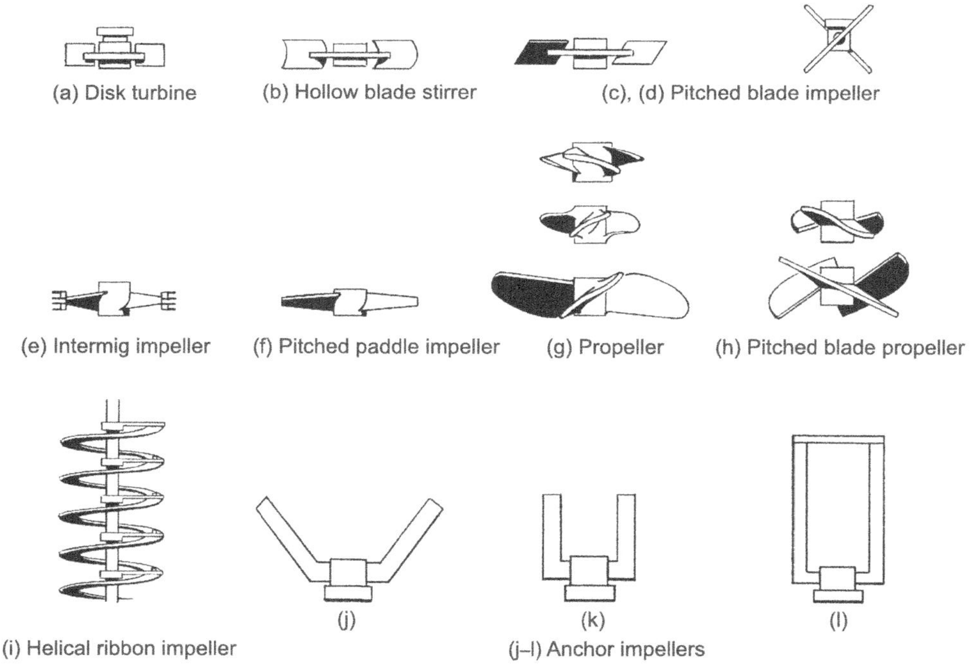

Figure 2.23 Different impellers. (Published with kind permission from BIOENGINEERING, copyright Bioengineering AG, Switzerland).

For improving mixing in bioreactors, one can opt the following:

(i) Different configurations

(ii) To mount impellers below the geometric center of the vessel (Doran, p155, 1995)

(iii) Using multiple impellers (Bader, 1987) (Fig. 2.24).

The mixing pattern in agitated system may be described as in Figures 2.25 (a) to (c).

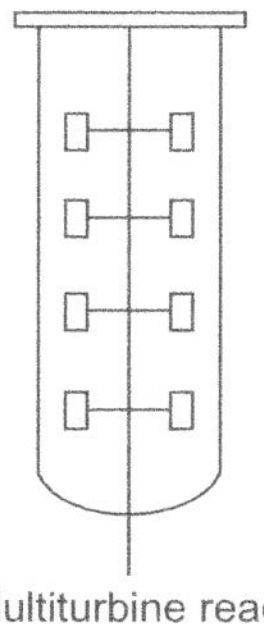

Figure 2.24 Multiple impellers.

2.5 OTHER BIOREACTOR CONFIGURATIONS

A few reactor configurations are mentioned here.

(i) Simple Vat Type Bioreactor (Fig. 2.26)

Cement or wooden tank with an agitator is the classical reactor.

Applications

- Distilleries
- Wine manufacturing industries

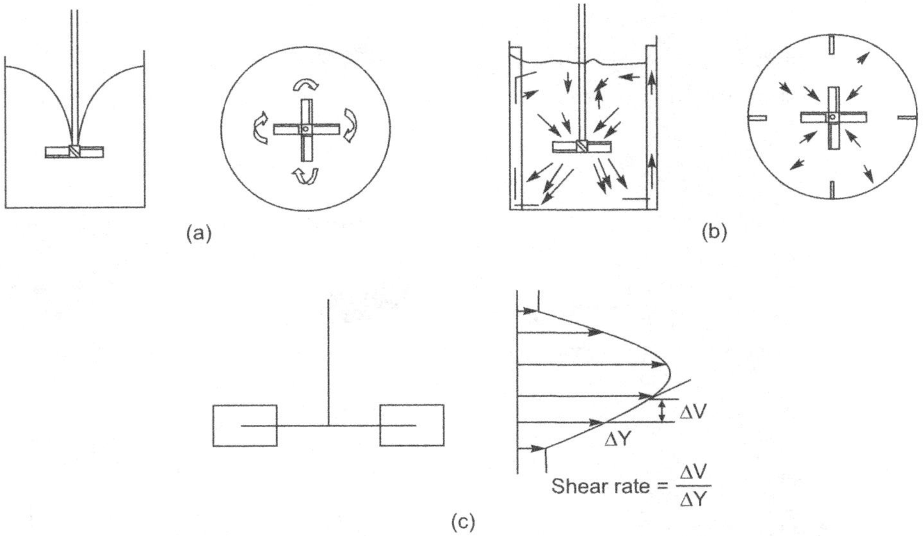

(a): Vortex flow pattern obtained with any type of impeller which is un-baffled
(b): Flow pattern obtained with any type of impeller with baffles
(c): Typical velocity pattern coming from the blades of a radial flow turbine showing calculation of the shear rate $\Delta v/\Delta Y$.

Figure 2.25(a)-(c) Flow pattern in agitated systems.

Disadvantages

This reactor is not suitable for strict aseptic processes.

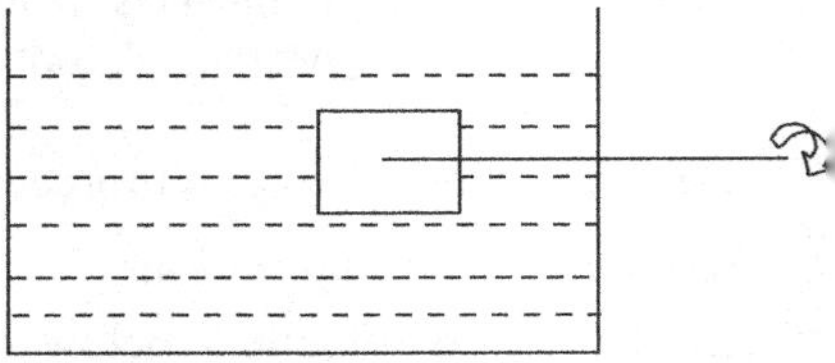

Figure 2.26 VAT bioreactor.

(ii) Bubble Column Bioreactor (Fig. 2.8)

Application

Shear sensitive system can be easily maintained.

Disadvantage

This reactor has inherent mass transfer problem and fails to provide uniform mixing.

(iii) Loop Bioreactor (Fig. 2.27)

They are of internal or external loop system.

Application

This is suitable for shear sensitive system.

Disadvantage

Mass transfer problem is associated with the reactor.

(iv) Pulsating Column Bioreactor (Fig. 2.28)

Application

This can find application for shear sensitive system.

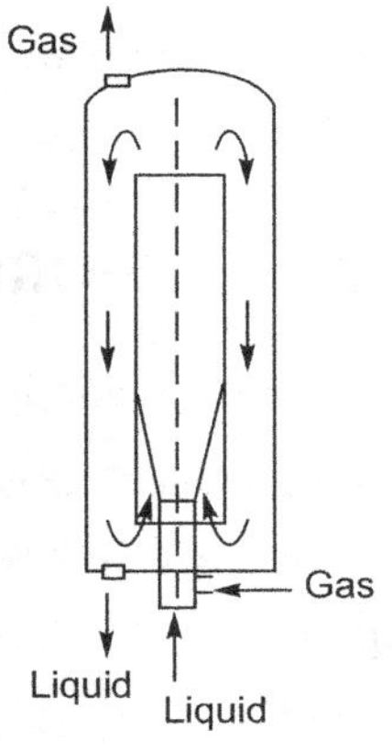

Figure 2.27 Internal loop bioreactor.

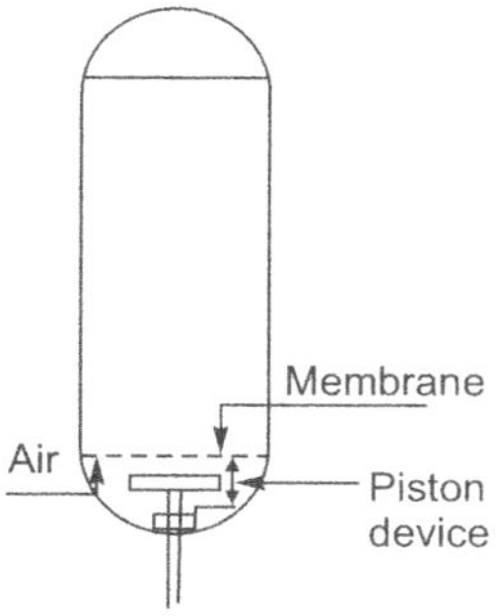

Figure 2.28 Pulsating column bioreactor.

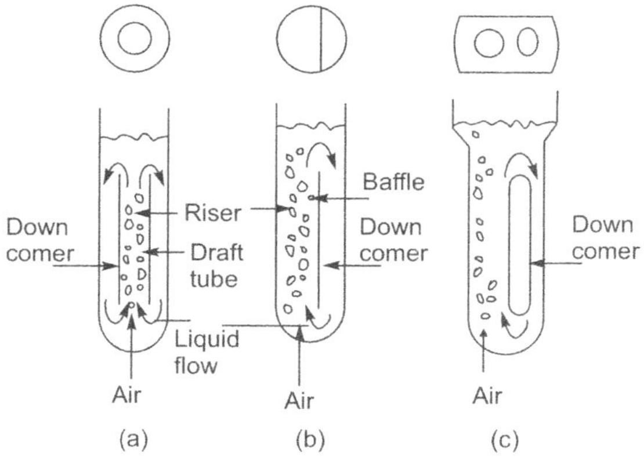

Figure 2.29 (a–c) Airlift bioreactors.

Disadvantage

Mass transfer problem cannot be avoided in this configuration of the bioreactor.

(v) Airlift Bioreactor (Fig. 2.29)

The fluid volume of the vessel is divided into two interconnected zones by a baffle or a draft tube.

- The zone sparged with air is called the riser.
- The other zone that receives no gas is called down comer.

There are internal and external loop systems. External loop is less common in practice. Internal loop configuration is divided into a concentric draft tube and a split cylinder type.

Application

This reactor can be used for shear sensitive system.

Disadvantage

One can find probable gas carryover to down comer.

(vi) Cell Holding Culture Bioreactor

They are generally of membrane device. Important classifications are:

(a) Membrane reactor
(b) Hollow fiber reactor
(c) Ultra-filtration reactor
(d) Multi-membrane reactor

(a) Membrane Bioreactors

Basic configurations will enable to separate cells from growth inhibiting metabolites and input medium. Two different configurations are:

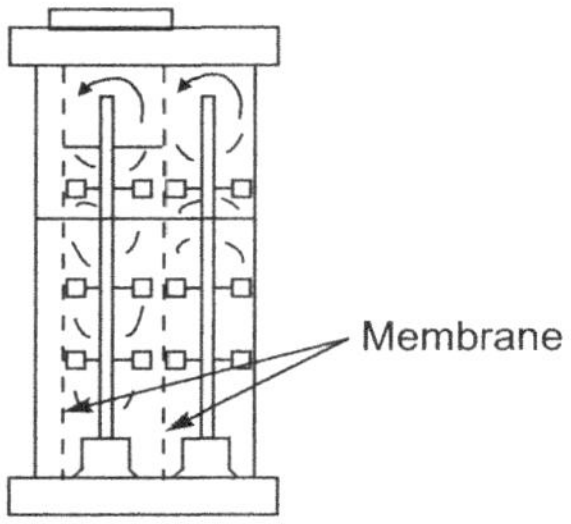

Figure 2.30 Membrane bioreactor.

(1) Two dialysis membranes separating stirred vessel: The smaller component of the reactor is integrated into the larger one (Fig. 2.30).

Application

It finds application in shear sensitive system.

Disadvantage

This prevents optimal mixing of the liquid phase and of the gas phase.

(2) Dialysis solid state bioreactor: This has a vibrating stirrer inside the inner dialysis membrane (Fig. 2.31).

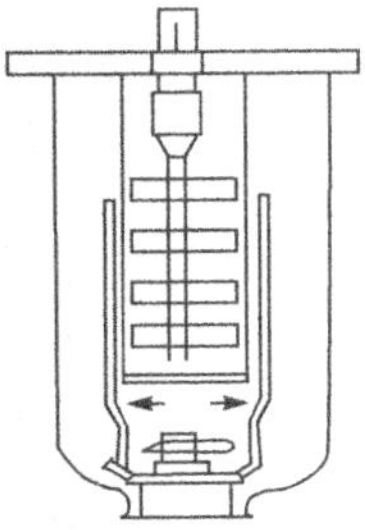

Figure 2.31 Dialysis solid-state bioreactor.

Application

Some applications has been observed in the literature for mammalian cells culture.

Disadvantage

This prevents optimal mixing of liquid and gas phases.

(b) Hollow Fiber Bioreactor

The membrane basket is fixed to the guide disk, which is suspended in the reactor cover through a connecting rod. The assembly is moved by the eccentric disk which is fixed to the controlled agitator shaft. The guide disk with membrane holding system carries out eccentric movement at a low rpm.

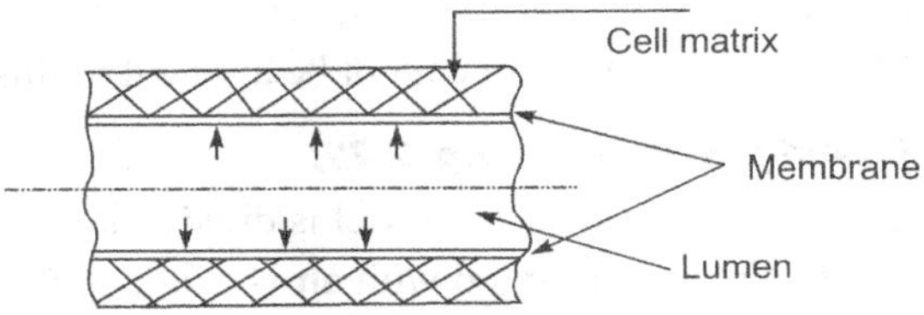

Figure 2.32 Hollow fiber bioreactor for human and mammalian cells, and plant cells culture.

Slowly moving eccentric sail-agitators with porous hollow fiber membrane (in the form of basket) are used (Fig. 2.32).

Application

For the growth of human and mammalian cells and plant cells, one can find application of this bioreactor.

Disadvantages

- Disturbances by foam flotation of micro-carriers
- Limited gas pressure within hollow fiber

(c) Ultra-filtration Bioreactor

This is an integration of a reactor and a cross-flow micro-filtration system into a single device (Fig. 2.33).

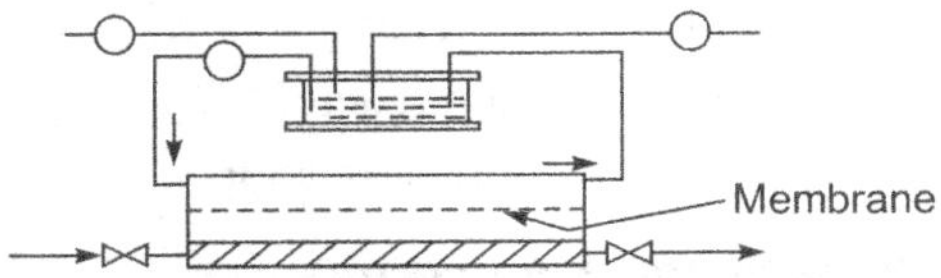

Figure 2.33 Ultra-filtration bioreactor.

Application

Following are the major applications of this reactor.

- Products synthesized in the stationary phase of growth
- Separation of inhibitory metabolites synthesized during reaction.

Disadvantage

It is difficult to separate dead cells during the process.

(d) Multi-membrane Bioreactor

Applications

One can find application in the following areas:

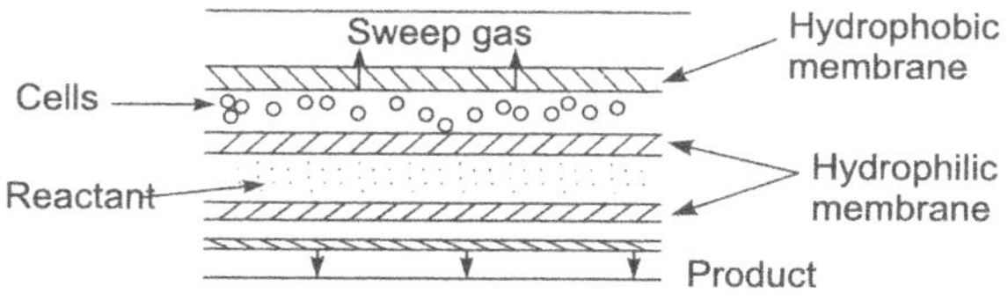

Figure 2.34 Multi-membrane bioreactor.

- Useful for bio-reactions that produce gas and an inhibitory compound
- Gas, cells, nutrients (reactants) and solvent are separated by membranes to prevent solvent emulsification in the fermentation broth
- Possibility of multistage conversion with multiple biocatalysts.

Advantages

This reactor has the following advantages:

- Low steric-hindrance and deactivation of enzymes
- No washout of cells
- Retention of higher concentration of cells
- Can achieve high volumetric productivity

(vii) Perfusion Bioreactor

The reactor consists of special porous silicon tube coiled around the draft tube and fed with air or pure oxygen. It has a rotating spin filter within the draft tube. The spin filter is a cylindrical sieve of stainless steel wire mesh with a pore size smaller than the cell diameter. The product and the cell free medium filter through the cylindrical sieve. The cells are retained within the reactor (Fig. 2.35).

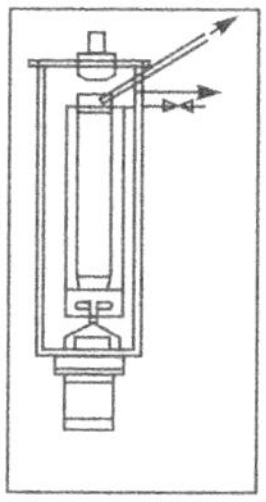

Figure 2.35 Perfusion bioreactor.

Applications

It finds application in bubble-free aeration system and for the growth of high density culture.

Disadvantage

Specific design is required to eliminate foam problem and enhanced gas transfer.

(viii) Photo-bioreactor

A photo-bioreactor is a system, which provides an artificial environment to grow phototrophic cell cultures. Photosynthetic culture requires sunlight or artificial illumination.

Construction

This consists of a reactor vessel, sterile medium delivery system, a head space gas circulation system, a light cabinet, a data communication interface, and a reactor control program.

The choice of photo-bioreactor design greatly impacts the overall efficiency. The design of this reactor has been the major problem in the research. It took many years to design a photo-bioreactor.

Open and closed systems are the two classes of photo-bioreactors.

(a) **Open type photo-bioreactor:** For example, outdoor ponds are considered as open type bioreactor. This type of bioreactors is often used to culture micro algae but mono-septic culture requires fully closed photo-bioreactors. It needs light. Photosynthesis can occur only at relatively shallow depths. However, too much light causes photo-inhibition.

The disadvantage of this type of bioreactor is that it has no definite control over process parameters. These systems offer little or no control over temperature, and incident light intensity and low utilization of CO_2 due to lack of turbulent flow. Contamination by other microorganisms influences the growth of the culture and decreases the quality of the product. Since the output rate per reactor volume is low, the production cost is high and hence we focus our attention on closed reactor.

(b) **Closed type photo-bioreactors:** These reactors are basically used for mono-culture (Fig. 2.36). The reactor consists of arrays of tubes that may be made of glass or a transparent plastic. A continuous single run tubular configuration may be used, or the tube may be wound around a cylinder. In addition to the tubes, flat or thin panels may be used for small scale operation (Tsoglin *et al.*, 1996).

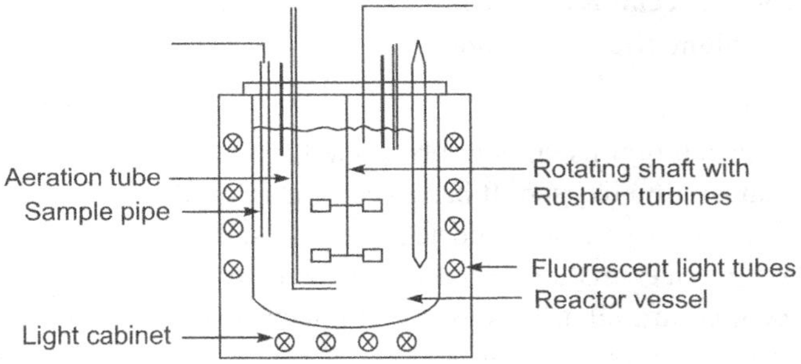

Figure 2.36 Photo-bioreactor.

(ix) Microgravity Bioreactor

Microgravity offers new ways of handling fluids, gases, and growing mammalian cells in efficient suspension cultures (Mausli, 1998). Some conceptual designs for a zero-g fermentor were developed in 1969. Unique suspension type cell culture systems could be developed for bioprocess technology experiments and production of biological products in microgravity condition. The first space bioreactor has been designed with microprocessor control having no gaseous headspace with circulation and re-supply of culture medium, and slow mixing in very low shear regimes. Various ground-based bioreactors are being used to test reactor vessel design, on-line sensors, effects of shear, nutrient supply, and waste removal from continuous culture of human cells attached to micro carriers.

(x) Special Bioreactors on the Basis of Process Requirement

A few important special reactors are mentioned below:

- Animal cell culture (See Section 2.8 – Reactors for animal cells cultivation)
- Plant cell culture (See Section 2.9 – Bioreactor for plant cell culture)

- Vacuum fermentation: To remove the inhibitory product as it is formed in the reaction phase. The difficulty of maintaining sterility in bioreactor at reduced pressure can be overcome by carrying out fermentation at atmospheric pressure and connecting an external vacuum system (Lee at al, 1981).
- Evaporation fermentation: This reactor has the combined fermentation and evaporation mechanisms (Li *et al.*, 2009).
- Adsorption fermentation/ Extractive fermentation: This is a fermentation process in which the product is removed from the reaction phase as it is formed by liquid-liquid extraction using an extractant for the product which is immiscible with water (Taya *et al.*, 1985). The product is, for example, ethanol. The extractant is generally unsaturated C_{14}- or higher aliphatic alcohols or saturated branched chain aliphatic alcohols (C_{14}- or higher).
- Dialysis fermentation: The continuous removal of product by dialysis and the intermittent replenishment of dialysate reservoir affecting cyclical changes in concentration of viable cells are the major feature of the reactor (Abbott and Gerhardt, 2004).

Microbioreactors: This is a polymer based bioreactors integrated with automated sensors and actuators for high-throughput bioprocess development. It has been demonstrated with 150 μL-volume in batch, continuous, and fed-batch mode using *E. coli* (Zhang *et al.*, 2007).

2.6 BIOREACTOR DEVELOPMENT FOR SOLID- STATE FERMENTATION (SSF)

So far we have discussed about the reactor development in SLF. A great variety of products are better obtained from SSF (**Table 1.1**).

2.6.1 Necessary Features of a Typical Bioreactor Used in SSF

(1) Containment of Reactant Bed

The following important points are considered:

- For metal support (for large-scale production)
  - Strength
  - Cost
  - Corrosion
  - Potential toxicity to process organisms
- For non-metallic support (for small scale production), viz., plastic, glass, wood are considered (Mitchell *et al.*, 2000).

(2) Support to Process Organisms

- It should not release organism or its parts (like spores) into the environment.

Question 2 ***Why should we not allow the organism to leave the bioreactor?***

Answer: Pathogenic organisms may cause allergic reactions and other health related problems to the support personals.

(3) Prevention of Contamination

The growth of undesirable organisms is avoided by controling water activity.

(4) Control of Environmental Conditions, viz.,

- pH
- Water activity
- Supply of oxygen (air)
- Temperature

(5) Maintenance of Homogeneous Condition in Reaction Bed

This is an uphill task in SSF.

Question 3 ***Why is it so?***
Answer: Organism does not like disturbance (mixing) during reaction.

Question 4 ***What will be the effect on process?***
Answer: Without mixing, it will cause thermal gradients and localization of gases (air/CO_2) in the reactor.

(6) Facilitate Upstream and Downstream Operations

They are listed below.

- Reactant preparation
- Inoculum preparation
- Aseptic measures (sterilization)
- Loading and unloading of bioreactor
- Product separation

(7) Mode of Operation

It is difficult to understand whether there are certain common things or not. There are certain doubts, which one can ask in this context.

- Can we categorize them in some groups?
- Can we analyze them effectively?

We have a long list of names of bioreactors. They are developed for specific use and application.

2.7 CLASSIFICATION OF BIOREACTORS

A number of general classifications of bioreactors have been proposed. In this context, we shall begin to classify the reactors based on more classical application-oriented processes (Fig. 2.37).

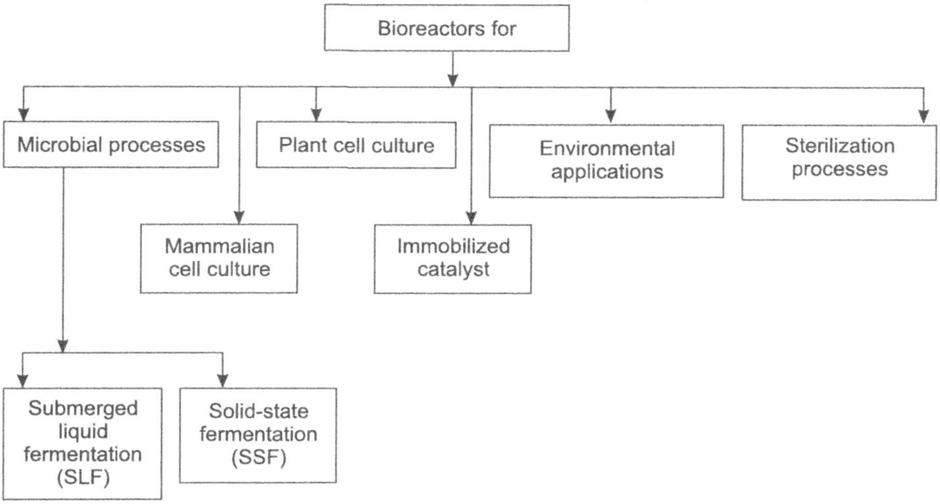

Figure 2.37 General classification of bioreactors based on applications to various processes.

2.7.1 Classification of Bioreactors in SLF

A considerable number of bioreactors differing in construction and operation have been used for various microbiological processes. Now to select and that too a practical one for a specific microbiological process is difficult and hence, there is a need for a rigorous system of classification. Various researchers have proposed separate approaches to classify the bioreactors. We shall refer a few important classification systems with relevant sketch of reactor. They are:

(A) Kafarov *et al.* classification
(B) Bungay and Belfort approach
(C) Schügerl's method of classification

Note: No single classification is unique.

(A) Approach of Kafarov et al.

Four different classes of bioreactors are suggested here (Kafarov *et al.*, 1988).

(1) Method of Supplying Energy and Aeration

They are of three different types.

(a) Energy supply in the gaseous phase The schematic diagrams of air-lift and bubble column reactors are given in Figures 2.38a and b.

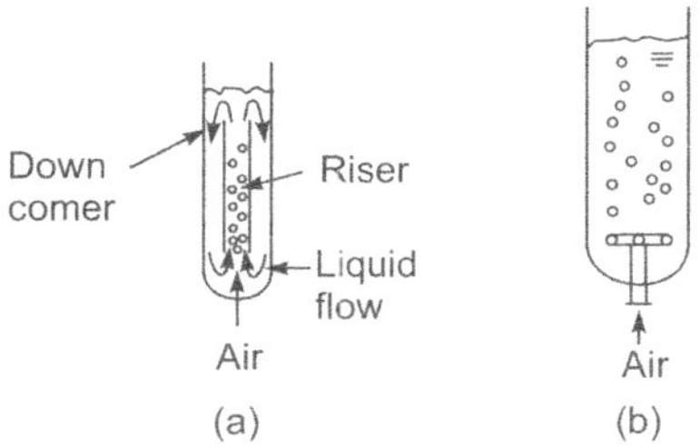

Figure 2.38 (a) Airlift bioreactor. (b) Bubble column bioreactor.

(b) Energy supply in the liquid phase Spouted bed reactor may be considered as an example (Fig. 2.39).

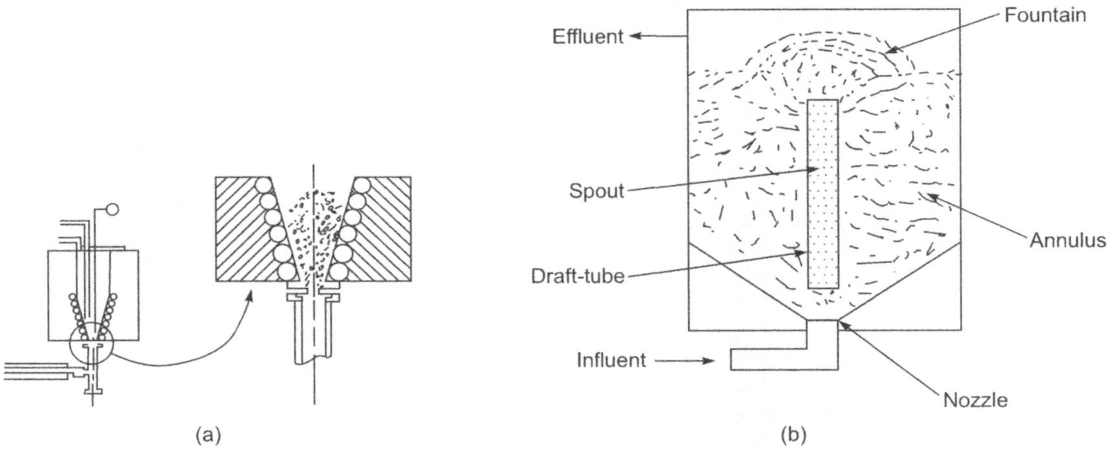

Figure 2.39 Spouted bed bioreactor.

(c) Combined energy supply Mechanically agitated reactor is a good example for this class (Fig. 2.40).

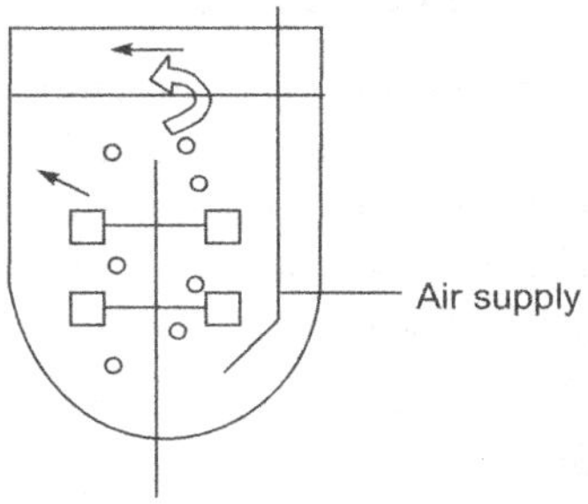

Figure 2.40 Bioreactor with mechanical agitation.

(2) Differences in Hydrodynamics

Based on hydrodynamics, bioreactors are sub-classified into three categories.

(a) Perfect mixing All ideal reactors like Figure 2.41(a) is given below.

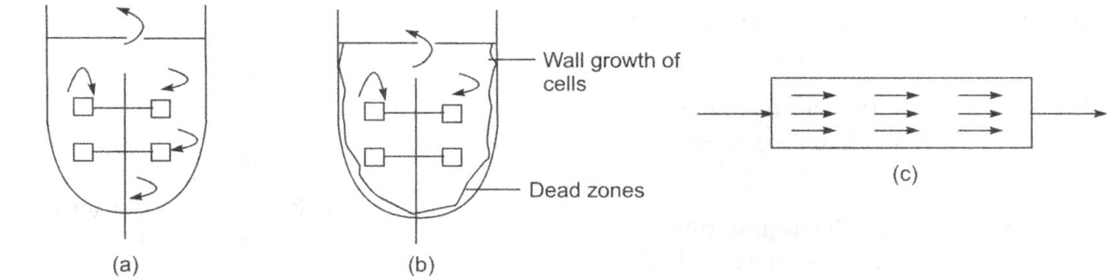

Figure 2.41 (a) Mechanically agitated batch bioreactor. (b) Mechanically agitated batch bioreactor with wall growth. (c) Ideal plug flow bioreactor.

(b) Partial mixing Non-ideal reactor is given in Figure 2.41b.

(c) Plug Flow Ideal PFTR, as per Figure 2.41c, indicates the concept.

(3) Principle of Operation

They are of three different types.

(a) Batch bioreactor All reactants are added to the reactor followed by proper sterilization and inoculum is added to it (Fig. 2.42). Reaction is followed as a function of time. At appropriate time, reaction is terminated. Then product synthesized in the process is isolated and separated.

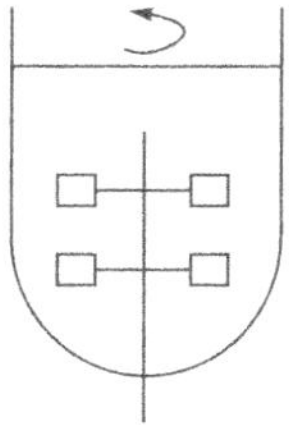

Figure 2.42 Batch bioreactor.

(b) Semi-continuous bioreactor Two different modes can be found, viz., batch-fed (Fig. 2.43) and fed-batch.

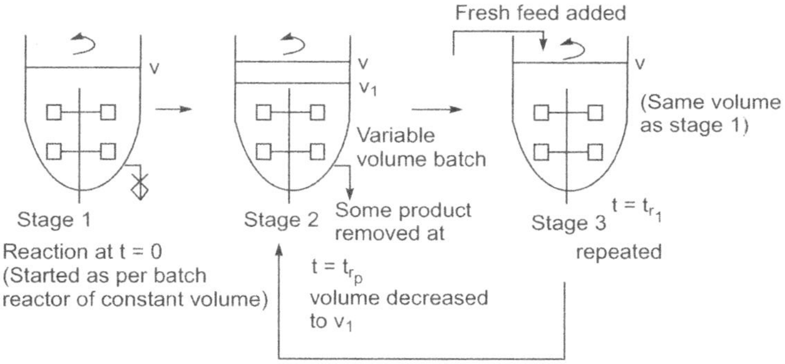

Figure 2.43 Semi-continuous (Batch-fed).

(i) Batch-fed This is an example of semi-continous mode of operation. A large number of examples are available in chemical reaction systems (Levenspiel, 1972).

(ii) Fed-batch mode Initially, it is started as variable volume batch till the desired volume is achieved. Then it is switched to continuous flow mode, but the input feed rates and output flow rates are not constant. These flow rates depend on the appropriate metabolic function, viz., respiratory quotient of the process organism. Detailed operation will be discussed in Chapter 3.

Question 5 ***What is the difference between a batch-fed and a fed-batch mode of operation?***

Answer:

Batch-fed mode	*Fed-batch mode*
1. Variable volume batch reactor with product output when it is formed	Initially variable volume batch reactor with no product output. Later variable flow rate, continuous flow, constant volume reactor
2. Feed is injected only when some product is removed	Feed rate depends on the metabolic function of the cell (i.e., respiratory quotient = RQ). If cells are at active growth phase, feed rate is high. When cells are at stationary phase, feed rate is low.

(c) Continuous bioreactor The examples are continuous flow stirred tank reactors (back-mix reactors-CFSTBR), plug flow reactors (PFTR), packed bed reactor (PBR), etc. (Fig. 2.44).

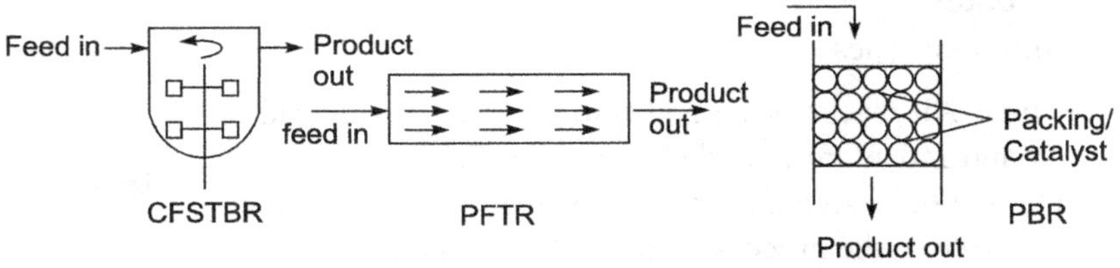

Figure 2.44 Continuous bioreactors.

(4) Conditions of Sterility for the Biosynthesis

(a) Sterile operation This is applied to the production of pharmaceutical and health related products.

(b) Aseptic operation This is applicable for fermentation industries.

(c) Non-sterile operation Reactors used for waste stabilization and treatment of waste are suitable examples.

(B) Bungay and Belfort Approach

Four ways of reactor classification has been described by Bungay and Belfort (1987).

(1) Combination of the mode of reactant feed to the reactor and reactor geometry, viz., batch (Fig. 2.45a), continuous flow (Fig. 2.45b), tubular packed bed (Fig. 2.45c), and fluidized bed bioreactor (Fig. 2.45d).

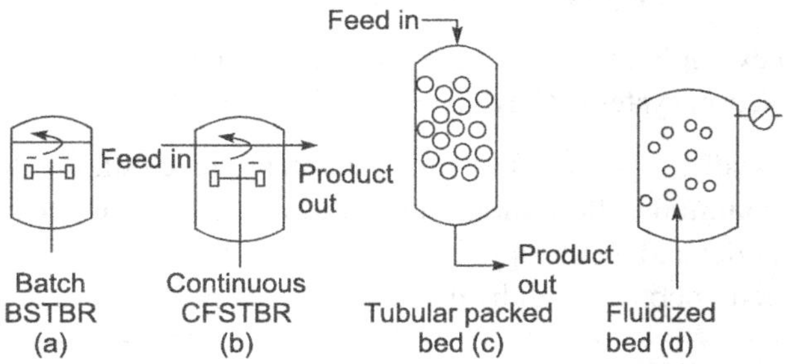

Figure 2.45 Combination of the mode of reactants feed.

(2) Configuration of biocatalysts within the reactor is:

(a) Free
(b) Suspended
(c) Immobilized

(3) Mode of providing mixing within the reactor: Mechanical systems, fluids and combination thereof are the various modes of mixing devices in reactor.

(4) Type of biocatalyst used in the reactor are enzyme, aerobic microbes, anaerobic microbes, etc.

This may include recent reactors developed for specific use and applications indicated in Section 2.5. The concept of classification is based on biocatalysts used (Table 2.2).

Table 2.2 *Bioreactors for different processes*

Type of phase / Catalyst	*Enzyme*	*Microbes*		*Immobilized biocatalyst*	*Entrapped biocatalysts*
		Aerobic	*Anaerobic*		
Homogenous liquid phase	BSTBR CFSTBR				
Two phase		BSTBR, CFSTBR, Bubble column, Airlift		BSTBR, CFSTBR, Air-lift, Packed bed, Fluidized bed, Hollow fiber, Membrane bioreactor	BSTBR, CSTBR, Hollow fiber, Membrane bioreactor
Multiphase			BSTBR, CFSTBR, Bubble column, Loop reactor		

(C) Schügerl's Method of Classification

There are many factors to be considered for bioreactor design (Schügerl, 1982) such as

- reduction of capital costs
- reduction of reactant evaporative losses
- reduction of energy input costs
- require to develop very large reactors, leading to additional design problems associated with increased power costs
- need for low shear reactors for shear sensitive cells, e.g., plant cell and mammalian cell culture. This problem is discussed under "Bioreactors for mammalian cell culture".
- need for increased reactant conversion.

To accommodate above problems, Schügerl's classification appears to be systematic.

Three categories of reactors are described in Tables 2.3-2.5.

Category	*Basic principle*	*Objective*
1.	Mechanically movable internals (Table 2.3)	Input energy for agitation and mixing
2.	Fluid circulation using external pump (Table 2.4)	Input energy through fluid circulation
3.	Compressed gas sparging (Table 2.5)	Input energy mediated by compressed gas

Conceptual drawings in the tables are based on Schügerl (1982).

Category (1) Mechanically Moved Internals in Bioreactors

In general, this category of reactors is of tank and column construction (co-current and counter current operations) with different inserts (agitators/spargers/stage separating trays/rotating internals).

We describe here the sub-classification with their advantages and disadvantages.

Table 2.3 *Bioreactors with mechanically moved internals*

Sub-class	*Conceptual figure*	*Description*	*Specific advantages (A) and disadvantages (DA)*
Fig. 2.46	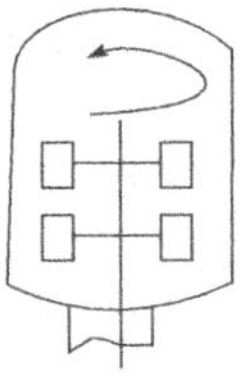	Mechanically stirred reactor with suspended baffles	A: Better mixing DA: Not suitable for shear sensitive cells
Fig. 2.47	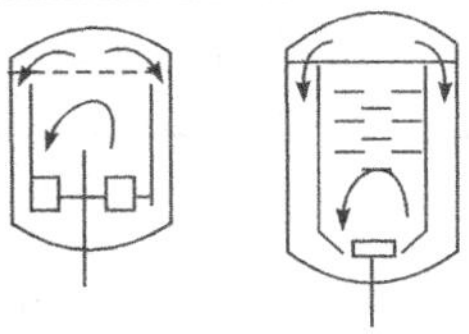	Mechanically stirred loop reactor	A: Reduced dead zone and shear DA: Reduced back mixing between loop and reactor wall which might increase the growth of cells on wall.
Fig. 2.48	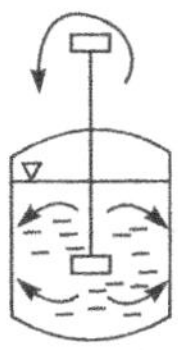	Self-aspirated, aerated reactors mechanically stirred	A: Reduced shear DA: Not suitable for fungal cultivation
Fig. 2.49	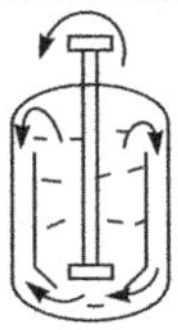	Mechanically stirred, aerated loop reactors	A: Reduced shear DA: Problem of wall growth
Fig. 2.50	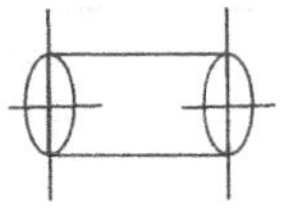	Horizontal loop reactors	A: Better for surface growth DA: Reduced mixing
Fig. 2.51	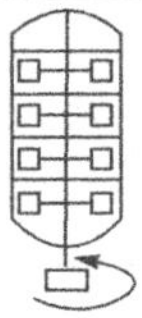	Cascade reactors with rotating mixing element	A: Better mixing, reduced dead space DA: More energy requirement

Contd.

Table 2.3 *Contd.*

Sub-class	*Conceptual figure*	*Description*	*Specific advantages (A) and disadvantages (DA)*
Fig. 2.52	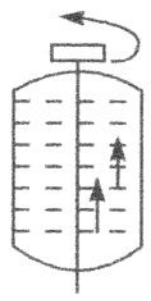	Cascade reactors with axial mixing element	A: Reduced shear and energy requirement DA: Mixing less uniform than stirred system
Fig. 2.53	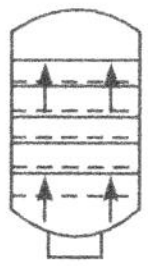	Cascade reactor with pulsating liquid	A: Shear sensitive cells DA: Scale-up problem
Fig. 2.54	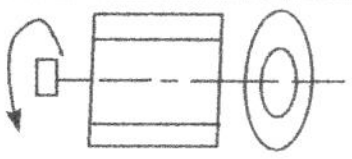	Thin layer	A: Higher surface area for reaction DA: Not suitable for suspended cells
Fig. 2.55	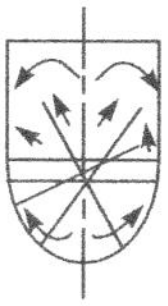	Paddle wheel reactor	A: Better mixing DA: Not suitable for shear sensitive system
Fig. 2.56	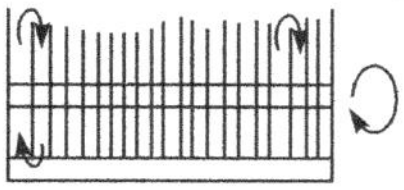	Disk reactor	A: Uniform mixing DA: High energy input, High shear

Category 2 Fluid Circulation Using External Pump

Energy input in this category of reactors is by means of an external pump. In general, reactors are also of "Tank" and "Column" construction (up-flow/down-flow/counter current). In addition to this, there are loops of different design (external loop/coaxial drafts/vertical partition wall) (**Table 2.4**).

Table 2.4 *Bioreactors with energy input by circulation with an external pump*

Sub-class	*Conceptual figure*	*Description*	*Specific advantages (A) and disadvantages (DA)*
Fig. 2.57	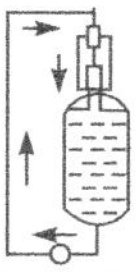	Plunging-jet	A: Mass transfer improved DA: Shear force increased which might affect shear sensitive cells
Fig. 2.58	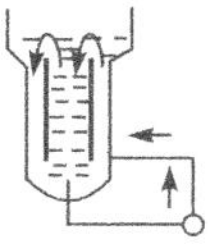	Jet-loop	A: Better gas dispersion DA: At increased velocity, a region is reached where jet approaches constant value due to air resistance

Contd.

Table 2.4 *Contd.*

Sub-class	*Conceptual figure*	*Description*	*Specific advantages (A) and disadvantages (DA)*
Fig. 2.59	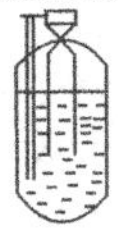	Plunging-channel	A: Combined advantages of Fig. 2.58 DA: Increased air forces
Fig. 2.60	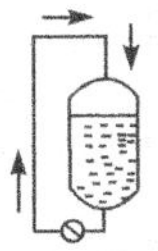	Nozzle-loop	Similar to Fig.2.58
Fig. 2.61	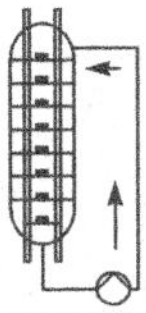	Multistage bubble column with counter flow	A: Reduced non-ideality DA: Difficult to operate with fungal system
Fig. 2.62	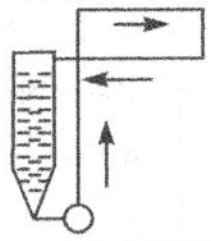	Tubular loop	Almost like nozzle-loop reactor
Fig. 2.63	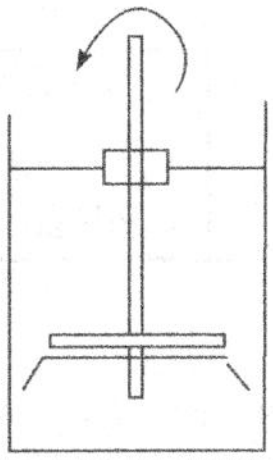	Rotating injector nozzle	A: Improved Mass transfer DA: Power consumption higher
Fig. 2.64	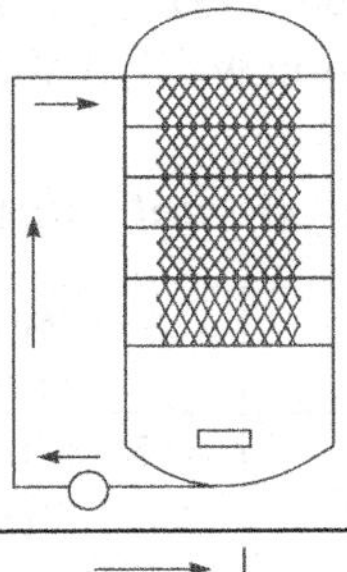	Packed bed reactor	A: Improved contact time DA: Higher pressure drop
Fig. 2.65	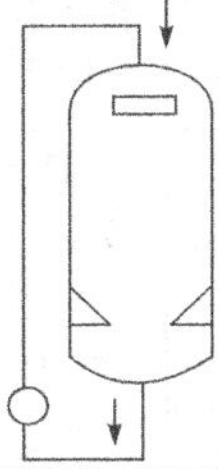	Bubble column down flow	A: Improved mass transfer than classical bubble column reactor DA: Not suitable for mycelial organisms

Category 3 Energy Supply by Compressed Gas Sparging

The emphasis is on the primary dispersion device for gas phase (Table 2.5). By developing two-phase dynamic aeration systems, various devices (venturi, nozzles, injectors, ejectors) are introduced to increase the specific interfacial area of the dispersed gas bubbles (Fig. 2.76). At the same time, control of foam formation is a difficult task.

Table 2.5 *Bioreactors with compressed gas sparging*

Sub-class	*Conceptual figure*	*Description*	*Specific advantages (A) and disadvantages (DA)*
Fig. 2.66		Single stage bubble column	A: low shear forces DA: Non-uniform mixing
Fig. 2.67		Single stage bubble column with control draft tube	Same as Fig. 2.66
Fig. 2.68	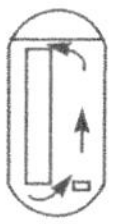	Internal tubular loop	Same as Fig. 2.66
Fig. 2.69		Vertical partition wall	Same as Fig. 2.66
Fig. 2.70	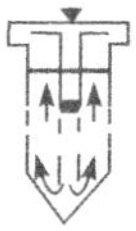	Downflow loop reactor	A: Low shear, mixing is marginally improved DA: Not applicable for fungal system
Fig. 2.71	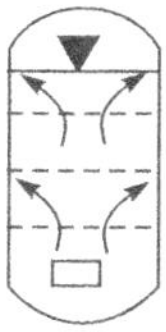	Co-current bubble column with stage separating trays	A: Improved mixing DA: Clogging by cells
Fig. 2.72	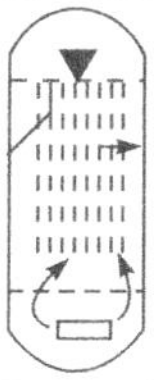	Co-current bubble column with static mixing level	A: Improved mixing DA: Clogging by cells, Difficult to scale up

Contd.

Table 2.5 *Contd.*

Sub-class	*Conceptual figure*	*Description*	*Specific advantages (A) and disadvantages (DA)*
Fig. 2.73	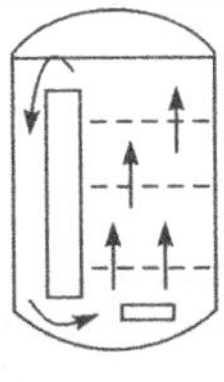	Loop with stage separating trays	A: Improved mixing DA: Clogging by cells, Difficult to scale up
Fig. 2.74	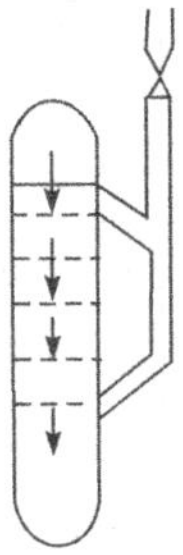	Bubble column with stage separating trays, external loop and pneumatic liquid pulsing	A: Suitable for shear sensitive cells DA: Clogging by cells, Difficult to scale up
Fig. 2.75	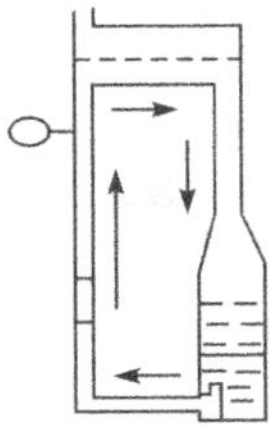	External tubular loop	A: Suitable for shear sensitive cells DA: High pumping capacity required

2.7.2 Classification of Bioreactors in SSF

Table 2.6 gives a comparative analysis of SSF and SLF with respect to the parameters influencing bioreactor design.

Two categories of reactors used in SSF processes are

(A) Laboratory scale
(B) Pilot/and industrial scale

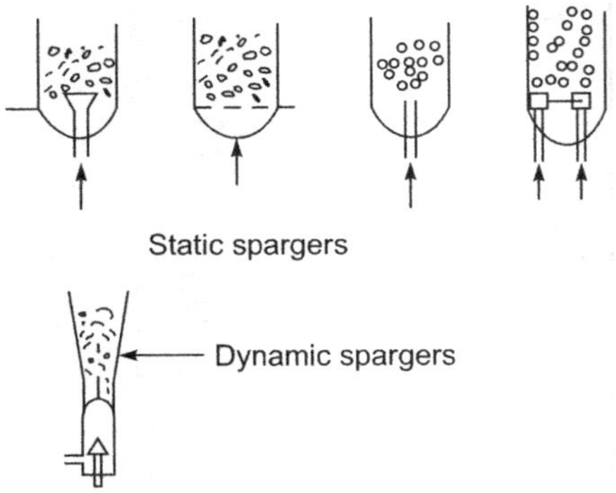

Figure 2.76 Configuration of spargers used to supply energy by compressed gas sparging in the bioreactors.

(A) Laboratory Scale

In earlier Section 2.3, use of slant and Petri-plate cultures have been explained elaborately. Further developments are given below.

(a) Without forced aeration and agitation

Relevant examples are jars, wide-mouth Erlenmeyer flasks, Roux bottles, roller bottles, etc.

Table 2.6 *Comparison of SSF and SLF*

S. No.	*SSF*	*SLF*
1.	Solid media contains *less water*.	Solids (either suspended or dissolved in liquid phase exist)
2.	*Impervious gas phase* between particles. This feature is important due to poor *thermal conductivity of air* compared to water.	This is a mixture of gas(G)-liquid(L) solid(S) phases.
3.	*Wide variety of matrices* used in SSF in terms of composition, size, mechanical resistance, porosity and water holding capacity (Durand, 2003)	This is a typical G-L-S multiphase system.
4.	*pH and temperature control* are difficult.	Classical pH and temperature control are followed.
5.	*Oxygen transfer* is a typical problem along with the complex control of temperature and water content in some design.	$k_L a$ is a typical quantity defined for all SLF systems.
6.	Morphology of the organisms, specially with fungal system are restored.	This is equally a complex problem to consider these effects on bioreactor design.
7.	Scale-up is difficult due to intense heat generation and heterogeneity of the system.	Scale-up is easier in this case. A few thumb rules in addition to chemical reactor scale-up may be followed.

Advantages

- It is a simple operation.
- Reduced power for aeration is observed.
- Large number of replications is possible.
- Temperature of the place, where they are kept, can be controlled.

Disadvantages

- Variation in production is observed.
- It is difficult to control parameters.

(b) With forced aeration

1. Without agitation
   - I. INRA – Small reaction chamber (developed around 2000)
   - II. ORSTOM (design developed during 1975–1980) (Durand, 2003)

Advantages

- It is possible to aerate culture.
- Measurement of respiratory rate of microorganism is possible.
- This is convenient for screening studies.

Disadvantages

- Sampling during fermentation is difficult.

I. INRA Small Reaction Chamber

This reactor is a modification of ORSTOM model. There are provisions for:

- measuring temperature during reaction by temperature probe
- controlling water activity in the reaction phase with water monitoring device

Advantages

- Regulation of water content is possible in the reaction phase
- Sampling during process is possible.
- Maintenance of water activity and temperature during process are possible.
- The range of air-inlet temperature and flow rate can be widened to get useful data.

II. ORSTOM Reactor

This is a typical packed bed reactor (Raimbault and Germon, 1976). Packing is solid reactant where humidity is controlled by passing moist air (Fig. 2.77). A series of similar units are arranged on a temperature controlled chamber. The device is provided with a vent at the top for release of CO_2 generated by respiration of the organism.

Advantages

- It is convenient for screening studies.
- Optimization studies are possible.
- Heat removal from reaction bed is easier.
- It is easy to fabricate and handle the system.
- Forced aeration is possible.

Disadvantage

There is no freedom for collecting sample during reaction.

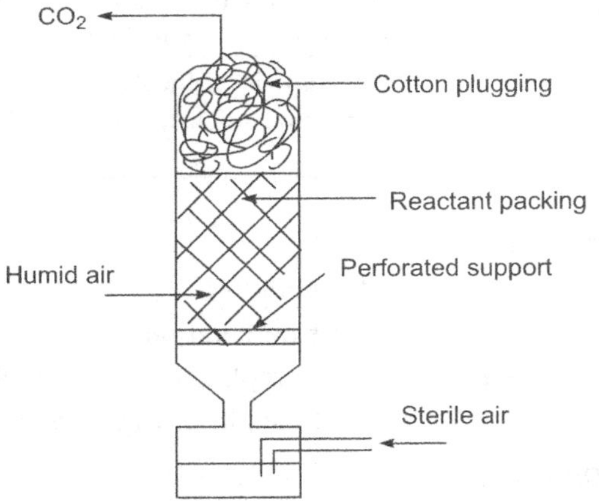

Figure 2.77 Typical laboratory scale bioreactor based on ORSTOM (not a real unit).

2. With agitation

   Rotating drum bioreactor of the following configurations is described in the literature.

   - with a perforating drum
   - with a horizontal paddle mixer (developed by Waginengen University, Durand, 2003) (Fig. 2.78).

Advantages

- Continuous mixing increases contact between reactor wall and solid medium.
- It facilitates oxygen transfer to microorganisms.

Disadvantages

- Agglomeration of reactant particles during the growth of mycelium increases difficulty in regulating temperature of solid medium.

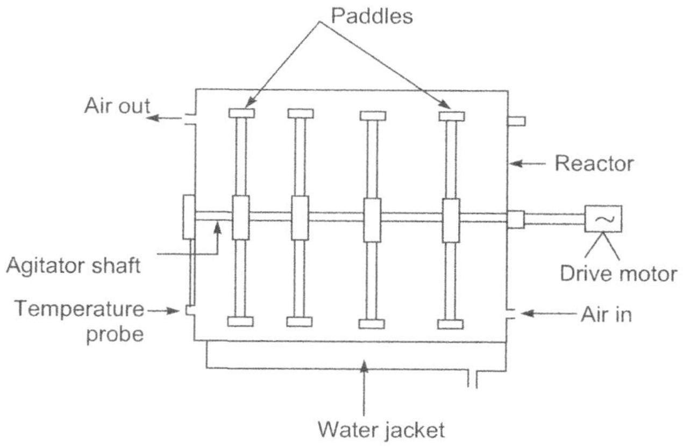

Figure 2.78 Conceptual drawing of horizontal paddle mixer.

- Oxygen transfer within agglomerated particles is very low or nil.
- Waterjacket on the working body of reactor is a serious problem that increases with scale.
- Due to continuous agitation, structure of solid modifies to a pasty mass and sometimes clay materials present cause abrasive action on mycelium and damage it.

3. Static bed: ORSTOM design is called Zymotis (Reussos *et al.*, 1993; Mitchell *et al.*, 2000) (Fig. 2.79).

   This consists of the following components.

   - Vertical internal heat transfer plates
   - Air passage through a temperature control unit

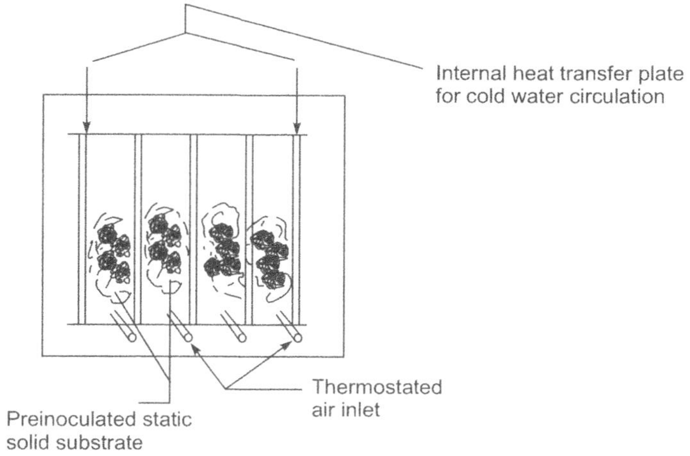

Figure 2.79 Conceptual drawing of Zymotis.

Disadvantages

- Shrinkage of medium volume that occurred during mycelial growth causes reduced contact with heat transfer plates.
- So poor heat transfer and air channeling are observed.
- Scale-up is difficult.

(B) Types of Pilot / Industrial Scale Bioreactors

A number of reactors have been efficiently developed for small scale operation. As we have observed with some of the designs, viz., 'Zymotis' and the model developed by INRA group, there are crucial problems of heat and mass transfer. However, one needs to address such situation and exploit it for large scale operation.

Question 6 ***Why do we have less number of options for large scale operations in SSF?***

Answer: Probable reasons may be considered here which will be seriously looked into the design of such bioreactor.

- Serious problem of shrinking of reaction bed affects heat and mass transfer.
- Shear sensitivity of microorganisms used is related to mechanical agitation. The agitation rate is correlated with the oxygen mass transfer and the distribution of temperature in the reaction bed.
- Degree of sterility desired for the process.
- Difficulties associated with charging reactants and emptying of reactor after reaction.
- Requirement of more manpower.
- Overall economics.

Let us review the varieties of bioreactors in this category.

Pilot / Industrial Scale Bioreactors

A probable classification is given in this context.

(a) Bioreactors without forced aeration
(b) Unmixed bioreactors with forced aeration
(c) Continuously mixed bioreactor with air circulation
(d) Discontinuously mixed bioreactor with forced aeration is classified as described in Figure 2.80.

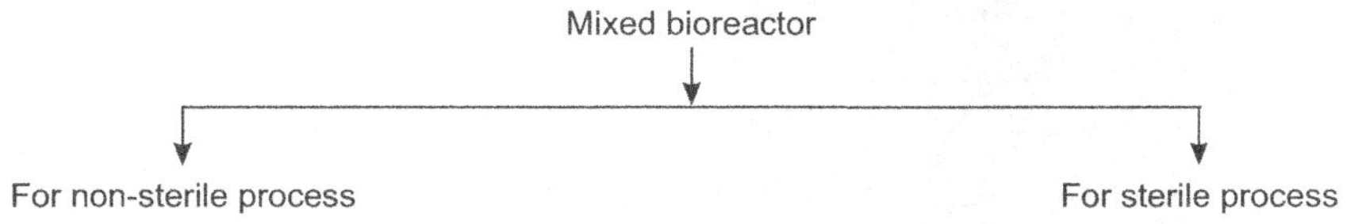

Figure 2.80 Scheme of probable classification of bioreactors used in Pilot / Industrial Scale (Durand, 2003).

Let us examine the advantages and disadvantages of each category of reactor.

(a) Bioreactors without Forced Aeration

This is a tray reactor (Durand, 2003). Koji process employs similar configuration (Fig. 2.81).

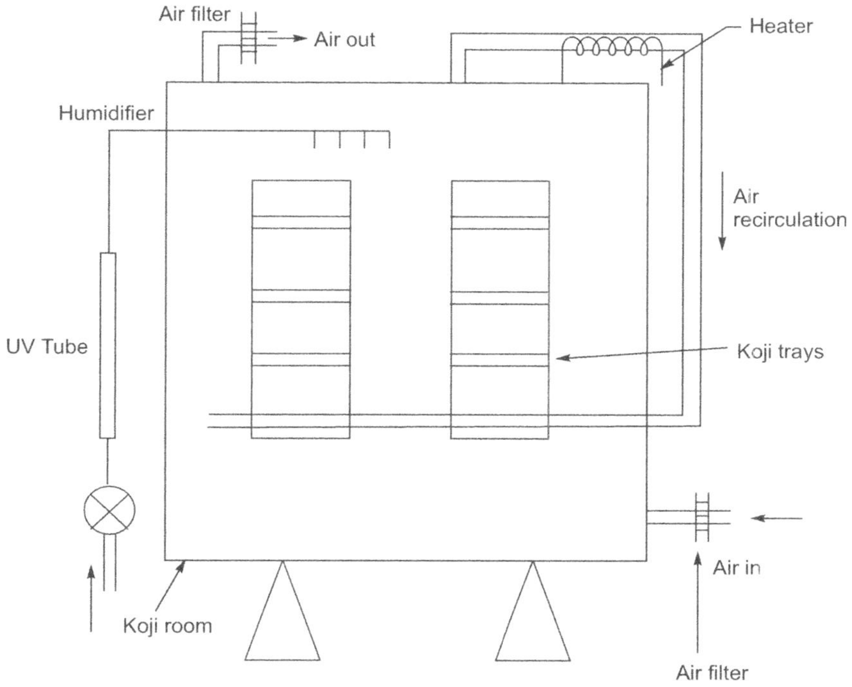

Figure 2.81 Koji type bioreactor.

Disadvantages

- More space is required for the bioreactor.
- It is labor intensive for the operation of this bioreactor.
- High capacity blowers are needed for air supply / circulation and for air exhaust.

(b) Unmixed Bioreactors with Forced Aeration

This is basically a packed bed reactor (Fig. 2.82).

Important features of this reactor are:

- No mechanical agitation is involved.
- Regulated air flow with controlled relative humidity creates agitation in the bed.
- Regulation of relative humidity and temperature is necessary in the reactor.

Disadvantages

- Control of metabolic heat production is a difficult job using this reactor.
- Temperature gradient along the bed is observed in the reactor.

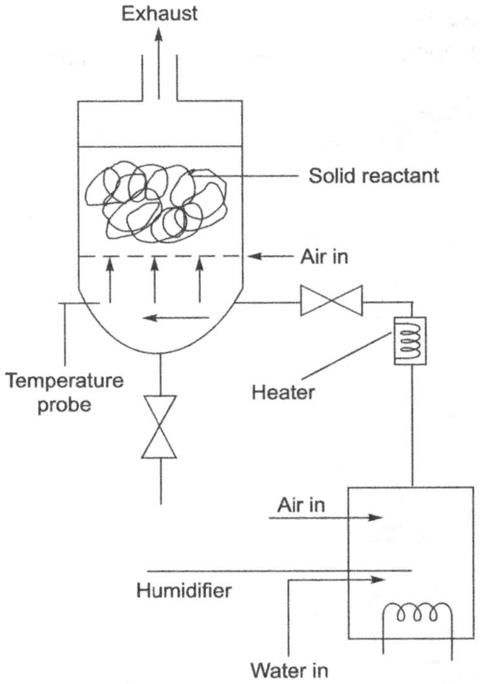

Figure 2.82 Conceptual drawing of packed bed bioreactor (Durand, 2003).

- Drying of bed is observed during the operation of this reactor.
- Improper distribution of water within the bed is a common problem in this reactor.

To reduce aeration, the following modified reactors are introduced. The facility for the supply of sterile air through each plate below heat exchangers is installed for the improved operation of this reactor.

For example:

- Prophyta – Used by a Company in Germany (Durand, 2003)
- Plafractor^{TM} – Developed by Biocon (India) (Suryanarayanan and Mazumdar, 2000)

Plafractor^{TM} reactor

The major features are listed below:

- Heat removal by conduction.
- Stacking and interconnecting individual modules.
- Non-communicating channels, deliver cooling and heating fluids, sandwiched between two sheets.
- Communicating channels deliver fluids for sterilization, for adjusting moisture, oxygen, and for extraction of compound of interest after reaction.
- Inside each module, there is a mixing arm revolving around the axis.

The schematic diagram of this reactor is described by Durand (2003).

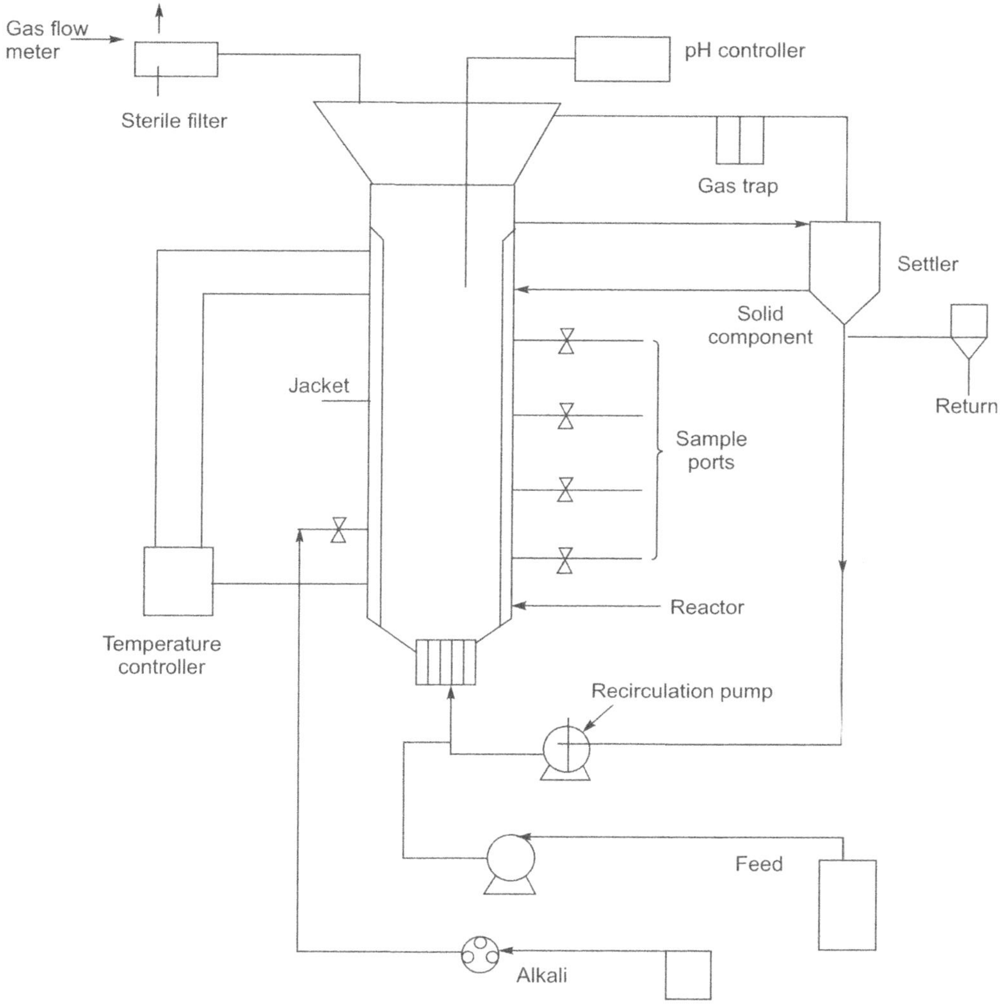

Figure 2.83 Typical fluidized bed bioreactor.

(c) Continuously Mixed Bioreactors with Air Circulation

Features are described below:

- Configuration includes rotating drum, fluidized beds (Fig. 2.83) and stirred drums
- Continuous mixing maximizes exposure of each reactant to thermostated air rotating drum reactors
- Good fluidization involves in fluidized bed reactor with a mixer at the bottom to break any substrate agglomeration and a space for disengagement of particles (Matsuno *et al.*, 1993).

Disadvantages

- Aggregates formed reduce heat, mass, and oxygen transfer.
- Shear caused by drum rotation affect mycelial growth.

(d) Discontinuously Mixed Bioreactor with Forced Aeration

Two varieties of reactors appear in this category depending on the processes, i.e., for non-sterile and sterile processes.

Common Features are listed below.

- They are packed bed reactors.
- Air passes through beds.
- Agitation device periodically operates to mix the reactants.
- Water sprayed to maintain water activity in the solid reactant bed.

1. Reactors for non-sterile processes

- Koji equipment with rotary disc (Reussos *et al.*, 1993) – Microcomputer controls temperature of inlet air, airflow rate, and agitation period.
- INRA Reactor (INRA group, Dijon, France) (Mitchell *et al.*, 2000)

Sensors fitted with the reactor are for the measurement of the following:

(i) Temperature
(ii) Moisture
(iii) Relative humidity
(iv) Inlet air flow rate

Other devices are like the previous configuration.

2. Reactors for sterile processes

Principle: This is described in reactors for non-sterile processes.

Question 7 ***Why do we need such a reactor?***

Answer:

- For applications, in pharmaceutical, food and health industries and for legislative reasons, product obtained in the process must be sterile.
- Microbial processes always call for sterile / aseptic environment.

(i) Reactor developed by Wageningen University and Hosokawa Micron (Rinzema *et al.*, 2000). Features of this reactors are:

- Conical vessel with a ribbon impeller
- *In situ* sterilization device
- Temperature sensor records at various heights of the reactor

(ii) Reactor devised by Durand *et al.*, (1994).
Features are listed below:

- Planetary mixing device

- Coupling with microcomputer to control and monitor functions during sterilization of empty bioreactor, sterilization of medium, control of parameters during reaction and data acquisition

The classification of reactors used for SSF and their scale-up appear to be inconclusive.

Reasons could be the following.

1. No proper quantitative approach has been established for heat and mass transfer aspects in SSF.
2. Development of probes is necessary for moisture measurement, control and regulation of process variables.
3. Development is necessary in the area of analytical procedures for reacting compounds, viz., solids and gases.
4. Problems are linked to the increase in volume / mass compaction / shrinkage, etc.

Question 8 ***Can we replace SLF by SSF?***

Answer: Volumetric productivity is higher in an equivalent SSF process compared to SLF. On that basis, one might think this question.

From economical consideration (Nigam and Singh, 1994), the change over from an existing established SLF process to SSF is not possible.

One thinks the next question.

Question 9 ***When can we consider the option for SSF?***

Answer: If we can improve the reactor and process control strategies, for deciding new investment, SSF can be the best alternative solution for a production process.

2.8 BIOREACTORS FOR ANIMAL CELL CULTIVATION

Generally, the bioreactors mentioned so far or will be discussed for other applications cannot be used as such for animal cell cultivation. In some cases, it requires modification or in majority of the cases cannot be used at all. The fundamental reason is that animal cells differ in structure and properties.

In most of the cases, the beginners are enthusiastic about the exploitation of any kind of animal cells. It is better to understand a little about the preferences of animal cell culture systems.

(a) Preferred Animal Cell Culture

They are:

- Animal cell properties
  - Physical characteristics: Animal cells have higher total dry weight per cell as they are larger than the microbial cells, for example: MRC-5 (human diploid fibroblast) has 28 times higher total dry weight per cell than *S. cerevisiae* cell.
  - Lower normal density is observed for animal cells than microbial cells.
  - A higher settling velocity is observed for animal cell than microbial cell. So it is easy to recycle animal cells.

- o Animal cells have a 10 to 100 times less surface area to volume ratio than microbial cells.
- o Animal cells display lower specific nutrients absorption rate (lower metabolic activities) than microbial cells
- o It has lower growth rate than microbial cells. CHO cells (Chinese Hamster Ovary cells) almost doubles in 18 h whereas *S. cerevisiae* cells doubles in approximately 1.5 h.

- Surface properties

Table 2.7 *Comparison of surface properties of animal cells and microbial cells*

Sl. No.	*Animal cells*	*Microbial cells*
1.	They do not have rigid cell wall or capsule. So they are sensitive to mechanical and chemical stresses.	They have rigid cell wall or capsule. So they are resistant to mechanical and chemical stresses.
2.	Cells are non-transformed, i.e., primary cells and diploid cell lines cannot grow in suspension. They need support.	Such problems do not arise here.
3.	Anchorage dependent cells cease to grow as they reach a particular cell density of a confluent monolayer culture.	This is a function of reactant.
4.	Cell surface has a variety of components needed for cellular communication, adhesion spreading, etc.	It is not that much prominent.
5.	For proper cell attachment, non-hydrophobic, positively or negatively charged components having high surface free energy is necessary. Classical glass or plastic surfaces are negatively charged.	This is visible in bio-film formation.

- Nutritional requirements
  The variation in growth yield of animal cells is due to the culture medium than operating conditions. On the other hand, microbial growth rate is quite high even on simple reactants. Animal cells require complex reactants including serum for growth.
- Control of process parameters and metabolic activities
  - o Temperature: Optimal growth of animal cells occurs between 36 °C and 38 °C.
  - o pH: Optimal values are between 6.8 and 7.8. Generally, $CO_2/NaHCO_3$ buffer system is used. It is better to use approximately 10 % CO_2 in the gas phase.
  - o Osmolarity: Narrow range of osmotic pressure is suggested for animal cells.
  - o Dissolved Oxygen: Optimal values lie between 30 and 40 % of saturation value with air.
- Metabolic activities
  Bioreactor for animal cell culture should be able to control metabolic activities. It is difficult to control metabolic activities in batch culture. Continuous or perfused culture is better for animal cells.

For varieties of cells or cell lines, mass production in reactors considers the following important parameters for bioreactor design:

- basic design of bioreactor and pipeline arrangements
- type of cells
- cell surface integrity and anchors for growth
- sterilization and operating temperatures
- rheological properties
- aeration type including bubble size and surface area / volume ratio.

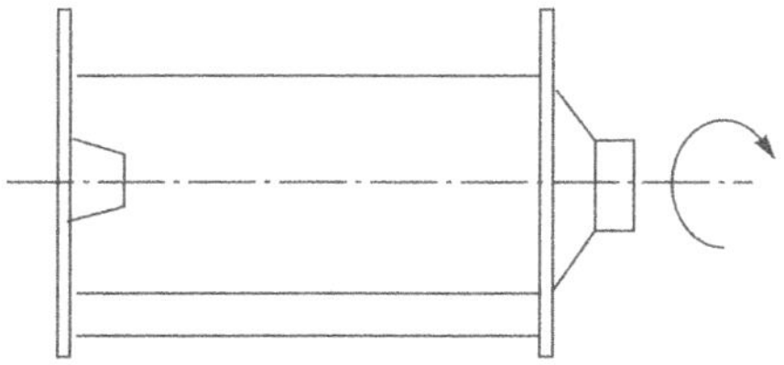

Figure 2.84 Roller bottle culture flasks.

2.8.1 Development of Bioreactors

(a) Conventional Laboratory Culture Reactors

Roller bottles to disposable plastics and larger bottles are the example of such reactors (Fig. 2.84). It has been changed since 1930.

Spinner flasks, after certain modification, take the concept of impellers and magnetic stirrers for better anchorage dependent growth.

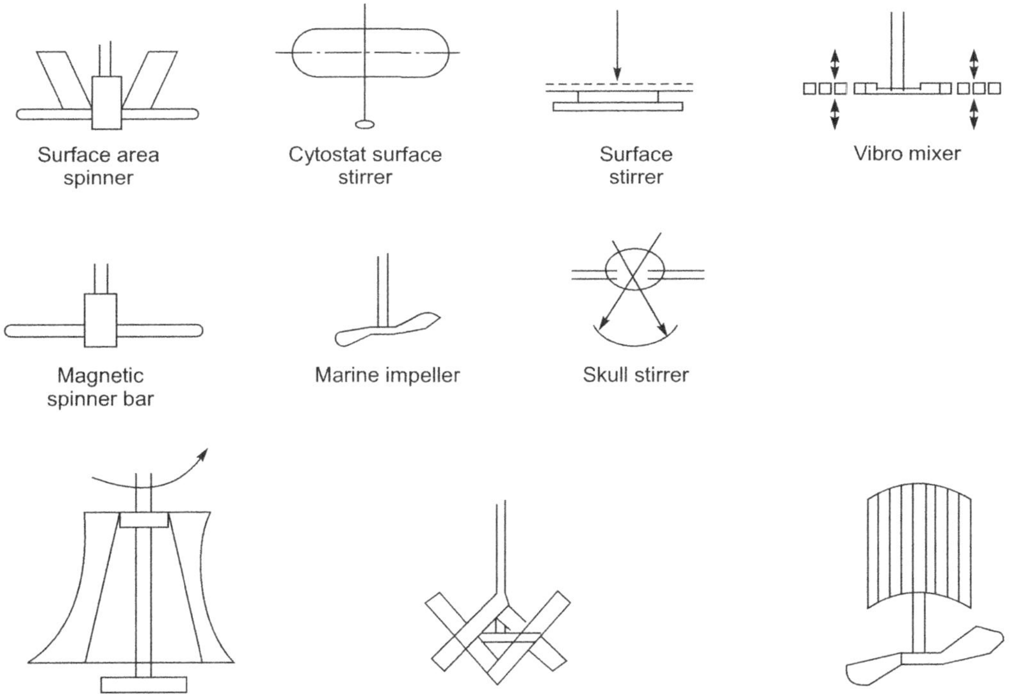

Figure 2.85 Various impellers.

Applications

Batch and extended batch operations are exploited in research laboratories, pharmaceutical industry, and production laboratories.

Advantage

- Replication of experiments is possible.

Disadvantage

- It is labor intensive for maintaining cultures.
- It has limited cell density, e.g., 10^6 cells / ml (Propst *et al.*, 1989).
- Spinner flasks are scaled up to 10 L or 20 L for batch cultures.

(b) Flexible Cell Culture Flasks

In this case, flasks are made of gas permeable plastics.

Advantage

- Closed system transfer (no pipetting) is possible.
- Improved $O_2 - CO_2$ gas exchange for high-density cultures is observed.
- Easy scale-up is possible within a conventional CO_2 incubator (Bacehowski *et al.*, 1990).
- It facilitates seeding of bioreactor through direct septum tubing connections, for reducing contamination.

Disadvantage

- It has limited size, i.e., maximum up to 3 L.

(c) Stirred Tank Bioreactors and its Developments

Initially, reactors used for microbial cultures were employed for mammalian cells. There are developments in this configuration (Fig. 2.86).

(d) Airlift Bioreactors and its Development

These reactors are evolved from aerobic microbial cultures (Fig. 2.87). However, the modifications are:

- vessel shape
- draft tubes (to improve circulation)
- efficient heat exchangers
- sophisticated process controls
- less damaging spargers

Advantages

- Fast conversion
- Cost-effective and efficient for scale-up of hybridoma cells

Disadvantages

- Cell densities greater than that obtained in roller bottle or spinner flask cannot be sustained in this reactor.
- Mass transfer limitation of O_2, nutrients / metabolites cause lesser concentration and purity of products.

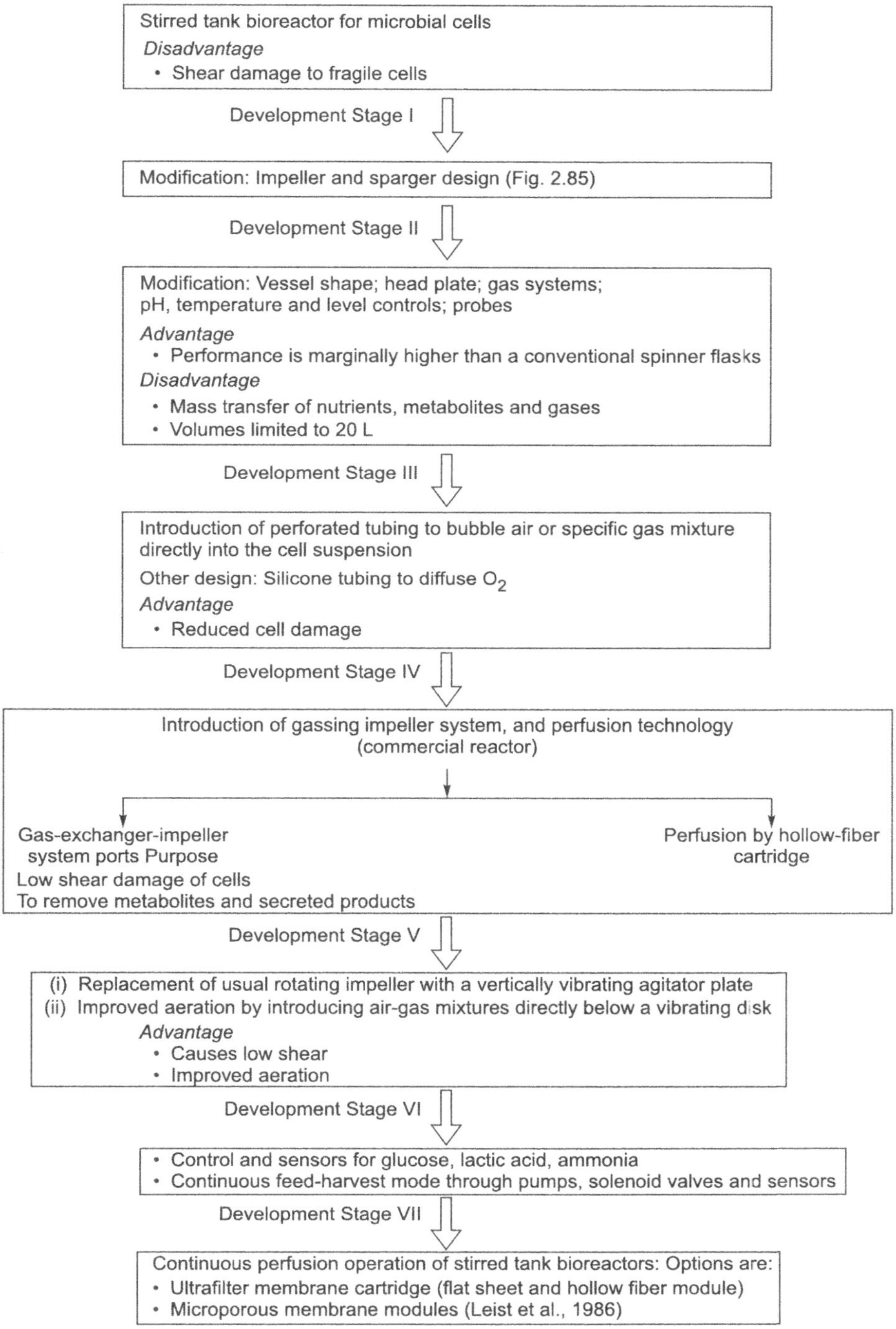

Figure 2.86 Stages of development of stirred tank bioreactors.

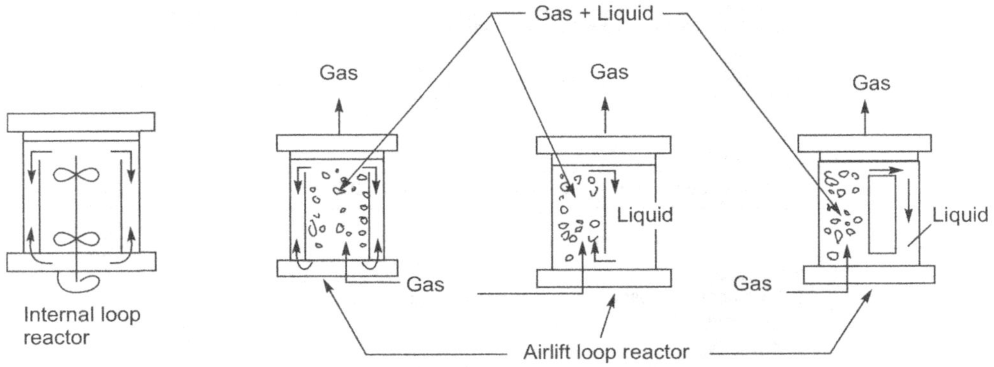

Figure 2.87 Bioreactors for animal cell culture.

- It needs help of perfusion technology for continuous operation, otherwise it is limited to batch mode

(Birch *et al.*, 1985; Cortessis and Proby, 1987)

Development 1

Combination of (i) cell immobilization or encapsulation; (ii) redesigned funnel-shaped glass bioreactor vessel; (iii) stainless steel frit at the base of tapered vessel

Development 2

- continuous feed-harvest and perfusion
- retrofitted with appropriate semi-permeable cartridge
- pumps
- sensors
- controls

(e) Membrane Perfusion Bioreactors and its Development

They maintain cells in a constant environment separated from a circulation medium by a semi-permeable membrane (Seaver, 1987). However, the basic configuration of bioreactors is of two types.

Basic configuration of bioreactor

⇓

Hollow fiber | **Flat sheet membrane**

Hollow Fiber Bioreactor

Materials of Construction

- Microporous membrane (e.g., 0.2 μm pores)—hydrophobic polysulfone, polypropylene, hydrophilic cellulose acetate-cellulose nitrate are used in this bioreactor.

- They are autoclavable.
- Aniosotropic ultrafilter membrane is also used in this application.They are not autoclavable or steam sterilizable.

Advantages are:

- Allows perpetual removal of waste products
- Uninterrupted replenishment of O_2, essential nutrients
- Maintenance of high cell density
- Less damage to cells

Disadvantages are:

- not effective for growing anchorage dependent cells for limited surface area
- incompatible surface chemistry
- diminished diffusion rates as cells clog membranes
- formation of dead zones in extra capillary space

Flat Membrane Sheet Bioreactor

Materials of Construction

This consists of alternating layers of micro-porous hydrophilic membranes.

Advantage

- Mass transfer is better.

Disadvantage

- No scale up module is available.

(f) Ceramic Matrix Bioreactor

This is a different approach to perfusion cell culture. The technique is to immobilize cells on a surface and then re-circulate the medium directly over the cells rather than relying on the diffusion of nutrients, metabolites and gases across a semi-permeable membrane.

Advantages

- It has no diffusion barrier.
- Informative process development tool for testing new culture conditions or cell lines can be done with this reactor.
- High cell density can be achieved in this reactor.

Disadvantage

- It is difficult to scale-up.

Materials of Construction

- Extruded ceramic cartridge (Lydersen *et al.*, 1985) is used for this reactor.

Question 10 ***What are the similarities and dissimilarities between ceramic matrix bioreactor and hollow fiber bioreactor?***

Answer:

Similarity

1. In both the reactors, suspension cells are retained while allowing the fluid phase to be operated in a continuous fashion.
2. Direct measurement of cell concentration is difficult.

Dissimilarity

Ceramic matrix bioreactor

- Cells are exposed to the bulk medium.
- It operates on plug flow behavior, i.e., nutrient and metabolic concentration gradients may have a significant effect on cell growth.

Hollow fiber bioreactor

- Liquid phase can be segregated into two streams by an ultra-filtration membrane and cells are exposed only to one stream.
- It does not operate on plug flow principle.

(g) Radial Flow Packed Bed Bioreactor

The concept is obtained from radial flow chromatography systems. Recirculation of culture media through a packed bed of cells (cells immobilized on glass bead / microcarriers / collagen matrices or stainless steel shavings) is removed through a port at the bottom of inner collection sleeves.

Advantages

- The system is suitable for operation at high flow rates.
- It is amenable to scale up.

(h) Other Reactors

Microsphere and fluidized bed bioreactors, static maintenance bioreactors, and microgravity bioreactors are also used for mammalian cell culture.

2.8.2 Classification of Bioreactors Used for Animal Cell Culture

Criteria considered in this case are:

1. Increase in total surface area (particularly important for anchorage dependent cells for compatible wettable surface requirement)
2. Low shear
3. Increased mass transfer of reactants and gases
4. Flow behavior

(1) Increasing Total Surface Area / Unit Reactor Volume

Small scale and large scale reactors are given below.

Small Scale are:

- Petri-dishes
- Roller bottles (Fig. 2.84)

Disadvantage

Monitoring and control of growth environment is difficult.

Large Scale are:

- Packed glass beads
- Stacked plate
- Rotating multiple tubes
- Roller bottles with multiple tubes inside
- Microsphere and fluidized bed reactor
- Hollow fiber bioreactor
- Ceramic matrix bioreactor

(2) Low Shear Causing Bioreactors in Different Scale

They are of small scale and pilot scale types.

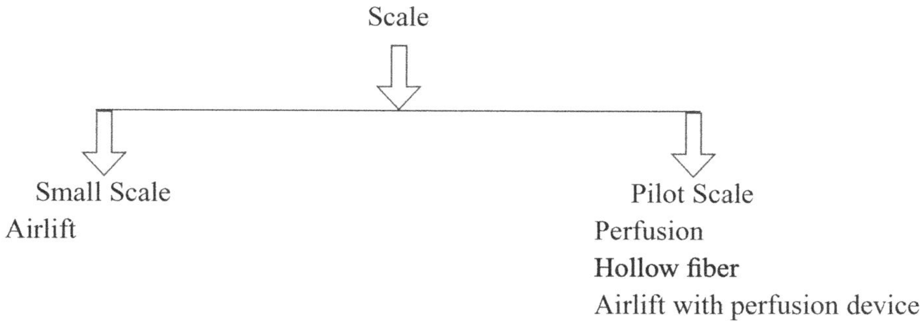

(3) Increased Mass Transfer of Reactants and Gases

Four different varieties are mentioned here.

- Conventional stirred tank bioreactor (STR)
- Modified STR with perfusion device
- Modified STR with variety of impellers (Fig. 2.85):
    - Marine type
    - Pitched blade
    - Sail type
    - "Skull"
    - Vibro-mixer

- Membrane stirrer
- Double screen annular cage impeller (made of two concentric cylindrical SS wire screen and fixed on to a hollow shaft) (Shi *et al.*, 1992) (Fig. 2.88).

• Bubble column

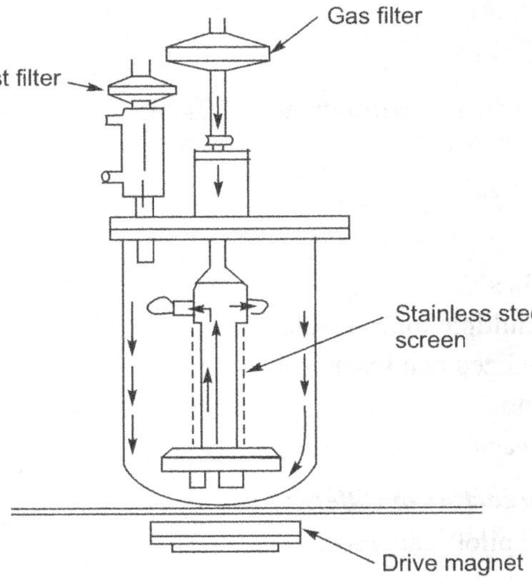

Figure 2.88 Double screen annular cage impeller.

(4) Flow Behavior

Reactor classification is based on flow behavior.

They are:

I Batch: Stirred tank reactor with different impellers

Disadvantage

Withdrawal of metabolite is not possible.

II Semi-continuous : Fed batch type

Disadvantage

This is same as mentioned in batch culture reactor.

III Continuous / back mix

- Stirred tank reactor with perfusion attachment
- Hollow fiber reactor with or without perfusion attachment
- Airlift reactor with or without perfusion attachment

Advantage

- High cell density in small size reactor

IV Plug flow: Ceramic matrix bioreactor (Marcipar *et al.*, 1983)

Detailed classification of bioreactor on flow behavior is discussed here.

- Batch culture systems—Liquid nutrient phase resides in the reactor. Only gas exchange can occur inside or outside the reactor.
- Continuous and perfusion culture systems—Liquid phase flows within or around the reactor allowing nutrient supply and product removal.

(I) Batch Culture Systems It can be classified as suspension culture, monolayer culture, airlift reactors, and micro-carrier culture.

Figure 2.89 T-flask.

- ***Suspension*** cultures: It can again be classified as
    - Static suspension in T-flasks (Fig. 2.89) multi tray units

  Example: anchorage dependent cells, hybridomas lymphoblasts
    - Spinner flasks – magnetic stirring (Fig. 2.90)
    - Stirred reactors (Fig. 2.91)

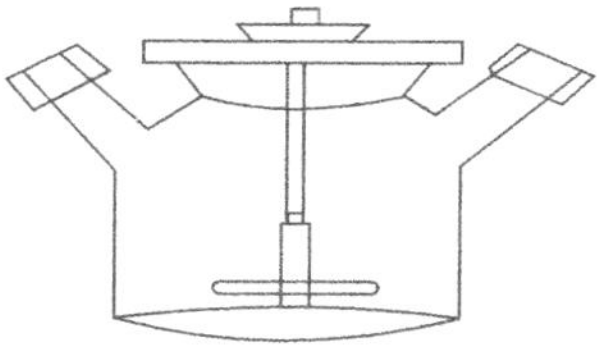

Figure 2.90 Spinner flask.

- ***Monolayer*** cultures: It can again be classified as
    - Stationary flasks and bottles, viz., T-flasks, Roux or Povitsky bottles, are used, but we can achieve only limited production.
    - Static or rotating plate type
    - Roller bottles and cylindrical containers type
    - Spiral films and plastic bags configuration (Fig. 2.92)
- ***Airlift reactors***

  It can be used for shear sensitive hybridomas, lymphoblastoid cells, etc.
- ***Micro carrier*** cultures

  It has a higher capacity, namely about 1000 m^3. It can be used for viral vaccines and interferon production. Anchorage dependent cells attach, spread, and grows to a confluent monolayer on simple solid spheres and is in suspension in liquid culture.

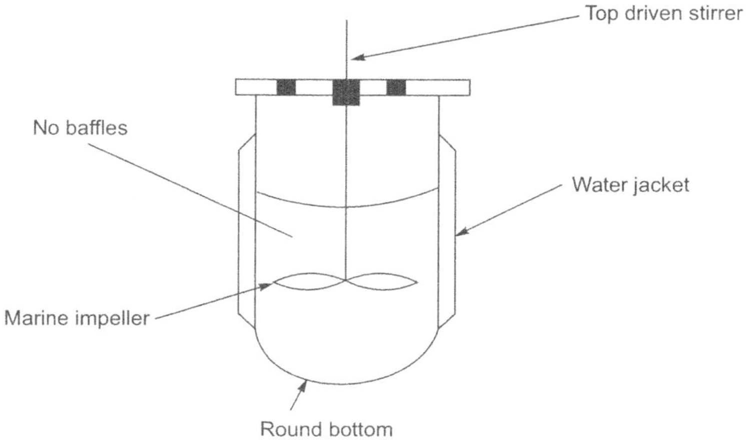

Note: No baffle to reduce turbulence

Figure 2.91 Stirred bioreactor.

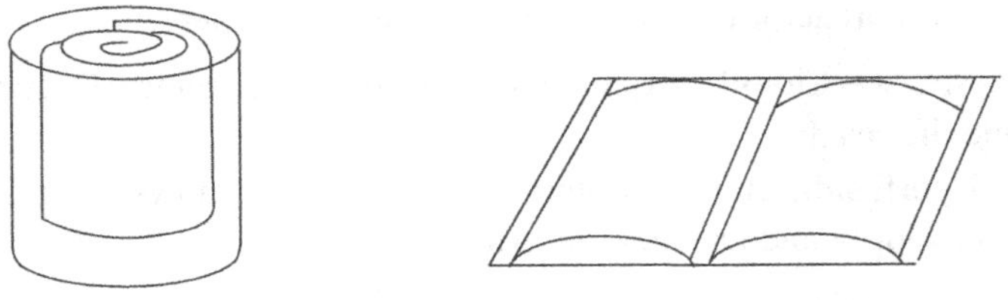

Figure 2.92 Spiral films and plastic bags.

Following are the properties of spherical micro carriers:

- They are non-porous, non-toxic, and rigid spheres.
- Dextran, collagen, gelatin, polyacrylamide, polystyrene, etc. are their matrices.
- Diameter of the sphere is approximately 300 μm.
- Concentration ranges between 5 and 15 kg dry weight /m^3.

Advantages

- High surface to volume ratio
- High cell density
- Homogeneous submerged culture
- Wide choice of micro carriers
- Easy monitoring and control, and scale up facility

Disadvantage

They cannot be applied for shear sensitive cells.

(II) Continuous Culture and Perfusion Systems

This system is better than the batch mode (Fig. 2.93). It gives adequate supply of oxygen and reactants. This configuration ensures removal of toxic metabolic components.

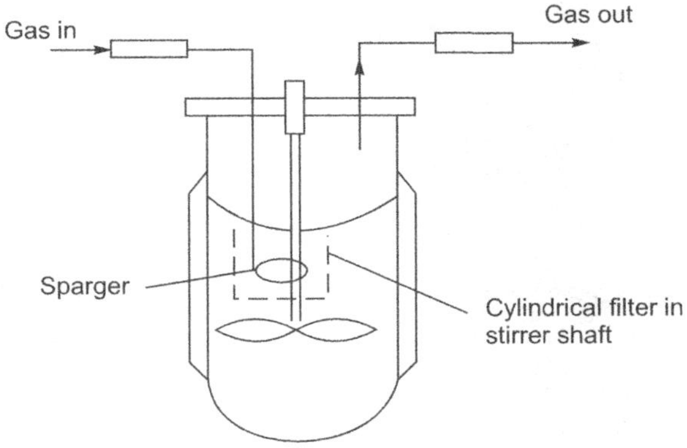

Figure 2.93 CFSTBR.

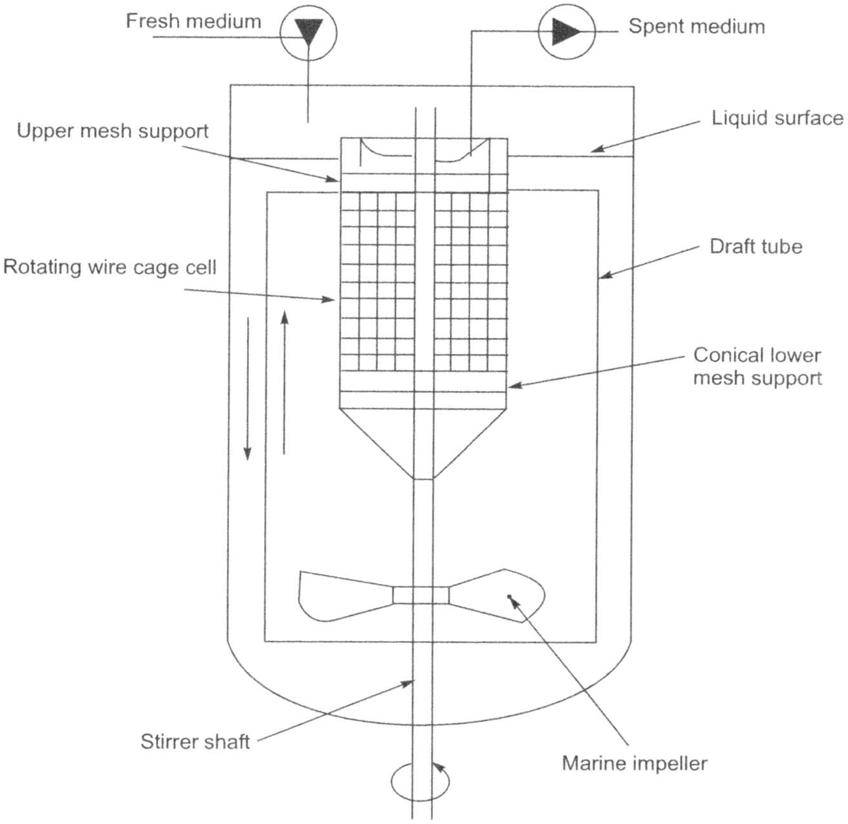

Figure 2.94 Perfusion system.

Classification of perfusion system

(A) ***Homogeneously*** distributed perfusion systems: This can be classified as

- Suspension
- Micro carrier
- Encapsulation
- Entrapment

(B) ***Heterogeneously*** distributed perfusion systems: This can be classified as

- Packed bed (Problems: Flow, degree of mixing, channeling, fouling, etc.) (Fig. 2.95)
- Ceramic (For high density cell culture) (Fig. 2.96)
- Membrane (Fig. 2.97(a))
- Hollow fiber (Fig. 2.97(b))

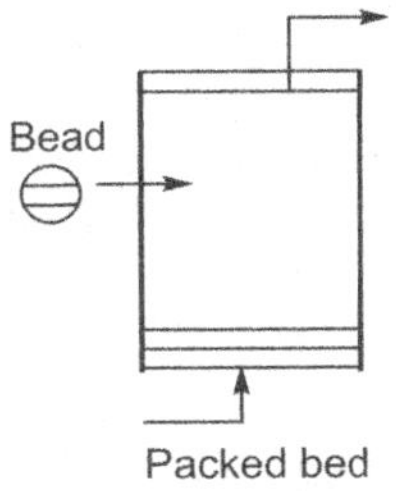

Figure 2.95 Packed bed configuration.

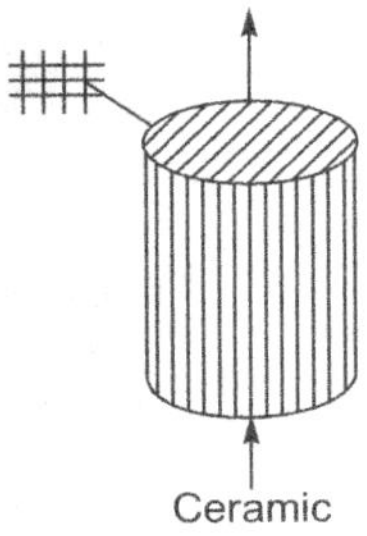

Figure 2.96 Ceramic bed reactor

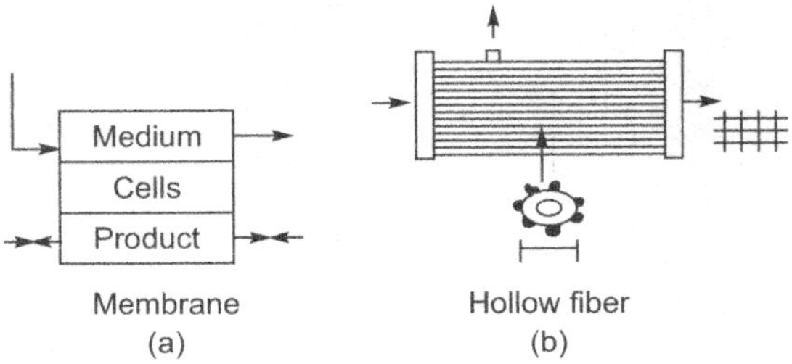

Figure 2.97 Hollow fiber and membrane reactors.

2.9 BIOREACTORS FOR PLANT CELL CULTURE

The classification of bioreactors depends on the choice of culture type used, viz., suspension culture and immobilized cell culture (Fig. 2.98). General survey of reactors for plant cell culture is reported by Panda *et al.*, (1987).

Bioreactor for

Suspension culture
- Shake- flask culture
- Stirred tank bioreactor
    - Batch
    - CFSTBR
- Bubble column
- Airlift
    - Internal loop
    - External loop
- Horizontal drum reactor
    - Static
    - Rotary

Immobilized cell culture

Figure 2.98 Classification of bioreactors for plant culture.

2.9.1 Reactors for Suspension Culture

Reactors used for suspension culture can operate either in batch, continuous flow or plug flow modes (Mukhopadhyay, 2004)

Shake culture is widely used for the initiation and serial propagation of plant cell suspension culture (Rajasekhar *et al.*, 1971).

Batch culture Borosilicate bottles of sizes between 4 l and 10 l having aeration device is used. Dispersion of cells is achieved by spinning culture vessel (80-120 rpm) fixed at certain angle to the horizontal (Lamport 1964, Wilson *et al.*, 1971).

Applications

- For scale-up
- Biochemical analysis

CFSTBR

A schematic of the reactor is described by Tsoglin *et al.*, (1996).

Application

- This reactor is particularly suitable for sequential treatments.

2.10 BIOREACTORS FOR IMMOBILIZED SYSTEM

For immobilized systems (microbial cells, plant cells, and enzymes) following different classes of reactors are used:

- Batch bioreactor
- Mixed bioreactor (Fig. 2.99)

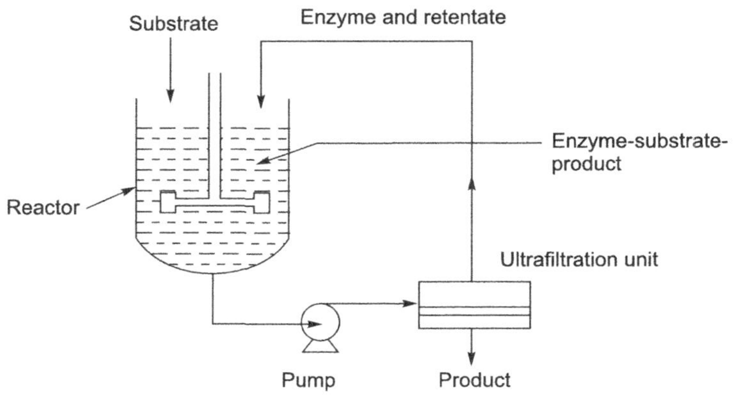

Figure 2.99 Mixed bioreactors for enzymatic reaction.

- Differential bioreactor
- Integral bioreactor

Besides these basic concepts, the typical classical contacting partitions of membrane bioreactors and airlift bioreactors, using immobilized catalysts are also used.

2.11 STERILIZATION BIOREACTORS

We consider here the use of sterilization for solid and liquid components. Even some of the liquid products or solutions are susceptible to heat. Let us see the application where the type of bioreactors is used for sterilization (Fig. 2.100). However, design of sterilization reactor is discussed by Aiba *et al.*, (1973) and Lee (1992).

Usually the components in liquids / aqueous solutions can be sterilized by radiation (UV or X-rays), microwave, sonication, filtration, heating, chemical addition, and high voltage electrical pulses. For large scale sterilization, heating, chemical addition and membrane filtration systems are almost standardized. High voltage electrical pulses and microwave are yet to prove their advantages with the established methods of sterilization.

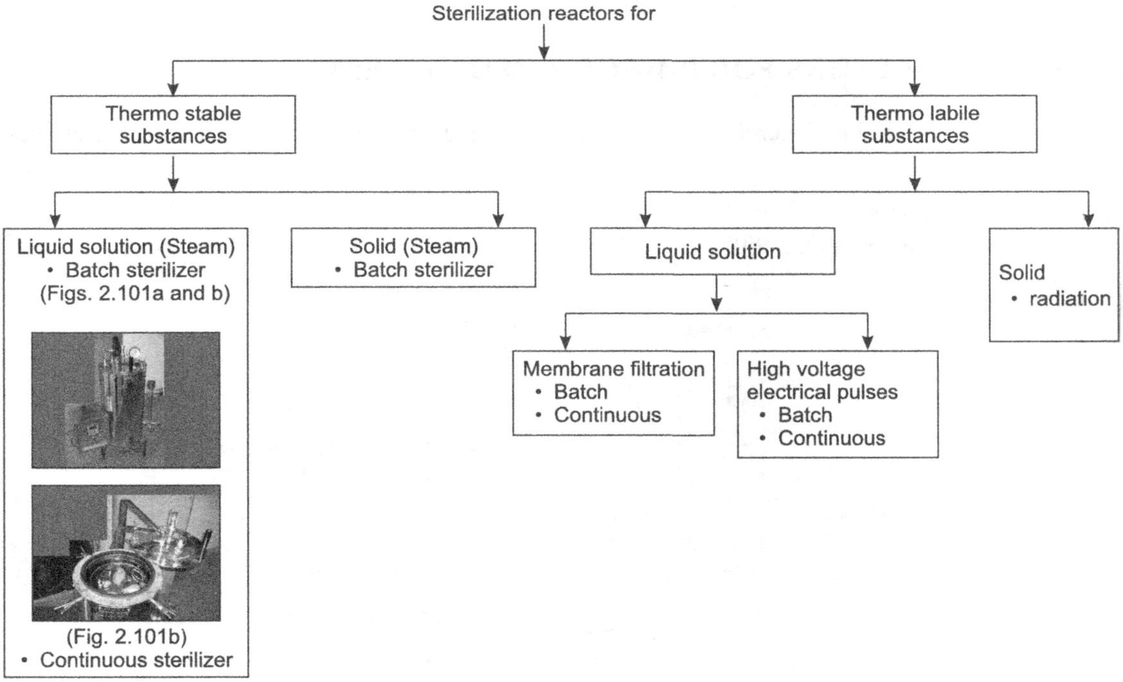

Figure 2.100 Sterilization bioreactors

2.12 BIOREACTORS USED IN DIFFERENT AREAS OF ENVIRONMENTAL CONTROL AND MANAGEMENT

In this case, the reactor configurations vary depending on the processes related to environmental issues. Waste management for environmental protection is given in various areas.

(i) Treatment of waste water
(ii) Digestion of organic slurries
(iii) Treatment of solid waste
(iv) Treatment of waste gases
(v) Soil remediation
(vi) Ground waste treatment

(i) Treatment of Wastewater

Treatment of wastewater is classified into two categories, viz., aerobic and anaerobic (Fig. 2.102).

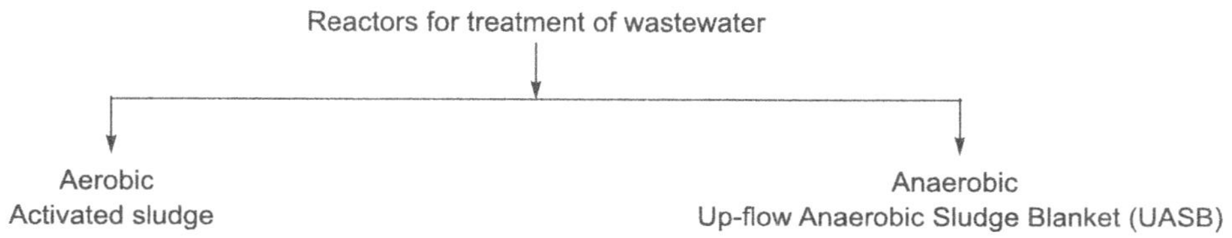

Figure 2.102 Classification of reactors for waste water treatment.

The lower energy costs and low sludge production are the important features of anaerobic wastewater treatment (Denac and Dunn, 1988) over the aerobic process. The detailed kinetic analyses for such processes are well appreciated, but it is not fully established. There are conventional reactors for wastewater treatment. They are not efficient for long residence times.

Bio-film reactor

The term "bio-film reactor" refers to that class of bioreactors where the biocatalyst exists in an anchored form, either on the surface of an inert "carrier" or attached to one another (Nicolella *et al.*, 2000). The carrier could be the wall of the reactor, baffles provided for this purpose or particles of some inert material. Biocatalysts such as microorganisms could also grow attached to one another, giving rise to "bio-granules".

The carrier or the bio-granule could be stationary as in a packed-bed or expanded bed system or mobile as in the case of a fluidized bed system. Typically, in such reactors, the rate of substrate conversion is limited by the rate of transport of substrate into the bio-film. In an anaerobic bio-film reactor, the biocatalyst will include all the different bacterial species responsible for the break down of complex or organic molecules to a final end product consisting of methane and carbon dioxide.

Bio-film formation

A bio-film is defined as a structured community of bacterial cells enclosed in a self-produced (Fig. 2.103) polymeric matrix and is adhered to an inert or living surface. In general, there are four stages for the development of a nature bio-film. They are initial attachment, irreversible attachment by the productions of extra cellular polymeric substances, early development, and maturation of bio-film architecture.

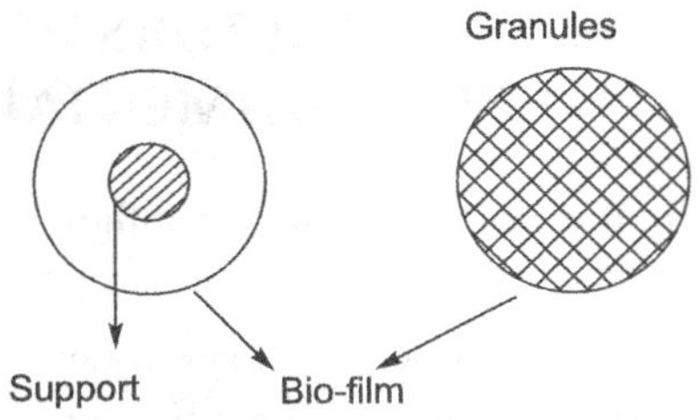

Figure 2.103 Bio-film.

Types of bio-film bioreactors

Bio-film reactors can be assembled in a number of configurations including batch, continuous stirred tank (CFSTBR; including agitating continuous reactors, and rotary continuous reactors), packed bed reactor (PBR), trickling bed reactor (TBR), fluidized bed reactor (FBR), airlift reactors (ALR), up-flow anaerobic sludge blanket (UASB), and expanded bed reactors. The operation of the above reactors changes from one to the reactor.

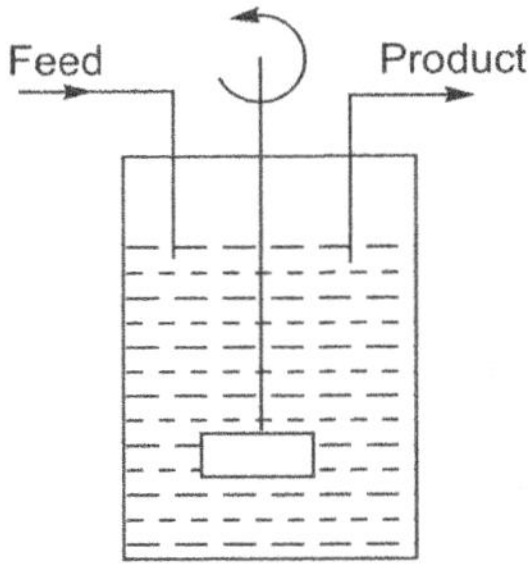

Figure 2.104 CFSTBR.

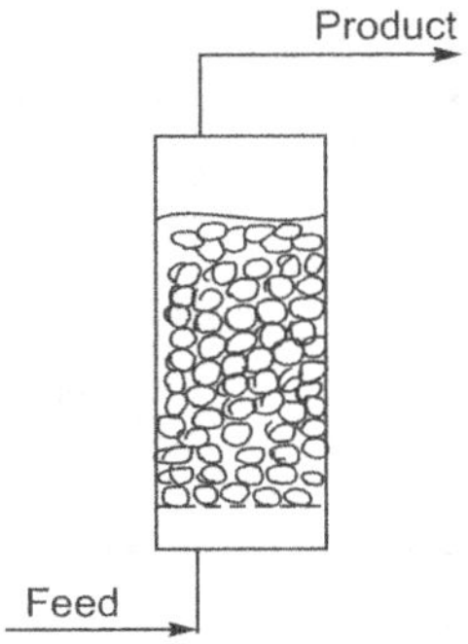

Figure 2.105 PBR.

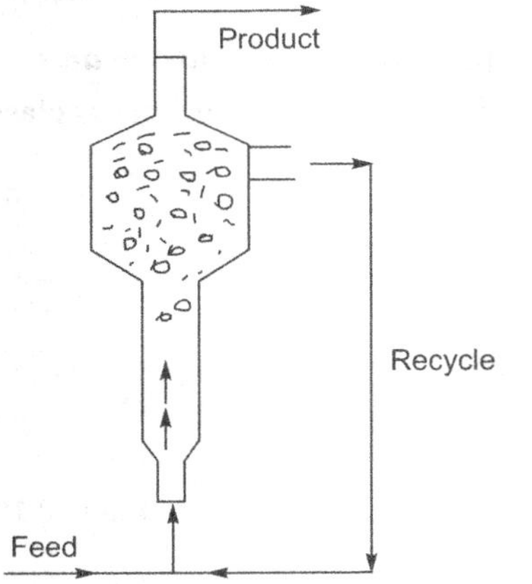

Figure 2.106 FBR.

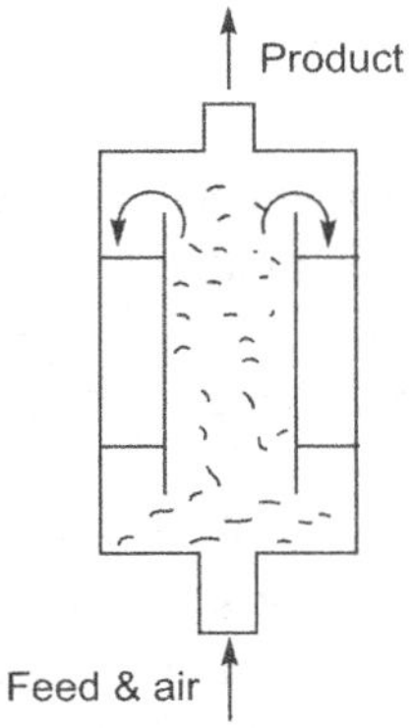

Figure 2.107 ALR.

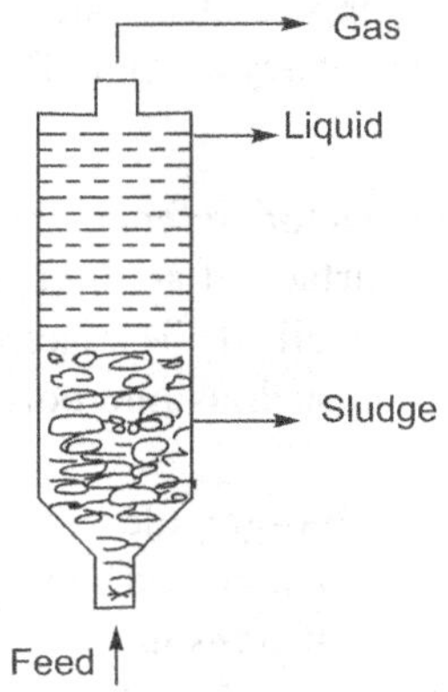

Figure 2.108 UASB.

(ii) Digestion of Organic Slurries

Anaerobic digestion

- Completely mixed anaerobic
- Up-flow anaerobic sludge blanket (UASB) bioreactor (Fig. 2.109)

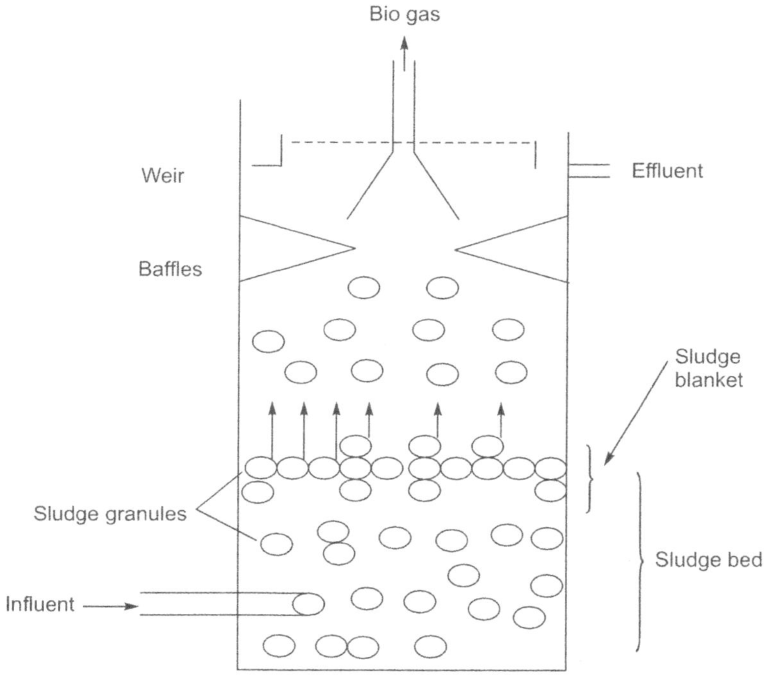

Figure 2.109 Schematic diagram of UASB reactor.

Reactors containing granular bio-film particles are used for anaerobic treatment of waste water or industrial effluents. As the name suggests, the flow in these reactors is in upward direction. At the top of the reactor, provisions are made for gases to escape and the sludge particles settle at the bottom part of the reactor. Reactor effluent is removed from the top of the reactor. UASB reactor was developed by Lettinga *et al.*, (1980).

- Solid-state reactor

Disadvantage

Nitrogen and phosphorus are minimized during recycle operation.

(iii) Treatment of Solid Waste

- Landfill bioreactor: It is associated with the problem of gas emission.
- Aerobic digesters
- Anaerobic digesters

(iv) Treatment of Waste Gases

Waste gas treatment can be done by bio-filter and bio-scrubber.

Bio-filter system

Advantages

- This is simple in construction and cheap in design.
- High internal surface area is suitable for recovering poorly soluble compound.

Disadvantages

- Large floor space is required for this system.
- It is not possible to control the process conditions.
- Support materials (compact) produce odor.

Bio-scrubber system

Advantages

- Less floor space is required for bio-scrubber.
- Sprayed liquid is a suspension of microorganism being cycled back and forth between the spray chamber and waste water treatment unit.
- Process parameters (nutrient supply rate, pH, etc.) are easily controlled during reaction.
- High air flow is necessary.
- Back pressure is low.

Disadvantages

- This is only for removal of gases which are sufficiently soluble since mass transfer rate is less.

(v) Reactor for Soil Remediation

Reports are available for continuous-time dynamic control method for optimizing remediation of ground water (Whiffen, 1998). Conceptual diagrams of reactors are given here.

It is difficult to design the reactor due to insufficient information on the following:

- Bio-availability: (To obtain good contact between contaminant molecule and microorganisms)
- Bio-stimulation (To supply microbes with stimulating agent)
- Bio-augmentation (how introduced microbes behave on the field)

Reactor

- Schematic of slurry phase bioreactor (used for land farming system) is given in Figure 2.110.
- Bio-ventor (for *in situ* soil bioremediation)

(vi) Ground Waste Treatment

- Reactors for (back mix) ‘pump and treat’ bioremediation can be used (Fig. 2.111).
- Schematic for reactor may be used for bio-fencing is given in Figure 2.112.

Advantages

- *In-situ* reaction is possible.
- No pumping cost is an added advantage over the reactor used for “pump and treat” technique.

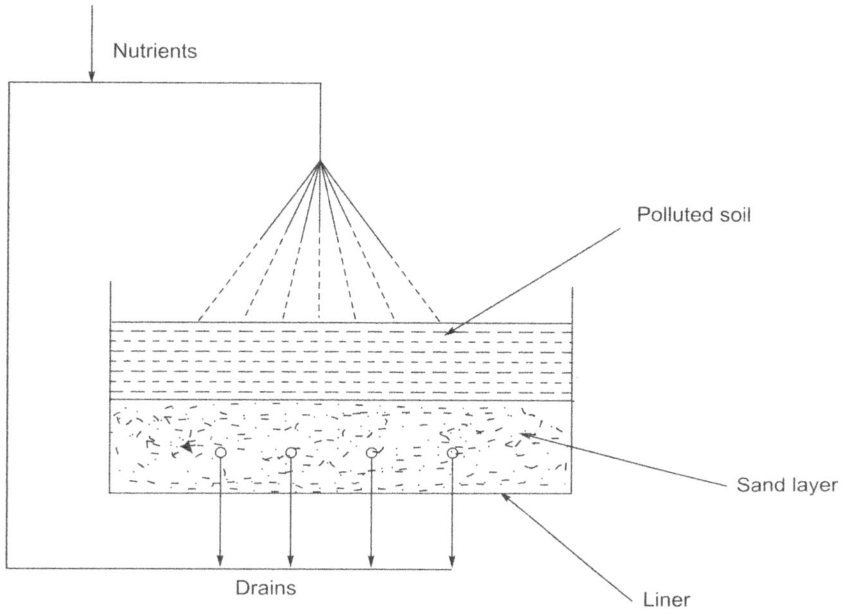

Figure 2.110 Slurry phase bioreactor.

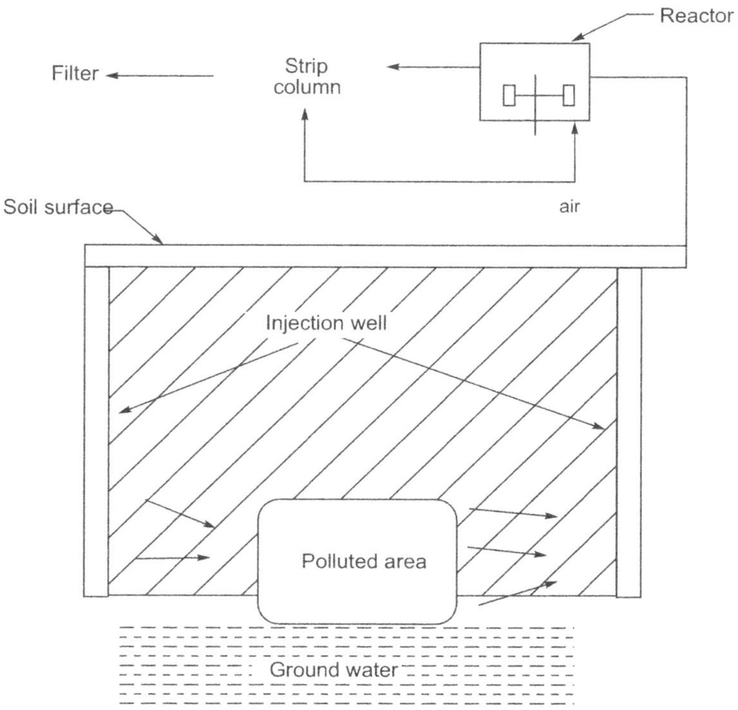

Figure 2.111 Pump and treat remediation bioreactor.

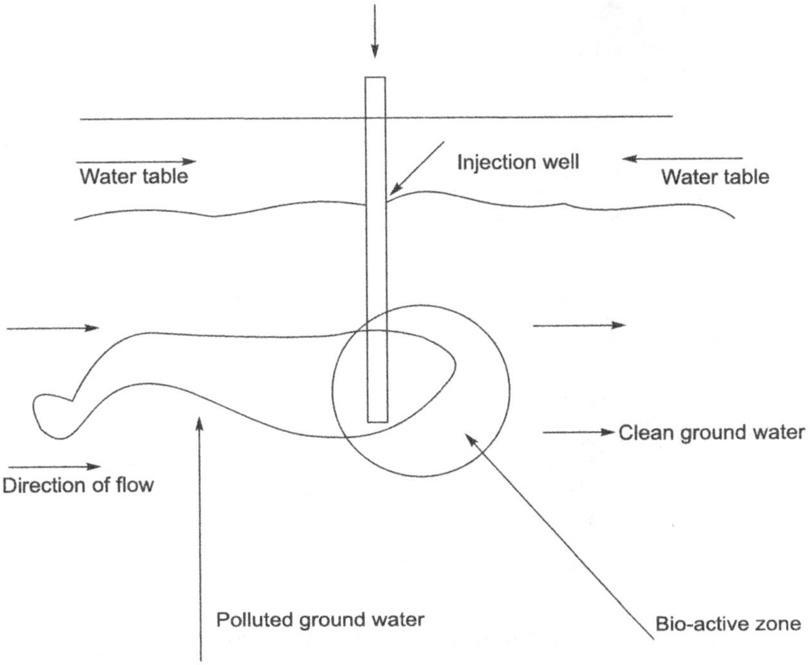

Figure 2.112 Reactors for bio-fencing.

Disadvantage

- No established kinetics and mass transfer data are available for this bioreactor.

2.13 BIOREACTORS USED FOR COMBINED REACTIONS AND SEPARATION

Membrane Bioreactors

Both municipal and industrial waste water treatment facilities benefit from membrane bioreactor (Fig. 2.113) by treating waste water stream from beverage, livestock, dairy, food, residential and in industrial sources.

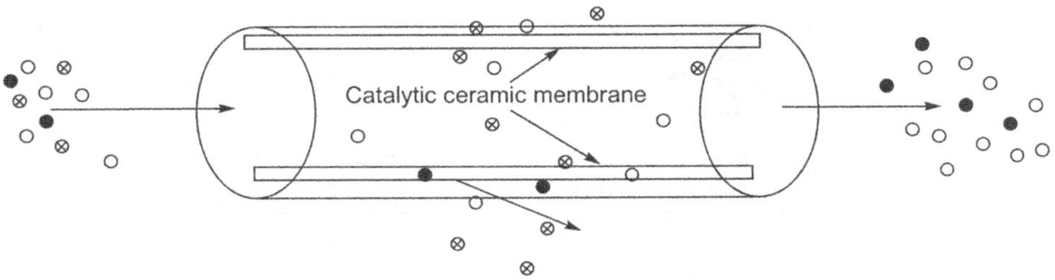

Figure 2.113 Membrane bioreactor

A membrane reactor is really just a plug flow reactor that contains an additional cylinder of suitable porous material within it. This is almost like the tube within the shell of a shell-and-tube heat exchanger. This porous inner cylinder is the membrane. The membrane is a barrier that allows only certain components to pass through it. The selectivity of the membrane is controlled by pore diameter.

Question 11 ***Why do we use a membrane reactor?***
Answer: Membrane reactors combine reaction with separation to increase conversion. One of the products of a given reaction is removed from the reaction through membrane forcing the equilibrium of the reaction to the right, so that more product will form. These are commonly used in dehydrogenation reactions (e.g., dehydrogenation of ethane), where only one of the products is small, which will pass easily through the membrane. This raises the conversion for the reaction, making the process more economical.

EXERCISES

2.1 Which of the following is not true?

I. Vortex formation can be prevented by:
(a) Off-centre location of the impeller on a shaft entering the vessel.
(b) Installation of baffles.
(c) Operation only in the laminar range for the impeller.

II. Bubble-cap bioreactors are not generally used now-a-days. The most important reason is:
(a) Problem of maintenance.
(b) The pressure drop is very high.
(c) The gas liquid contact is not at all good.
(d) Initial cost is very high.

2.2 You are supplied with a classical stirred tank bioreactor for the growth of cells. How will you convert this reactor for the use of mammalian cell? Explain with reasons.

2.3 The organism is slow-growing, shear sensitive, and highly aerobic. It requires inducer to be fed intermittently during fermentation. What will be your recommendation for a bioreactor to be used in this process?

2.4 Inidicate the application of the following impellers :
(a) Vaned disk
(b) Rushton turbine
(c) "skull"
(d) paddle wheel
(e) screw impellers

2.5 Give the sketches for fluid motion for the individual impellers in a bioreactor.

2.6 In "VAT" type reactor, molasses fermentation is carried out with *Saccharomyces cerevisiae*. How does it control contamination?

2.7 In an aqueous-two phase fermentation using *Trichoderma harzianum* (a fungus), what type of bioreactor configuration will you suggest to separate endoglucanse (an enzyme) in the top phase and mycelia retained in the reactor?

2.8 Categorize the following configuration of reactor into basic reactor concept of differential, integral, backmix and batch.

(i) Bubble column
(ii) Airlift
(iii) Pressure cycle
(iv) Hollow fiber
(v) Tubular

2.9 In continuous sterilizer, what could be possible configurations?

2.10 A varieties of bottom configuration of reactor are given in this chapter.

(a) Draw the possible fluid flow diagram with the disc turbine impellers.
(b) Where will you achieve more ideal situation and why?

REFERENCES

Abbott BJ, Gerhardt P (2004) Dialysis fermentation. I. Enhanced production of salicylic acid and naphthalene *P. fluorescence. Biotechnology and Bioengineering*, **12**, 577-589.

Aiba S, Humphrey AE, Millis NF (Eds) (1973) *Biochemical engineering*, 2nd Edn. University of Tokyo, Tokyo, Japan.

Bader FG (1987) “Modelling mass transfer and agitator performance in multiturbine fermentor”, *Biotechnology and Bioengineering*, **30**, 37-51.

Bacehowski DV, Brellatt Jr. JP, Kolanko W, Smith T (1990) “Adherent cell culture flask”, US Patent, No. 4,939,151, July 03.

Bungay HR and Belfort G (Eds) (1987) *Advanced Biochemical Engineering*, John Wiley & Sons, Singapore.

Denac M, Dunn J, *Packed-and-fluidized-bed biofilm reactor for anaerobic waste water treatment*, 1988.

Dillion CP, Rahoi DW, Tuthill AH (1992a) Stainless steel for bioprocessing, Part 2 classes of alloys, *Biopharm*, **5**, 32-39.

Dillion CP, Rahoi DW, Tuthill AH (1992b) Stainless steel for bioprocessing, Part 3 classes of alloys, *Biopharm*, **5**, 40-44

Doran PM (Ed) (1995) *Bioprocess Engineering Principles*, Academic Press, London.

Durand A (2003) “Bioreactor designs for solid state fermentation”, *Biochemical Engineering Journal*, **13**, 113-125.

Durand A, Renand R, Maratra J, Almanza S, Pelletier A (1994) Reactor for sterile solid fermentation methods. World Pat. No. WO 94 18306.

Kafaov VV, Vinarov AY, Gordeev LS (1988) “Modelling of bioreactors”, *International Chemical Engineering*, **28**, 14-35.

Kundu S, Panda T, Majumdar SK, Guha B, Bandyopadhyay KK (1984) Pretreatment of Indian cane molasses for increased production of citric acid, *Biotechnology and Bioengineering*, **26**, 1114-1121.

Lamport DTA (1964) Cell suspension culture of higher plants: Isolation and growth energetics. *Experimental Cell Resarch*, **33**, 195-206.

Lee JH, Woodward Jc, Pagan RJ, Rogers PL (1981) Vacuum fermentation for ethanol production using strains of *Zymomonas mobilis,* Biotechnology Letters, **3**, 177-182.

Lee JM (Ed) (1992) *Biochemical Engineering*, Prentice-Hall, Int. Series.

Leist C, Meyer HP, Fiechter A (1986) Process control during the suspension culture of a human melanoma cell line in a mechanically stirred loop bioreactor. *Journal of Biotechnology*, **4**, 235-246.

Li H, Chai X-S, Deng Y, Zhan H, Fu S (2009) Rapid determination of ethanol in fermentation liquor by full evaporation head space gas chromatography, *Journal of Chromatography* A, **1216**, 169-172.

Lettinga GA, Van-Velsen FM, Hobma SW, De-Zeeuw WJ, Klapwijk A (1980) Use of the upflow sludge blanket (USB) reactor concept for biological waste water treatment, *Biotechnology and Bioengineering*, **22**, 699-734.

Levenspiel O (Ed) (1972) *Chemical Reaction Engineering*, 2nd edn., John Wiley & Sons.

Lydersen BJ, D'Elia NA, Nelson KL (Eds) (1994) *Bioprocess Engineering: Systems, Equipment and Facilities*, John Wiley & Sons, Inc., New York.

Lydersen BK, Pugh GG, Paris MS, Sharma BP, Noll LA (1985) Ceramic matrix for large scale animal cell culture, *Biotechnology*, **3**, 63-67.

Matsuno R, Adachi S, Uosaki H (1993) Bioreduction of prochiral ketones with yeast cells cultivated in a vibrating air-solid fluidized bed fermentor, *Biotechnology Advances*, **11**, 509-517.

Maeusi P-A (1998) Bioreactor in particular for microgravity, US Patent, Number 5,846,817, December 08.

Marcipar A, Henno P, Lentowojt, E, Roseto A, Broun G (1983) Ceramic-supported hybridomas for continuous production of monoclonal antibodies, *Ann. N.Y. Acad. Sci.*, **413**, 416–420.

Mitchell DA, Kriegar N, Stuart DM, Pandey A (2000) New developments in solid state fermentation. Part II Rational approaches to design, operation, and scale-up of bioreactors, *Process Biochemistry*, **35**, 1211-1225.

Mukhopadhyay SN (Ed) (2004) *Process Biotechnology Fundamentals*, Viva Books Pvt Ltd, New Delhi.

Nicolella C, van Loosedrecht MCM, Heijnen SJ (2000) Particle-based biofilm reactor technology, *Trends in Biotechnology*, **18**, 312-320.

Nigam P, Singh D (1994) Solid-state (substrate) fermentation systems and their applications in biotechnology, *Journal of Basic Microbiology*, **34**, 405-423.

Panda AK, Mishra S, Bisaria VS, Bhojwani SS (1989) Plant cell reactors – A perspective. *Enzyme Microbial Technology*, **11**, 386-397.

Panda T, Kundu S, Majumdar SK (1984) Studies on citric acid production by *Aspergillus niger* using treated Indian cane molasses, *Process Biochemistry*, **19**, 183-187.

Propst CL, VonWedel RJ, Lubinisecki, AS (1989) *Using mammalian cells to produce products in Fermentation Process Development of Industrial Organisms* (Neway, JO) (Ed), Marcel Dekker Inc. New York/Basel, Vol. 4, Chapter 5, pp. 221-276.

Raimbault M, Germon JC. French Patent No. 76-06-677, 1976.

Rajasekhar EW, Edwards M, Wilson SB, Streact HE (1971) Studies on the growth in culture of plant cells XI. The influence of shaking rate on the growth of suspension cultures. *Journal of Experimental Botany*, **22**, 107-117.

Reussos, S, Raimbault M, Prebers I, Lonsane BK (1993) Zymotis, a large scale solid state fermenter, design and evaluation. *Applied Biochemistry and Biotechnology*, **42**, 37-51.

Rinzema A, Oostra J, Timmer HR, Sijitsma L, Vander Wel P, Tramper J (2000) Mixed fermenters for solid state cultivation of *C. minitans* for biological pest control, *Third Eur. Symp. Biochem. Eng. Sc.*, Department of Biotechnology, Technical University, Denmark, 11-13 September.

Schügerl K (1982) New bioreactors for aerobic processes, *International Chemical Engineering*, **22**, 591-610.

Seaver SS (Ed) Culture method affects antibody secretion of hybridoma cells. In: Commercial production of monoclonal antibioties. Marcel Dekar, New York, pp. 49-71, 1987.

Shi Y, Ryu, DY, Park SH (1992) Performance of mammalian cell culture bioreactor with a new impeller design. *Biotechnology and Bioengineering*, **40**, 260-270.

Solomon GL (Ed) (1968) *Materials and methods in fermentation*, I&L Press, Oxford.

Suga K, Waki T, Kumano M, Chimange P, Shin SB, Ichikawa K (1980) Production of cellulose in fed-batch system, pp 371-392, *Proceedings of Bioconversion and Biochemical engineering*, Symposium 2, vol II, Ghose TK (ed) IITDelhi, New Delhi.

Suryanarayanan S, Mazumdar K (2001) US Patent 6,197573.BI.

Taya M, Ishii S, Kobayashi T (1985) Monitoring and control for extractive fermentation of *Clostridium acetobutylicum*. *Journal of Fermentation Technology*, **63**, 181-187.

Tsoglin LN, Gabel BV, Fal'kovich TN, Semenenko VE (1996) Closed photobioreactors for microalgal cultivation, *Russian Journal of Plant Physiology*, **43**, 131-136.

Venkataraman A, Kumari JA, Babu PSR, Panda T. (1991) Critical analysis of process development on *in vitro* growth of growth of chick-embryo, Animal cell culture and productivity biologicals, *Proc. of* 13th *Annual Meeting of the Japanese Assoc. for Animal Cell Technology*, Kyoto, Dec. 11-13, 1990 (Eds) Sasaki R and Ikura K, Kluwer Academic Publisher, Dordrecht, pp. 47-52.

Whiffen G (1998) Piecewise continuous control of ground water re-mediation, US Patent, Number 5,813,798, September 29.

Wilson SB, King PJ, Street HE (1971) Studies on the growth in culture of plant cells XII. A versatile system for the large scale batch or continuous culture and plant cell, *Journal of Experimental Botany*, **22**, 177–207.

Zhang Z, Perozziello G, Bocazzi P, Sinskey AJ, Geschke O, Jensen KF (2007) Microbioreactors for bioprocess development, JALA, June, 143-151.

FURTHER READING

Birch JR, Boraston R, Wood L. Bulk production of monoclonal antibodies in fermenters. *Trends in Biotechnology*, **3**, 162-166, 1985.

Cortessis GP, Proby CM (1987) Airlift bioreactors for production of monoclonal antibodies. *Biopharmaceutical. Manufacture*, **1**, 30-33.

Hu W.-S, Dodge TC (1985) Cultivation of mammalian cells in bioreactors. *Biotechnology Progress*, **1**, 209-215.

Mitchell DA, Berovic M, Krieger N (2000) Biochemical Engineering aspects of solid-state bioprocessing, In New products and new areas of bioprocess engineering (Berovic *et al.*,) (Eds) *Advances in Biochemical Engineering / Biotechnology*, Vol. 68 (T. Scheper, Managing Ed) Springer Verlag, Berlin, pp. 54-136.

Raghavararao KSMS, Ranganathan TV, Karanth NG (2003) Some engineering aspects of solid-state fermentation, *Biochemical Engineering Journal*, **13**, 127-135.

Robinson T, Nigam P (2003) Bioreactor design for protein enrichment of region cultural residues by solid state fermentation, *Biochemical Engineering Journal*, **13,** 197-203.

Shrikumar S (2003) Current industrial practice in solid state fermentations for secondary metabolite production: the Biocon India experience, *Biochemical Engineering Journal*, **13**, 189-195,

Schügerl K (Ed) (1991) *Bioreaction engineering Characteristics features of bioreactors*, Vol. 2: John Wiley & Sons, Chichester.

Vorlop J, Lehman J (1988) Scale up of bioreactor for fermentation of mammalian cell cultures with special reference to oxygen supply and mixing, *Chemical Engineering and Technology*, **11**, 171-178.

Viniegra – Gonza'lez G, Favela-Torres E, Aguilar CN, Romero–Gomez SdeJ, Dia'z–Goinez G, Angur C (2003) Advantages of fungal enzymes production in solid state over liquid fermentation systems, *Biochemical Engineering Journal*, **13**, 157–167.

Vogel HC (Ed) (1983) *Fermentation and biochemical engineering handbook—Principles, process design and equipment*, Noyes Publications, Park Ridge.

CHAPTER 3

Bioreactor Operation

OBJECTIVES

- To learn the operations of basic reactors
- Focus will be given on common operations. Specific and special operations will be mentioned for different type of catalysts used in bioreactors.

3.1 INTRODUCTION

The contacting patterns in bioreactors for various reactions are mentioned in the Chapter 2. The reactors, in general, can be classified into three basic operating bioreactors:

- Batch
- Mixed
- Semi-batch

The operation of these basic reactors requires some knowledge, which will be discussed in this chapter. However, there are some common operations irrespective of the type of catalysts (organism/cell/enzymes) used. Specific and special operations will be indicated for different type of catalysts.

3.2 COMMON OPERATIONS OF BIOREACTOR

3.2.1 Setting up of Bioreactor for Submerged Liquid Fermentation (SLF)

The following are the common operations that need to be performed to carry out reactions in submerged liquid state.

Step (a) Cleaning of All Components of the Bioreactor

This must be done thoroughly and carefully. It will give a chance to check the health conditions of all gaskets, connectors, filters, pipelines, pressure gauges, probes, etc.

Step (b) Arrangements of All Components In-place

Components of the bioreactor mentioned in Chapter 2 (Fig. 2.2) need to be arranged in a proper fashion. A few things require information in detail.

- Greasing the seal, O-rings, gaskets
- Thorough checking of sparger
- Checking of membrane, O-rings, electrolyte, and connector settings to the controller
- Standardization of probes
- Checking of flow meters, air filter connections and gas-exhaust connections
- Checking of response for pressure gauge
- Checking of heat exchangers and steam line connections for any scale formation and leakage, etc.
- Special protection for glass vessel

Step (c) Arrangement of Accessories

- *Air supply* Oil-free compressed and dry air supply is required in the process. So the suitable oil-free compressor plus the supply line should be thoroughly checked before operation.
- *Air supply line pressure* Supply line pressure is generally 1 bar.
- *Float-type flow meter* This meter is designed with a housing needle valve and flange connection to the air piping. A laminated safety cover protects the operating personnel in case of an accident.
- *Sterile air inlet filter* The filter needs to have a pore size of about 0.2 micrometers or less.
- *Filter housing* Sterilizable-in-place and easily removable inlet air filter is desirable as the reactor accessory.
- *Air line with sparger* Easy removal and cleaning of sparger is desired in the reactor.
- *Exhaust gas cooler* To minimize liquid loss and evaporation of culture medium, effluent gas needs to pass through a water-cooled exhaust air condenser.
- *Temperature trap* To prevent contamination over exhaust gas piping, steam conversion is required preferably through a double jacket system.
- *Exhaust gas filter* Easily removable and sterilizable filter is connected to the gas line to prevent contamination.
- *Exhaust gas analyzer* If provision is available, it can be connected to the off-line gas analyzer.
- *Peristaltic pumps* One pump each for the addition of acid, base, and antifoam are necessary for the bioreactor. In addition to this, if one needs to add reactants, additional peristaltic pump for each flow is necessary. Pumps with proper silicone tubing are desired. Pumps are calibrated prior to the operation. Tubing, connector, and flows used in this regard must be sterile as desired by the process.
- *Chilling unit* Generally, water at around 4°C is used to control the temperature of the reaction. This is pumped through a pipeline connection to the jackets/heat exchange device in bioreactor.
- *Sampling device* For collecting representative samples from the bioreactor, a connection is placed inside the bioreactor which is further connected through a tube to a peristaltic pump. For large fermentor, sampling device may be more than one, including the gravity sampling system.
- *Steam generator /boiler* For all large bioreactors and in some small scale bioreactors, sterilization device is an in-built heat exchanger. In some cases external steam is supplied through the steam line connections to the heat exchanger of the bioreactor.

Care must be ensured to avoid pressure more than 35 psig. Safety valve in the steam line or reactor is an additional precaution.

- *Autoclave* For small bioreactor without *in-situ* sterilization device, sterilization of the bioreactor, its accessories, fluids, connectors are done by placing them in the autoclave.

Step (d) Handling of Bioreactor

I. Mount the reactor vessel on the bottom support. In some cases, the reactor vessel itself is molded with the bottom plate (Fig. 2.2).
II. Install the head-plate (top-end plate) on to the reactor vessel with proper greasing and a gasket.
III. For *in situ* system, the entire unit is installed on the driving shaft properly.
IV. For autoclavable reactor system step (III) is not necessary. The top end plate with necessary fitting for sterilization must be fixed.
V. Installation on the top-end plate (common for all reactor configurations):
   (i) Fix the aeration system with proper filter unit (Fig. 2.25).
   (ii) Connect the filter device to the sparger connection.
   (iii) Connect the exhaust gas condenser with filter unit.
   (iv) Connect the thermo-trap at the condenser. Connect it together with safety valve and pressure gauge plus pressure control valve.
   (v) Install the temperature sensor (for example, Pt 100). Then connect it to the water bath (before sterilization, this is not required for autoclavable reactor).
   (vi) Install the pO_2 electrode (in some cases externally calibrated) and connect to the amplifier.
   (vii) Install pH electrode (externally calibrated) and connect to the amplifier. Connect the over pressure line. Note: Connection to the amplifier is made after autoclaving the sterilizable components.
   (viii) Connect sampling device and close it with sterile sleeve.
   (ix) Connect the tubes for addition of acid, base, and antifoam. (Note: for *in-situ* reactor system, these connectors are separately sterilized and fitted through the peristaltic pump and required connector under aseptic conditions.)
   (x) Inoculum transfer line is prepared similar to step (d-V-i).
   (xi) Flange the drive and connect it to the stirrer shaft. (Note: This connection to the stirrer shaft is made after sterilization for autoclavable reactor.)

Step (e) Filling of the Bioreactor

(i) Fill the reactor to maximum of about 80% of the total reactor volume with proper medium (reactants in proper form).
(ii) Adjust the pH to desired level and check the desired pH after sterilization and even immediately after the addition of inoculum.

Step (f) Testing the Leak-proof of the System

(i) Close all the extra ports with proper fittings.
(ii) Operate the reactor at some desired set points for temperature, stirrer speed, pH, and air flow.
(iii) Close the overpressure valve (i.e., exit gas valve) and increase the pressure to 1 bar. Then stop supply of air and check the variation in pressure.

(Note: In some autoclavable bioreactor, pressure gauge is not fixed to the top-end plate. In this case, make some provision to fit the pressure gauge and do the similar operation stated in (f-iii). For autoclavable bioreactor, exhaust gas line is immersed in water through a flow meter).

Step (g) Sterilization

(A) For *in situ* sterilizable bioreactor (operations (g) to (i) to be followed)

(i) Open the outlet gas valve slowly to release the over-pressure.
(ii) Take off the heat exchanger tubes and connect the steam supply and condense tubes to the baffle arrangement for indirect sterilization.
(iii) Close the exhaust gas valve when sufficient steam comes out. This indicates expulsion of air from the bioreactor enclosure.
(iv) Start the agitator at low speed during sterilization cycle.
(v) Begin with direct steam sterilization through the sampling pipe.
(vi) Stop indirect sterilization (Note: Change the condensate tube to the thermo-trap).
(vii) Start the direct steam supply through the aeration pipe line and open the condensate drain valve.
(viii) At 1 bar gauge pressure, open slowly the exhaust gas valve and regulate the pressure inside the reactor (Note: Sterilization temperature of 121°C adjusts itself at 1 bar).
(ix) During the sterilization, temperature-pressure regulation is carried out by the exhaust gas valve.

Step (h) Cooling of the Bioreactor

(i) Connect the cooling water supply line and the water drain connection to the heat exchanger (for direct water piping connection).
(ii) Close the exhaust valve after the sterilization time.
(iii) Begin with cooling.
(iv) End the steaming through sample pipe.
(v) When the pressure inside the reactor vessel is between 0.2 and 0.4 bar, stop steam supply and slowly supply sterile air.
(vi) Select the flow rate of air to have an over-pressure of about 0.2 bar.
(vii) When the temperature drops due to cooling and is about to reach cultivation temperature (2°C to 5°C above cultivation temperature), stop the direct tap water cooling.
(viii) Connect the thermostat system and begin with water bath temperature control.
(ix) Open carefully the exhaust gas valve.
(x) Connect the tap water supply to the exhaust condenser and then to the stirrer shaft and begin with cooling.

Step (i) Control Parameter Adjustment

(i) Adjust the parameter set-points
(ii) Connect the storage bottles for aseptic addition of acid, alkali, and antifoam.
(iii) Do the first sampling (by creating over pressure in the vessel and rejecting the dead volume).

(iv) Do the external measurement of the pH value and check the medium microscopically for any presence of organism (i.e., for contaminating organisms, if any).

(v) Calibrate p_{O_2} sensor by passing nitrogen to 0% value followed by saturation with air to 100% value.

Step (j) Inoculation

If after step (*i*) all operations are found alright, the next step is to add the desired organism to the bioreactor vessel in the form of inoculum.

(i) Fill the inoculum into sterile inoculum bottles in a laminar flow chamber.

(ii) First fill sterilized media into the bioreactor and then add the inoculum from steps (j)–(m) through a sterile sleeve.

Step (k) Closing and Opening the Sleeve

(i) Before opening the sleeve, keep the flow rate at a position of inoculation set point.

(ii) During this opening of sleeve, care must be ensured to protect the system from contamination.

(iii) Keep these connection parts nearer to flame.

(iv) After this opening of sleeve operation, sleeve is closed with sterile blind component.

(v) Then adjust the flow rates to process set point.

Step (l) Sterilization

(B) For Autoclavable Bioreactor (operations (l) i-xx are to be followed).

(i) Fill the bioreactor vessel with medium (approximately up to 80% of reactor volume).

(ii) Place the head plate with proper greasing of gasket on the top end of the reactor vessel.

(iii) Close all necessary blinds.

(iv) Place the externally calibrated pH probe and p_{O_2} probe through the respective ports.

(v) Then tighten the respective nuts.

(vi) Attach silicone tubes for acid, alkali, antifoam additions (to be operated through the peristaltic pumps), inoculums addition and sample withdrawal connections to the respective ports on the top end plate of the bioreactor. All tubes and connector openings should be closed with proper cotton plugging and cover with aluminum foil. All tubes should be properly clipped so that the medium should not come out during autoclaving.

(vii) Remove connectors to pH and p_{O_2} sensors, which are inserted into the reactor. Cover the top of the sensors with proper cotton and aluminum foil cover.

(viii) Disconnect the driving shaft connection.

(ix) Place the complete bioreactor with all attachments as discussed above in the autoclave.

(x) Operate the autoclave and maintain at 121°C for 20 minutes.

(xi) After sterilization, release slowly the pressure of autoclave with care.

(xii) Cool the autoclave.

(xiii) Then take out the reactor with all accessories and connect it to the driving shaft.

(xiv) Connect the sensors to the controllers and switch them on.

(xv) Connect air supply line and adjust the air flow rate.

(xvi) Exhaust line port should be immersed in 4M NaOH solution. Check if bubbles of air come out from the exhaust line.

(xvii) In the presence of gas flame, do the aseptic connections for acid, alkali, antifoam, inoculum addition, and sample withdrawal, etc.

(xviii) Connect thermostatic system to circulate cold water through the jacket of the bioreactor.

(xix) Measure all initial parameters and adjust it to the desired level before addition of inoculum.

(xx) Make sure that the bioreactor operates with satisfaction.

Step (m) Addition of Inoculum

Inoculum is developed separately in a shake flask or in a smaller dimension bioreactor using proper growth medium and suitable conditions for growth.

For example: Production of chitinase by *Trichoderma harzianum* (Felse, 1999) is discussed here.

One hundred milliliter of sterile seed medium (composition in (kg/m^3) glucose monohydrate – 10, $(NH_4)_2SO_4$ – 1.40, KH_2PO_4 – 2.0, NaH_2PO_4 – 6.9, $MgSO_4$ – 0.3, citric acid monohydrate – 10.5, peptone – 1 and urea – 0.3) contained in 500-cm^3 Erlenmeyer flask needs to be inoculated with *Trichoderma harzianum* from a fresh 105 h old working slant. Conidia concentration of 10^5 spores/cm^3 is generally used. Culture, after inoculation, is incubated on a rotary shaker maintained at 160 rpm at 30°C for 43 h to obtain the seed culture. Approximately 1.3 g cell dry weight/l from seed culture is transferred to chitinase production medium.

Cells are transferred aseptically to the bioreactor. The type of inoculums and its size depend on the organism which will be discussed later in this chapter ("Inoculum development" section).

3.2.2 Inoculum Development for Bioreactor Operation

Preparation of inoculums is necessary for bio-reaction to be carried out in large scale using cells. The condition and methods of development differ greatly from organism to organism. General information is highlighted here. For specific organism one can get detailed information from the published literature.

Step 1: Right Source for Organisms

(a) For microorganism: Micro-organisms are generally obtained in pure culture from culture center (Table 3.1). Sometimes researchers isolate microorganisms from natural sources, viz., soil, air, water or from the growth on natural substances and then characterize them as per standard microbiological procedures. One may consider the books on microbiology and manual (e.g., Bergye's manual). However, the organisms are obtained from the culture collection centers in the form of a lyophilized powder. It is necessary to revive the organism by the standard procedures.

Step 2: Revival of Organism

Cells are obtained from **Step (1)** at low temperature frozen conditions (at –196°C in liquid nitrogen), freeze-dried (lyophilization) or liquid dried (L-drying) forms.

From this stock, cells are revived aseptically by transferring them on to a suitable growth medium and at culture conditions. It is, generally, grown on a slant culture or on Petri-plate culture or in shake culture (Chapter 2: Sections 2.3 (a) and (b)). Those cultures are incubated at most suitable growth

Table 3.1 *Source of organism, cells, and cell lines*

Abbreviation	*Full form*	*Address/Company*
NCIM	National Collection of Industrial Microorganisms	National Chemical Laboratory, Pune, India
MTCC	Microbial Type Culture Collection	Institute of Microbial Technology, Chandigarh, India
ATCC	American Type Culture Collection	www.atcc.org
DSM	Deutsche Sammheng von Mikroorganismen	DSM und Zellkulturen, GmbH, Mascheroder Weg, Braunschweig, Germany (Malik and Claus, 1987)

conditions (like temperature: 30°C for yeast and fungus, 37°C for mesophilic bacteria and 60°C to 65°C for thermophilic organisms; agitation at preferred shaker speed). They are called working culture. Proper growth of working culture is necessary. It is preferred to optimize the biological parameters, viz., cell age in slant or plate growth, cell number density and period during inoculum preparation (Dasu *et al.* 2003).

In most of the cases, few serial transfers on slants in the beginning are preferred to get proper growth phase. The revived organism is preserved in specified conditions till further use. A common practice is to keep the slant growth at 4°C in refrigerator. It is also recommended that unnecessary frequent transfer of organisms may be avoided as the organism will lose its proper activity. To start with a reaction, the organism at proper growth phase is transferred to suitable sterilized growth medium (Appendix 1 in Chapter 1).

Step 3: Transfer to Liquid Culture

After revival stage, cells are counted (if possible) by hemocytometer and transferred to growth medium in the shake-flask culture. Organism is grown under specific growth conditions. An example is given in the previous section (Section 3.2.1 (m)).

For large scale bioreactor operations, inoculum is developed in several stages. Every stage follows proper growth and maintenance conditions. A schematic diagram is given in Figure 3.1.

Step 4: Separation of Cells

In some cases, only cells are used as inoculum, probably to avoid any metabolites produced in the

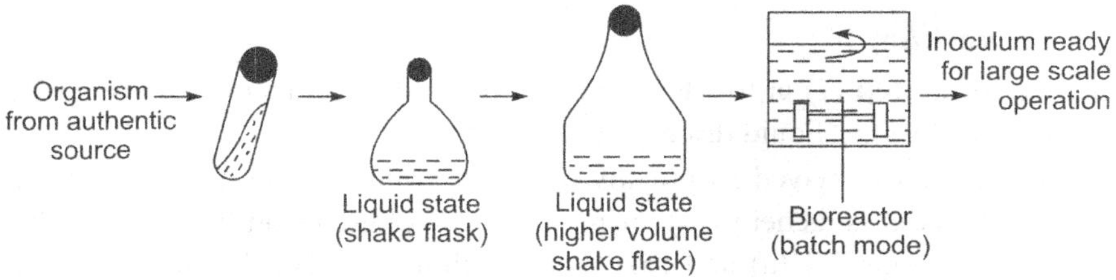

Figure 3.1 Schematic flow diagrams for large scale inoculum preparation.

supernatant during inoculums development. Cells are separated by centrifugation usually at 4000 x g for 15-20 minutes, from the culture filtrate.

In this case, cells are suspended in an aliquot volume of sterilized production medium and transferred to the bioreactor for the immediate start of reaction.

A few specific examples are given here to help the reader to practice by themselves.

(i) Process with bacteria

Aerobic bacteria

Steps

- *Selection of organisms and procurement from standard culture collection centers*: For example, penicillin acylase synthesis by *Escherichia coli* may be selected as an example (Babu and Panda, 1991).
- *Culture maintenance:* The organism *E.coli*, in this case, can be maintained on the medium given in Appendix 1 in slants by serial transfer every month and incubation at suitable temperature for a specified time till appreciable growth is visible. Then it is stored at 4°C till further use. Using *E.coli*, growth conditions are ideal at 30°C for 24 h.
- *Development of inoculum:* Cells from suitable growth slants are dispensed in a measured volume of suitable medium aseptically. Measurement of cells can be made in a counting chamber (hemocytometer) before it is transferred to a growth medium. The inoculum is prepared from these cells by growing in a shake flask condition at specified temperature of growth and shaking conditions.
- *Inoculation:* The inoculation for small scale laboratory experiments is carried out in a laminar flow chamber. For reactor operation, the inoculum is transferred with the help of a peristaltic pump.
- *Sampling:* For laboratory scale experiments, a small aliquot is collected in the laminar flow chamber under strict aseptic conditions.
- *Analysis of samples:* Each sample collected at a specified time is analyzed for cell (by gravimetric or spectrophotometric or turbidimetric analysis), product following standard analytical procedures and residual reactant (for reducing sugars by standard analytical procedures available for complex reactants, total carbon, nitrogen and other elements are estimated by standard procedures).

Anaerobic bacteria A few modifications of the steps are followed for aerobic bacteria.

Modifications:

- *Culture maintenance:* The organism (e.g., *Clostridium* sp.) is maintained by transferring liquid culture into freshly prepared special medium (for example, CM3 medium in Appendix 1) in a serum bottle every fortnight and the incubation is made at 60°C. After this, the organism can be stored for a month at 4°C. Every operation should be carried out in a CO_2 incubator.
- *Maintenance of anaerobic conditions during reaction:* For strict anaerobic conditions in the reactor, oxygen free nitrogen is sparged throughout the reaction. Before entering into reactor, nitrogen is passed through alkaline pyrogallol solution and Fildes and McIntosh indicator solution (Kundu 1983).
- *Inoculation:* This is done by puncturing the septum of the serum bottle or through a reaction inlet with the help of a hypodermic syringe.
- *Sampling:* Aliquot of sample is collected with the similar principle of inoculation by a hypodermic syringe.

(ii) Process with yeast Similar steps as in the case of bacterial process with a few modifications need to be followed for the process using yeast.

Modifications

- *Culture maintenance:* The organism (e.g., *Saccharomyces cerevisiae* – Anjani Kumari and Panda, 1992) is maintained on MGYP slants (refer: Appendix 1 (c)). This is sub-cultured after a regular interval of 15 days. Ideal growth temperature is between 28°C and 30°C.

(iii) Process with fungus A few modifications of the aerobic bacterial process is followed for this organism.

Modifications

- *Culture maintenance:* The organism (e.g., *Trichoderma* sp.- Théodore and Panda, 1995) is maintained on potato-dextrose-agar (PDA) slants or in some cases on Czapek-Dox medium (refer Appendix 1 (d)-(e)).
    - Ideal growth temperature is between 28°C and 30°C.
    - Spores from a well-developed slant growth are suspended in sterile distilled water containing traces of Tween 80.
    - Development of inoculums: Inoculum is developed from the spores in the form of suspended mycelium or tiny pellets (e.g., *Aspergillus niger* in most cases develops pellets (Nair and Panda, 1996)).
    - Dry cell mass equivalent is the basis for transfer of mycelium from inoculum to a reactor for further fermentation.

3.3 SELECTION/IDENTIFICATION OF OTHER COMMON FACTORS NECESSARY FOR SMOOTH OPERATION OF BIOREACTORS

Bioreactor operation is certainly different from chemical reactor operation. For biological reactions in a bioreactor, one can find a few variables which is not good to mix-up with other variables to be evaluated during the reactor analysis.

They are:

- Microbiological parameters: Age of the organism in slant growth, age of the inoculum in pre-stage shake culture, inoculum level to begin the reaction, etc.
- Physical parameters: Temperature of the reaction, shear forces (agitator speed / flow of air for aerobic reaction or flow of inert gases like nitrogen in anaerobic reaction), etc.
- Chemical Parameters: The pH of the reaction, reactant composition – in a typical biological reaction, viz., reaction involving Czapek-Dox (refer Appendix 1) medium have five constituents. They are considered as five different variables.

In most of the cases, biological reactions in a reactor are set right with these three classes of variables a priori to study the reactor analysis in detail.

One should not forget about the interaction of these variables. For the simplicity of understanding, a few examples are worked out to study them.

Methods of analysis are either statistical or mathematical or both. Mathematical optimization of such variables is complex at this stage. Statistical analysis is considered here.

Example 3.1 Development of production medium for the production of extracellular enzyme by an organism. The organism is reported to produce the product using a medium given in Table 3.2.

Table 3.2 *Reactant level reported or assumed*

Constituents	*Concentration (kg/m^3)*
A	30
B	5
C	3
D	1
E	0.5
F	0.5
G	0.01

Theory for this solution The development of medium for maximum production by the organism is carried out initially in two stages.

Stage I: A first order design, viz., Plackett–Burman design (Plackett and Burman, 1946) is employed to screen the important medium constituents, which significantly influence production.

Stage II: The reactant screened by the Plackett-Burman design is studied by a second order design, viz., the central composite design (Box and Wilson, 1951). The complex algorithm of Box is employed to determine the levels of reactants required for a production.

Stage III: Central composite design, for example, deals with single output response (i.e., product produced). A number of output responses (for example, cell concentration, products synthesized, etc.) are considered to get in-depth analysis. The solution is called distance function approach (Gowrishanker and Panda, 2008).

Stage I: Plackett–Burman Screening Design

Plackett and Burman (1946) introduced fractional two level factorial designs in k-variables where the number of design points, N, is equal to $(k + 1)$. These designs are available only when N is a multiple of 4.

To construct a Plackett–Burman design in k variables, a row consisting of elements equal to +1 or –1 such that the number of positive one is $(k + 1)/2$ and the number of negative ones is $(k – 1)/2$ is selected. This row is chosen as the first row in the design. The next $(k – 1)$ rows are generated from the first row by shifting it cyclically one place $(k – 1)$ times, that is, the second row is obtained from the first row by shifting it one place. The third row is obtained from the second row by shifting it one place, and so forth. Finally, a row of negative ones is added to the previous k-rows producing a design with $N = k + 1$ rows.

The Plackett–Burman screening design is based on probabilistic model

$$Y' = \beta_0 + \beta_i x_i + \varepsilon \tag{3.1}$$

where

Y' is predicted response

β_0 is offset term

β_i are linear effects

x_i are independent variables

ε is error

with assumption that no interaction occurs among factors.

The method of evaluating the coefficients (β_i)is illustrated below. The Plackett-Burman design is, generally, best suited for efficient screening and accuracy in picking out the important variable in comparison with a random balance design or a fractional design (Williams, 1963).

Each medium constituent is treated as a variable in the Plackett–Burman design. Sometimes to satisfy the requirement of the matrix, dummy variables(s) are used for analysis. The matrix showing these variables with two levels is called Hadamard matrix. The experimental units of the Hadamard matrix is established, generally in a region of 25% from the center of the experimental domain which is the reported or assumed growth medium composition (Table 3.2). For example, the sign (–) represents the coded levels of unoptimal medium constituents, whereas the sign (+) represents 125% of the unoptimal medium. All the experiments are performed in duplicate at least. The average of the maximum production obtained in duplicate sets is the response.

The effect of each variable on the response for organism is given by the Equation (3.2).

$$\text{Effect} = \frac{\text{Response at (+) level}}{\text{4 or multiple of 4}} - \frac{\text{Response at (−) level}}{\text{4 or multiple of 4}} \tag{3.2}$$

The t-values for the null hypothesis parameter, $H_0 = 0$ and the probability of finding the effect by chance is also determined. The confidence levels given by 100 × (1-probability due to chance) are calculated.

Variables with confidence levels greater than 80% (for example) are considered to have significant effect on production. The choice of confidence levels depends on the problem where majority of the variables are screened for a particular problem. The example in Table 3.2 will give some deviation in this case.

Estimation of model coefficients, β_i, in a regression model

Estimation of the model coefficients, β_i, in regression model is illustrated for the second order polynomial model used for fitting the data obtained in the optimization of microbiological parameters (slant age, inoculum age, and inoculum level) and screened medium constituents for the maximum production using the central composite design, viz.,

$$Y = \beta_0 + \sum_{i=1}^{m} \beta_i X_i + \sum_{i=1}^{m} \beta_{ii} X_i^2 + \sum_{i=1}^{n-1} \sum_{j=2}^{m} \beta_{ij} X_i X_j \tag{3.3}$$

where Y denotes the observed response. X_i and X_j represent the levels of the real value of the variables.

The random errors are assumed to be independently distributed as normal variables with a zero mean and a common variance, σ^2.

In matrix notation, the model Equation (3.3) over N observations, is

$$\mathbf{Y} = \mathbf{X}\beta + \varepsilon \tag{3.4}$$

For a specific example,

$$Y = \beta_0 + \beta_1 x_1 + \beta_2 x_2 + \beta_{11} x_1^2 + \beta_{22} x_2^2 + \beta_{12} x_1 x_2 + \in$$

where

$$\underset{N \times 1}{Y = \begin{bmatrix} Y_1 \\ Y_2 \\ . \\ . \\ . \\ Y_N \end{bmatrix}} \qquad \underset{N \times (5+1)}{X = \begin{bmatrix} 1 & X_1 & X_2 & X_1^2 & X_2^2 & X_1 X_2 \\ 1 & X_{11} & X_{12} & X_{13} & X_{14} & X_{15} \\ 1 & X_{21} & X_{22} & X_{23} & X_{24} & X_{25} \\ . & . & . & . & . & . \\ . & . & . & . & . & . \\ . & . & . & . & . & . \\ 1 & X_{N1} & X_{N2} & X_{N3} & X_{N4} & X_{N5} \end{bmatrix}}$$

$$\underset{(5+1) \times 1}{\boldsymbol{\beta} = \begin{bmatrix} \beta_0 \\ \beta_1 \\ \beta_2 \\ \beta_{11} \\ \beta_{22} \\ \beta_{12} \end{bmatrix}} \qquad \underset{N \times 1}{\varepsilon = \begin{bmatrix} \varepsilon_1 \\ \varepsilon_2 \\ . \\ . \\ . \\ \varepsilon_N \end{bmatrix}} \tag{3.5}$$

The normal equations are

$$\mathbf{X'Xb} = \mathbf{X'Y} \tag{3.6}$$

where $\mathbf{X'}$ is the transpose of matrix $\mathbf{X}$ and $\mathbf{b}$ is the matrix of coefficient estimates.

The solution to these normal equations is

$$\mathbf{b} = (\mathbf{X'X})^{-1}\,\mathbf{X'Y} \tag{3.7}$$

where $(\mathbf{X'}X)^{-1}$ is the inverse of $(\mathbf{X'X})$. Both $(\mathbf{X'X})$ and $(\mathbf{X'X})^{-1}$ are symmetric matrices.

The fitted second order model in coded variables is

$$\mathbf{Y'} = \beta_0 + \beta_1 \mathbf{x}_1 + \beta_2 \mathbf{x}_2 + \beta_{11} \mathbf{x}_1^2 + \beta_{22} \mathbf{x}_2^2 + \beta_{12} \mathbf{x}_1 \mathbf{x}_2 \tag{3.8}$$

The predicted values of the response under study (Y') are obtained using Equation (3.8).

Estimation of *t*-values in Plackett–Burman design

The *t*-values and the confidence levels are determined as follows:

$$E = \frac{\sum D_i^2}{n} \tag{3.9}$$

E is variance of the concentration effect

D is concentration effect for the dummy variables and

n is number of dummy variables.

The student t-value, $t(x_i) = E(x_i)/\text{S.E.}$ (3.10)

where $E(x_i)$ is the effect of variable x_i,

$$(x_i) = \frac{\text{Response at (+) level}}{\text{number of runs}} - \frac{\text{Response at (−) level}}{\text{number of runs}}$$

S.E. is standard error of the concentration effect which is square root of E.

The P-value of each concentration effect (confidence level) is then calculated using the $t(x_i)$ values obtained.

Solution of example for stage I (Plackett–Burman screening)

Steps

Step 1: Variables and their levels employed in the Plackett–Burman design for the screening of medium constituents are given in Table 3.3. It is assumed that the reported values in Table 3.2 is considered as "– Level". "+ Levels" may be considered 25% more than the corresponding value in the "– Level."

Table 3.3 *Actual (–) and (+) levels of variables*

Variables	*Levels (kg/m³)*	
	+	–
A	37.5	30
B	6.25	5.0
C	1.25	1.0
D	3.75	3.0
E	0.625	0.5
F	0.625	0.5
G	0.0125	0.01

Step 2: Hadamard matrix for screening of reactant is given in Table 3.4

Table 3.4 *Hadamard matrix*

Run #	*Variables with their levels*						
	A	*B*	*C*	*D*	*E*	*F*	*G*
1	+	+	+	–	+	–	–
2	–	+	+	+	–	+	–
3	–	–	+	+	+	–	+
4	+	–	–	+	+	+	–
5	–	+	–	–	+	+	+
6	+	–	+	–	–	+	+
7	+	+	–	+	–	–	+
8	–	–	–	–	–	–	–

Step 3: One needs to carry out the experiments with the composition stated in each run of Table 3.4. For example

Run # 1 Composition of medium constituent

Component	*Concentration (kg/m^3)*
A	37.5
B	6.25
C	1.25
D	3.0
E	0.625
F	0.5
G	0.01

Same run should be carried out at least three times keeping all other variables of the experiment constant.

Step 4: Samples for each experimental run are regularly taken and should be analyzed for the desired product without any bias following standard estimation technique. Results need to be tabulated in the following Table 3.5.

Step 5: Product obtained against different runs are given in Table 3.5.

Table 3.5 *Production against each run in Table 3.4*

Run	*Product (kg/m^3)*
1	0.9
2	0.6
3	0.5
4	0.8
5	0.7
6	0.7
7	0.7
8	0.7

Step 6:

$$\text{Effect of each response is} = \frac{\text{Response at (+) level}}{4} - \frac{\text{Response at (−) level}}{4}$$

The model coefficients, *t*-values, probability due to chance, and confidence level are calculated (Table 3.6). In Table 3.6, values are assumed here. One needs to calculate from the output of real experimental runs. Khuri and Cornell (1987) have explained the detailed methods of calculation.

Table 3.6 *Confidence levels for variables*

Variable	*Confidence level (%)*
A	80
B	60
C	70
D	75
E	60
F	60
G	70

Step 7: One can assume confidence level of 65% and above to significantly influence the production. Then variables *B*, *E*, and *F* do not influence production. If one assumes 75% and above confidence level for similar study, then *B*, *C*, *E*, *F*, and *G* do not influence the production. One may consider confidence level of 70% to screen the variables.

Step 8: Hence, the composition of screened medium is given in Table 3.7.

Table 3.7 *Composition of screened medium*

Variables	*Level (kg/m³)*
A	+
B	-
C	+
D	+
E	-
F	-
G	+

Step 9: In the stage II, optimization study will optimize levels of variables *A*, *C*, *D*, and *G* only. For future experiments, *B*, *E*, and *F* variables will be at (–) level composition throughout the study of the optimization.

Stage II: Optimization of concentration of screened medium constituents (reactants) using second order design (example, central composite design).

Theory for the second order design (CCD):

For example

A coded 2^4 - factorial central composite experimental design with two axial (α) points at a distance of α from the design centre and four replicates about the center point making a total of 30 experiments is employed to study the screened variables grouped as x_1, x_2, and x_3 and x_4. The experiments are performed at least in duplicate. The design matrix and the levels of the independent variables are made for the purpose of experiments. The average of the maximum product obtained in the duplicate sets is taken as the response, *Y*. The acquired data are fitted into an empirical and full second order polynomial model (Equation 3.11).

$$\hat{Y} = \beta_0 + \beta_1 x_1 + \beta_2 x_2 + \beta_3 x_3 + \beta_4 x_4 + \beta_{11} x_1^2 + \beta_{22} x_2^2 + \beta_{33} x_3^2 + \beta_{44} x_4^2 + \beta_{12} x_1 x_2 + \beta_{13} x_1 x_3 + \beta_{23} x_2 x_3 + \beta_{14} x_1 x_4 + \beta_{24} x_2 x_4 + \beta_{34} x_3 x_4 \quad (3.11)$$

where,

$\hat{Y}$	is predicted response
β_0	is offset term
$\beta_1, \beta_2, \beta_3, \beta_4$	are linear effects
$\beta_{11}, \beta_{22}, \beta_{33}, \beta_{44}$	are squared effects
$\beta_{12}, \beta_{13}, \beta_{23}, \beta_{14}, \beta_{24}, \beta_{34}$	are interaction effects
x_1, x_2, x_3, x_4	are independent variables .

The method of evaluating the coefficients, β_i, is illustrated in Plackett–Burman design analysis. Iso-response contour plots are made for the variable studied.

The complex algorithm of Box (1965) is a sequential search technique which has proven effective in solving problems with nonlinear objective functions subject to nonlinear inequality constraints. This procedure finds the maximum of a multivariable, nonlinear function subject to nonlinear inequality constraints. Detailed procedure is discribed in Appendix 3.1.

Solution to Example 3.1 for stage II optimization by central composite design

After screening the variables *A*, *C*, *D*, and *G* for the production of a desired product by Plackett and Burman design, the variables are optimized by a 2^4-factorial central composite experimental design with α and six replicates at the centre point ($n_0 = 6$), leading to a total 30 experiments. The α-values can be suitably assumed to get a broadened experimental space. The levels of the independent variables chosen and design matrix (experimental plan) are given in Tables 3.8 and 3.9, respectively.

Table 3.8 *Variables and their levels employed in the central composite experimental design for their optimization*

Variables	Coded levels				
	$-\alpha$	-1	0	$+1$	$+\alpha$
	Actual values (kg/m^3)				
A	$-\alpha_1$	20	37.5	55	$+\alpha_1$
C	$-\alpha_2$	0.75	1.25	1.75	$-\alpha_2$
D	$-\alpha_3$	2	3.75	5.5	$+\alpha_3$
G	$-\alpha_4$	0.0075	0.0125	0.0175	$+\alpha_4$

With the above composition as per the design plan given in Table 3.9, experiments in at least triplicate are conducted for every individual run. The product values obtained are called experimental response.

By applying multiple regression analysis on the experimental data, the second order polynomial Equation (3.11) have been found to explain the production.

The regression Equation (3.11) is optimized by iteration method of Rosenbrock to determine the optimal concentration of each medium constituent. The optimal production of the desired product can be experimentally achieved by conducting experiments using optimal level of variables. From Equation (3.11), one can get the theoretical production, which is called theoretical response. Both the theoretical and

Table 3.9 *Central Composite Design Matrix employed for the optimization of variables A(= x_1), C(= x_2), D (= x_3), and G (= x_4)*

Run#	x_1	x_2	x_3	x_4
1	+1	– 1	+1	+1
2	+1	+1	+1	–1
3	0	0	0	0
4	–1	–1	+1	–1
5	+1	–1	–1	–1
6	–1	+1	+1	+1
7	–1	+1	–1	–1
8	+1	+1	–1	+1
9	–1	–1	–1	+1
10	0	0	0	0
11	–1	–1	–1	–1
12	–1	+1	+1	–1
13	+1	+1	–1	–1
14	0	0	0	0
15	0	0	0	0
16	+1	–1	–1	+1
17	+1	–1	+1	–1
18	+1	+1	+1	+1
19	–1	–1	+1	+1
20	–1	+1	–1	+1
21	0	0	$+\alpha_3$	0
22	$+\alpha_1$	0	0	0
23	0	$+\alpha_2$	0	0
24	0	0	$-\alpha_3$	0
25	0	0	0	0
26	0	0	0	0
27	0	$-\alpha_2$	0	0
28	0	0	0	$+\alpha_4$
29	0	0	0	$-\alpha_4$
30	$-\alpha_1$	0	0	0

experimental responses are necessary to be compared and analyzed for the verification and validation of the experimental plan.

In this regard, the values of R (coefficient of correlation) and R^2 (coefficient of determination) of the production are calculated to check whether there exists a good agreement between the experimental and the predicted values of product synthesis. Statistical testing of the model is done by the Fisher's statistical test for analysis of variance (ANOVA).

Details of the ANOVA table is given below.

Construction of the Analysis of Variance (ANOVA) Table

Experiments are carried out according to the central composite experimental plan. Data are analyzed and shown in a tabular form, where the table is called an analysis of variance (ANOVA) table (Table 3.10). The content of the table gives the information regarding the separate sources of variation in the data.

Table 3.10 *Analysis of variance (ANOVA) table*

Source of variation	*Degree of freedom*	*Sum of squares (SS)*	*Mean square (MS)*	*F-value*	*Probability >F*
Due to regression (fitted model)	$p - 1$	SSR	SSR/$(p - 1)$	MSR/MSE	
Residual (error)	$N - p$	SSE	SSE/$(N - p)$		
Total	$N - 1$	SST			

The total variation in a set of data is called the total sum of squares (SST). The quantity SST is the sum of the squares of the deviations of the observed response (Y_m) about their average value ($\bar{Y}$) for N observations.

$$\bar{Y} = (Y_1 + Y_2 + Y_3 + \ldots\ldots\ldots\ldots\ Y_N)/N, \tag{3.12}$$

$$SST = \sum_{u=1}^{N} (Y_m - \bar{Y})^2$$

The quantity SST is associated with $(N - 1)$ degrees of freedom since the sum of deviations $(Y_m - \bar{Y})$ is equal to zero. The total sum of squares can be divided into two parts: the sum of squares due to regression (or the sum of squares explained by the fitted model) and the sum of squares unaccounted for in the fitted model. The sum of squares due to regression (SSR) is given in Equation (3.13).

$$\text{SSR} = \sum_{m=1}^{N} \{Y(x_m) - \bar{Y}\}^2 \tag{3.13}$$

The deviation $(Y(x_m) - \bar{Y})$, is the difference between the value predicted by the fitted model for the mth observation and the overall average of the Y_m's. If the fitted model contains 'p' parameters, the number of degrees of freedom associated with SSR is 'p-1'. The sum of squares unaccounted for by the fitted model, also called the sum of squares of the residuals or the sum of squares of the errors (SSE) is given in Equation (3.14).

$$\text{SSE} = \sum_{m=1}^{N} \{Y_m - Y(x_m)\}^2 \tag{3.14}$$

The number of degrees of freedom for SSE is the difference: $(N - 1) - (p - 1) = N - p$. The usual test of significance of the fitted regression equation is a test of the null hypothesis, H_0: all values of β_i (excluding β_0) are zero. The alternative hypothesis is H_a: at least one value of β_i (excluding β_0) is not zero. Assuming normality of the errors, the test of H_0 involves first calculating the value of the F-statistic from Equation (3.15).

$$F = \frac{\text{Mean Square regression}}{\text{Mean Square residual}} = \frac{\text{SSR}/(p-1)}{\text{SSE}/(N-p)} \tag{3.15}$$

If the null hypothesis is true, the F-statistic in Equation (3.15) follows an F-distribution with $(p-1)$ and $(N-p)$ degrees of freedom in the numerator and in the denominator, respectively. The next step in the test of H_0 is to compare the value of F in Equation (3.15) with the table value, $F_{\alpha, p-1, N-p}$ which is the upper 100 α% point of the F-distribution with $(p-1)$ and $(N-p)$ degrees of freedom, respectively. If the F-value exceeds $F_{\alpha, p-1, N-p}$, the null hypothesis is rejected at the α level of significance. It is concluded that the variation accounted for by the model is significantly greater than the unexplained variation. Another accompanying statistic is the multiple coefficient of determination (R^2),

$$R^2 = \text{SSR/SST} \tag{3.16}$$

The value of R^2 is a measure of the proportion of total variation of the values of Y_u about the mean $\hat{Y}$ explained by the fitted model. It is often expressed as a percent. This implies that approximately $100 \times R^2$ % of the total variation in product synthesis is explained by the fitted model.

3.4 SPECTRUM OF BASIC BIOREACTOR OPERATIONS

Specific cases of start-up of bioreactor will be discussed for various bioreactor operations.

(a) Overall Unit Operations for Bioreactors

Following unit operation may be taken for a batch reactor, but they are basic for all the reactors (Fig. 3.2).

1. Blank sterilization: Pre-batch cleaning
2. Medium formulation: Statistical experimental design followed
3. Medium batching: Medium injection
   - All connection closed
   - Probes calibrated & inserted
4. Medium sterilization: *In situ* sterilization by steam from boilers or steam generated in the systems
   - Autoclaved
5. Medium hold:
   - To maintain reaction compounds in a safe state post-sterilization
   - Final adjustment of reactor setting can be made prior to inoculation
6. Reaction inoculation: To add inoculum to sterilized reactor in a controlled aseptic way
7. Incubation /Reaction: Necessary valves and connectors must be operated carefully in a controlled aseptic manner to record all possible control variables. Sampling operations at regular intervals.
8. Termination of reaction /harvest: Adjustment of pH or lower down temperature to facilitate product recovery
9. Cleaning: Options are available, viz., full sterilization or heating in the presence of caustic detergents to fully automated clean in place (CIP) systems with complex valve operations of their own.

Figure 3.2 Flow sheet for sequence of unit operations required for the reactor operation.

Figure 3.3 is the summary of basic ideal bioreactors that differ on various modes of operation. It is necessary to design such an equipment and experiment that will generate accurate and meaningful data. However, one can find no unique laboratory bioreactor, which can be used for all types of reactions. In this chapter, various types of bioreactors are discussed that can be chosen to obtain kinetic parameters for a specific reaction system.

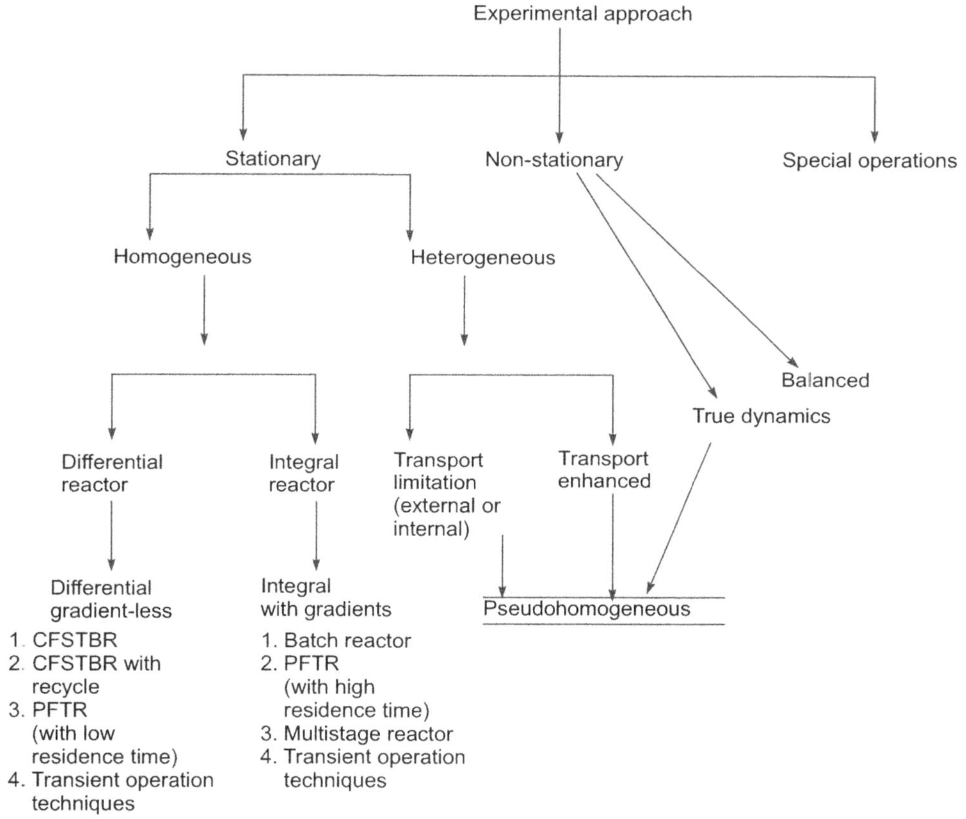

Figure 3.3 Spectrum of bioreactors operation.

The following general criteria can be used to evaluate various types of laboratory reactors:

- Ease of operational flexibility and construction,
- Cost of bioreactor,
- Ease of sampling, parameter monitoring, analysis and control,
- Effectiveness of contact between reactants and catalyst,
- Catalyst retention, and
- Degree of isothermality.

3.4.1 Experimental Laboratory Bioreactors

Homogenous Reactions

This is limited to soluble enzymes reacting with soluble reactants. Two major classes of reactors are differential and integral reactors. A differential reactor is used to determine the rate of reaction as a function of concentration. The reactor is considered to be gradientless. The design equation is similar to the CFSTBR design equation.

(a) Integral bioreactors with Gradients

I. Stirred batch reactor The catalyst is mixed initially with the reactants. Progress of reaction is followed as a time function. If one needs to know only the concentration of reactant or concentration of product without disturbing the catalyst, the sample usually pass through the membrane filters to separate catalyst and fluid. This will stop the reaction. If the catalyst decays, the activity and selectivity will vary during the course of data collection.

II. PFTR The operation is similar to the batch reaction initially, but later on reactants alone are pumped into the reactor. At the outlet, a membrane filter retains the catalyst in the reactor. Flow rate of reactants is limited so that no back mixing occurs in the reactor.

(b) Differential Bioreactors

I. CFSTBR Initial stages are similar to batch stirred tank reactor. The feeding of reactant and retention of catalyst in the reactor are similar to the PFTR.

II. Recycle stirred reactors The operation is similar to CFSTBR. The catalyst in the concentrated form in the outlet is recycled back to the bioreactor.

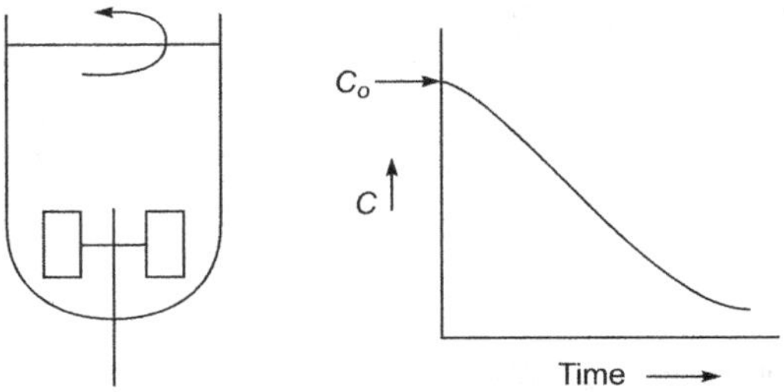

Figure 3.4 Batch (Discontinuous reactor).

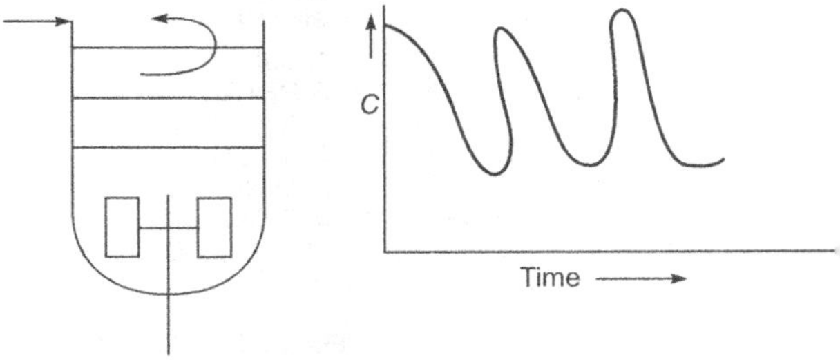

Figure 3.5 Semi-continuous reactor.

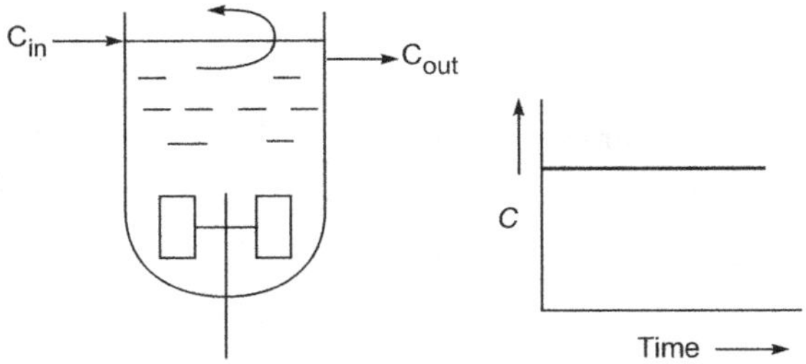

Figure 3.6 CFSTBR (Continuous flow).

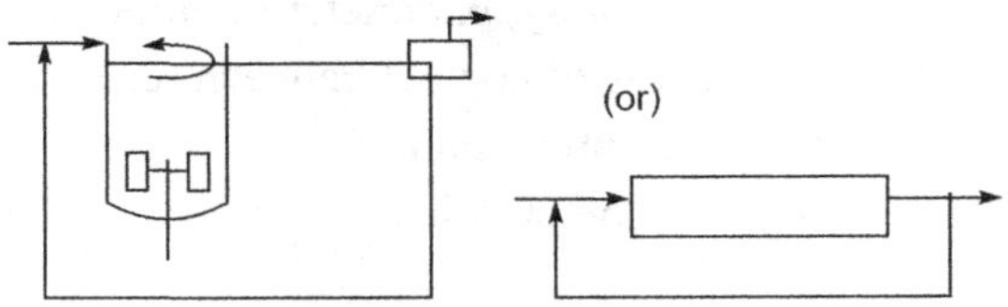

Figure 3.7 Recycle reactor.

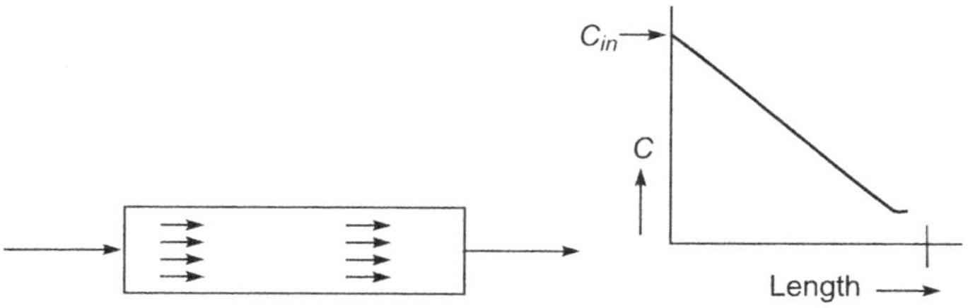

Figure 3.8 Continuous plug flow reactor (PFTR).

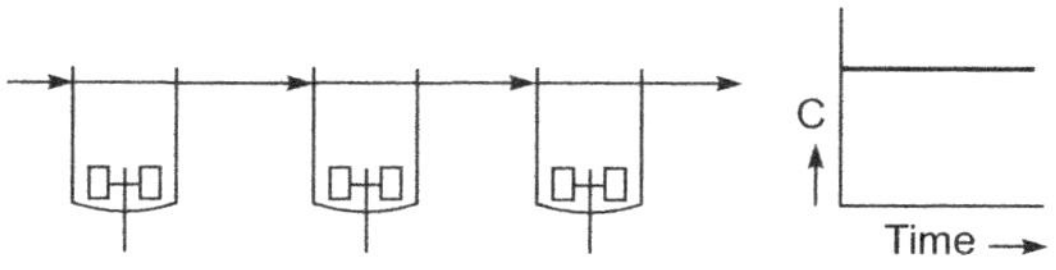

Figure 3.9 Multistage CFSTBR.

Figures 3.4–3.9 Schematic representation of bioreactors used to carry out homogenous reactions.

Heterogeneous Reactions

Immobilized enzymes and cellular reactions can be classified in this group. To simplify such a difficult situation, the theory in general is explained for stirred batch reactor, integral reactor and differential reactors. Then separate discussion is made on microbial free cells and cellular reaction, reactions involving immobilized systems using animal cells, reactions involving plant cells and special operations. General theory is discussed here.

I. Stirred Batch Reactor

In a stirred batch reactor, the catalyst is dispersed as slurry (schematic presentation in Figure 3.4). The reactor has better contact between the catalyst and the reactants than either the differential or integral reactors.

It has sampling problem. Samples are usually passed through the separating unit or withdrawn through filters to separate the catalyst and other components in fluids, there by stopping the reaction. Its isothermality is good. The contact time is known since the catalyst and reactants are fed at the same time. If the catalyst decays, the activity and selectivity will vary during the course of data collection.

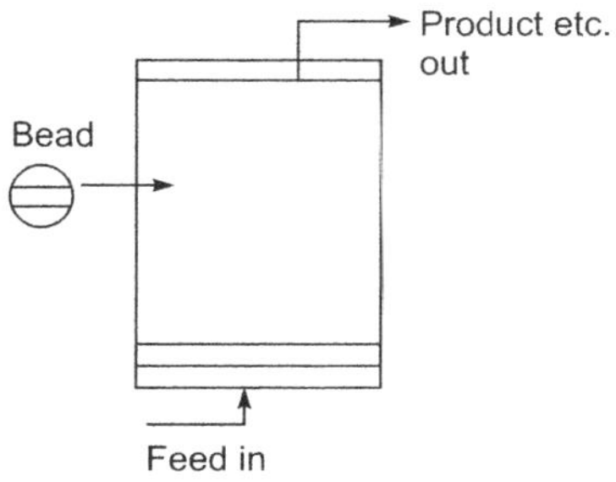

Figure 3.10 Packed bed reactor.

II. Integral Bioreactors

(a) Fixed bed reactor: The construction is simple (Fig. 3.10). Channeling or by passing of some of the catalysts by the reactant stream is a problem. There is more contact between the reactant and catalyst in the integral reactor than in the differential reactor. This is probably due to greater length. This reactor is best used for immobilized cells (Babu, 1991).

Case (i) For immobilized cells: If a reaction is highly endothermic, or exothermic, significant axial and radial temperature gradients can result. If the reaction follows different paths (metabolic cycles) with different activation energies, different products will be obtained at various temperatures. This makes it more difficult to evaluate the reaction rate constants because the cellular mechanism changes with changing temperature along the length of the bioreactor.

Case (ii) For deactivation of cells/enzymes: If catalysts decay during the experiment, the reaction rates will be significantly different at the end of the reaction. Particularly for immobilized cells, the reaction may follow different reaction paths as the deactivation of cells occurs. Hence, the selectivity to a particular product will vary during the course of the experiment. Again it will be difficult to sort out various rate law parameters for the different reactions. So this bioreactor is not suitable for reactions where cell /enzymes deactivation occurs. Other integral reactors are PFTR with high residence time.

III. Differential Bioreactor

(a) Tubular bioreactor: A schematic arrangement is shown in Figure 3.11.

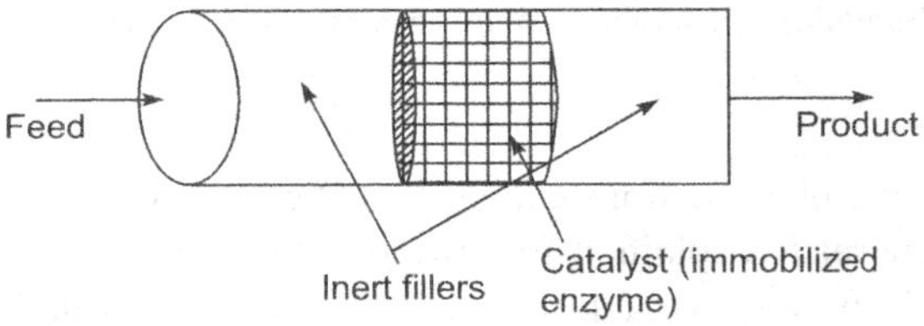

Figure 3.11 **Schematic of experimental differential reactor (an example of tubular bioreactor).**

A differential reactor is normally used to determine the rate of the reaction as a function of concentration for heterogeneous systems. It consists of a tube containing small amount of catalyst arranged in the form shown in Figure 3.11.

As small amount of catalyst is used, the conversion of reactants in the bioreactor is extremely small. The reactant concentration through the reactor is essentially constant and approximately equal to inlet concentration, i.e., the bioreactor is considered to be gradient less. The reaction rate is considered uniform within the bioreactor. Care must be ensured to check the reactant that should not bypass or channel through the packed catalyst.

Problem (i): If the catalyst decays during investigation, this reactor is not a good choice. The reaction rate parameters at the start of the run will be different from those at the end of the run.

(ii) In some cases, sampling and analysis of product will be difficult for small conversion in multi-component systems.

(b) CFSTBR: In this bioreactor, sterile feed is pumped into the bioreactor whereas immobilized catalysts are initially present in the reactor. Immobilized particles are not allowed to leave the bioreactor. Similar kind of problems like in tubular bioreactor is encountered in this case.

(c) Catalyst Contained CFSTBR (CCCFSTBR): A number of designs are available for immobilized systems. A typical example is given in Figure

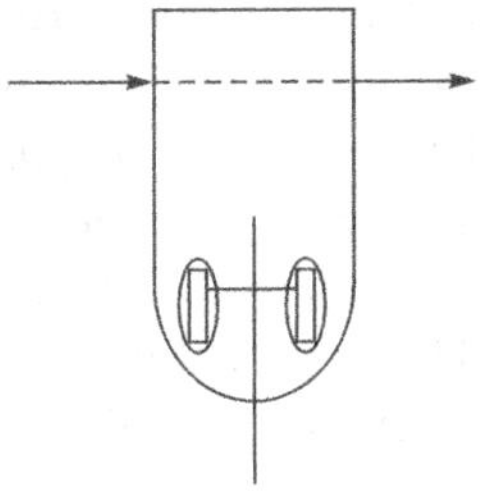

Figure 3.12 CCCFSTBR.

3.12 where catalyst particles are contained in the impellers. Impellers with catalyst rotate at high speed to minimize external mass transfer (Fogler, 1992).

3.4.2 Microbial Free Cells and Cellular Reactions in Basic Bioreactor

Based on the environment, type of cell, and cellular product, the mode of reactor operation is selected. The reactor can be batch, continuous flow, and semi-continuous (fed-batch and batch-fed) mode.

(A) Operation of Batch Bioreactor

All reactants (substrates) are added in the beginning of reaction following proper aseptic conditions. Then proper cells are added as inoculum to initiate the reaction. No product(s) and cells are withdrawn or no inoculum is added during the reaction. At a specified time, the reaction is terminated. This is confirmed by analyzing the samples during the course of reaction. Generally, ideal reactor operates at constant reaction volume, Flow rate $_{input}$ = Flow rate $_{output}$ = 0. All shake-flask experiments are generally batch. For statistical experimentations, in majority cases, shake flask studies are usually done to achieve replica of the experiments.

Exception There are some exceptions to ideal batch reactor operation.

(i) For mixed culture fermentation, in some cases a specific organism is added after a certain span of time of reaction. This is called phasing of the organism (Panda, 1986).

(ii) Typical inducers or controlling agents are added during different intervals of batch operation. For example, phenylacetic acid is added during batch production of penicillin acylase by *Escherichia coli* (Babu, 1991). Antibiotic pressure is required for certain antibiotic marker resistant plasmids. Otherwise, loss of plasmid occurs at a rapid rate (Chen *et al.*, 2002).

Advantages of batch reactor

- Lower investment costs as not much control is used here.
- Greater flexibility is achieved by using a bioreactor for various products and product specification.
- Higher conversion level is achieved as a result of well-defined cultivation period.
- Less risk of contamination and cell mutation happens due to relatively shorter cultivation periods.

However, disadvantages are the following.

- Non-productive ideal time for filling of medium to reactor, heating, sterilizing, cooling, emptying of reaction product, and cleaning the reactor.
- Greater stress on measuring instruments due to frequent sterilization.
- Higher expenditure due to preparation of several subcultures for inoculum.
- High expenses as more personnel or process computers are needed to control the non-stationary process (this can affect the advantage of lower investment cost).
- Greater risk to service personnel from possible contact with some pathogenic microorganisms or toxic products.

(a) Where shall we use the batch bioreactor?

Batch reactor is used in the following applications.

- To produce small amounts of products.

- For high risk of microbial mutation.
- For discontinuous separation of product from the cultivation medium.

Batch reactions are carried out in

- Shake-flask
- Stationary flask
- Serum bottle
- Petri-plate
- Fabricated and commercial bioreactors

(b) How shall we start a batch reaction?

(i) Shake flask experiments for aerobic microorganisms: Fermentation is carried out in various capacities of flasks (Erlenmeyer or conical flasks).

Steps

1. Medium (reactants) as per desired composition and/or optimal requirement (as suggested in statistical optimization procedure) is prepared for the proposed reaction. The pH (initial) for growth of organism is adjusted by standard HCl or KOH or NaOH solution.
2. The required volume of medium (for example, 100 ml in 500ml–Erlenmeyer flask) is dispensed in the flask. The mouth of the flask is closed with non-absorbent cotton in such a way that purified gases can only be exchanged through the cotton filter.
3. Over the cotton plugging, aluminum foil is placed to protect the cotton plugging from direct contact with steam.
4. Similarly some amount of distilled water in a flask, plugged as per steps 2 and 3, glass pipettes and rods are wrapped individually with aluminum foils and are placed in a stainless steel pipette sterilizer, pipette tips for micro pipettes placed inside a beaker and covered with aluminum foil, and a few glass test tubes, plugged with cotton and covered with aluminum foils are kept in an autoclave along with the flasks containing medium (in step 3).
5. Sterilization in an autoclave (check before for sufficient water level inside the autoclave) is done at 121°C (equivalent gauge pressure of 1.5 kgf/cm^2) for 15 minutes. (Caution: After autoclaving do not open the steam pressure hurriedly or try to open the lid. Normal cooling of sterilizer is desired in all the cases).
6. After cooling, one flask containing sterilized medium is inoculated by selected organism from a slant growth using a Pt–nichrome–loop. The aseptic operation is done inside a laminar flow chamber which must be previously cleaned and the platform is sterilized by isopropanol or suitable reagent followed by the illumination of UV-light in the laminar flow chamber. The illumination for about 5-10 minutes is sufficient to sterilize the inner chamber of laminar flow system. The transfer of organism to culture medium is strictly done inside laminar flow chamber in the presence of a flame.

   (Caution: (i) Do not expose organism or yourself (the experimenter) to UV-rays. The UV-source must be switched off before the inoculation. (ii) While handling LPG-flame, care must be taken to avoid any kind of damage to organism and the person).
7. The inoculated flask is kept on an orbital shaker maintained at a particular temperature for the growth of the organism and at a particular rpm. Inoculum will be ready by 12–24 hours of growth.

This may vary depending on the requirement of the experiments (for example, if one wants to study the microbiological parameters, like inoculum age, the inoculum is incubated for a desired period). Now the inoculum is ready for further study. One may be interested to know about the number of cells / unit volume of inoculum, cell dry weight equivalent, any product synthesis and residual reactants left. They can be estimated by standard published literature (Panda and Gowrishankar, 2008).

8. After the inoculum preparation, simultaneously suitable production medium is prepared and sterilized following the steps (1) to (4) stated above.
9. The suitable amount of inoculum expressed as, for example 10^4 to 10^8 cells/ cm^3 of production medium or 0.01 to 0.02 kg dry cell weight equivalent/m^3 of production medium, is transferred from steps (7) to step (9).
10. The production flasks are incubated in a temperature controlled incubator shaker for a specific time where maximum product is obtained. Samples are taken regularly to analyze cells, product, and reactants.
11. (a) After the reaction, if the product is extracellular, culture filtrate obtained after centrifugation is used to recover the product. Cells are destroyed by sterilizing in an autoclave and discarded following the laboratory hygiene procedure.
    (b) If the product is intracellular or associated with the cell, cells obtained after centrifugation are homogenized by ultrasonic disintegrator or by mechanical procedure with pestle-mortar or by chemical action. The culture filtrate is sterilized and discharged as per steps in 11(a).
12. Used empty flasks, glass pipettes, glass rods (which are reusable) are sterilized and washed with recommended detergents.
13. All disposable materials, like pipette, plastic tips, used aluminum foil and cotton, etc. are disposed following the procedure of laboratory hygiene and maintenance.

(ii) Fabricated or commercial bioreactor for aerobic microorganisms: One can expect variation of reaction conditions in shake-flask experiments. General problems associated with shake-flask experiments are the following.

- Control of pH, homogeneity, and dissolution of gases are difficult.
- Sample volume and sampling strategies are not satisfactory.
- Variation in conditions of fermentation in flask is prominent and is inevitable.

So the bioreactor is used to avoid such bottleneck in the experiments.

Steps for Operation

1. A standard bioreactor has the following assembly:
    - pH probe, measuring and control devices
    - Temperature sensing, measuring, and controlling devices
    - Dissolved oxygen sensing, measuring, and controlling devices
    - Antifoam control and monitoring accessories

    Other control and accessories are given in details in Chapters 2 and 8.

    Such system is used for the kinetic study.
2. Medium (reactants) is prepared as per the method discussed in Section 3.3.

3. Preparation of inoculum for reaction and reactor sterilization is done simultaneously.
4. Inoculum is prepared following the steps of shake flask experiments, i.e., steps 1 to 4.
5. (a) Bioreactor components are checked and verified for fit to use. Except the probes, all parts of the reactor are cleaned following standard procedures. The gaskets, O-rings, etc. are cleaned and greases are applied. They are placed on to the respective parts of the reactor. Details are as per Section 3.2.1.
   (b) Air filter is checked for the membrane. The air-line connection is properly checked till the sparger.
   (c) Steam connection line (for *in situ*) sterilization system is carefully checked for any clogging or hindrances.
   (d) Temperature control section is checked for proper display of temperature. Jacket / cooling coil is checked for smooth control of temperature.
   (e) The pH probe is cleaned gently and carefully. This is standardized with standard buffer before it is inserted in to the reactor.
   (f) The dissolved oxygen probe is cleaned properly. The membrane on the probe is checked for any fouling or damage. Placing the probe in water, which is first de-aerated by passing nitrogen, standardizes the dissolved oxygen probe. Then saturated air is passed through the water in which the probe is immersed. After standardization, the probe is inserted into the proper place on the reactor so as to immerse in the reaction fluid.
6. (a) The reactor, after proper cleaning and greasing of the components, is filled up to approximately 60% capacity with standard reaction medium designed by statistical procedure described in section 3.3.
   (b) Then the probes are inserted gently.
   (c) The connectors of the probe to the controller are removed before sterilization.
   (d) The outer parts of the probe (outside the reactor) are packed nicely with non-absorbent cotton and aluminum foil so that steam does not enter into the connection side. A hand pump, before sterilization, should pressurize the pH probe.
   (e) The air filter unit is separately covered with aluminum foil.
   (f) Acid- and base-containing flasks are properly connected with silicone tubes which is provided with a pinch-clip to avoid spilling of acid / base during sterilization.
   (g) All outside connections are properly capped for avoiding contamination.
7. (a) For *in situ* reactor, components in 6(f) and (g) plus other additional materials are separately sterilized as per the procedure of sterilization desired in shake-flask experimentation.
   (b) The reactor containing reactants and fitted with probes are sterilized by passing steam at very low agitator speed. The steam release valve initially controls escape of steam. After sufficient release of steam, the exhaust valve is closed and the pressure builds up in the reactor. After desired sterilization, cooling water is switched on.
   (c) After proper cooling, the components and connectors are attached to the reactor under a flame to ensure proper sterilization.
   (d) Positive pressure is ensured by passing sterile air through the air filter connected to the reactor.

(Caution (i) During the entire period of sterilization till the cooling to a desired temperature, the reactor vessel (for glass vessel) should be covered with a protective cover. (ii) All necessary connections should be effectively sterilized. (iii) Steam connection should be turned off carefully after sterilization.)

8. For autoclavable reactors:

   Steps 1-6, described for fabricated or commercial bioreactor operation, are the same. Additional steps are described below.

   (a) This is applicable for small laboratory scale operation. The medium (reactants) is poured into the reactor and other fittings are as per step 5(a) above. The entire reactor assembly is kept in an autoclave for sterilization. All tubes and connections with the reactor are properly clipped to avoid drainage/spill over of medium during sterilization. The sterilization is carried out at 121°C for 20 minutes. After proper cooling, the reactor is connected to the original reactor position and the agitator drive assembly is connected. Other connections are also made in the presence of a flame.

9. Inoculation is done in the presence of a flame by adding required inoculum through the inoculation port or it may be pumped by peristaltic pumps.
10. As the reaction starts, sample is taken immediately and during proper intervals.
11. The reaction is terminated when the desired product concentration is reached. This decision is taken after analyzing all representative samples.
12. All other steps are similar to the steps 11 to 13 of shake-flask experiments.

(iii) For Anaerobic Fermentation – Stirred Flask System

(a) *Preparation of oxygen- free nitrogen gases:* Nitrogen is passed though alkaline pyrogallate solution followed by Fildes and McIntosh indicator and silica gel column. This oxygen- free nitrogen is connected to pre-sterilized air filter assembly. This gas is bubbled to the reaction medium (Kundu *et al.*, 1984).

(b) *Control of off-gas:* The pre-sterilized air filter is attached to the exit gas stream, which is connected prior to the condenser to prevent evaporative loss during reaction. Finally, the exit gas is allowed through formalin solution to avoid contamination problem. The system has the provision for adding inoculum. Magnetic stirrer provides the agitation. Provision for temperature control and monitoring devices are attached to the assembly.

(c) Medium in the designed system is sterilized separately in an autoclave.

(d) It is connected later to the nitrogen supply and off-gas connection lines.

(e) Inoculation is done with the help of hypodermic syringe through the inoculation port.

(f) Samples are withdrawn with the help of a peristaltic pump attachment.

(g) Termination of reaction, disposal of cell, and culture filtrate are as per the procedure suggested for the shake-flask experiments.

(iv) Bioreactor Operations for Anaerobic Culture

The bioreactor operation described for aerobic processes is followed, but throughout the sterilization and during reaction vigorous sparging of oxygen-free nitrogen is done to ensure proper anaerobic environment. Additional redox measurement and control devices are required for anaerobic operation. Other steps and arrangements are exactly the same as those described for basic bioreactor operation for aerobic culture.

(v) Common Study in a Bioreactor

After formulating optimal medium components and microbiological parameters, the role of pH, agitation, and air flow rates for aerobic fermentation are studied to achieve near optimal values for the bioreactor operation. Two popular techniques are used: (i) Self-directing optimization (SDO) and (ii) Taguchi's orthogonal design.

They are discussed here with examples.

Example 3.2 Batch bioreactor study is necessary to obtain most suitable reaction conditions, i.e., pH for reaction, airflow rate, and agitation rate using limited run in a bioreactor. The desired product level is improved by this operation. Suggest a suitable solution to this study.

Solution: This problem is solved using the self-directing optimization technique of Hendrix (1980). Let us discuss the theory for the same followed by a numerical example.

Theory to this solution: It is difficult to conduct many experiments simultaneously in bioreactor systems. Self-directing optimization (SDO) is a non-statistical technique employed to determine the levels of variables when experiments cannot be run simultaneously (Panda *et al.*, 1997). SDO is called the rotating simplex method of optimization. It is self-correcting if an error causes the simplex to move in the wrong direction. It begins with patterned set of experiments in all of the concerned variables. When this pattern of experiments has been run, the experiment that gave the worst result is identified. This experiment is then discarded and replaced by a new experiment according to the rule.

Pattern of the simplex and its application: The shape of the simplex is described depending on the number of variables. For example, if there are two variables, the simplex is a triangle.

- For three variables, the simplex is a tetrahedron.
- For more variables, the simplex is a multidimensional geometry.

The movement of the simplex is visualized in the response plane. For example, the simplex is a tetrahedron for three variables. The variables are represented by four sets of combinations. The levels of variables for these four combinations are fixed randomly or from earlier reported values.

After the pattern of experiments is done, the worst run is discarded. This experiment is replaced by a new combination, which is the reflection of the worst point on the response plane.

Determining the reflection of a tetrahedron, for example, in a three dimensional plane is difficult. For this reason, a rule is employed to identify the new combination of variables. The rule is twice the average of the best points minus the worst point. The new experiment is performed and the worst point is again identified and replaced. This iteration is repeated till no further improvement in response values is observed.

Drawbacks of SDO

- Experiments must be run sequentially.
- Important independent variables and their range of interest must be known before hand.

Numerical Solution to Problem 3.2: Table 3.13 shows initial range of variables chosen as required by the SDO for the production of compound (P).

Table 3.11 *Initial range of variables*

Run #	***pH (Controlled)***	***Aeration (m^3)/ (m^3)(min)***	***Agitation (revolutions/ minutes***
1	(H)	(L)	(L)
2	(H)	(H)	(L)
3	(L)	(L)	(H)
4	(L)	(H)	(H)

Steps

(i) Four separate runs are carried out in a bioreactor.

(ii) Product, cell, and reactant concentrations are evaluated regularly.

(iii) At a particular point of reaction (e.g., 120 h of reaction) the product levels are compared for the four different runs.

(iv) Data are tabulated

Run #	***pH (Controlled)***	***Aeration (m^3)/ (m^3) (min)***	***Agitation revolution/ minutes***	***Product (kg/m^3)***
1	H	L	L	P_1
2	H	H	L	P_2
x3	xL	xL	xH	xP_3
4	L	H	H	P_4

'x' represents worst run.

The experiment that gave the worst result is assumed to be Run # 3.

(v) Consider only Run # 1, 2, and 4.

Calculate the following:

1. Sum of best points
2. Average of best points
3. Two times the average of best points
4. Two times the average of best points minus the worst points (Run #3 in this case). This step gives the new combination of run.

Detailed calculation is given by Felse and Panda (1999).

*For similar kind of situation, such approximations are made to ensure the sensitivity of the available measurement devices fitted with the bioreactor.

(vi) Again with the new set of variables, the experiment is carried out and the product concentrations will be measured in the experiment. Similar analysis like step (v) will be made with Run # 1, 2, 4 and the new run.

For bioreactor analysis, one needs to start with optimal medium constituents and 'near optimal' values of the variables stated here. The analysis for reactor design then appears meaningful. In this book, the reactor analysis in Chapter 4 has been done following these procedures.

Another example of optimization following Taguchi's design is considered here. The relative influence of chemical parameters (concentration of screened medium constituents) and physical parameters, viz., controlled pH, agitation, aeration rate on product synthesis in batch bioreactor can be determined using Taguchi's method (Taguchi, 1986). Taguchi's method is applied in industrial process design principally in the development trials to generate enough process information to establish the optimal conditions for a particular process using minimum number of experiments possible.

The screening of medium constituents done in shake flask experiments are considered here as chemical components along with the other variables pH (controlled), agitation, and aeration.

Example 3.3 Further optimization of chemical components screened by Plackett and Burman design is done using Taguchi's method rather than to study them in second order design. Reactions are performed in a bioreactor to see the combined influence of controlled pH, aeration rate, and agitation rate.

Solution: Experiments, as per orthogonal array procedure, have the pair wise balancing property where every test setting of a design parameter occurs with every test setting of all other design parameters the same number of times. Orthogonal array experiments minimize the number of test runs while keeping the pair wise balancing property (Byrne and Taguchi, 1989).

The medium constituents which are screened by Plackett-Burman experimental design (variables *A*, *C*, *D*, and *G*) are taken into consideration for this study. The levels of these chemical parameters (the medium constituents) are chosen based on the results obtained in the shake flask fermentation. The levels of physical parameters, viz., controlled pH, agitation, and aeration, are chosen based on the earlier reports. According to Taguchi's technique, for 7 variables 8 runs is used to evaluate the influence of various parameters on the production. The levels 1 and 2 represent the coded levels of parameters at lower and higher levels. The variables and their levels employed in the Taguchi's experimental design are given in Table 3.12. Experiments have been performed according to the experimental plan given in Table 3.13.

Table 3.12 *Variables and their levels employed in the Taguchi's Robust Design method for optimal production*

Variables	**Levels**	
	1	**2**
A (kg/m^3)	37.5	50.0
B (kg/m^3)	1.25	1.5
C (kg/m^3)	2.25	3.75
D (kg/m^3)	0.005	0.0125
E (pH controlled)	5.5	7.5
F (Agitation rpm)	150	7.5
G (Aeration m^3/(m^3) (min)	4.0	6.0

Table 3.13 *Taguchi's Robust design matrix for optimal production.*

Run #	*Variables and their levels*						
	A	*B*	*C*	*D*	*E*	*F*	*G*
1	1	1	1	1	1	1	1
2	1	1	1	2	2	2	2
3	1	2	2	1	1	2	2
4	1	2	2	2	2	1	1
5	2	1	2	1	2	1	2
6	2	1	2	2	1	2	1
7	2	2	1	1	2	2	1
8	2	2	1	2	1	1	2

Variables B, C, and D are originally marked as variables C, D, and G in the Plackett and Burman design.

Samples are analyzed from the independent experiments for specific product.

After performing experiments, analysis of control factor for production has been done by Phadke (1989). The relative influence of each parameter on product synthesis in a batch bioreactor has been determined using the following equations.

$$\text{SS} = \frac{\sum_{i=1}^{n} (\text{sum of yields in the } i \text{ level of } A \text{ factor})^2}{\text{The number of yields in the } i \text{ level of } A \text{ factor}} - \frac{(\text{Total sum of yields})^2}{\text{total number of yields}} \quad (3.17)$$

$$\%\ \text{SST} = (\text{SS/SST}) \times 100 \quad (3.18)$$

where,

SS means sum of squares

SST stands for total sum of squares.

The sum of product synthesized for individual levels for a particular experiment has been calculated from Table 3.13. Detailed numerical calculation is described by Dasu *et al.* (2000).

(B) Operation of Continuous Flow Bioreactor (CFBR)

In the continuous mode of operation, reactant is continuously added to the reactor and products plus unconverted reactants and cells are removed simultaneously from the system (Schematic presentation in Fig. 3.13). All reaction variables and control parameters remain constant with time. Time constancy in parameters is observed for productivity and output, i.e., steady state.

Generally, CFBR is carried out in a well-controlled bioreactor. Those will avoid the limitation of shake-flask experimentation as well as batch bioreactor studies.

A typical operation of CFBR is designed in such a way that a typical reactant is limiting in the reaction. The addition of reactant (feed flow) to the reactor during CFBR operation changes the environmental conditions which will influence cellular physiology. The product concentration is less than the batch-fed reactor. One needs to operate in such a way that the productivity is maximal in CFBR. In general,

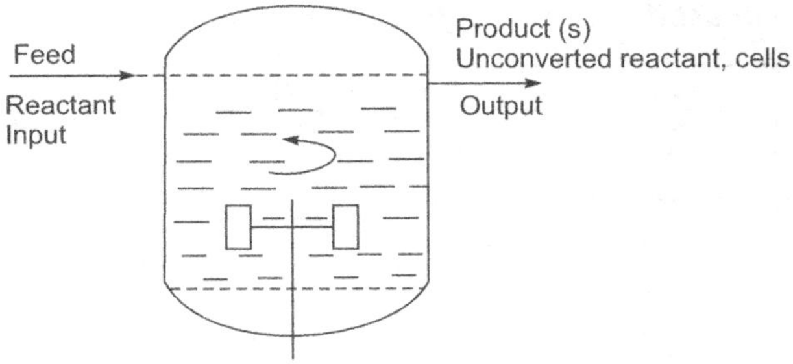

Figure 3.13 Continuous flow bioreactor-showing input and output streams.

in the literature of biological reactions, CFBR is called "chemostat". Chemostat means "chemeo" plus "statis", i.e., chemical composition throughout the reactor is constant.

On the other hand, in a batch reactor, the chemical composition is always different which is a function of time.

In summary, CFBR is also called as: (a) pH stat – In CFBR feed flow is adjusted to maintain pH constant in the reactor.

(b) Turbidostat – Feed flow is adjusted to maintain the turbidity constant. This concept is not practical if insoluble reactants along with cells are present. This concept may not be suitable for enzyme reactions with or without soluble reactant, plant cell, and animal cell reactions.

So in a simplified way,

$$\text{Input feed flow rate } (F_{in}) = \text{output flow rate } (F_{out}) \neq 0 \qquad (3.19)$$

Therefore, the volume of reaction is constant in ideal CFBR operation.

Some requirements for continuous flow bioreactor operation are almost the same as described in batch bioreactor operation.

Additionally, one should consider the following requirements.

- Additional feed and product delivery line with peristaltic pumps.
- Large volume of sterile feed.
- Product collection should be made under rigorous aseptic conditions.

Of course, there are certain advantages associated with the operation of continuous flow bioreactor.

Advantages:

- Large scope for mechanization and automation.
- Lower reactor volumes, i.e., no unproductive idle time.
- Low wages bill.
- Product quality is constant as operating conditions are invariant.
- Less possible danger to service personnel from pathogenic microorganisms or toxic materials due to improved mechanization.
- Less wear and tear on instruments from sterilization.

However, one cannot ignore the disadvantages of this reactor.

Disadvantages:

- Low flexibility in operation.
- Raw material quality must be uniform.
- High investment cost.
- Continuous removal of insoluble substances can be very expensive.
- High risk of mutation due to long cultivation period.

So, in general, continuous flow operation is **preferred** with:

- High production rates.
- Microorganisms with high mutation stability.

There are different modes of continuous flow bioreactor operations: They are:

(I) Continuous Flow Stirred Tank Bioreactor (CFSTBR)

- Single input single stage continuous flow bioreactor (SISSCFSTBR)
  - with sterile feed (no inoculum or cell added with the feed).
  - with sterile feed but dosing of a critical compound.

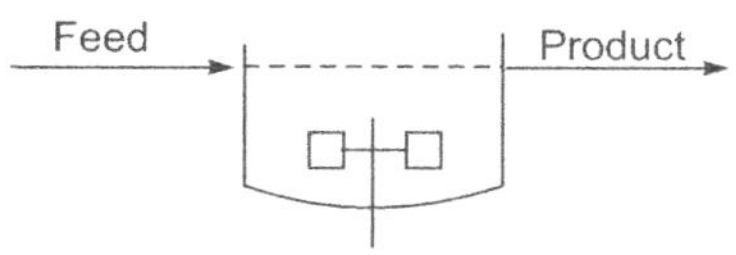

Figure 3.14 SISSCFSTBR without dosing.

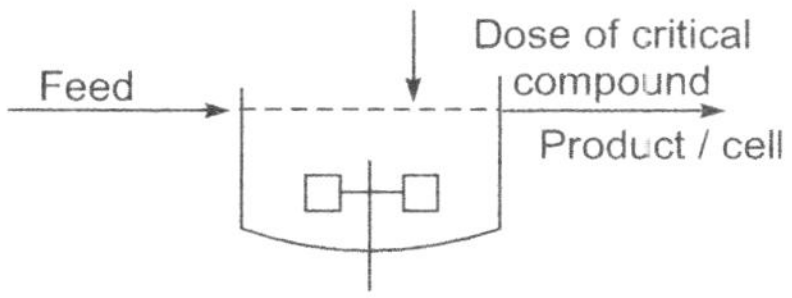

Figure 3.15 SISSCFSTBR with dosing.

- Single stage continuous flow with cell recycle.

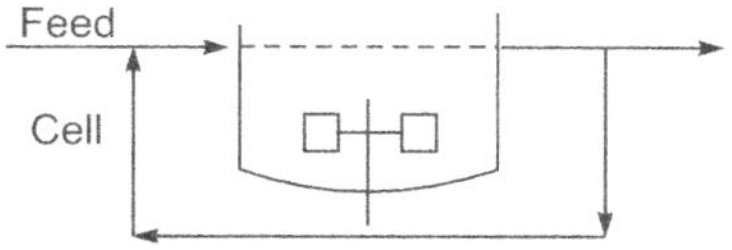

Figure 3.16 SSCFSTBR with recycle.

- Multistage continuous flow bioreactor
  - With sterile feed (Fig. 3.17)
  - With intermittent feed at any stages of train of bioreactor (Fig. 3.18)
  - With cell recycle (Fig. 3.19)

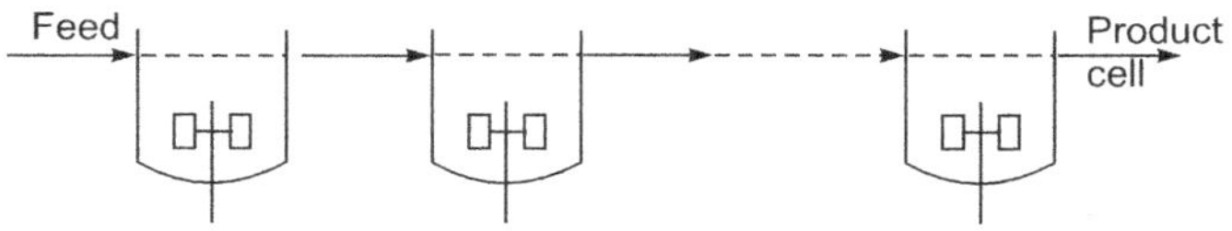

Figure 3.17

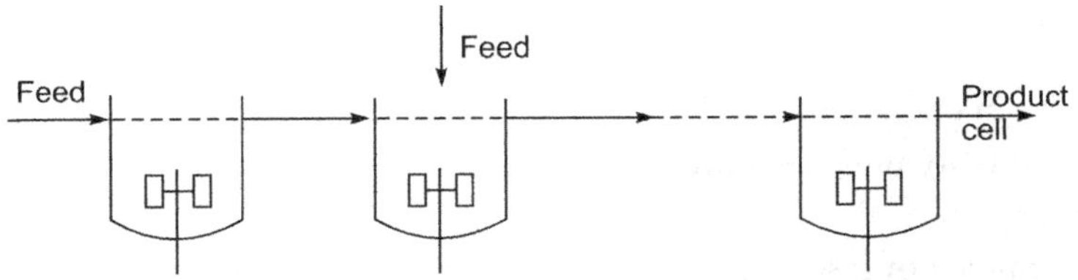

Figure 3.18

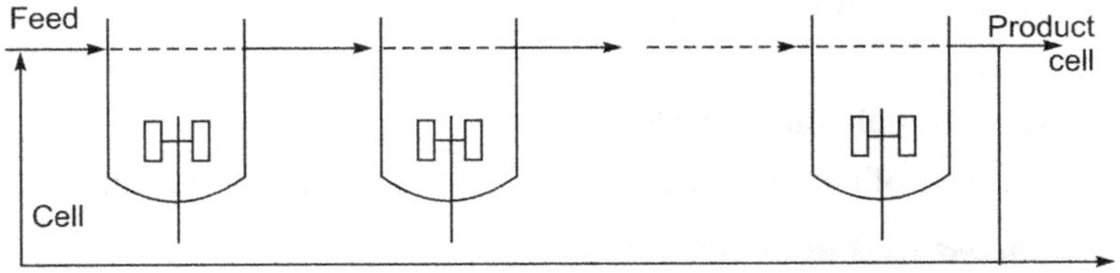

Figure 3.19

(II) Continuous Plug Flow Reactors (PFTR)

This is described in Figure 3.20.

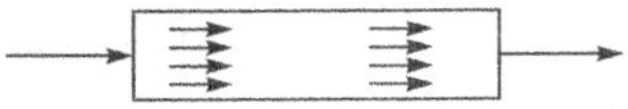

Figure 3.20

How do they work?

1. Single Stage CFSTBR

The preparation for the start-up of the reactor is almost the same as discussed for batch bioreactor operation. There is some modification.

- **Addition of inoculum:** Amount of cell added to initiate the reaction is like batch bioreactor. However, the feed, without any cell which is used in the reaction, is pumped aseptically to the reactor and simultaneously output stream (flow rate is same for maintaining constant reaction volume) is switched on. Flow rate is maintained initially in such a way that cell division rate matches with the flow rate. Otherwise all cells without multiplication will come out from the reactor. Feed of reactant alone without cell is called sterile feed ($C_{X_F} = 0$). Flow rate of reactant is changed to suit the requirement for production. At a very low flow rate, the reaction approaches batch reaction behavior.

At high flow rates of reactant, the conversion of reactant will be reduced and more and more cell will come out from the reactor.

At one stage of feed flow rate, one can achieve 100% discharge of cell in the output stream. The phenomenon is called **washout of cells.** No conversion of reactant is observed in this case.

One can study the following:

- Kinetic parameter estimation for the organism, as there is no reactant limitation.
- Validation of mathematical expression using this reactor.
- Optimization of production.
- Set up the goals for cell, product, and reactant flow rates.

Some deviations

(a) Additional feed: Single stage CFSTBR sometimes is associated with additional intermittent feed of a critical reactant. For example, the addition of a typical inducer is necessary when there is depletion of

such component. Other example is the addition of an antibiotic to keep the antibiotic pressure at constant level for a plasmid-bearing cell.

(b) Recycling: Single stage recycle reactor is used when high cell concentration is necessary in the reactor. Product stream is attached to a settler or a filter. The concentrated cell slurry in certain ratio (called recycle ratio) is mixed to the feed stream before it enters the reactor. It is necessary that toxic components have to be avoided in the recycle stream.

2. Multistage Continuous Culture Operation

A few important variations in operation are indicated here. This subject has been elaborately discussed by Moser (1985).

(a) Single input multistage operation (SIMO) Single stream system of multistage CFSTBR is associated with a single medium input. This flow rate is constant through all other following stages. The important **points** are the following.

- First stage behaves like a CFSTBR.
- Following stages cannot affect the previous stages.
- The space velocity or dilution rate (D) cannot be changed in one stage as it is interconnected.
- The flow rate is constant in all the stages i.e.,

Flow rate in the input (F) = Volume (V) × Dilution rate (D) (3.20)

Therefore, $F = V_1D_1$(1st stage) $= V_2D_2$ (2nd stage) $= \ldots\ldots\ldots\ldots = V_nD_n$(nth stage). (3.21)

The typical relation of $C\acute{x}$ (steady state cell concentration) and D is given here for two- stage operation (Fig. 3.21).

(i) This appears that $C\acute{x}_1$ (first stage SS C_x) and $C\acute{x}_2$ (second stage SS C_x) become zero at the same dilution rate (i.e., D_c).

(ii) $C\acute{x}_1 << C\acute{x}_2$

(iii) This is true that any unreacted reactant in first stage will enter into the second stage where it is used completely. The output of the second stage contains negligible reactant.

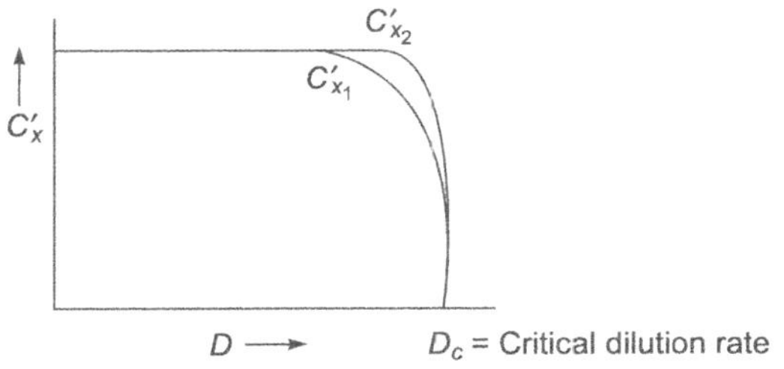

Figure 3.21 $C\acute{x}$ vs D.

This suggests the following.

I. If all the stages contain same reaction volume, all of them will washout at the same critical dilution rate.

II. For stages containing unequal reaction volume, the whole system will washout unless there is at least one reactor whose dilution rate is less than D_c, though in later stages the dilution rate may be greater than D_c without these reactors washing out.

III. D_c for the entire system is considerably lesser than that for a single reactor.

Question 1 ***Is multiple stage beneficial?***

Answer: As the number of stages increases there will be gradual depletion of reactant and endogenous metabolism in cell continue in later stages. This leads to change in chemical composition and physiological state of cell.

(b) Multi-input multi-stage operation (MIMO): More possibilities are offered in MIMO than SIMO. There is multiple feed input and all of them are independent variables. Each feed input may contain a variety of reactants. This is a complex experimental situation. The behavior of a simplest two input–two stage system can be visualized by Figure 3.22.

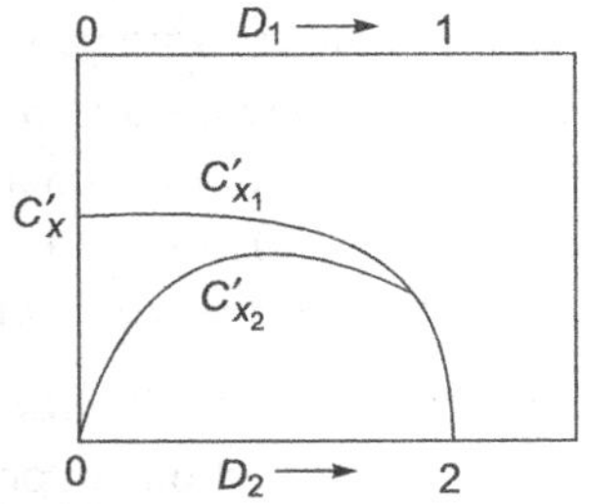

Figure 3.22 Two-input two-stage system.

Relative insensitivity of the second stage happens to change in the first stage. So the higher flexibility of the multi-input multistage systems is to operate at high flow rates. This quality makes MIMO superior to SIMO.

Importance of multi-stage systems

- Maximum conversion can be achieved by complete utilization of reactant in subsequent stages by maximizing productivity in the first stage. This indicates higher reaction order for better results. Example: costly reactant and environment reactors.
- Physiological status of cells can be maintained by MIMO. Example: secondary metabolite synthesis.
- Long residence time can be maintained without reducing dilution time, i.e., without affecting productivity.
  Example: for reactions at stationary growth phase and complex reactants
- Demand for optimal operating conditions.
- Assurance for high quality products.

Difficulty Complex operating conditions and kinetic expression for C_X and C_S in later stages are prominent.

(c) Operation of continuous plug flow bioreactor (PFTR): An ideal PFTR is a tube reactor through which undisturbed liquid flows uniformly (Fig. 3.20).

The composition of the fluid varies from one position to the other along the length of the reactor. So this reactor is also called "integral reactor" or "gradient reactor" or "reactor with distributed parameters".

The ideal PFTR operation is not possible with reactions involving cells, free enzyme plus insoluble reactants, and immobilized cells or enzyme with reactants. They cause back mixing in PFTR.

(a) Difference between CFSTBR (gradient less reactor) and PFTR (gradient reactor)

1. If the objective of a process is cell production, there is a little difference.
2. If the objective is reduction in effluent reactant concentration, PFTR is the optimal choice.
3. MIMO is an equivalent to PFTR or tubular reactor.

(b) Application of PFTR

1. For first order reaction, conversion in PFTR is better than in CFSTBR. Example: waste water treatment, sterilization reactor.
2. Immobilized cell bioreactor.
3. Conceptual tubular reactor for solid reactant processing and shear sensitive cell culture.

(c) Advantages: Tubular reactor has a number of advantages over CFSTBR.
- Mixing is more uniform by eliminating dead spaces. This facilitates easy scale-up.
- Surface area to volume ratio is sufficiently high which offers better transfer operations.
- Higher productivity and optimal conversion can be achieved in PFTR.

(D) Operation of recycle reactors They offer advantages compared to non-recycle continuous flow reactors (CFSTBRs, PFTR – including tubular reactor):

- Improved treatment rates
- Smaller vessel size
- Higher conversion.

Recycling of cells provides a way for proper maintenance of cells in reactor. A basic concept of PFTR with recycle is shown in Figure 3.23.

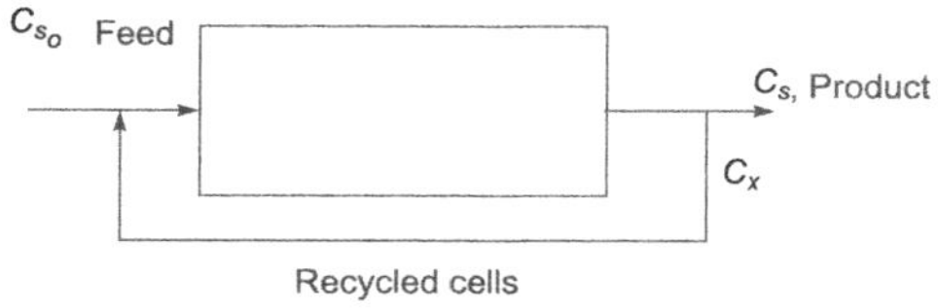

Figure 3.23 PFTR with recycle.

(E) Semi continuous bioreactor operations: There are two basic classes of semi-continuous reactors.

- Batch-fed bioreactor and
- Fed-batch bioreactor

This is a combination of batch and continuous mode of operation. It has a programmed reactant addition for secondary metabolite production, i.e., cell growth and product formation often occurs in separate phases.

There are certain advantages over the other processes.

Advantages

- High level of flexibility of operation
- Semi-stationary method of operation even in the case of slightly mutating microorganisms
- High yield as a result of well defined cultivation time(no cells are added)
- Environmental conditions for microbes may be optimized to suit the growth phase or production phase and age of culture.

However, there are a few disadvantages as well.

Disadvantages

- Non-productive idle time
- Higher wages bill due to greater staff requirement or use of process computers
- More wear and tear on measuring instruments

There are a lot of controversies over the use of these terminologies. In this book, the idea of fed-batch reaction is considered based on Suga *et al.*, (1980).

E (i) Batch-fed bioreactor: It is a variable volume batch reactor with product output. Feed is injected only when some part of the reaction mixture is removed from the reactor. Cells are not removed completely. For a new batch, one needs to add inoculum. In this case, there is no need to add fresh inoculum.

In batch-fed cultivation, intermittent feeding of nutrients or substrate or reactant is used to supplement the reactor contents and provide control over the reactant (substrate) concentration. High substrate concentration may be inhibitory or may switch on to undesirable metabolic pathways (Doran, 1995). Batch-fed operation shows the same characteristics as pure batch operations, in that concentration levels change with time and that some down time occurs during initial charging of reactant and final discharge of products. The ability to manipulate concentration levels within the bioreactor by an appropriate feeding strategy confers a high degree of flexibility to batch-fed operation. The important characteristics of batch-fed operation are the following (Dunn *et al.*, 1992):

- Extension of batch growth or product formation by additional substrate feeding.
- Possibility of operating with separate conditions for production and growth phases.
- Development of high biomass and product concentration.
- Alleviation of substrate and product inhibition.
- High control of fermentation process due to nutrient control.
- Possibility to maintain any nutrient limitation condition.

Applications

- Foreign protein synthesis and stability improvement (Chirulova et al, 1997; Ramirez and Bentley, 1995).
- High cell density cultivation of hybridoma cells (Zhou *et al.*, 1995).
- Increased ethanol productivity by recombinant *E.coli* (Hilaly *et al.*, 1995).
- Secretion of biologically active–anti trypsin (Kang *et al.*, 1996).
- Improved thringensin production by *Pichia pastoris* (Jong *et al.*, 1995).
- Recombinant protein production by *Pichia pastoris* (Vijay *et al.*, 1997).
- Simultaneous production of ginseng-saponin and polysaccharides by high cell density (Zhang and Zhong, 1997).
- Single chain Fv antibody fragment (ScFv) by recombinant *E.coli* (Harrison *et al.*, 1997).
- Human anti-thrombin III production by BHK cells (Honda *et al.*, 1998).

E (ii) Fed-batch bioreactor: The reactor starts with a little volume of reaction medium with inoculum in a batch mode. Gradually, the volume increases stepwise till the final volume of operation is achieved in the reactor. At that point, feed and output streams are functional to keep the volume of the reaction constant. Feed rate is guided by some important metabolic parameter so that the growth rate remained constant during the reaction.

To control the fed-batch culture, a suitable control parameter should be chosen that expresses the microbial activities accurately and rapidly. Suga *et al.*, (1980) has introduced this concept for fed-batch culture to produce cellulase by *Trichoderma reesei.* They have suggested CO_2 evolution rate as the control parameter. The fed-batch culture could provide a constant environment like a steady state continuous culture. To maintain the specific growth rate constant, CO_2 concentration in the effluent gas must be controlled in the bioreactor operation.

Steps are

1. The specific growth rate (μ) for maximum productivity is selected for a reaction.
2. The change of CO_2 concentration should be kept to the specific value determined by Suga *et al.*, (1980).

$$C_{CO_2/\text{set point}} = \left(m_{CO_2} + \frac{\mu}{Y_{CO_2}}\right)\frac{C_x}{F_g} V_0 (\exp(\mu t)_{\text{set point}}) + C_{CO_2/\text{intial}} \quad (3.22)$$

where,

$C_{CO_2/\text{set point}}$ is set point CO_2 concentration (kmol/m^3)
$C_{CO_2/\text{initial}}$ is initial CO_2 concentration
C_x is cell concentration (kg cell /m^3)
μ is specific growth rate (h^{-1})
F_g is gas flow rate,
m_{CO_2} is CO_2 evolution rate for maintenance (kmol/kg cell/(h)$^{-1}$)
Y_{CO_2} is the yield constant for CO_2 evolution (kg-cell/kmol)
V_o is initial working volume in fed- batch culture, m^3
t stands for time, h

3. The concentration of CO_2 obtained from Equation (3.22) is called set point CO_2.
4. If the CO_2 evolved from the reactor is lower than the set point, the feed-rate of reactant could be increased automatically.
5. The feed-rate is controlled by comparing the analyzed CO_2 from the reactor with the set-point (Fig. 3.24).

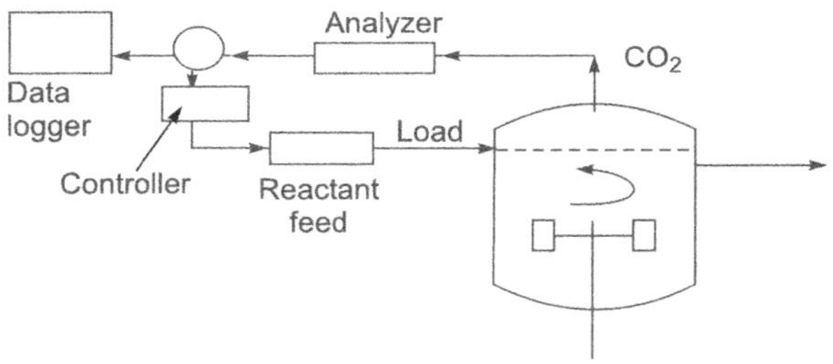

Figure 3.24 Fed-batch reactor control system.

Above process is based on single major C-source. If two major C-sources are present in the reaction (i.e., reaction involving mixed C-sources), following steps are desirable which are explained in Example 3.3.

Example 3.4 Let us assume that starch and amylose are the C-sources. How will you control both the C-sources in the reactor?

Solution: The feed rate of starch, F_S, is set to be almost the same value as the degradation rate of starch by enzyme produced in the reaction.

Feed rate of amylose (F_A) is set so as to get the expected specific growth rate in addition to batch feed rate (F_S).

The F_{AS} is the feed rate of amylose as controlled by the CO_2 evolution rate.

This is achieved by experiment using repeated fed-batch culture in a similar analogy by Suga *et al.* (1980).

(F) Operations of other special reactors

F(i) Fed-batch under vacuum cycling: To remove toxic metabolite from the reaction, this technique is exercised to run the reactor without the effect of inhibitions. The scheme of operation is given below.

To the regular fed-batch reactor system, vacuum system is connected with the aid of a vacuum pump. The vacuum pump is operated intermittently and the pressure is controlled manually or automatically. As the liquid level in the bioreactor (under vacuum) is dropped for boil-off of vapor, sterile feed is automatically added into the reactor to maintain constant working volume. The pressure in the reactor is slowly decreased till the reaction fluid begins to boil. Condensate is collected, by keeping the receiver at cold condition. The pH of the reaction fluid is maintained constant by acid / alkali addition. The reaction is carried out under quasi-adiabatic condition (Kundu, 1983).

F(ii) Dialysis reactor operation: This is a specialized reactor used for cells and immobilized system. A filtration unit is placed inside the reactor (Fig. 3.25).

Reactant enters the reaction zone by diffusing through a dialysis membrane from the reactant (medium) zone. Reaction takes place and product diffuses back through the membrane into the recovery zone. Cells cannot escape the reaction zone. So high cell density can be maintained in the reactor.

Drawback

Membrane fouling is a serious problem. Stirred coupled membrane reactor reduces fouling.

F(iii) Fluidized-bed bioreactor (FBR) operation: Fluidized beds are very common in the chemical

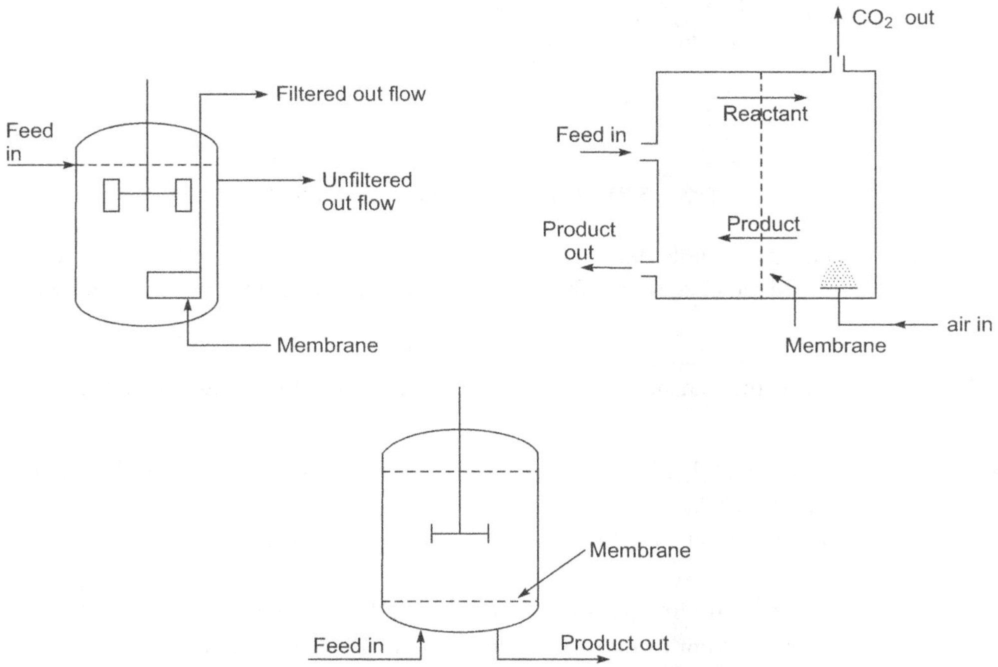

Figure 3.25 Schematic of dialysis bioreactor.

industry. Fluidized beds constitute a relatively new concept in the field of biotechnology. The three most important applications for bio-fluidization are:

- Enzymes immobilized on a solid matrix,
- Pure cultures of cells immobilized on a solid matrix,
- Various types of effluent treatment.

Cells / enzymes in immobilized form are held in the vessel by a mesh. Reactant solution (medium) is circulated by a pump. This can be used as a semi-continuous or continuous operation to allow the trapped cells to effect chemical changes on constituents of the medium without being washed out along with the reacted medium. This is suitable for animal cells in small-scale operation and large-scale operation for effluent/decontamination process.

Multi-stage fluidized bed with immobilized – chymotrypsin to separate D- and L-phenylalanine has also been reported in the literature. For the production of alcohol, Margaritis and Wallce (1995) have set up a fluidized bed of immobilized *Zymomonas mobilis* cells.

A schematic of a gas-liquid-solid FBR for waste-water treatment is shown in Figure 3.26.

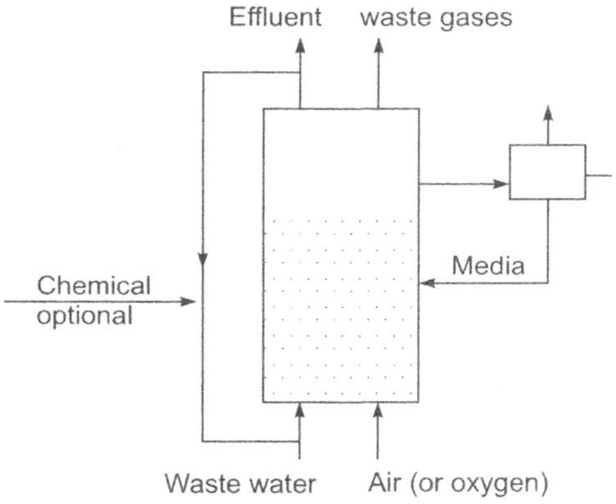

Figure 3.26 Three phase fluidized bed bioreactor.

In a bioreactor with heavy particles (density of matrix particles larger than that of liquid) biomass support (e.g. silica, sand, coal), fluidization is commonly conducted with an upward co-current flow of gas and liquid through a bed of particles. Under fluidization conditions, the bed is fluidized with an upward flow of liquid counter to the net gravitational force of the particles.

Once fluidized, each particle provides a large surface area for bio-film formation and growth. The support media eventually becomes covered with bio-film and the vast available growth surface afforded by the media results in increased biomass concentration approximately to an order of magnitude greater than that maintained in a suspended cell growth system.

Gas fluidized bed systems are often used in the chemical industry, but are rarely found in biotechnological processes. They are sometimes used for drying the biomass.

A practical problem which can occur in the operation of an FBR is the over expansion of fluidized bed due to excessive growth of biomass on support media. This can lead to washout of bio-film coated particles from the bioreactor. The problem of the over expansion has, generally, been solved by the removal of biomass-laden particles from the bioreactor.

F (iv) Bubble-driven bioreactor operation: Sparging without mechanical agitation can also be done for simultaneous aeration and agitation. Two classes of bubble driven bioreactors are bubble column bioreactors and air-lift bioreactors (Figure 3.27).

Bubble driven bioreactors are commonly used in the culture of shear sensitive organisms such as molds and plant cells. An airlift bioreactor differs from bubble column bioreactors by the presence of a draft tube, which provides better means of mass and heat transfer.

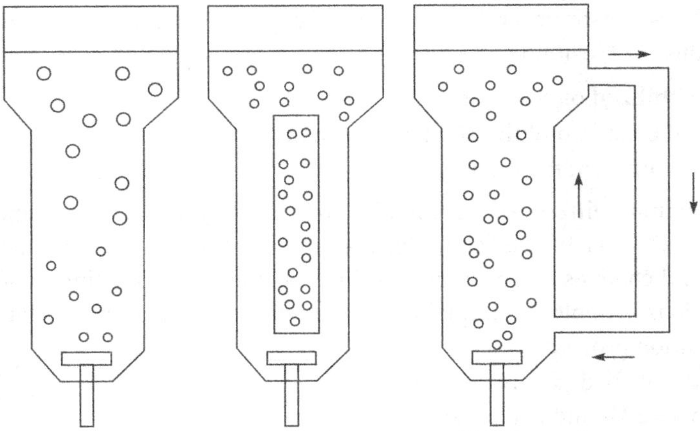

Figure 3.27 Bubble driven bioreactors.

The vessel design eliminates the need for a stirrer system. A tall and thin vessel is the best shape with high aspect ratio. Sometimes a conical section is used in top of the reactor to facilitate gas exchange.

Bubble driven bioreactors are generally tall with liquid height to base diameter ratios of between 8 : 1 and 20 : 1 (aspect ratios for microbial cell culture is higher than for animal cell culture). The tall design of these bioreactors leads to the following:

- High gas hold-ups,
- Long bubble residence times,
- A region of high hydrostatic pressure near the sparger, i.e., at the base of the reactor.

Bubble Column Bioreactor: Usually the bubble column reactors are cylindrical in shape with an aspect ratio between 4 and 6 (height to diameter ratio). Gas is sparged at the base of the column through perforated pipes, perforated plates, sintered glass or microporous spargers. Oxygen transfer and mixing are influenced mainly by the gas flow rate and the rheological properties of the fluid. Internal devices such as horizontal perforated plates, vertical baffles, and conjugated sheet packing may be used in the vessel to improve mass transfer and modify the basic design.

Characteristics of basic bubble column reactor are the following.

- No moving parts
- No sealing problems
- Absolute sterile operation
- Air sparger supplies air bubbles distributed at the reactor bottom and the bubbles move upward in the form of bubbles concurrently or counter currently to the liquid, and
- Diameter of the reactor is enlarged at the top for facilitating bubble disengagement and foam destruction.

Advantage

High degree of flexibility in construction and design accommodates several modifications, e.g., introduction of internal and external loops, gas distribution system, etc.

Airlift Reactor Operation: Many biorcator designs have been proposed for increasing the oxygen transfer rate and for minimizing the power consumption. One of the most promising configuration is found to be the airlift bioreactor (Chisti, 1989).

Airlift bioreactors are comprised of four distinct zones, each with its own distinct flow pattern (Fig. 3.28).

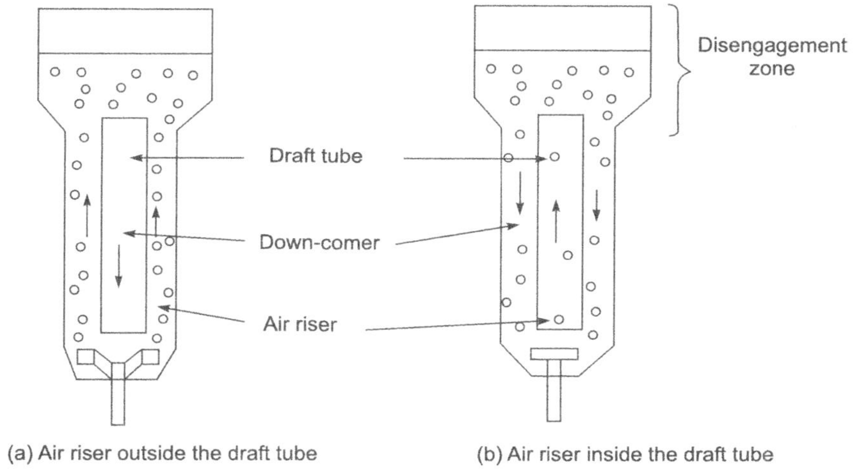

Figure 3.28 Location of air riser.

The three major configurations of airlift bioreactors are

- Split cylinder internal loop reactor
- Draft-tube internal loop reactor
- External loop reactor

Components in Airlift Reactor

Riser The region into which bubbles are sparged is called the air riser. The air riser may be on the inside or outside of the draft tube. The latter design is preferred for large scale fermentors as it provides better heat transfer.

Gas-liquid Separator The liquid leaving the top of the riser enters the gas disengagement zone, where, depending on its specific design, some or most of the dispersed gas is removed from the reactor.

Down-comer The rising bubbles in the air riser cause the liquid to flow in the vertical direction. To counteract these upwards forces, liquid will flow in a downward direction in the down-comer.

Bottom The liquid flows through the down-comer to the base of the device (the bottom) where it re-enters the riser.

The sudden widening at the top of the reactor slows the bubble velocity and thus disengages the bubbles from the liquid flow. Carbon dioxide-rich bubbles are thus prevented from entering the down-comer.

The reduced bubble velocity in the disengagement zone also leads to a reduction in the loss of medium due to aerosol formation.

The increase in area will also help to stretch the bubbles in foam, causing the bubbles to burst. The axial flow circulation caused by the draft tube also helps reducing foam formation.

Airlift bioreactors are more expensive to construct than bubble column reactors. There are several designs existing for airlift bioreactors. The most commonly used design is one with a central draft tube.

Main **functions of draft tube** are the following.

1. The draft tube enhances the axial mixing through out the whole reactor.
2. The draft tube reduces bubble coalescence. This occurs due to circulatory effect that the draft tube induces in the reactor. Circulation occurs in one direction and hence the bubbles also travel in one direction. Small bubbles lead to an increased surface area for oxygen transfer. The magnitude of liquid circulation is one of the most important design and scale-up parameters for airlift reactors.
3. It equalizes shear forces through out the reactor. This is believed to be the major reason why airlift bioreactors have higher productivities than stirred tank reactors. In stirred tank reactors, there will be high shear near impeller and low shear near the surface. The draft tube distributes shear evenly throughout the reactor.

F(v) Bio-film reactor operation: Basic concept of biofilm-reactor is described in Chapter 2. However, the properties of bio-films and their potential advantage are important for the operation of biofilm reactors.

Properties of bio-films are continuous mode of operation, easy control of process and heterogeneous kinetics having transport problems: for gas/liquid, liquid, liquid/solid and solid (due to disintegration of film). It is difficult to control the shape of the particles.

Flocs are encountered in this reaction. Thus the process is more complex and is difficult to control.

Potential advantages of bio-film are the following.

- Highest productivity can be achieved as no washout occurs.
- Highest cell concentration can be achieved.
- Principles of fluidization can be applied to this process.

General theory of Bio-film Bio Reactor operation: Microorganisms grow on the packing surface in the reactor. Strictly they are microbial film (bio-film) reactors and belong to the surface reactor group. The reactant is added at the reactor head, trickles over the microbial film and down to the bottom (percolating reactors). Air is added at the bottom and moves up in the counter flow.

Probable sequence of the reaction is given below.

- Adsorption of dissolved organic compounds on the moist support surface,
- Transport of microbial cells to the surface,
- Adhesion of cells to the surface,

- Bio-film formation by cell growth, and
- Sloughing of bio-film by shear or lack of oxygen in the case of thick layers and cell autolysis on the support surface.

(i) In a **batch bio-film reactor,** the immobilized cells have to be used for repeated batches. However, it is likely that during the late stationary phase culture production would experience inhibition and thus reduces productivity.

(ii) In a **CFSTBR,** the feed medium is fed to the reactor and product is withdrawn at the same rate as feed. They are stirred using a mechanical device such as impeller. CFSTBRs cannot be packed with the adsorbent support covered by bio-films. However, they can be used if fibrous bed support is used for adsorption of cells. In that case, cells can grow and form a bio-film on the fibrous bed. Agitation can be provided in such a case.

(iii) Suitable support materials are packed in packed bed reactor (PBR) followed by inoculation with the culture to form bio-film. The reactor is supplied with a feed that is not deficient in nutrients. Such reactors are usually fed at the bottom, thus getting product at the top of the reactor.

(iv) In **tubular bed reactors (TBR)** some of the bio-films may not get sufficient feed thus affecting reactor efficiency/productivity adversely. In gaseous fermentations, gas may occupy significant space in the reactor and may form stagnant pockets. This may affect the efficiency of the reactor. These reactors have been used at large scale successfully in anaerobic wastewater treatment and acetic acid production.

(v) **Airlift** reactors contain two concentric tubes, a riser (an inner tube) and a down comer (an outer tube). In these reactors, mixing is achieved by circulating air at the bottom of the reactor. As a result of force applied by the air (at the bottom of the inner tube), the liquid in the inner tube moves up which then overflows (the inner tube) downward thus creating eddies to mix the liquid.

(vi) **Operation of up-flow anaerobic sludge blanket (UASB)**

Wastewater enters the bottom of the reactor through the inlet liquid distribution system and passes upward through the dense anaerobic sludge bed. Because of the high biomass concentration, it has been demonstrated that volumetric organic loading rates as high as 50 kg chemical oxygen demand (COD) per m^3 per day could be employed to the reactor. The liquid velocity inside the reactor is usually in the range of 0.5-1.0 m/h. This reactor consists of a sludge bed, a sludge blanket, and a clarifier zone supplemented with a physical device called the gas-solid separator (Fig. 2.109).

F(vi) Operation of photobioreactor: Basic concept in photobioreactor system has been described in Chapter 2 (Fig. 2.36).

Open type bioreactors are often used to culture micro algae, but mono-septic culture requires fully closed photobioreactors because it needs light.

Closed photobioreactors are basically used for monoculture. It consists of arrays of tubes that may be made of glass or transparent plastic (Tsoglin *et al.*, 1996). Continuous single run tubular configuration or helical wound configuration may be used (Fig. 3.29).

In addition to the tubes, flat or thin panels are used for small scale operation. The enclosed systems offer better control over process variables, greater CO_2 utilization efficiency, and reduced contamination. Photoreactors are normally operated in continuous mode.

Factors Affecting Selection and Design of the Light Source(s) in Photobioreactor

The design of an efficient light source for a photobioreactor requires knowledge of the various aspects of light that influences micro algae to be developed. Some factors that influence the light requirements of an algal culture are:

- Type of algal culture.
- Type of light source.
- Intensity of light source.
- Dark period requirement of the algal culture.

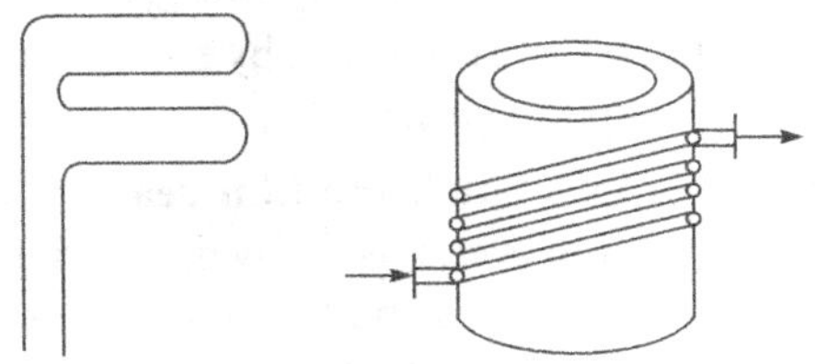
(a) Continuous run (b) Helical wound

Figure 3.29 Closed loop photobioreactor components.

Type of Algal Culture

Different types of algae cultures need different light and nutrient sources (Tsoglin *et al.*, 1996). The light requirement for algae depends on the major pigments present in the algal cell. The algae of interest are blue-green and green algae. Various pigments present in blue-green and green algae are chlorophyll and β-carotene. The main pigment in these two types of algae is chlorophyll a. Chlorophyll a is located as a part of core reaction center protein complexes and in the light-harvesting antenna. Different pigments absorb / harvest different regions of visible light energy.

Type of Light Source

After identifying the type of algal culture to be grown, it is important to identify the right type of light source (appropriate wavelengths) to achieve a high level of photosynthetic efficiency. Electricity is used in closed loop photobioreactors to produce light, making it essential to ensure that the light source is optimized relative to algae and cost of production so that a high level of electrical efficiency is obtained. The efficiency of converting electricity into light varies with different light source, which is further compounded when wavelength of light is considered for photobioreaction. Light sources with descending order of efficiency are light emitting diodes, grow flux/fluorescent light and incandescent/ halogen lamps. Essentially any type of light source that produces light between 400 nm and 500 nm and between 525 nm and 680 nm supports the growth of blue-green algae. Higher energy photons are absorbed from light with wavelengths of 400 nm to 500nm and 525 nm to 630 nm. It has to lose energy as heat before they can be used in photosynthetic reaction center that needs 680 nm and 700 nm. The best way to achieve such high concentration of photons of red light close to 680 nm and 700 nm is by using light emitting diodes with peak wavelengths close to 680 nm.

Intensity of Light Source

Intensity of a light source gives the number of photons that are available for photosynthetic process. The energy associated with photons of wavelength of 680 nm is the energy level required by chlorophyll a to initiate photosynthesis. Light with a wavelength of 680 nm is near the longest wavelength of visible light. Therefore, most of the visible light has sufficient energy to support photosynthesis. However, if the wavelength is small, the energy associated with the wavelength is high. Once cell growth starts, higher light intensities result in increased cell growth up to a light intensity where growth stops. The light intensity at which cell growth begins is known as the compensation light intensity, while the light intensity at which no further increase in growth takes place with increase in light intensity, is known as saturation light intensity. Further increase in light intensity does not increase the specific growth rate, but at the same time it does not hinder growth. The point at which increased light intensity decreased the specific growth rate is the point where photo-inhibition begins.

Dark Period

Very little information is available as to how much time the cell should be exposed to light. There are reports on variable dark periods. Long dark periods generally resulted in biomass loss as well as a decline in growth rates because the algae undergo photorespiration and consume oxygen and carbohydrates.

F(vii) Operation of membrane bioreactor: Membrane bioreactors resemble a plug flow reactor that contains an additional cylinder having porous material in it. This is conceptually a shell and tube heat exchanger configuration. The membrane is a barrier that allows certain components to pass through it. The selectivity of the membrane is controlled by pore diameter. In fact, membrane reactor combines reaction and separation so that the conversion of the reactants is improved in the reactor.

Hollow fiber bioreactor (HFBR): Cells are embedded in fibers contained in a cartridge, which is immersed in circulating culture medium (Cadwell, 2004). Extensions of this method are to use cartridges containing two different bundles of fibers as a separation membrane between a pair of reactor vessels and allow gases to dissolve and permit metabolites to exchange without cells.

Conventional Axial Flow Configuration

Individual hollow fiber in the HFBR contains three regions. They are lumen region (inner tube), permeable membrane (wall of the inner tube), and spongy matrix (annular region) inside which an active biocatalyst in the form of either enzymes or live cells is immobilized. Cells are injected through the sample ports into this region. The substrate (reactant) solution is fed through the lumen and diffuses inside in the outward direction through the permeable membrane to react with active enzymes or live cells supported on the spongy matrix. The product diffuses back to the lumen and flows downstream (Fig. 3.30).

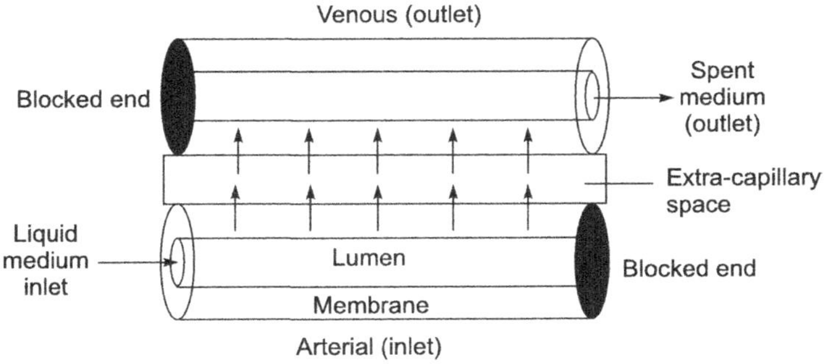

Figure 3.30 Schematic diagram of HFBR.

The lumen pressure is greater than the pressure in extra-capillary space only in the half of the inlet of the bioreactor. A convective secondary flow is induced by the trans-membrane flux. In the distal half of the axial flow bioreactor, flow re-enters the fiber lumen form the extra capillary space. The non-uniform substrate (reactant) delivery can lead to non-uniform cell growth within axial-flow bioreactors.

Intercalated Alternate Dead Ended Configuration

In this bioreactor, two micro porous fiber sets are laid out as sheets and blocked at opposite ends. The fiber sets are stacked and rolled together to form two intercalated spirals. Liquid medium containing a limiting substrate (reactant) such as dissolved oxygen or glucose enters the arterial fiber set. Because of the blocked end, the liquid medium flow is forced across the fiber membrane into the extra-capillary space. The liquid medium flows across the extra capillary space through the venous fiber membranes and finally it is drawn through the venous luminary to the bioreactor exit. With this configuration, the largest pressure drop occurs across the fiber membranes. The promise of this design is to better utilize convective dominant transport, and hence, the construction should rely on micro porous rather than ultra-filtration hollow fibers. This configuration is intended for the culture of mammalian cells.

The goal of the alternate dead-ended design is to minimize axial gradients along the bioreactor and to enhance convective transport across the extra capillary space. Under idealized conditions, the axial pressure gradients are negligible within fiber lumen and extracapillary space. The convective flux is predominantly radial and nutrients are delivered uniformly to the cell growth region.

Multi-coaxial Configuration

It is composed of four tubes of increasing diameter placed one inside the other, creating four spatially isolated compartments (Fig. 3.31).

Cells are sandwiched in the space between the two innermost semi-permeable tubes or hollow fibers. A radial flow of media from an outer compartment occurs through the cell mass compartment and to an inner compartment. Gas-permeable tubes creates the outermost compartment. This housing is used to oxygenate the perfusion media to periportal levels in the outer compartment. The coaxial hollow fiber bioreactor performance is four-fold better than conventional hollow fiber bioreactors. This novel multi-coaxial bioreactor is mainly used for three-dimensional cultures of adherent cell types (Wolfe *et al.*, 2002).

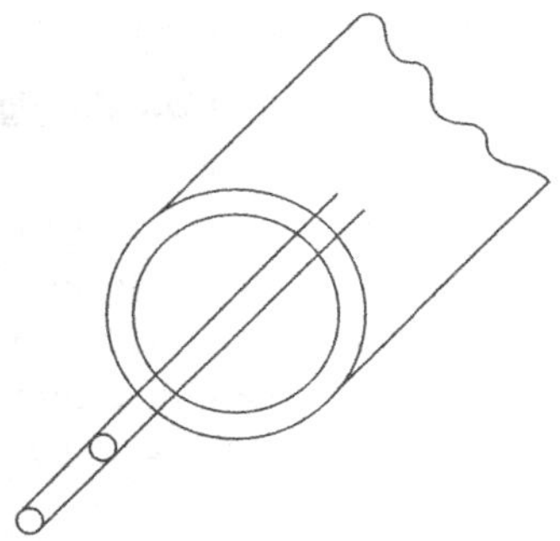

Figure 3.31 Multi-coaxial bioreactor.

3.5 REACTOR OPERATION FOR IMMOBILIZED SYSTEMS

A few reactors have already been discussed which are used for immobilized catalysts. They are:

Batch reactors

- Continuous stirred tank reactor
- Expanded bed bioreactor

Continuous flow reactors

- Continuous flow stirred tank reactors
- Expanded bed continuous flow reactors
- Fluidized bed reactors
- Packed bed reactors

- Membrane reactors
- Other heterogeneous transport reactors

The reactors used with free microbial cells are also described here. In addition to this, packed bed and expanded bed reactors are of special considerations. It is also considered that shear forces are not excessive to damage the immobilized catalyst.

For conceptual purpose, the reader will be interested to know the typical example of reactor operation involving immobilized systems.

For immobilized systems, it is necessary to prepare the catalysts of particular interest either using enzymes or using cells. For whole cells immobilized catalyst, Figure 3.32 gives the details of the process.

Organism grown in shake-flask to obtain sufficient amount of active cells for a particular period

⇩

Centrifugation

⇩

Residue as cells

⇩

Washing of cells with buffered solution

⇩

Cell slurry

⇩

Cells mix with proper immobilizing component (pectin / agar / polyacrylamide or agarose, etc.)

⇩

Extrusion: Cells and immobilizing agents extrude through a hypodermic needle and the resulting drops are collected in a mixture of solidifying agent (viz., $CaCl_2.2H_2O$, $Na_2B_4O_7.10H_2O$, etc.) with gentle stirring at ~ 30°C

⇩

Curing of immobilized catalysts

⇩

Prepared immobilized beads are cured in solidifying agents for ca. 24 h at 4°C

⇩

To use stored beads, steps like incubation, washing as indicated in the figure are followed

⇩

Incubation : To a preferred fermentation medium immobilized beads are added for a short period

⇩

Washing with a buffer to eliminate free and weakly attached cells from the immobilized beads

⇩

Immobillized catalysts are ready for reaction

Figure 3.32 Process for immobilization for cells.

(a) Batch Expanded Bed Bioreactor Operation

The expanded bed bioreactor is sterilized empty. This is then filled with sterile medium. A known quantity of immobilized whole cells is added to the medium. Later the circulation pump is switched on to start the operation of expanded bed bioreactor. Other conditions of fermentation like temperature and pH need to be controlled during the process. The time of fermentation depends on the experiment. Speed of the pump is maintained at a particular rate.

How will you calculate bed expansion characteristics of expanded bed bioreactor? The expanded bed bioreactor is filled with the medium and a known amount of immobilized whole cells are added to the medium aseptically. Initial height of the bed is measured. The circulation pump is then switched on for the expansion of the bed. Expansion of the bed (in percentage) is measured at different flow rates.

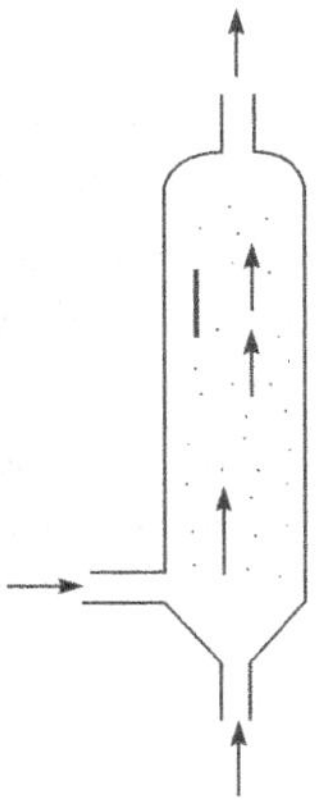

Figure 3.33 Straight-through transport reactor.

(b) Heterogeneous Immobilized Catalyst Transport Reactors

The theory of heterogeneous transport reactors is described here.

(i) Straight-through Transport Reactors

For the production of methane from methanol using immobilized cells, catalyst particles are driven away by the reaction solution (Figure 3.33). This eliminates catalyst decay because the catalysts and reactants are fed continuously (Fogler, 1992). Reactant and catalyst contact time will be difficult to evaluate.

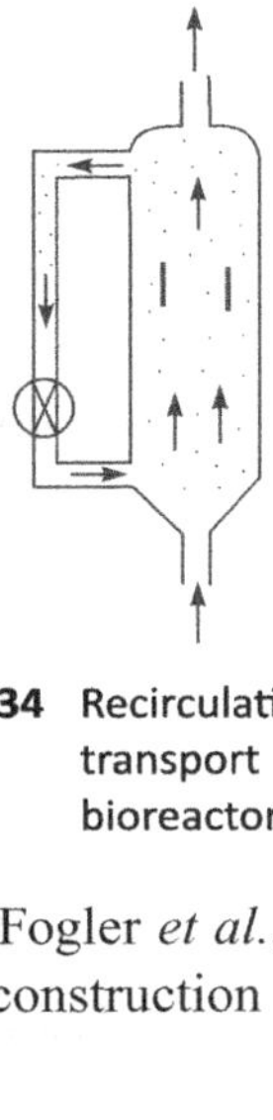

Figure 3.34 Recirculating transport bioreactor.

(ii) Recirculating Transport Reactor

This is a modification of the previous reactor to achieve well-mixed condition. Since the reactor is operated at steady state, the kinetic parameters measured at the start of the experiment will be same as those measured at the end. However, since partially deactivated catalyst will be mixed with the other catalyst, the product distribution and the kinetic parameters might not be same as those in the straight-through transport reactors (Fogler *et al.*, 1992). The incorporation of a recirculation system adds a degree of complexity to the construction (Figure 3.34).

3.6 OPERATION OF ANIMAL CELL BIOREACTORS

In Chapter 2, animal cells used in the reaction have been dealt with different types of reactors. However, to start the reactions with animal cells, some typical important steps under strict aseptic condition are necessary (Frommer *et al.*, 1989).

This section explains the recommended procedures for the operation of animal cell bioreactors and precautions generally followed to carry out the reaction with animal cells.

3.6.1 Methods of Preparation of Culture of Animal Cells

(a) Primary Cell Culture

For *in vitro* culture of animal cells, pieces of tissues are treated with trypsin solution to obtain single cells. A drop of this suspension is placed on the Petri dish or any flat surface. Cells adhered to surface will grow if proper reactants containing amino acids, growth factors or vitamins plus salts, glucose as C-source and bicarbonate solution in equilibrium with 5% CO_2 in the gas phase are supplied. Sometimes calf-or fetal calf-serum is added to the growth medium

To avoid bacterial contamination of cells, penicillin and streptomycin are added to the medium. Virus contamination can be detected by abnormalities in the growth of cells. Cells multiply until they occupy the whole surface. A few cells are transferred to a sterile container along with fresh sterile medium. One should be careful about the cells growing as a monolayer on the surface of culture vessel. Cells might undergo mutation. Cells derived from normal cells cannot be sub-cultured indefinitely. Generally, 40-50 transfers appear to be critical.

(b) Malignant Tissue Cells

These cells give rise directly to cell lines of indefinite life span. They grow in suspension culture having high cell density. Cell density is higher than that obtained from normal cells.

(c) Antigen-specific Hybridomas

These are hybridomas obtained by fusing myeloma (tumor) cells with lymphocytes. This is repeatedly immunized with an antigen of interest. Hybridomas can grow in suspension culture.

(d) Vaccine-specific Cells

Growing viruses on normal cells having finite life and non-tumorigenic characteristics produces specific viral vaccines. For example, hamster kidney cells and human embryo fibroblasts are used for the production of vaccines against Foot-and-mouth and human viral diseases, respectively.

3.6.2 Sources of Contamination

(a) Explants and primary cell cultures are sometimes contaminated with microorganisms (e.g., viruses and mollicutes) originating from donor animals.
(b) Cell cultures are also contaminated during manipulation or via contaminated media components.

3.6.3 Safety Precautions for Animal Cell Cultures

(a) For Primary Cell Culture

(1) Contamination in primary cell cultures are not known specifically. Work related to primary cell culture must be done under Containment Category 1(CC1) (Formmer *et al.*, 1989).
(2) If there are medium risk or high risk contaminants, the work should be performed under Containment Category 2(CC2) or Containment Category 3(CC3). This type of requirement is related to the work with retroviruses.
(3) Cells of specific pathogen free laboratory animals may be used to avoid pathogenic contaminants.

(b) For Handling Established Cell Lines

All established cell lines should be considered with potential low risk and should be handled in CC1, otherwise with CC2 or CC3.

(c) For Handling Cell Products

This is followed as per the procedures described for primary cell culture or cell lines culture.

(d) For Cell-free Products

If pathogen is eliminated from products or inactivated, physical contaminant for further handling of products is not required. For every batch, it is necessary to assay reverse transcriptase activity.

For details, the readers may consult a standard practice book on "animal cell culture".

Infection status of the donor should be assessed prior to the work related to animal cell culture. Non-human cells or cell lines have less danger.

In addition to this, toxic chemicals are used in tissue culture (e.g., 12-O-tetradecanoylphorbol 13-acetate or dimethylsulfoxide).

3.6.4 Basic Precautions

(1) Correct containment category should always be used. If the experimenter does not know, safety officer may be consulted in this regard.
(2) A protective laboratory coat should always be worn.
(3) Gloves should be used for all culture work.
(4) Masks are also advised.
(5) Care should be taken for contaminated sharps.
(6) All liquid and solid waste should be disposed as per established procedure.
(7) For cell culture work in a hood, a quantity of 10% hypochlorite solution should be placed in the hood.
(8) All work should be subjected to local and national regulations.

3.6.5 Batch Reactor Operation

The reaction is carried out in the reactor as described in Chapter 2 (under mammalian cell bioreactor). During the course of reaction, used medium is periodically removed and then replaced with a fresh one.

3.6.6 Continuous Flow (CHEMOSTAT) Culture Operation

Continuous or chemostat culture is a homogeneous system with a fixed culture volume. The reactor is fed with fresh medium in a constant rate and used medium containing products and cell leave the system at the same rate.

A steady state is reached at which all the state variables of the culture are independent of time. The rate of cellular growth is determined by the medium reactants flow rate.

Chemostat cultures will ensure high growth rates and low residual concentrations of metabolic wastes, and remain low density culture systems because cell densities cannot reach a high value.

3.6.7 Perfusion Culture Operation

Perfusion culture systems allow cell retention at a very high concentration in the bioreactor. Cells are continuously perfused with fresh medium (Lehman *et al.*,1988). Products and/or metabolic wastes are continuously removed without cells. Perfusion systems can reach a steady state when cells cease to grow, that is, in a density limited or confluent culture in contrast to chemostat culture, the steady state is characterized by a very low rate because growth can only occur when replacing dead cells. Perfusion cultures can either be a homogeneously or heterogeneously distributed system (Looby and Griffiths, 1988).

(a) Operation of Membrane Aerated and Perfused Animal Cell Bioreactor

Stirred tank bioreactors are equipped with a moving micro porous membrane (e.g., polypropylene hollow fiber membranes) for aeration and additional membrane of same type for perfusion of medium (Lehman *et al.*, 1987). Generally, 3 m of membrane per liter reactor volume are used for aeration and for medium perfusion. For optimization of perfusion system, several other hydrophilic or hydrophilized micro porous membranes are used in the reaction. The length of the membrane is kept within a range between 20 and 40 cm to shorten the time until membrane fouling starts to influence flux and transmembrane pressure.

The lengths of different hollow fiber membranes are adjusted to get comparable membrane surface area. Transmembrane pressure is measured with autoclavable peizo-resistive pressure gauges. The schematic diagram of the system is given in Figure 3.35. Normally, perfusion rates are not increased to the theoretical maximum more than 5 reactor volumes per day. Membrane life time could be significantly increased to better perfusion.

(b) Application

Hybridomas of different origin, insect cell lines in suspension cultures, and recombinant cell lines in suspension or micro-carriers or as spheroids can be used in the perfusion reactor.

(c) Factors Responsible for Membrane Lifetime of Perfusion System

Following measures may be adopted for longer life of the membranes.

1. Restoration with proteases
2. Characterization of fouling materials: Proteins (6800 mg/m^2), RNA (6 mg/m^2), and phospholipids (400 mg/m^2).

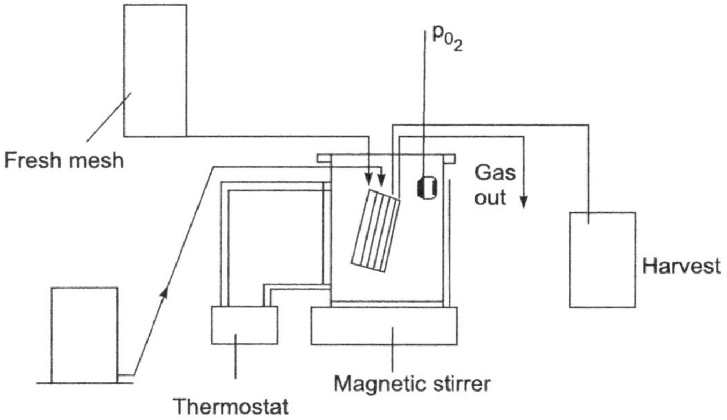

Figure 3.35 Perfusion bioreactor.

3. Screening for optimum perfusion membrane: A special polypropylene membrane is generally used for the perfusion purpose.
4. Influence of cell culture media and supernatants: Reduced protein in media increases the lifetime of membrane. However, a compromise may be made since most cell lines require several proteins as growth factors, nutrient carriers, attachment factors, etc. In addition to this, the cell derived proteins, phospholipids, and membrane vesicles, cellular debris, etc. also reduce membrane lifetime.
5. Role of perfusion modes
   (i) Supernatant is filtered out of the bioreactor without back-flushing and fresh medium is added directly.
   (ii) Pumps run alternatively as described in 5(i). Fresh medium is used for back flushing of the membrane.
   (iii) Harvest pump runs continuously and the fresh medium is added directly into the bioreactor. A third pump runs in cycles back-flushing the membrane with filtered supernatants.
6. Optimal back flushing
   In order to achieve a positive effect, back-flushing volume and velocity need to be optimized for this reactor operation. To keep the volume, which is filtered through the membrane as small as possible, the back-flushing volume should be small. The back-flushing velocity (e.g., 20-25 l/ (m^2) (min)) should be much higher than the velocity of filtration.
7. Membrane pretreatment
   Nontoxic (ionic and non-ionic) polymers are used for this purpose.

3.6.8 Operation of Hollow Fiber Bioreactor for Hybridoma Culture

A few additional steps are necessary for hybridoma culture and their use in reactor (Hara *et al.*, 2003).

(a) Preparation of Hybridoma Cells

A typical example for the production of hybridoma cells by fusion techniques is given in Fig. 3.36.

(b) Operation of Hollow Fiber Bioreactor

(i) How Shall We Start the System?

(1) In laminar flow hood, remove one of the reservoir side arm caps. Flame the side arm and pour known quantity of sterile basal medium into the reservoir. Replace the cap.
(2) Start the pump at a very low speed. Draw medium slowly, about 5 cm^3/ min, into and through the tube. Make sure that the moving liquid front will push trapped gas ahead of it.
(3) Hold the in-line filter upright to vent all gas and to fill the housing completely with medium. Then replace it in a horizontal position.
(4) In a similar fashion, hold the cartridge assembly upright to vent all gas and to fill the hollow fibers completely with medium, then replace it in a horizontal position.
(5) Re-circulate the medium slowly. Check for leaks and tighten fittings whenever necessary.

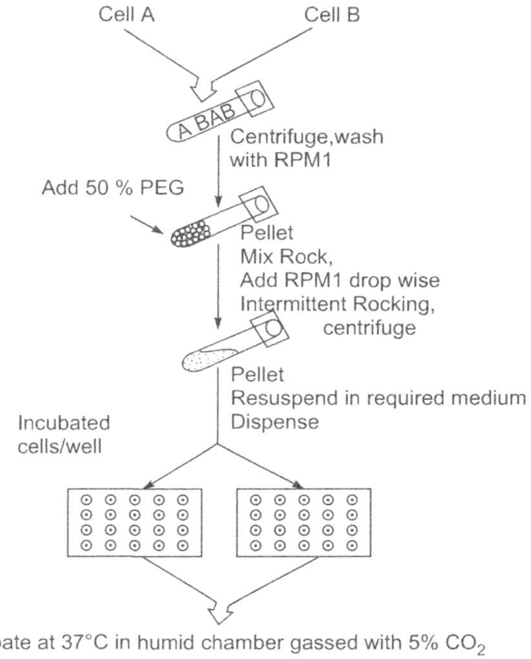

Figure 3.36 Schematic diagram to illustrate the formation of hybridomas.

(ii) How Will You Fill the Extracapillary Space?

(1) In a laminar flow hood, open the stopcock on the extra capillary space port. Remove the filling bell cover. This procedure may be sufficient to permit the extra capillary space to fill over a period of several hours. If not, partially close the adjustable tube, clamp distant medium outlet port as well.

(2) Tap and tilt the cartridge to dislodge all bubbles. Vent the gas through the extra capillary spaceport tube and filter. When all the gas has been vented, close the stopcock on the extra capillary space port.

(3) Open the adjustable clamp completely in the recirculating loop tube. Recap the filling bell on the filter unit.

The extra-capillary space should remain filled without any further manipulations. During normal operation, most commonly start-up and medium-change times, gases may come out of the solution at suitable temperature and pressure. This can be minimized by reducing the time for which the system is outside the incubator or warm room, and by keeping the reservoir slightly above the cartridge. If gases do accumulate in the extracapillary space, vent the gas as described above.

(iii) How Will You Prepare the System?

Operation without cells should continue for several days to assure proper performance of the pump, sealing of the connections, removal of extraneous substances, and sterility of the system. The basal medium used initially should be replaced at least once. Serum containing medium should be used during the final change.

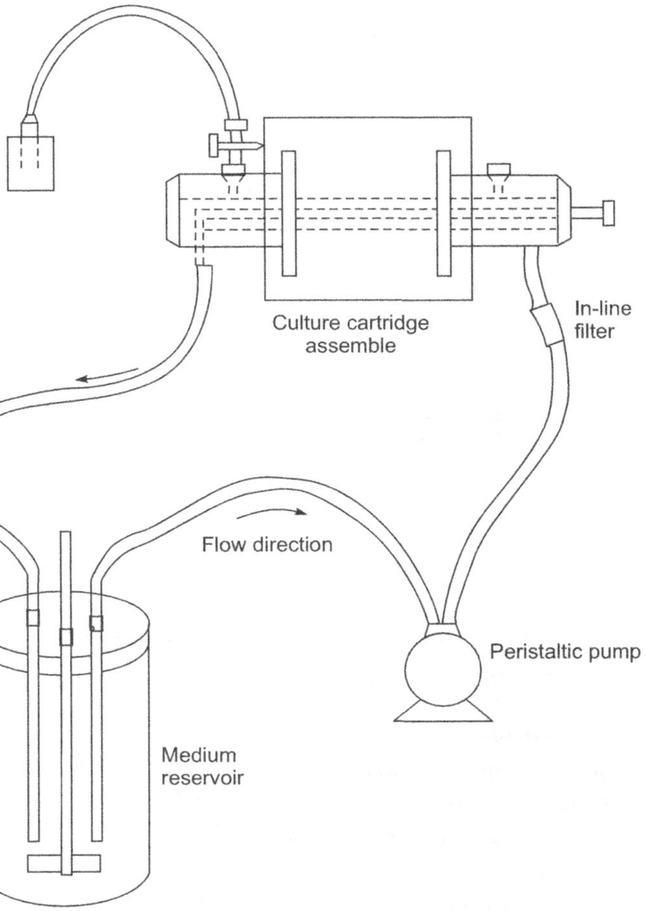

Figure 3.37 Hollow fiber reactor assembly.

(iv) Inoculating Cells

(1) Turn off the pump and place the system in a laminar flow hood. Remove the filling bell cover from filter unit. Then open the stopcock on the extra-capillary space port.

(2) Fill a lock syringe with a few cm^3 of serum containing medium. Remove the lock cap from the 3-way stopcock on the inoculation port. Insert the syringe into the opening. Open the stopcock channel leading into the cartridge and inject the entire contents of the syringe. This procedure will purge the cell inoculation tube off any trapped bubbles. Fill it with serum-containing medium, which will force several volume of medium from the extra-capillary space (Fig. 3.37).

(3) Fill the syringe with cells in a small volume of serum-containing medium. Close the stopcock and replace the empty syringe with cell suspension syringe. Open the stopcock and inject the entire cell suspension into the cartridge. This procedure will force most of the cells into the center of the hollow fiber, as well as some volume of medium from the extra-capillary space. Some cells will, however, remain in the cell inoculation tube.

(4) Fill a syringe with a small volume of serum containing medium. Close the stopcock and replace the empty syringe with the syinge filled with medium. Open the stopcock and inject the entire contents of the syringe. Repeat this process with another small volume of serum containing medium. This procedure will displace all cells from the cell inoculation tube into the hollow fiber. This will force several small volumes of medium from the extra-capillary space.

(5) Close the stopcock on the extra-capillary space port and cover the filling bell. Remove the stopcock from the cell inoculation port and cover the opening with a suitable cap.

(6) With pump still off, allow some time for the cells to settle onto the hollow fibers. Periodic rotation and gentle rocking of the cartridge will aid the distribution of cells throughout the hollow fiber bundle.

(v) How Shall We Operate the System?

Depending upon the cell type and number of cells inoculated, rapid proliferation will continue for a few weeks. Eventually, cells will come to fill all the space available in the cartridge and will remain in a viable, stable state for weeks or months thereafter. The medium flow rate can be slower during the early phase of culture growth when the numbers are small, but it must be faster when the culture will mature with large number of cells.

(1) Replace some volume of medium every 1-3 days, depending upon the number of cells and metabolic rate of the cells in the culture. Aseptic medium exchange is most easily accomplished by pouring and refilling through a side arm as described above or by siphoning and refilling through a tube inserted in a top cap port.

(2) Continue perfusion of the culture during this process. Work quickly since cells will begin to die off within minutes without constant supply of nutrients.

(3) Serum levels may be reduced when the rate of proliferation subsides and the culture reaches maturity.

3.7 OPERATION OF BIOREACTORS FOR PLANT CELL CULTURE

Plant cell culture can be divided into two main groups, viz., callus culture and cell suspension culture. The basic growth reaction components are inorganic micro-and macronutrients, carbon source, vitamins and plant growth hormones (Murasighe and Skoog, 1963). For bioreactor operations, following preparations are needed to initiate the reaction in the presence of plant cell cultures.

Method of Preparation of Plant Cells

Explants –isolation and maintenance of callus

A section of plant organs is excised by a knife. The isolated explants are placed into medium consisting of mineral salt mixture, a carbon source, vitamins, and phytochromes (cytokinins /auxins) to initiate cell division.

(a) Sub-culture of callus Explants / callus are preserved in the test tubes from which aseptically a few explants are transferred to Petri dish. Usually after 4 weeks, the explants incubated on medium with (2, 4-D) formed a substantial callus (Reinert and Yeoman, 1982). Cells on the lower side of explants towards the center are frequently necrotic in appearance and they are not suitable for sub-culture. The necrotic area should be removed and discarded. Transfer the developing explants / callus to a fresh medium in a culture tube.

(b) Cells for suspension culture Cells grown in the test tube culture are aseptically removed and placed on to a Petri dish. A 250 ml conical flask containing 60 ml growth medium with 2, 4-D previously sterilized and a few pieces of callus are transferred to the medium in the conical flask. The culture is placed on the shaker at a suitable temperature. A few transfers are made in liquid culture to obtain a suitable cell density.

Reactor Operation

Batch, repeated batch (semi-continuous), and continuous flow reactor operations are common for plant cell culture. Photo-bioreactor is also used which is discussed earlier in this chapter. The basic practice of operation is almost like microbial systems. However, in this case, strict aseptic conditions should be maintained.

3.8 REACTORS FOR WASTE MANAGEMENT

In a strict sense, reactors used for waste management do not require strict aseptic conditions. The mixed microbial and aquatic systems stabilize the waste. In a batch growth curve, the reaction might proceed after stationary phase of growth. A few classical reactors are trickling filters, rotating biological contactors, etc.

EXERCISES

3.1 Write the applications of Plackett and Burman design in medium formulation for the growth of animal cells?

3.2 State and calculate the elements in a typical ANOVA table. Following relevant information is:

Y_u	is	observed response
N	is	number of observations
$\tilde{Y}$	is	average response
p	is	probability
β_i	is	coefficients in the polynomial assumed
H_o	is	null hypothesis
H_a	is	alternate hypothesis
F	is	F-statistic
α	is	level of significance

3.3 The central composite experimental design was suggested for the optimization of pH (initial) and temperature for the production of carboxymethylcellulase (a product) synthesized by a fungal strain. The range of pH (initial) and temperature of fermentation was between 4 and 6, and between 25°C and 35°C respectively. You need to answer the following:

(a) Code pH (initial) and temperature of fermentation as per CCD.

(b) Write the matrix for experimental purposes.
(c) How many factorial points are indicated in the matrix? What are they?
(d) How will you decide the number of centre points? Express your answer on the basis of total design points.
(e) What are the criteria for rotatability?

3.4 The response equation for $\beta - 1$, 3-glucanase production in submerged culture is expressed by the following equation.

$$\tilde{Y} = 0.4015 + 0.0043\,x_1 + 0.0242\,x_2 - 0.0402\,x_1^2 - 0.0419\,x_2^2 - 0.00415\,x_1x_2$$

where $\tilde{Y}$ = predicted response
x_1 and x_2 are coded values of the variables

Answer the following:

(a) Write the values of coefficients of offset term, interaction terms, squared effect terms and linear effect terms.
(b) Calculate the optimal values of x_1 and x_2.

3.5 What are the differences between self-directing optimization and Taguchi's experimental design plan?

3.6 (a) The response equation for esterase production in submerged culture is expressed by the following equation.

$$\tilde{Y} = 5.87 + 0.036\,x_1 + 0.16\,x_2 - 0.64\,x_1^2 - 1.35\,x_2^2 + 0.31\,x_1x_2$$

where $\tilde{Y}$ = predicted response
x_1 and x_2 are coded values of the variables
Answer the following:
Calculate the optimal values of x_1 and x_2.

(b) Following polynomial expression represents the predicted response for chitinase production.

$$\tilde{Y} = 2.148 \times 10^{-3} - 1.134 \times 10^{-5}\,x_1 + 3.714 \times 10^{-4}\,x_2 - 2.807 \times 10^{-4}\,x_1^2 - 2.863 \times 10^{-4}\,x_2^2 - 2.512 \times 10^{-4}\,x_1x_2$$

Write the values of coefficients of offset term, interaction terms, squared effect terms and linear effect terms.

3.7 State the criteria for Plackett-Burman design. You have variables sucrose, yeast extract, K_2HPO_4, $NaNO_3$, KCl, $MgSO_4$ and $FeSO_4$. Code them properly and write the Hadamard matrix.

3.8 What are the differences between Box and Behnken design and factorial experimental design plans?

3.9 Comment on the following statements:
(a) Challenges to chemical engineers in biotechnology
(b) Mammalian cell cultures vs. transgenic animals for the production of pharmaceutical proteins.

3.10 (a) Why do you need bioreactor for the fermentation process?

(b) In a chemostat operation, imperfect mixing caused by pellets, calculate at steady state the substrate concentration, consider Han and Levenspiel model to describe μ (specific growth rate).

(c) What is the advantage of perfusion airlift reactor over ultra filtration coupled reactor?

REFERENCES

Anjanikumari J and Panda T (1992) Studies on critical analysis of factors in influencing increased production of protoplasts from *Trichoderma ressei* mycelium. *Enzyme Microbial Technology*, **14**, 241-248.

Babu PSR (1991) "Studies in biosynthesis of penicillin amidase in *E.coli*, stabilization and immobilization of the enzyme associated with whole cells." Ph.D. Thesis, IIT-Madras, India.

Babu PSR and Panda T (1991) The role of phenylacetic acid inbiosynthesis of penicillin amidase in *Escherichia coli. Bioprocess Engineering*, **6**(1-2) 71-74.

Box MJ (1965) A new method of constrained optimization and a comparison with other methods. *Computer Journal*, **8**, 42-52.

Box GEP and Wilson KB (1951) On the experimental attainment of optimum conditions. *Journal of Royal Statistics Society*, **B 13**, 1-38.

Byrne DM and Taguchi S (1987) The Taguchi approach to parameter design, *Quality Progress*, Dec. 19-26.

Cadwell JJS (2004) New developments in hollow fiber cell culture, *American Biotechnology Laboratory*, July 2004: www.fibercellsystems.com/documents/ABL-Cadwell.pdf

Chen J, Leblanc DJ, and Galli DM (2002) DNA inversion on conjugative plasmid pVT745, *Journal of Biotechnology*, **184**, 5926-5934.

Chirulova V, Cregg JM and Meagher MM (1997) Recombinant protein production in an alcohol oxidase-defective strain of *Pichia pastoris* in fed batch fermentations. *Enzyme and Microbial Technology*, **21**, 277-283.

Chisti Y (Ed) (1989) *Airlift Bioreactors*, Elsevier, London.

Dasu VV and Panda T (2000) Optimization of microbiological parameters for enhanced griseofulvin production using response surface methodology, *Bioprocess Engineering*, **22**, 45-49.

Dasu VV, Panda T, and Chidambaram M (2003) Determination of major influencing parameters for improved griseofulvin production in a batch bioreactor by Taguchi's Method, *Process Biochemistry*, **38**, 877-880.

Doran PM (Ed) (1995) *Bioprocess Engineering Principles*, Academic Press, London.

Dunn IJ, Heinzle E, Ingham J, and Prěnosil J (1992) "Bioreactor Modelling," pp 97-111, In: *Biological Reaction Engineering* (Second Edition), VCH Publishers, New York, USA.

Felse PA (1999) "Process development for extracellular chitinase production by *Trichoderma harzianum*." Ph.D. Thesis, Indian Institute of Technology Madras, pp 77–78.

Felse PA and Panda T (1999) Self-directing optimization of parameters for extracelluar chitinase production by Trichoderma harzianum in batch mode, Process Biochemistry, 34, 563-566.

Frommer W, Ager B, Archer L, Brunius G, Collins CH, Donikian R, Frontali C, Hamp S, Houwink EH, Kuenzi MT, Kramer P, Lagast H, Lund S, Mahler JL, Normant-Plessier F, Sergeant K, Muijs GT, Vranch SP, and Werner RG (1989) Safe biotechnology III. Safety precautions for handling microorganisms of different risk classes. *Applied Microbiology and Biotechnology* **30**, 541-552.

Gowrishankar BS and Panda T (2008) Critical analysis of application of generalized distance function for optimization of important variables for esterase synthesis by *Saccharomyces cerevisiae*. *Bioresources Technology*, **99**, 5545-5555.

Hara M, Kanemura Y, Miyake J, Okano H, Yamasaki M, Nakamura Y, Yamamoto M, Oda E, and Kodama E (2003) Monoclonal antibodies, hybridoma, cell isolation method, isolated cells and immunological method, European Patent, Number 03251974.6 dt. 28.03.2003.

Harrison JS, Keshavarz-Moore E, Dunnill P, Berry MJ, Fellinger A and Frenken L (1997) Factors affecting the fermentative production of a lysozyme-binding antibody fragment in *Escherichia coli*, *Biotechnology and Bioengineering*, **53**, 611-22.

Hendrix C (1980) Through the response surface with test tube and pipe wrench, *Chemtech*. **10**, 488-497.

Hilaly AK, Karim MN, and Linden JCI (1995) A study on real-time optimization of a fedbatch recombinant *Escherichia coli* fermentation, *Control Engineering Practice*, **3**, 485-493.

Honda H, Lenas P, Watanabe H, Kitade T, and Kobayashi T (1998) Human antithrombin III variant production from recombinant BHK cells in a fed-batch culture with on-line control of glucose and glutamine concentrations, *Journal of Fermentation Bioengineering*, **85**, 532-535

Jong JZ, Hsiun DY, Wu Wt, and Tzeng YM (1995) Fed-batch culture of *Bacillus thuringiensis* for thuringiensin production in a tower type bioreactor, *Biotechnology and Bioengineering*, **48**, 207-213.

Kang HA, Nam SW, Kwom KS, Chung BH, and Yu MH (1996) High-level secretion of human alpha 1-antitrypsin from *Saccharomyces cerevisiae* using inulinase signal sequence. *Journal of Biotechnology*, 48, 15-24.

Khuri AI and Cornell JA (Eds) (1987) *Response surfaces designs and analyses*, Marcel Dekker, Inc., AQSC Quality Press, New York.

Kundu S (1983) "Microbial conversion of cellulose to ethanol," Ph.D. Thesis, IIT-Delhi, India.

Kundu S, Ghose TK, and Mukhopadhyay SN (1983). Bioconversion of cellulose into ethanol – Product inhibition. *Biotechnology and Bioengineering*, **25**(4) 1109-1126.

Lehman J, Piel GW, and Schulz Z (1987) Bubble culture aeration with porous moving membranes, *Development Biology Standard*, **66**, 227-240.

Lehman J, Vorlop J, and Buentemeyer H (1988) "Bubble free reactors and their development for continuous culture with cell recycle" in RE Spier, JB Griffiths (Eds) *Animal cell Biotechnology*, vol-3, Academic Press, London, pp 221-237.

Looby D and Griffiths JB (1988) Fixed bed porous glass sphere (porosphere) bioreactors for animal cells, *Cytotechnology*, **1**, 339-346.

Malik KA and Claus D (1987) Bacterial culture collections: their importance to biotechnology and microbiology, *Biotechnology and Genetic Engineering Reviews*, **5**, 137–197.

Margaritis A and Wallace JB (1982) The use of immobilized cells of *Z. mobilis* in a novel fluidized bioreactor to produce ethanol, *Biotechnology and Bioengineering Symposium*, No.12, 147–159.

Murashige T and Skoog F (1962) A revised medium for rapid growth and bioassay with tobacco tissue culture, *Physiologia plantarum*, **15**, 473-497.

Nair SR and Panda T (1997) Statistical optimization of medium components for improved synthesis of pectinase by *Aspergillus niger*, *Bioprocess Engineering*, **16**(3), 169-173.

Panda T (1986) "Studies on biosynthesis of cellulase and xylanase by the mixed culture of *Trichoderma reesei* D1-6 and *Aspergillus wentii* Pt2804," Ph.D. Thesis, IIT-Delhi, India.

Panda T, Babu PSR, Kumari JA, Rao DS, Théodore K, Rao KJ, Kesava SS, Kapat A, Nair SR, Sinha J, Sreenivas R, Lakshmi Prasanna G, Venkata Dasu V, Pazouki M, Felse PA, Naidu GS, Gokul B, Uma S, Srividya K, Muralidhar RV, Balamurugan K, and Chandrasekhar K(1997) Bioprocess optimization – A challenge. *Journal Microbiology and Biotechnology*, **7**(6) 367-372.

Phadke SM (1989) *Quality Engineering Using Robust Design*, Englewood Cliffs, New Jersy, Prentice Hall.

Plackett RL and Burman JP (1946) The design of optimum multifunctional experiments. *Biometrika*, **33**, 305-325.

Ramírez DM and Bentley WE (1995) Fed-batch feeding and induction policies that improve foreign protein synthesis and stability by avoiding stress responses, *Biotechnolology and Bioengineering*, **47**, 596–608.

Reinert J and Yeoman MM (1982) *Plant cell and tissue culture: A laboratory manual*, New York, Springer-Verlag.

Suga K, Waki T, Kumano M, Chimange P, Shin SB and Ichikawa K (1980) Production of cellulose in fed-batch system, pp 371-392, *Proceedings of Bioconversion and Biochemical engineering*, Symposium 2, vol II, Ghose TK (ed) IITDelhi, New Delhi.

Taguchi G (1986) *Introduction to Quality Engineering*, Tokyo, Asian Productivity Organisation.

Théodore K and (1995) "Studies on optimization of β-1,3-glucanase production by *Trichoderma harzianum* NCIM 1185," Ph.D Thesis, Indian Institute of Technology Madras, Chennai, India.

Théodore K and Panda T (1994) Statistical optimization of production medium for β-1, 3-glucanase synthesis by *Trichoderma harzianum*, *Biotechnology Techniques*, **8**, 381–384.

Théodore K and Panda T (1995) Application of response surface methodology to evaluate the influence of temperature and initial pH on the production of β1-3 glucanase and carboxymethylcellulase from *Trichoderma harzianum*. *Enzyme and Microbial Technology*, **17**, 1043-1049.

Tsoglin LN, Gabel BV, Fal'kovich TN, and Semenenko VE (1996) Closed photobioreactors for microalgal cultivation, *Russian Journal of Plant Physiology*, **43**, 131-136.

Vijay C, Cregg JM and Meagher MM (1997) Recombinant protein production in an alcohol oxidase-defective strain of *Pichia pastoris* in fedbatch fermentations, *Enzyme and Microbial Technology*, **21**, 277-283.

Williams KR (1963) Comparing screening designs. *Industrial Engineering Chemistry*, **55**,29-32.

Wolfe SP, Hsu E, Reid LM, and Macdonald JM (2002) A novel multi-coaxial hollow fiber bioreactor for adherent cell types: Part I hydrodynamic studies, *Biotechnology and Bioengineering*, 77, 83-90.

Zhang Y-H and Zhong J-J (1997) Hyper-production of ginseng saponin and polysaccharide production by high-density cultivation of *Panax notoginseng* cells. *Enzyme and Microbial Technology,* **21**, 58-63.

Zhou W, Rehm J, and Hu W-S (1995) High viable cell concentration fed-batch cultures of hybridoma cells through on-line nutrient feeding. *Biotechnolology and Bioengineering*, **46**, 579–587.

APPENDIX 3

The Complex Algorithm of Box

For example, the maximization of F $(X_1, X_2, X_3 \ldots\ldots, X_N)$ subjects to $G_k \le X_k \le H_k$, $k = 1, 2, \ldots\ldots$M. The implicit variables $X_{N+1}, \ldots\ldots, X_M$ are dependent functions of the explicit independent variables $X_1, X_2, \ldots, X_N$. The upper and lower constraints H_k and G_k are either constants or functions of the independent variables. No derivatives are required in this example. The procedure should tend to find the global maximum due to the fact that the initial set of points are randomly scattered throughout the feasible region. The algorithm proceeds as follows:

1. An original complex of $K \ge N + 1$ points is generated consisting of feasible starting points and $(K - 1)$ additional points generated from random numbers and constraints for each of the independent variables:

$$X_{ij} = G_i + r_{ij}(H_i - G_i) \qquad (3.23)$$
$$i = 1, 2, \ldots\ldots, N \text{ and } j = 1, 2, \ldots\ldots K - 1$$

   where r_{ij} are random numbers between 0 and 1.

2. The selected points must satisfy both the explicit and implicit constraints. If at any time the explicit constraints are violated, the point is moved a small distance δ inside the violated limit. If an implicit constraint is violated, the point is moved one half of the distance to the centroid of the remaining points.

$$X_{ij}(\text{new}) = (X_{ij}(\text{old}) + \bar{X}_{i,c})/2 \qquad (3.24)$$
$$i = 1, 2, \ldots N.$$

   where the coordinates of the centroid of the remaining points $\bar{X}_{i,c}$ are defined by

$$\bar{X}_{i,c} = \frac{1}{K-1}\left[\sum_{j=1}^{K} X_{ij} - X_{ij}(old)\right], i = 1, 2, \ldots N. \qquad (3.25)$$

   This process is repeated as necessary until all the implicit constraints are satisfied.

3. The objective function is evaluated at each point. The point having the lowest function value is replaced by a point which is located at a distance α times as far from the centroid of the remaining points as the distance of the rejected points on the line joining the rejected points and the centroid:

$$X_{ij}(\text{new}) = \alpha(X_{ic} - X_{ij}(\text{old})) + \bar{X}_{i,c} \qquad (3.26)$$
$$i = 1, 2 \ldots\ldots\ldots\ldots N$$

   Box recommended a value of $\alpha = 1.3$.

4. If a point repeats in giving the lowest function value on consecutive trials, it is moved one half the distances to the centroid of the remaining points.
5. The new point is checked against the constraints and is adjusted as before if the constraints are violated.
6. Convergence is assumed when the function values at each point are within β units for "r" consecutive iterations. Iteration is defined as the calculations required selecting a new point, which satisfies the constraints and does not repeat in yielding the lowest function value.

CHAPTER 4

Biochemical Aspect of Bioreactor Design

OBJECTIVES

- To analyze the basic SLF bioreactors considering reactions separately involving cells and enzymes
- To calculate important parameters for bioreactor designs

4.1 INTRODUCTION

Reactor operations vary among each other. One class of reactor for different catalysts (microbial cells-free and immobilized form, plant cells and animal cells) requires separate unit operations to start the reactor. For example, a batch reactor with aerobic microorganisms in free form calls for entirely different upstream operations and control than an anaerobic microorganism in free form under the same conditions. Similarly, batch reactor operations for aerobic organism in free form are not same for the same aerobic organism in immobilized form. Analysis of each individual reactor will be cumbersome in this book. This book, therefore, deals with the basic reactor configurations.

For submerged liquid fermentation (SLF), the following configurations can be applied to free and immobilized microbial cells, and free and immobilized enzymes:

- Batch bioreactors
- Flow reactors
  - Continuous flow stirred tank bioreactors (CFSTBR)
  - Plug flow (Tubular) reactors (PFTR)
  - Combination of reactors
- Semi-continuous reactors
  - Batch-fed
  - Fed-batch

Initially, ideal reactors will be discussed in detail followed by the factors which are responsible for non-ideal behavior.

Analysis of reactors for microbial cells in SLF using special reactors, viz., fluidized bed, hollow-fiber, packed bed, airlift, will be discussed separately in Chapter 9.

For better understanding, the organization of this chapter is given below. Let us discuss the reactors using microbial systems in submerged liquid fermentation.

4.2 ORGANIZATION OF THIS CHAPTER

Section A: Reactors for submerged liquid fermentation for microbial cells

Part 1: Batch reactor

Part 2: Continuous flow reactor

Part 3: Semi-continuous reactor

Part 4: Combination of reactors

Section B: Reactors for enzyme reactions and immobilized cells

Some common analysis is indicated here to model and to clearly understand biological process. This gives an opportunity to combine the kinetic model with the model for the bioreactor analysis. A bioreactor model is represented by a set of dynamic mass balances for reactant, cells, and products which describe the change in concentration of these state variables with time. The bioreactor may be any type of device ranging from a test tube or a shake flask to a well-mixed bioreactor. The feed is normally assumed to be sterile, i.e., cell concentration in the feed is zero.

We need to arrive at the general model expression for bioreactor through the following discussion:

- General growth reaction
- Rate laws
- Stoichiometry
- General mass balance
- General energy balance

4.2.1 General Growth Reaction

For example, the reaction with a typical aerobic organism can be written as per Equation (4.1).

$$C_{\text{carbon source}} + C_{\text{N-source}} + C_{O_2} + C_{\text{minerals}} + C_{\text{micronutrients}} + \ldots\ldots\ldots + C_{\text{cell}} \text{ (as inoculum)}$$

$$\xrightarrow[\text{Temperature, Do level, etc.}]{\text{Reactant medium conditions, Microbial parameters and other conditions-pH,}} C_{\text{cell}} \text{ (more cells)} + C_{\text{product}} + C_{H_2O} + C_{CO_2} \tag{4.1}$$

A more simplified form of cell reaction is given by Han and Levenspiel (1987) by the Equation (4.2).

$$\text{Reactants} + \text{Cells (inoculum)} \longrightarrow \text{More cells} + \text{Products} \tag{4.2}$$

If one looks at the left hand side of either Equation (4.1) or (4.2), it is mandatory that for biological reactions involving cells, minimum amount of live and active cells should be added at the start of the reaction. Those active cells are known as inoculum. As the reaction progresses, cells in inoculum multiply by consuming reactants and form more cells as indicated in the right hand side of the equations (4.1) and (4.2). During multiplication of cells, they form products. This reaction is influenced by a large number of factors. Some of the important factors are reactant medium composition, feeding strategy, microbial parameters (viz., slant age, inoculum age, inoculum level, phase of growth, physiological conditions, etc.) and other conditions like pH of the reaction (whether controlled or uncontrolled), temperature, ionic strength, shear forces (fluid forces in bubble driven systems, mechanical forces in mechanically agitated systems).

4.2.2 Rate Laws

Many laws exist for the cell growth reaction [Equation (4.2)].

A few of them are given in Chapter 1. Among them, Monod's equation is the simplest one. However, the formal kinetic approach is based on the following (Moser, 1985).

- Real conditions are obtained from experiments.
- Analogy concept is chosen for mathematical functions (Monod's equation has analogy with Henri–Michaelis–Menten equation).
- Suitable quantification of process is necessary.
- The meaning of the parameters is required for this purpose.

Detailed information on the classes of the mathematical models is given by Thilakavathi *et al.*, (2007). Let us consider Monod's equation to simplify the situation.

$$\text{Cell growth rate} = \frac{dC_x}{dt} = \mu C_x \tag{4.3}$$

where, C_x is cell concentration (kg/m^3 or g/l)
μ is specific growth rate (h^{-1} or time^{-1})
$\frac{dC_x}{dt}$ is expressed in (kg/(m^3)(h)) or (g/(l)(h))

Specific cell growth rate

$$\mu = \mu_m \frac{C_s}{K_s + C_s} \tag{4.4}$$

where, μ_m is the maximum specific growth reaction rate, h^{-1} (time^{-1})
K_s is the saturation parameter (g/l or kg/m^3) (analogous to Michaelis-Menten constant, K_m, in enzyme kinetics)
C_S stands for reactant concentration (g/l or kg/m^3).

So the growth rate based on Equation (4.2) can be written as

$$\frac{dC_x}{dt} = \mu_m \frac{C_s C_x}{K_s + C_s} \tag{4.5}$$

Growth rate depends on a large number of factors:

- Reactant composition and concentration
- Status of cells in inoculum
- Product and its concentration
- New cells formed and its concentration
- Physical factors: temperature, shear forces
- Physiochemical factors like pH, etc., and many more.

If the factors are not favorable for cells, reduction in growth rate happens. This is called inhibition (affected by chemical/biochemical components), inactivation (affected by C_{H+}), and deactivation (affected by physical factors).

The generalized rate law inclusive of inhibition by reactants, cells, and products is suggested by Han and Levenspiel (1987).

$$r_g = \frac{r_{max}\left(1-\frac{C_i}{C_i^*}\right)^n C_s}{K_s\left(1-\frac{C_i}{C_i^*}\right)^m + C_s} C_x \tag{4.6}$$

where,

$$r_g = \frac{dC_x}{dt}$$

r_{max} is analogous to μ_{max} (time^{-1} or h^{-1})
C_i is inhibitor concentration (g/l or kg/m^3)
C_i^* is critical inhibitor concentration beyond which growth stops (g/l or kg/m^3)
n and m are dimensionless constants.

4.2.3 Temperature Dependence of Rate Law for Growth

Every organism possesses one typical growth temperature, below and above which growth rate slows down.

For example, *Escherichia coli* strain produces penicillin acylase. It has been reported that total reaction time (fermentation time) increases by 2 h at 28°C than at 30°C **(Babu, 1991).**

Further, the lag time for growth and total time for fermentation decrease with an increase in growth temperature of 33 °C beyond which those parameters increase with an increase in temperature. A typical plot (Fig. 4.1) is given for analyzing this situation.

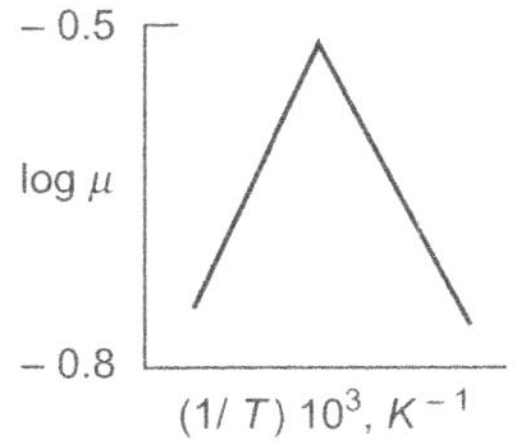

Figure 4.1 Temperature dependence of μ.

The typical equation is in analogy with the Arrhenius equation in chemical reaction engineering (Levenspiel, 1990).

$$\log \mu = \log A - \frac{E}{2.303\,RT} \tag{4.7}$$

where

μ is specific growth rate, (time^{-1})
A is a constant depending on the frequency of formation of activated complexes of reactants.
E is a constant known as activation energy.
R is the ideal gas law constant.
T is the temperature, K

Figure 4.1 is different from the plot of ln k vs. 1/T using classical Arrhenius equation (cf. Figure 25, p77 of Levenspiel's book: Levenspiel, 1972). Here, one can calculate both activation and deactivation energies for cell.

4.2.4 Stoichiometry

Stoichiometry is a branch of science that deals with the quantitative composition of chemical compounds and quantitative conversion in chemical reactions. This is also accompanied by thermodynamics and kinetics. Like kinetics, stoichiometry also requires detailed knowledge of mechanism of reactions and the balance of significant compounds.

Complex multiple reactions occur in microorganisms. So the stiochiometry for cell growth is very complex and varies with microorganisms, reactants, and environmental factors. Hence, in case of multiple reactions, the general form of the principle of conservation of component species is expressed by

$$\sum_i v_i r = 0 \tag{4.8}$$

where

v is stoichiometric coefficient

r is general form of reaction rate, kg/m^{3}s

It is difficult to apply Equation (4.8) in bioprocesses, as large numbers of compounds are involved in cellular reactions (reactants and metabolites). One can approximate gross-stoichiometric analysis, viz., the application to multiple reactions at steady state.

$$\sum_i \nabla n_i = -\sum_i v_i r \tag{4.9}$$

where n is the mass flux per unit volume, mol/m^{3}s.

The balance equation for conservative properties has no production term, so Equation (4.9) becomes

$$\sum_i \nabla n_i = 0 \tag{4.10}$$

This equation is limited to steady state condition and applies to more simplified situation. For biological reactions, it has limited application. For biological reaction, we recall Equation (4.2). This is often in the form of

$$v_s S + \text{cells} \longrightarrow v_x \text{ more cells} + v_p \text{ products} \tag{4.11}$$

This equation suggests the definition of yield coefficients

$$Y_{i/j} \approx \frac{v_i}{v_j} \tag{4.12}$$

For example,

$Y_{X/S}$ is called yield factor (macroscopic variable). This is not a constant.

So $Y_{X/S}$ = mass of new cells formed/mass of reactant consumed to produce new cells

$$= \frac{\Delta C_x}{\Delta C_s} = \frac{C_{x_1} - C_{x_0}}{C_{s_0} - C_{s_1}} = \frac{r_x}{r_s} \text{ (kg cells formed/kg reactant consumed)} \tag{4.13}$$

where

C_{x_1} is cell concentration at time t_1,
C_{x_0} is cell concentration in inoculum,
C_{s_0} is initial reactant concentration,
C_{s_1} is reactant concentration at time t_1.

Actual amount of reactant consumed for new cell generation is less as some reactants are used for the maintenance of cells. Maintenance of cell is defined as the repairment of DNA and RNA in the cells.

The rate of reactant consumption for maintenance irrespective of formation of new cells is

$$r_{s_{\text{maint}}} = mC_x \tag{4.14}$$

where m is maintenance coefficient for cell.

We can define m as in Equation (4.15)

$$m = \text{mass of reactant consumed for maintenance/(mass of cells) (time)} \tag{4.15}$$

So the actual yield coefficient for cell growth accounting for maintenance is $Y_{X/S}$ in Equation (4.16).

$$Y_{X/S} = \text{mass of new cells formed/(mass of reactant consumed for cell growth only)} \tag{4.16}$$

This is graphically evaluated as in the Figure 4.2.

Table 4.1 gives the definition of different yield factors (Moser, 1985).

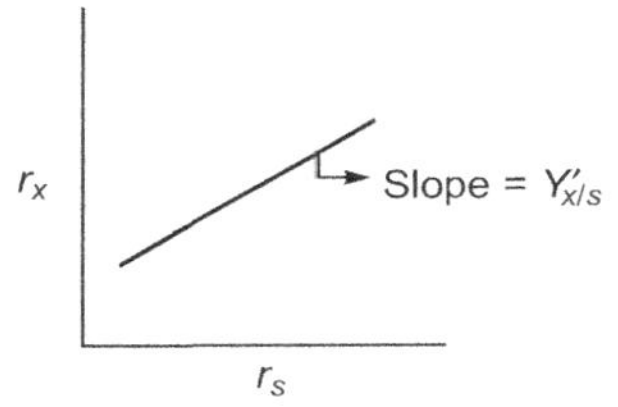

Figure 4.2 Graphical method of calculation of $Y'_{X/S}$.

4.2.5 Application of Yield Factors

It is necessary to correlate reactant utilization, cell growth, cell maintenance, and product synthesis. Yield factors are used in this difficult situation of complex reactions (cell growth reactions). In general, let us consider the reactant auditing.

Net rate of reactant consumption $(-r_s)$ = rate of reactant consumed by cells $(Y_{S/X}r_g)$
+ rate of reactant consumption for product synthesis $(Y_{S/P}r_P)$
+ rate of reactant consumed for cell maintenance (mC_x)

$$-r_s = Y_{S/X}r_g + Y_{S/P}r_P + mC_x \tag{4.17}$$

where $Y_{S/X} = 1/Y_{X/S}$ and $Y_{S/P} = 1/Y_{P/S}$ are defined earlier.

In majority of the cases, it is difficult to separately account for the reactant consumed for cells and products. Even, in general, it is difficult to say in a quantitative way the complete range of products in cellular reactions. One may be afraid that recent development in biological reactions has not come out with the complete answer to such a vital bio-processing problem.

Present situation is to simplify this complex problem of reactant distribution into a lump yield factor that accounts for both cell and product synthesis. Equation (4.17) is simplified to Equation (4.18).

$$-r_s = Y_{S/X}r_g + mC_x \tag{4.18}$$

Table 4.1 *Yield factors used in biological reactions and stoichimetric analysis*

S.N.	*Basis/ concept*	*Definition of yield factor*
1.	Reactant consumption	$Y_{X/S}$ = mass of cells formed/mass of reactant consumed $Y_{P/S}$ = mass of product formed/ mass of reactant consumed to form product.
2.	Mole units	$Y_{X/S(mol)}$ = mass of cells formed / moles of reactant consumed $Y_{X/S(atom)}$ = mass (g) cells formed/g-atoms of reactant consumed
3.	ATP synthesis as basis	Y_{ATP} = mass of cell formed/(mol of ATP formed during growth)
4.	C-mole of cell	Y_S^c = g-atoms of cell C formed /moles of reactant consumed Y_C^c = g-atoms of cell C formed/g-atom of reactant–C consumed Y_{ATP}^c = g-atoms of cell C formed/moles of ATP formed Y_O^c = g-atoms of cell C formed/g-atoms of O_2 consumed
5.	Available electrons in reactants	Y_{ave^-} = g of reactants/electrons $= \frac{Y_{X/S}}{a}$ where a is number of carbon atoms in one mole of reactant. or $Y_{ave^-} = \frac{Y'_{X/S}}{\gamma_s} M_x$ γ_s is degree of reduction of reactant. M_x is the molecular weight of cell.
6.	Heat of combustion	Y_{kJ} = g of cell dry matter/kJ energy of combustion of reactant For aerobic reactions: $Y_{kJ} = \frac{Y_{X/S}^{mol}}{\Delta H_s}\left(\frac{g/mol}{kJ/mol}\right)$ Some times $Y_{kJ} = \frac{Y_{X/S}}{\Delta H_S - (Y_{X/S} \cdot \Delta H_X)}$ where ΔH_X is the heat of combustion of cell(kJ/mol or kJ/g) and ΔH_S is the heat of combustion of reactant (kJ/mol).

The corresponding rate of product formation is

$$r_P = (-r_s)Y_{P/S} = r_g Y_{P/X} \tag{4.19}$$

For the simplest form of this relation, one can visualize the stationary phase of growth, where no cell growth occurs. Rate of product formation takes the Monod's relation [Equation (4.20)] (Fogler, 1992).

$$r_P = \frac{k_P C_{S_2} C_x}{K_{S_2} + C_{S_2}} \tag{4.20}$$

where

k_P is the specific rate constant for product (h^{-1})
C_{S_2} is the concentration of second reactant (kg/m^3)
K_{S_2} is the saturation constant with respect to secondary reactant (kg/m^3)

Similar to Equation (4.19), for secondary reactant case:

$$r_P = (-r_{S_2})Y_{P/S_2} \tag{4.21}$$

Therefore, the net rate of reactant consumption during stationary phase is

$$-r_{S_2} = Y_{S_2/P} + mC_x \tag{4.22}$$

From Equations (4.20) and (4.22), Equation (4.23) is obtained below.

$$-r_{S_2} = Y_{S_2/P}\frac{k_p C_{S_2} C_x}{K_{S_2} + C_{S_2}} + mC_x \tag{4.23}$$

It is always preferred that the concentration of product achieved at a particular time of reaction (i.e., C_P) should be correlated with reactant gradient through the yield factor. A typical example with batch reaction is:

$$C_P = Y_{P/S}(C_{S_0} - C_S) \tag{4.24}$$

In biological reactions, all phases of growth of microorganisms influence the kinetics and stoichiometry. Above treatments are limited to specified situation of cell growth, viz., stationary phase. We would be interested to know the dynamic effect in bio-reactions. This prevails in the accumulation term of mass balance in the reactor system, which will increase or decrease with time of reaction. A physical concept is given below (Fig. 4.3).

4.2.6 The Mass Balance

In bio-reactions, three major components, viz., cells, reactants, and products, are considered for mass balance analysis. Particularly for microbial cells, number of living cells (except mycelial growth systems) or mass of living cells is used in mass balance expression.

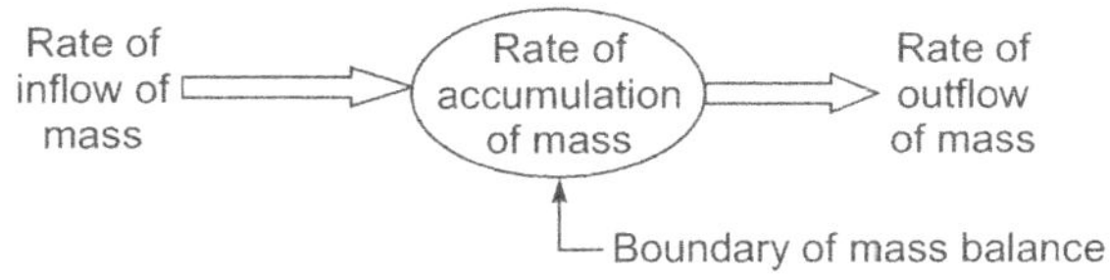

Figure 4.3 Understanding the mass balance.

The total mass balance means a statement that in a chemical reactor or in a bioreactor, mass is neither created ncr destroyed.

From Figure 4.3, within the balance region, we can write

$$\begin{aligned}(\text{Rate of accumulation of mass}) &= (\text{Rate of mass flow entering the balance region}) \\ &\quad - (\text{Rate of mass flow leaving the balance region})\end{aligned} \tag{4.25}$$

More precisely, in biological reaction, component balance is described in the following form [equation (4.26)].

$$\text{Accumulation} = \text{Net formation rate} + \text{Input} - \text{Output} \tag{4.26}$$

The accumulation specifies the rate of change of a compound in the bioreactor. On the other hand, net formation rate is given by a reactant uptake rate (negative being withdrawn from the system) and for metabolic products and cells, this term is the rate of formation of these variables.

The boundary region is chosen in such a way that the region is essentially homogeneous or well mixed. Equation (4.25) is expressed in terms of density (ρ), system volume (V), and volumetric flow rate (F) as per Equation (4.27).

$$\frac{d}{dt}(\rho V)_S = (\rho F)_i - (\rho F)_o \tag{4.27}$$

where '*s*' stands for the system, '*i*' and '*o*' for input and output, respectively and expressed in units.

$$\frac{(\text{kg/m}^3)\text{m}^3}{s} = (\text{kg/m}^3).(\text{m}^3/\text{s}) - (\text{kg/m}^3).(\text{m}^3/\text{s})$$

For biological systems, we consider density change to be negligible:

$$\frac{dV}{dt} = F_i - F_o \tag{4.28}$$

This signifies the change in volume of reaction in a reactor. The significance of Equation (4.28) is:

(1) If $\frac{dV}{dt} = 0$, $F_i = F_o$, continuous flow reactor.
(2) If $F_i = F_o = 0$, batch reactor operation
(3) If $F_i > F_o$, start up of reactor or filling
(4) If $F_i < F_o$, emptying of reactor
(5) If

$$\left.\begin{array}{c} F_i > O, \\ \text{and} \\ F_o = 0 \end{array}\right\} \quad \text{semi-batch operation}$$

For generalized component balances in biological reaction, we look for Equation (4.26) in the form of Equation (4.29).

(Rate of accumulation of component j within the boundary region)
= (rate of inflow of component *j*)
– (rate of outflow of component *j*)
+ (production rate of component *j* within the region) (4.29)

having units in (kg/s) or (mol/s). Therefore, Equation (4.29) is expressed as Equation (4.30).

$$\frac{d(VC_{S_j})}{dt} = (FC_{S_j})_{\text{in}} - (FC_{S_j})_{\text{out}} \pm (r_j V)_{\text{out}} \tag{4.30}$$

So, accumulation = input – output ± production (4.31)

The subscript 'out' on accumulation and production terms refers to the well-mixed assumption. $r_j V$ has the unit of mol/s. The sign before r_j designates whether the component is produced (+ ve sign) or consumed (– ve sign). Equation (4.30) refers to a well-mixed balance region whose volume might change.

4.2.7 General Energy Balance in Bioreactors

Energy balance is needed when the heat of bio-reaction causes a variation in bioreactor temperature. Energy balance is written following the same set of rules as given above for mass balances (Dunn *et al.*, 2003).

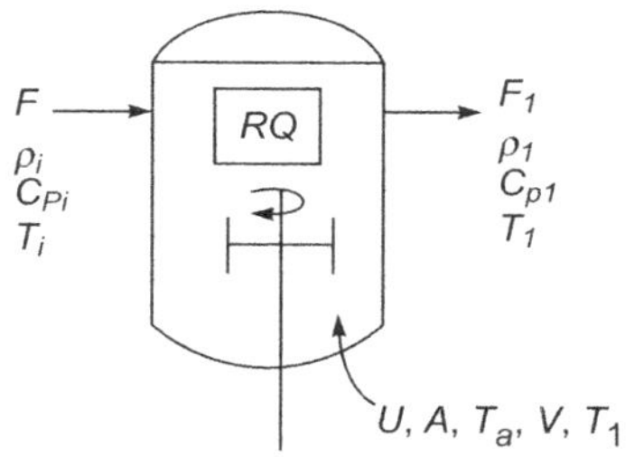

Figure 4.4 A continuous bio reactor showing only the energy related variables.

Rate of energy accumulation = rate of energy in by flow – rate of energy out by flow
– rate of energy out by transfer
+ rate of energy generation by reaction
+ rate of energy added by agitation
+ rate of energy lost due to evaporation of fluid
from the reaction medium (4.32)

An exact derivation of the energy balance was given by Aris (1969).

$$\sum_{i=1}^{S}(n_{i1} C_{Pi1})\frac{dT_1}{dt} = F_{in}\sum_{i=1}^{S} C_{iin}(h_{i0} - h_{i1}) + UA(T_a - T_1) + r_Q V + \Delta H_{\text{agit}} \tag{4.33}$$

where n_i are the number of moles of components, i
C_{Pi} are the partial molar heat capacities, and
h_i are the partial molar enthalpies.

r_Q is the rate of heat production at T_1. If the heat capacities, C_{Pi}, are independent of temperature, enthalpies at T_1 can be expressed in terms of heat capacities as

$$h_{i1} = h_{iin} + C_{Pi}(T_1 - T_{in}) \tag{4.34}$$

and with

$$\sum_{i=1}^{S} n_{iin} C_{Piin} = \sum_{i=1}^{S} n_i C_{Pi} = V\rho C_p$$

Thus with these simplifications,

$$\overbrace{V\rho C_P \frac{dT_1}{dt}}^{\text{accumalation}} = \overbrace{F_{in}\rho C_P (T_1 - T_{in})}^{\text{flow term}} + \overbrace{UA(T_a - T_1)}^{\text{heat transfer term}} + \overbrace{r_Q V}^{\text{heat of bioreaction}} + \overbrace{\Delta H_{\text{agit}}}^{\text{heat due to agitation}} \tag{4.35}$$

The units of each term in the equation are energy per time (kJ/h or kcal/h). Densities and heat capacities of liquids can be taken as essentially constant. The term $r_Q V$ gives the rate of heat released by the bioreaction.

The rate term $r_Q V$ can be written in different forms.

In terms of substrate uptake and a substrate-related heat yield,

$$r_Q = r_S Y_{Q/S} \tag{4.36}$$

In terms of oxygen uptake and an oxygen – related heat yield,

$$r_Q = r_{O_2} Y_{Q/O_2} \tag{4.37}$$

where r_{O_2} is the rate of oxygen uptake, Y_{Q/O_2} is the yield factor $\left(\frac{\text{kJ}}{\text{moles of } O_2 \text{ consumed}}\right)$

and $Y_{Q/S}$ is the yield factor $\left(\frac{\text{kJ}}{\text{gm of reactant}}\right)$

In terms of heat of reaction, mol of substrate, and a substrate uptake rate,

$$r_Q = r_S \Delta H_{r,\,S} \tag{4.38}$$

r_s is the substrate uptake rate and $\Delta H_{r,\,s}$ is the heat of reaction for the substrate.

Other Terms

The heat of agitation may be the most important heat effect for slow growing cultures, particularly with viscous cultures. Other terms, such as heat losses from the reactor due to evaporation, are also important.

SECTION A: BIOREACTORS FOR SUBMERGED LIQUID FERMENTATION OF MICROBIAL CELLS

Part 1: Batch Bioreactors

4.3 INTRODUCTION

From the knowledge of batch bioreactor operation we find certain differences from the knowledge of batch chemical reactor. Two different batch reactor operations are closely observed in this section.

Fundamental commonalities of ideal batch bioreactor and batch chemical reactors are:

- Feed flow rate $_{\text{input}}$ = Product flow rate $_{\text{output}}$ = 0
- Basically, we say constant volume reaction.
- The design criterion is **total batch time**.

Differences in these two reactors are given in Table 4.2. This is considered with respect to the evaluation of total batch time.

Hence, biological batch process takes longer time than the chemical batch process. Again t_f, t_d, and t_c are usually less in chemical reactor operation than the biological reactor system.

Profiles of Batch Reaction

We consider for chemical reaction,

$$\text{R} \rightarrow \text{P}$$

C_r = Concentration of reactant

C_P = Concentration of product.

For bioreactor (Fig. 4.6),

$$R \xrightarrow{\text{cells}} P$$

C_X = Cell concentration

Table 4.2 *Differences in batch bioreactor and batch chemical reactors*

Criteria (Time required for)	*Chemical reactor*	*Bioreactor*
(1) Sterilization of empty reactor	–	Sterilization time for reactor (t_{er})
(2) Sterilization of reactants	–	Sterilization time for reactants (t_{sr})
(3) Filling the reactor with reactants	Filling time (t_f)	Filling time (t_f)
(4) Inoculum preparation	–	Inoculum development time (t_i)
(5) Inoculation	–	Inoculation time (t_{ino})
(6) To carry out reaction for desired conversion	Reaction time (t_r)	Reaction time (t_r)
(7) After reaction, discharge of contents	Discharge time (t_d)	Discharge time (t_d)
(8) Cleaning the reactor to start the next batch	Cleaning time (t_c)	Cleaning time (t_c)
Total batch time = t_{batch}	$(t_{batch})_{ch} = t_f + t_r + t_d + t_c$	$(t_{batch})_{bio} = t_i + t_f + t_{er} + t_{sr} + t_{ino} + t_r + t_d + t_c$

For chemical catalytic reaction, the catalyst is unchanged whereas in biochemical reaction the catalyst (cell) undergoes a number of changes. So cell is included in the reaction scheme:

Reactants + Cell (inoculum) → More cells + Products.

So, we cannot find elementary reactions in cell mediated systems.

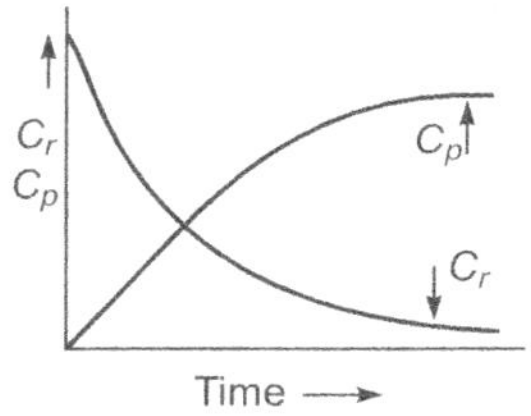

Figure 4.5 Profiles of batch reaction for chemical reactor.

4.3.1 Calculation of Total Batch Time

From the previous discussion, we find total batch time expression in Equation (4.39).

Total batch time for bioreactor

$$= t_i + t_f + t_{er} + t_{sr} + t_{ino} + t_r + t_d + t_c \quad (4.39)$$

Calculation of different time components in Equation (4.39) is done in the following discussion.

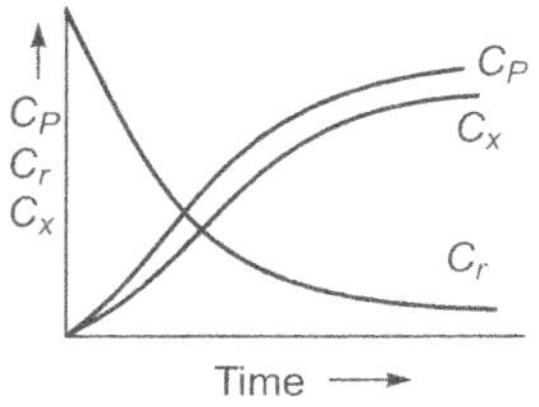

Figure 4.6 Profiles of batch bioreactor.

(a) Inoculum Development Time (t_i)

This time includes time required for the preparation of medium for slant culture, slant culture preparation including sterilization, growth time of organism (which depends on microorganisms and as per the requirement of the experiments, if statistical experimental design is followed for slant growth, etc.), transfer of slant growth to liquid culture in shake flask, and inoculum age in shake flask. For large scale inoculum development, cells are transferred from shake flask culture to bioreactor in small scale. This is associated with additional time for sterilization of reactants and reactor cleaning time.

(b) Filling Time (t_f)

This is approximately 30 minutes.

(c) Sterilization Time for Empty Reactor (t_{er}) and Reactants (t_{sr})

t_{er} is a summation of three different times, viz.,

$$t_{er} = t_{er1} + t_{er2} + t_{er3} \tag{4.40}$$

where

t_{er1} = time required to raise the temperature of the vessel at $T = T_0$ to a specific value of T for sterilization.

t_{er2} = period of time during which reactor is kept at constant T, which can be taken as 30 minutes.

t_{er3} = time required for vessel to cool down after cessation of steam supply.

We shall find relevant literature on sterilization reactors to calculate these parameters (Lee, 1992).

(d) Inoculation Time (t_{ino})

This time is approximately between 30 minutes and 45 minutes.

(e) Reaction Time (t_r)

Calculation of t_r in a bioreactor is difficult. One needs to consider cell and product concentrations. It is interesting to note that cell and product concentration generally do not maximize at one time. If it is growth-associated product synthesis, the problem is simple, i.e., cell and product concentrations maximize at the same time of reaction (t_r). Otherwise, it is difficult to calculate. A general procedure is devised for batch reaction using ODE (ordinary differential equation).

For the beginners, let us consider a simple example of batch cell growth where cell growth is an important consideration. We need to calculate the time at which the cell concentration is maximal. Here, we can consider this time as batch reaction time for cell growth (t_r).

For ideal batch bioreactor simple equation for cell growth (i.e., Monod's equation) is assumed for this purpose. Other assumptions are mentioned below.

(1) All cells in inoculum are live cells.
(2) There is no phase difference in cells in inoculum.
(3) Reaction is well mixed and no unwanted products are synthesized.
(4) Cell is the only controlling factor.

4.3.2 Calculation of Batch Reaction Time from Ideal System

We assume Monod's equation for quantifying growth.

(a) Monod's Kinetics

$Y_{X/S}$ is considered as overall value $\left(\dfrac{C_X - C_{X_0}}{C_{S_0} - C_S}\right)$ obtained and being treated as a constant.

From mass balance of cells and reactants

$$\frac{dC_X}{dt} = \mu C_X = \frac{\mu_m C_S}{K_S + C_S} C_X \tag{4.41}$$

$$\frac{dC_S}{dt} = -\frac{1}{Y_{X/S}} \frac{\mu_m C_S}{K_S + C_S} C_X \tag{4.42}$$

Adding Equations (4.41) and (4.42)

$$\frac{d(C_X + Y_{X/S} C_S)}{dt} = 0$$

Boundary conditions are

$$\text{at } t = 0,\ C_X = C_{X_0},\ C_S = C_{S_0}$$

$$\text{at } t = t_r,\ C_X = C_X,\ C_S = C_S$$

$$C_X + Y_{X/S} C_S = \text{constant}$$

i.e.,

$$C_X + Y_{X/S} C_S = C_{X_0} + Y_{X/S} C_{S_0}$$

$\therefore$

$$C_X = C_{X_0} + Y_{X/S} C_{S_0} - Y_{X/S} C_S \tag{4.43}$$

Therefore,

$$\frac{dC_S}{dt} = -\frac{1}{Y_{X/S}} \frac{\mu_m C_S}{K_S + C_S} [C_{X_0} + Y_{X/S} C_{S_0} - Y_{X/S} C_S] \tag{4.44}$$

On separating the variables,

$$\frac{(K_S + C_S)}{C_S\left[C_{X_0} + Y_{X/S} C_{S_0} - Y_{X/S} C_S\right]} dC_S = -\frac{1}{Y_{X/S}} \mu_m dt \tag{4.45}$$

On integrating, Equation (4.45) from C_{S_0} to C_S and t from 0 to t results in Equation (4.46).

$$\int_{C_{S0}}^{C_S} \frac{(K_S + C_S)}{C_S\left[C_{X_0} + Y_{X/S} C_{S_0} - Y_{X/S} C_S\right]} dC_S = -\int_0^{t_r} \frac{1}{Y_{X/S}} \mu_m dt \tag{4.46}$$

LHS of Equation (4.46) is

$$\int_{C_{S0}}^{C_S} \frac{(K_S + C_S)}{C_S\left[C_{X_0} + Y_{X/S} C_{S_0} - Y_{X/S} C_S\right]} dC_S$$

$$= \int_{C_{S0}}^{C_S} \frac{K_S}{C_S\left[C_{X_0} + Y_{X/S} C_{S_0} - Y_{X/S} C_S\right]} dC_S + \int_{C_{S0}}^{C_S} \frac{dC_S}{\left[C_{X_0} + Y_{X/S} C_{S_0} - Y_{X/S} C_S\right]}$$

Let

$$C_{X_0} + Y_{X/S} C_{S_0} = a \tag{4.47}$$

$\therefore$

$$\text{LHS} = \int_{C_{S0}}^{C_S} \frac{K_S}{C_S\left[a - Y_{X/S} C_S\right]} dC_S + \int_{C_{S0}}^{C_S} \frac{dC_S}{\left[a - Y_{X/S} C_S\right]}$$

Upon integration by parts, results the LHS

$\therefore$

$$\text{LHS} = \frac{K_S}{a} \int_{C_{S0}}^{C_S} \frac{dC_S}{C_S} + \frac{K_S Y_{X/S}}{a} \int_{C_{S0}}^{C_S} \frac{dC_S}{\left[a - Y_{X/S} C_S\right]} + \int_{C_{S0}}^{C_S} \frac{dC_S}{\left[a - Y_{X/S} C_S\right]}$$

From Equation (4.46)

$$\frac{K_S}{a}\left[\ln\frac{C_S}{C_{S_0}} - \ln\frac{a - Y_{X/S} C_S}{a - Y_{X/S} C_{S_0}}\right] - \left[\frac{1}{Y_{X/S}} \ln\frac{a - Y_{X/S} C_S}{a - Y_{X/S} C_{S_0}}\right] = -\frac{\mu_m}{Y_{X/S}} t_r \tag{4.48}$$

From Equation (4.47)

$$\therefore \qquad a - Y_{X/S}C_{S_0} = C_{X_0} \tag{4.49}$$

Substituting Equation (4.48) in Equation (4.49) and rearranging

$$(C_{X_0} + Y_{X/S}C_{S_0} + Y_{X/S}K_S)\ \ln\left[\frac{C_{X_0} + Y_{X/S}C_{S_0} - Y_{X/S}C_S}{C_{X_0}}\right] - Y_{X/S}K_S \ln\frac{C_S}{C_{S_0}} = \mu_m(C_{X_0} + Y_{X/S}C_{S_0})t_r \tag{4.50}$$

where $t = t_r$ is the batch reaction time. Equation (4.50) is a function of C_{X_0}, C_S and C_{S_0}. We need to accurately measure C_S values during the reaction. This will avoid the need for measurement of cell concentration.

(b) Limitation of Equation (4.50)

If the reaction involves complex reactants (for example: yeast extract, malt extract, peptone, meat extract) or polymeric substances (for example: cellulose, lignin, pectin, starch, and natural complex materials like rice straw, wheat straw, etc.), estimation of reactant is a serious problem.

In this case, following procedure may be followed.

Considering cell synthesis:

As
$$\mu = \frac{1}{C_X}\frac{dC_X}{dt} \tag{4.51}$$

$$\mu C_X = \frac{\mu_m C_S}{K_S + C_S}C_X$$

$$Y_{X/S} = \frac{C_X - C_{X_0}}{C_{S_0} - C_S}$$

$$C_S = \frac{1}{Y_{X/S}}(C_{X_0} - C_X + Y_{X/S}C_{S_0}) \tag{4.52}$$

$$\frac{dC_X}{dt} = \mu C_X = \frac{\mu_m C_S}{K_S + C_S}C_X$$

$$= \frac{\mu_m \dfrac{1}{Y_{X/S}}\left(C_{X_0} - C_X + Y_{X/S}C_{S_0}\right)}{K_S + \dfrac{1}{Y_{X/S}}\left(C_{X_0} - C_X + Y_{X/S}C_{S_0}\right)}C_X \tag{4.53}$$

On integration of Equation (4.53)

$$\int_{C_{X_0}}^{C_X} \frac{K_S + \dfrac{1}{Y_{X/S}}\left(C_{X_0} - C_X + Y_{X/S}C_{S_0}\right)}{\left(C_{X_0} - C_X + Y_{X/S}C_{S_0}\right)C_X}\,dC_X = \int_0^{t_r}\frac{\mu_m}{Y_{X/S}}\,dt$$

Separating the variables and substituting from Equation (4.47)

$$K_S \int_{C_{X0}}^{C_X} \frac{dC_X}{(a - C_X)C_X} + \frac{1}{Y_{X/S}} \int_{C_{X0}}^{C_X} \frac{dC_X}{C_X} = \int_0^{t_r} \frac{\mu_m}{Y_{X/S}} dt$$

$$\therefore \quad \frac{K_S}{C_{X_0} + Y_{X/S}C_{S_0}} \ln\left[\left(\frac{C_X}{C_{X_0}}\right)\left(\frac{Y_{X/S}C_{S_0}}{C_{X_0} + Y_{X/S}C_{S_0} - C_X}\right)\right] + \frac{1}{Y_{X/S}} \ln\frac{C_X}{C_{X_0}} = \frac{\mu_m}{Y_{X/S}} t_r \qquad (4.54)$$

Equation (4.54) requires initial reactant concentration (C_{S_0}), cell concentration in the inoculum (C_{X_0}) and C_X. So C_X measurement should be accurate during the reaction.

Example 4.1 A typical example for the calculation of batch time

A batch reactor needs to be designed for an organic acid production. It is required to produce 100 kg/day of organic acid.

Available data for this fermentation are given below.

$Ks = 0.12$ kg/m^3 $\qquad (Y_{X/S})_{max} = 0.63$

$C_{S_0} = 4.2$ kg/m^3 $\qquad \mu_m = 0.08$ h^{-1}

$C_{X_0} = 0.2$ kg/m^3 $\qquad C_S = 5\%$ of C_{S_0}

The reaction is growth associated in nature. Assume that the density of the medium is 1000 kg/m^3. Also assume that the reaction mixture occupies 60 % of the total reactor volume. Take height to diameter ratio of the reactor vessel as 3:1.

Solution:

$$C_S = 5\% \text{ of } C_{S_0} = 0.21 \text{ kg/m}^3$$

Reaction is growth associated in nature

$$\therefore \quad Y_{X/S} = Y_{P/X} = 0.63$$

$$Y_{P/X} = \frac{C_P - C_{P_0}}{C_X - C_{X_0}}, \quad C_{P_0} = 0$$

$$C_X = C_{X_0} + Y_{X/S}C_{S_0} - Y_{X/S}C_S$$

$$\therefore \quad C_X = 2.7137 \text{ kg/m}^3$$

$$Y_{P/X} = 0.63$$

$$C_P = 1.583 \text{ kg/m}^3$$

The time required to attain this product concentration can be calculated using the following expression derived from Monod's equation

$$(C_{X_0} + Y_{X/S} C_{S_0} + Y_{X/S}K_S) \ln\left[\frac{C_{X_0} + Y_{X/S}C_{S_0} - Y_{X/S}C_S}{C_{X_0}}\right] - Y_{X/S} K_S \ln\frac{C_S}{C_{S_0}} = \mu_m(C_{X_0} + Y_{X/S} C_{S_0})t$$

$$t = 34.4575 \text{ h} = 1.4357 \text{ days}$$

It is given that required production is 100 kg/day.

So from 1.4357 days 143.57 kg of product should be produced. Volume of the reaction medium to get 143.573 kg of product = 143.573/1.5836 = 91 m^3

Therefore, total volume required = 91/0.6 = 152 m^3. The total volume required to carry out the reaction is very large, hence we use multiple tanks.

Taking single tank volume as 1 m^3, number of tanks required is 152.

Given height to diameter ratio of tank is 3:1

Therefore,

Diameter of a single tank = 0.75 m

Height of a single tank = 2.25 m

Number of tanks required = 152.

Suppose we are interested in the product along with the cell produced. We need to calculate the time of reaction when the product concentration is maximum. In such a situation, t_r is the time when product concentration is maximum. Product synthesis is a complex phenomenon in biological system.

4.3.3 Calculation of t_r for Simultaneous Synthesis of Cells and Products

In this case, we assume Luedeking and Piret expression (Luedeking and Piret, 2000).

$$\text{Luedeking – Piret expression is } q_p = \alpha\mu_g + \beta \tag{4.55}$$

where $q_p = (1/C_x)\, dC_p/dt$ = specific product formation rate,
α is growth associated constant and
β is non-growth associated constant.

Therefore,

$$\frac{dC_P}{dt} = \alpha\frac{\mu_m C_S}{K_S + C_S}C_X + \beta C_X$$

$$\frac{dC_S}{dt} = -\frac{1}{Y_{X/S}}\frac{\mu_m C_S}{K_S + C_S}C_X - \frac{1}{Y_{P/X}}\left(\alpha\frac{\mu_m C_S}{K_S + C_S}C_X + \beta C_X\right) \tag{4.56}$$

Let

$$Y_1 = Y_{X/S} \text{ and } Y_2 = Y_{P/X}$$

$$\frac{dC_S}{dt} = \left(-\frac{1}{Y_1}\frac{\mu_m C_S}{K_S + C_S} - \frac{\alpha}{Y_2}\frac{\mu_m C_S}{K_S + C_S} - \frac{\beta}{Y_2}\right)C_X \tag{4.57}$$

$$C_X = C_{X_0} + Y_1 C_{S_0} - Y_1 C_S \tag{4.58}$$

Let

$$C_{X_0} + Y_1 C_{S_0} = a$$

$$b = \frac{1}{Y_1} + \frac{\alpha}{Y_2}$$

That is

$$C_X = a - Y_1 C_S$$

$$\therefore \quad \frac{dC_S}{dt} = -\left(\frac{\mu_m C_S b}{K_S + C_S} + \frac{\beta}{Y_2}\right)(a - Y_1 C_S)$$

$$= -\left(\frac{\beta K_S + (\beta + b\mu_m Y_2)C_S}{(K_S + C_S)Y_2}\right)(a - Y_1 C_S)$$

Let $e = (\beta + b\mu_m Y_2)$

$$\therefore \quad \frac{dC_S}{dt} = -\left(\frac{\beta K_S + eC_S}{(K_S + C_S)Y_2}\right)(a - Y_1 C_S) \tag{4.59}$$

$$\int_{C_{S0}}^{C_S} \frac{(K_S + C_S)dC_S}{(\beta K_S + eC_S)(a - Y_1 C_S)} = -\left(\frac{1}{Y_2}\right)\int_0^t dt \tag{4.60}$$

Integrating Equation (4.60) gives Equation (4.61).

$$\frac{A}{e}\ln\left(\frac{\beta K_S + eC_S}{\beta K_S + eC_{S_0}}\right) - \frac{B}{Y_1}\ln\left(\frac{a - Y_1 C_S}{a - Y_1 C_{S_0}}\right) = -\frac{1}{Y_2}t \tag{4.61}$$

where

$$A = \frac{(e - \beta)K_S}{(ae + \beta Y_1 K_S)} \quad \text{and} \quad B = \frac{Y_1 K_S + a}{(ae + \beta Y_1 K_S)} \tag{4.62}$$

Example 4.2 General steps for a batch bioreactor design

Detailed steps are given below.

I. Mass balances

For cell

We recall the generalized mass balance Equation (4.25)

Rate of accumulation of cells = Rate of cells entering – rate of cells leaving + net rate of generation of living cells

As there is no input and output in this bioreactor, the above equation simplifies to

$$\frac{VdC_X}{dt} = 0 - 0 + (r_g - r_d)V \tag{4.63}$$

where r_d is the death rate of cells

$$\frac{VdC_X}{dt} = (r_g - r_d)V \tag{4.64}$$

where

$$r_g = \frac{\mu_m C_S}{K_S + C_S}C_X, \text{ assuming Monod's equation}$$

For reactant

Similarly from general mass balance equation,

$$\frac{VdC_S}{dt} = -r_s V, \text{ where } r_s \text{ has the unit kg/m}^3\text{.h}$$

Reactant disappearance is due to cell synthesis and due to the maintenance of the cell.

So we get

$$V\frac{dC_S}{dt} = Y_{S/X}(-r_g)\,V - mC_xV \tag{4.65}$$

or

$$\frac{dC_S}{dt} = Y_{S/X}(-r_g) - mC_x \tag{4.66}$$

For product balance

$$V\frac{dC_P}{dt} = r_pV = Y_{P/S}(-r_s)V \tag{4.67}$$

where r_P has unit kg/m^3.h.

Also

$$V\frac{dC_P}{dt} = Y_{P/X}(r_g)V \tag{4.68}$$

II. Rate equations

For r_g = rate of growth. We can assume Monod's model or depending on the reaction scheme we can consider the equation in Table 1.3 of Chapter 1.

For death rate of cells, assuming first order kinetics,

$$r_d = k_dC_x \tag{4.69}$$

and maintenance rate of cell $= mC_X$ (4.70)

III. From stoichiometry, we can get a correlation between r_g and r_p. The simplest model is $r_P = Y_{P/X}r_g$ (4.71)

However, the following variations will influence Equation (4.71).

- Different phases of growth in batch reactor
- Growth-associated and non-growth associated behavior
- Diauxic behavior

Different phases of growth of cell in a batch bioreactor

There are a typical few phases of cell in batch bioreactor (Fig. 4.7).They are lag phase(1), accelerated growth phase (2), exponential growth phase (3) , post logarithmic growth phase (4),stationary phase (5), accelerated death phase (6) and exponential death phase (7). This may be assumed for bacterial growth. r_g is different in each phase which will be addressed in this chapter.

Growth-associated and non-growth associated behavior of product synthesis

The kinetic relationship between growth and product formation depends on the role of the product in cell metabolism. This suggests the micro kinetics in the metabolic level to determine the shape of the

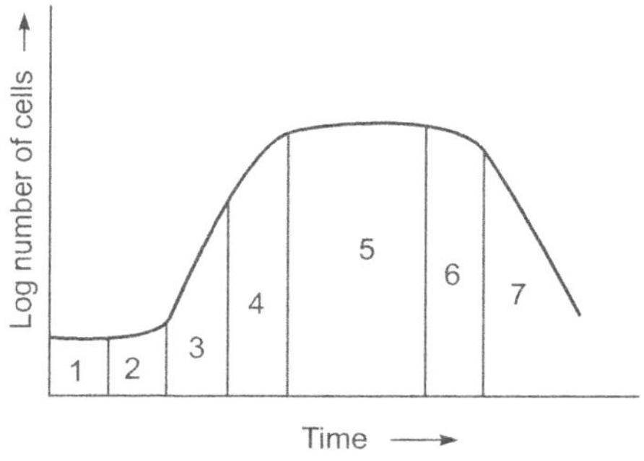

Figure 4.7 Phases of growth of microorganisms in batch culture.

curve of formal kinetics. Growth associated products are directly engaged in catabolic pathway (for example, the yeast fermentation for ethanol production).

Non-growth associated products have no association with primary metabolism. These products are called secondary metabolites (for example, antibiotics and toxin production).

In mixed-growth associated product formation, the products are indirectly connected to energy production pathways and are the result of a characteristic genetically manipulated metabolism (for example, lactic acid, citric acid, etc.)

Diauxic growth behavior

This is a phenomenon in the growth of micro organisms where two exponential growth phases are separated by a lag phase (Fig. 4.8).

Microorganism displays this biphasic growth or "diauxic" behavior when grown on two different C-sources. The first growth phase corresponds to exclusive utilization of one of the compounds followed by a period of adaptation before the other reactant is metabolized and growth recommences. In this example, the organism (*Escherichia coli*) does not have enzymes to metabolize lactose initially (Fig. 4.8). So it grows first on glucose, which is easily transported to the cell and rapidly metabolized.

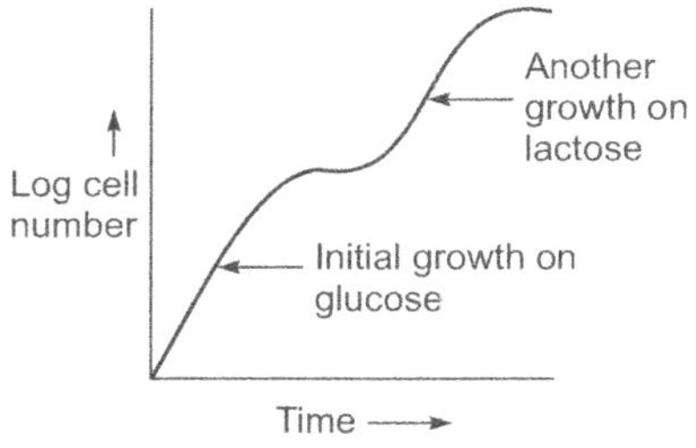

Figure 4.8 Diauxic growth behaviors.

IV. Combining equations

For cells

Equations (4.64), Monod's equation and Equation (4.69) give Equation (4.72).

$$\frac{dC_X}{dt} = \frac{\mu_m C_S}{K_S + C_S} C_X - k_d C_X \tag{4.72}$$

For reactants

Equation (4.64) and Monod's equation give Equation (4.73)

$$\frac{dC_S}{dt} = -Y_{S/X} \frac{\mu_m C_S}{K_S + C_S} C_X - mC_X \tag{4.73}$$

For products

Equation (4.68) and Monod's equation give Equation (4.74).

$$\frac{dC_P}{dt} = +Y_{P/X}\frac{\mu_m C_S}{K_S + C_S}C_X \tag{4.74}$$

Equations (4.72)–(4.74) are solved on an ODE equation solver. The parameters are either calculated or known from certain experiments.

Example 4.3

Rate equations for the growth of eukaryotic cells, the reactants consumption and products formation are given below.

$$\frac{dC_X}{dt} = \mu_m\left(1 - \frac{C_P}{C_{P_{\max}}}\right)^m\left(\frac{C_S}{K_S + C_S}\right)C_X$$

$$r_S = -\frac{1}{Y_{X/S}}\frac{dC_X}{dt}$$

$$r_P = \frac{1}{Y_{X/P}}\frac{dC_X}{dt}$$

Data given

$$\begin{aligned} K_s &= 1.6\ \text{kg/m}^3 & Y_{X/S} &= 0.06\ \text{kg/kg} \\ C_{X_0} &= 0.1\ \text{kg/m}^3 & \mu_m &= 0.24\ \text{h}^{-1} \\ C_{P_0} &= 0\ \text{kg/m}^3 & (Y_{X/P}) &= 0.16\ \text{kg/kg} \\ C_{P_{\max}} &= 100\ \text{kg/m}^3 & m &= 2 \end{aligned}$$

(1) Calculate the change of C_X,C_P, and C_S as a function of time when $C_{S_0} = 100$ kg/m^3.
(2) What is the role of reactant concentration on specific growth rate of cell?
(3) If we increase the reactant concentration, check whether rate of cell growth increases or not.
(4) Show the effect of initial reactant concentration on μ_{net}.

Solution

$$\frac{dC_X}{dt} = \mu_m\left(1 - \frac{C_P}{C_{P_{\max}}}\right)^m\left(\frac{C_S}{K_S + C_S}\right)C_X$$

$$= \mu_m\left(1 - \frac{C_{P_0} + \left(\dfrac{C_X - C_{X_0}}{Y_{X/P}}\right)}{C_{P_{\max}}}\right)^m\left(\frac{C_{S_0} - \left(\dfrac{C_X - C_{X_0}}{Y_{X/S}}\right)}{K_S + C_{S_0} - \left(\dfrac{C_X - C_{X_0}}{Y_{X/S}}\right)}\right)C_X$$

Let

$$a = Y_{X/P}C_{P_{\max}} - Y_{X/P}C_{P_0} + C_{X_0}$$

$$b = C_{X_0} + Y_{X/S}C_{S_0}$$

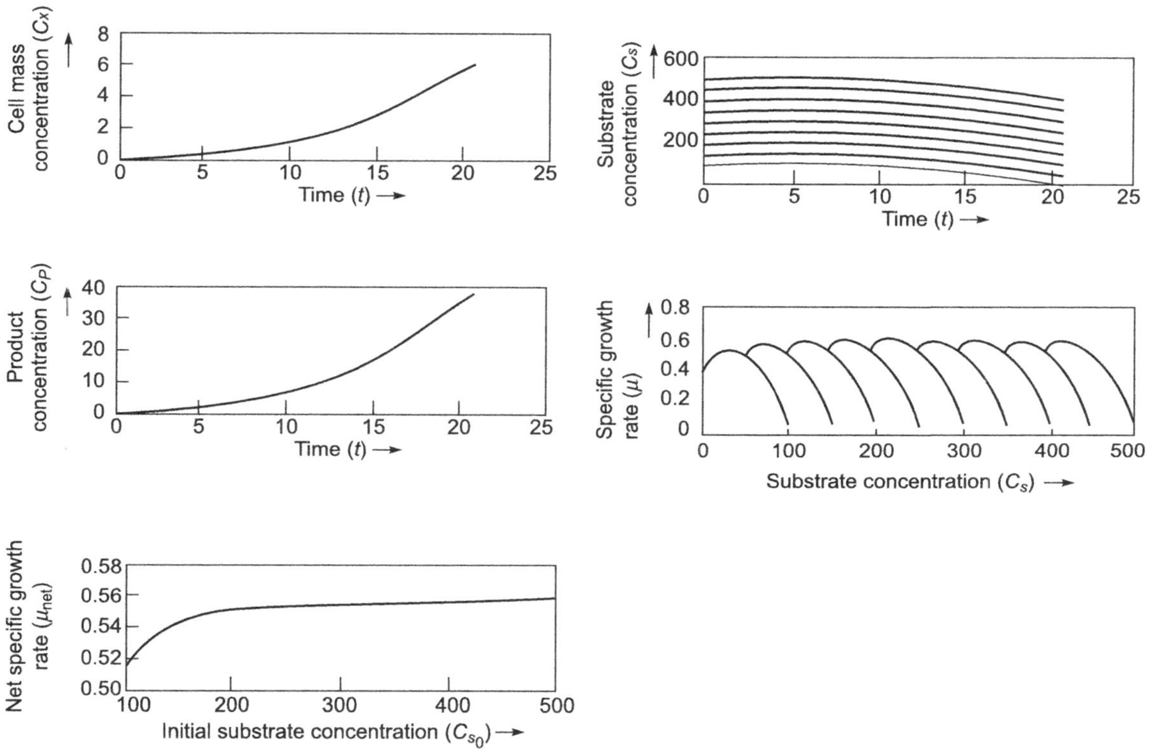

Figure 4.9 Solutions for C_X, C_S, C_p vs. t.

$$c = C_{X_0} + Y_{X/S}C_{S_0} + Y_{X/S}K_S$$

$$\frac{dC_X}{dt} = \frac{\mu_m}{\left(C_{P_{\max}} Y_{X/P}\right)^2}(a - C_X)^2\left(\frac{b - C_X}{c - C_X}\right)C_X$$

First integrate this equation from C_{X_0} to C_X and t from 0 to t. Then one can solve the equation in MATLAB for different C_X and C_{S_0}. We get different plots for C_X, C_S, and C_P vs. t as given in Figure 4.9.

4.3.4 Non-Ideality in Batch Bioreactor

Sources of non-ideality in batch reactor are categorized in two ways:

1. Microbiological factors and
2. Physical factors

They influence the batch reaction time (t_r).

Microbiological sources of non-ideality are the following.

(a) Different growth rates in various phases

(b) Presence of dead cells

(c) Inhibition caused by biochemicals and chemicals
(d) Conversion of productive cells to unproductive cells
(e) Growth of cells in the reactor walls
(f) Age of cell population

Physical factors which cause non-ideality are the following:

1. Flow patterns, turbulence, and gas dissipation.
2. Gas retention
3. Recirculation
4. Mixing

Let us discuss first the microbiological sources of non-identity.

(a) Different Growth Rates in Various Phases

Various growth phases are taken into consideration. It is assumed that μ is constant with in the same phase. For reference we recall Figure 4.7.

Let us assume that the rate of growth of cell is expressed by Equation (4.75)

$$\frac{dC_X}{dt} = \mu \phi_w C_x \tag{4.75}$$

where ϕ_w is apparent growth constant suggested by Schügerl (1985).

Let us consider the duration of phase and ϕ_w in different phases in the Table 4.3.

The product formation rate is considered in terms of modified Luedeking and Piret Equation (4.76).

$$\frac{dC_P}{dt} = \alpha \mu \phi_w C_x + \beta(1 - \phi_w) C_x \tag{4.76}$$

Table 4.3 *Duration of phase and corresponding values of ϕ_w.*

Sl. no.	Phase	Duration of phase	ϕ_w
1	Lag phase	0 to t_0	0
2	Transition phase	t_0 to t_L	φ_w
3	Exponential phase	t_1 to t_c	1
4	Post exponential phase	t_c to t_d	$\left(\frac{C_{X_C}}{C_{X_m} - C_{X_C}}\right)\left(\frac{C_{X_m} - C_X}{C_X}\right)$
5	Stationary phase	t_d to t_f	$\frac{C_{X_D}}{C_X}$

Similar phase consideration is also discribed by Schügerl (1985).

Growth rate expressions are considered first in different phases (Schügerl, 1985).

(1) Lag Phase

From Equation (4.75), $\frac{dC_X}{dt} = \mu \phi_w C_x$

For lag phase $\phi_w = 0$

$\therefore$ $$\frac{dC_X}{dt} = 0$$

$$C_x = \text{constant} = C_{X_0}$$

$$\boxed{\therefore\ C_X = C_{X_0}} \qquad (4.77)$$

(2) Transition Phase

From Equation (4.75) $$\frac{dC_X}{dt} = \mu\phi_w C_X$$

For transition phase $\phi_w = \varphi_w$

$\therefore$ $$\frac{dC_X}{dt} = \mu\varphi_w C_X$$

$$\int_{C_{X_0}}^{C_{X_L}} \frac{dC_X}{C_X} = \int_{t_0}^{t_L} \mu\varphi_w dt$$

$$\boxed{C_{X_L} = C_{X_0}\ \exp\left(\mu\varphi_w\left(t_L - t_0\right)\right)} \qquad (4.78)$$

(3) Exponential Phase

From Equation (4.75) $$\frac{dC_X}{dt} = \mu\varphi_w C_X$$

For exponential phase $\phi_w = 1$

$$\int_{C_{X_L}}^{C_{X_c}} \frac{dC_X}{C_X} = \int_{t_L}^{t_c} \mu dt$$

$$\boxed{C_{X_c} = C_{X_0}\ \exp\left(\mu\varphi_w\left(t_L - t_0\right)\right)\exp\left(\mu\left(t_c - t_L\right)\right)} \qquad (4.79)$$

(4) Post-exponential Phase

From Equation (4.75) $$\frac{dC_X}{dt} = \mu\phi_w C_X$$

For post-exponential phase $$\phi_w = \left(\frac{C_{Xc}}{C_{Xm} - C_{Xc}}\right)\left(\frac{C_{Xm} - C_X}{C_X}\right)$$

$$\frac{dC_X}{dt} = \mu\left(\frac{C_{Xc}}{C_{Xm} - C_{Xc}}\right)\left(C_{Xm} - C_X\right)$$

$$\frac{dC_X}{dt} = \mu\left(\frac{C_{Xc}C_{Xm}}{C_{Xm} - C_{Xc}}\right) - \mu\left(\frac{C_{Xc}C_X}{C_{Xm} - C_{Xc}}\right)$$

Let

$$A = \left(\frac{C_{Xc}C_{Xm}}{C_{Xm} - C_{Xc}}\right)$$

$$B = \left(\frac{\mu C_{Xc}}{C_{Xm} - C_{Xc}}\right)$$

$\therefore$
$$\frac{dC_X}{dt} = \mu A - BC_X$$

On integrating from C_{X_C} to C_{X_D} and from t_C to t_D

$$\boxed{C_{X_D} = \frac{1}{B}\left[\mu A - \left(\mu A - BC_{X_c}\right)\exp\left(-B\left(t_D - t_c\right)\right)\right]} \tag{4.80}$$

(5) Stationary Phase

From Equation (4.75)
$$\frac{dC_X}{dt} = \mu\phi_w C_x$$

For stationary phase
$$\phi = \frac{C_{X_D}}{C_X}$$

$\therefore$
$$\frac{dC_X}{dt} = \mu C_{X_D}$$

On integrating from C_{X_D} to C_{X_f} and from t_D to t_f

$$C_{X_f} = \frac{1}{B}\ [\mu A - (\mu A - BC_{Xc})\exp(-B(t_D - t_c))]\ (1 + \mu(t_f - t_D)) \tag{4.81}$$

where

$$C_{X_c} = C_{X_0}\exp(\mu\varphi_w(t_L - t_0))\exp(\mu(t_c - t_L))$$

$$A = \frac{C_{X_m}C_{X_c}}{C_{X_m} - C_{X_c}}$$

$$B = \frac{\mu C_{X_c}}{C_{X_m} - C_{X_c}}$$

Practical fermentations are carried out maximum up to the stationary phase. If cells begin to lyse, it causes serious problem in downstream operations of product. In waste treatment reactors, reactions are continued with cells beyond the stationary phase.

Now the product synthesis in different phases are considered using the generalized expression of Equation (4.76).

$$\frac{dC_P}{dt} = \alpha\mu\phi_w C_X + \beta(1 - \phi_w)C_X$$

(1) Lag Phase

$$\phi_w = 0$$

$$\frac{dC_P}{dt} = \beta C_{X_0}$$

On integration

$$\boxed{C_P = C_{P_0} + \beta C_{X_0} t_0} \tag{4.82}$$

(2) Transition Phase

For transition phase $\phi_w = \varphi_w$

$$\frac{dC_P}{dt} = \alpha\mu\varphi_w C_X + \beta(1 - \varphi_w)C_X$$

Let

$$a = \alpha\mu\varphi_w$$

$$b = \beta(1 - \varphi_w)$$

$$\frac{dC_P}{dt} = (a + b)C_{X_L}$$

Substitute for C_{X_L} from Equation (4.78).

On integrating from C_{P_0} to C_{P_L} and from t_0 to t_L

$$\boxed{C_{P_L} = C_{P_0} + (a+b)\,C_{X_0}\,(t_L - t_0)\exp\big(\mu\varphi_w\,(t_L - t_0)\big)} \tag{4.83}$$

(3) Exponential Growth Phase

For exponential phase $\phi_w = 1$

$$\frac{dC_P}{dt} = \alpha\mu C_{X_c}$$

On integrating from C_{P_L} to C_{P_c} and from t_L to t_c

$$C_{P_c} = C_{P_L} + \alpha\mu(t_c - t_L)C_{X_0}\exp[\mu\{\varphi_W(t_L - t_0) + (t_c - t_L)\}] \tag{4.84}$$

(4) Post-exponential Phase

For post-exponential phase $\phi_w = \left(\dfrac{C_{X_c}}{C_{Xm} - C_{X_c}}\right)\left(\dfrac{C_{X_m} - C_X}{C_X}\right)$

$$\therefore\ \frac{dC_P}{dt} = \alpha\mu\left(\frac{C_{X_c}}{C_{X_m} - C_{X_c}}\right)\left(C_{X_m} - C_{X_D}\right) + \beta\left(\frac{C_{X_D} - C_{X_C}}{C_{X_m} - C_{X_C}}\right)C_{X_m}$$

$$C_{P_D} - C_{P_C} = \left\{\begin{aligned} &\alpha\mu\frac{C_{X_L}\exp(\mu(t_c - t_L))}{C_{X_m} - C_{X_L}\exp(\mu(t_c - t_L))}\times \\ &\left(C_{X_m} - \frac{1}{B}[\mu A - (\mu A - BC_{X_L}\exp(\mu(t_c - t_L))\exp(-B(t_D - t_L)))]\right) \\ &+\beta\left(\frac{\frac{1}{B}[\mu A - (\mu A - BC_{X_L}\exp(\mu(t_c - t_L))\exp(-B(t_D - t_L)))] - C_{X_L}\exp(\mu(t_c - t_L))}{C_{X_m} - C_{X_L}\exp(\mu(t_c - t_L))}\right)C_{Xm}\end{aligned}\right\}(t_D - t_c) \tag{4.85}$$

where $C_{X_L} = C_{X_0} \exp(\mu\varphi_w (t_L - t_0))$

$$A = \frac{C_{X_0} \exp[\mu\{\varphi_w (t_L - t_0) + (t_c - t_L)\}] C_{X_m}}{C_{X_m} - C_{X_0} \exp[\mu\{\varphi_w (t_L - t_0) + (t_c - t_L)\}]}$$

$$B = \frac{\mu C_{X_0} \exp[\mu\{\varphi_w (t_L - t_0) + (t_c - t_L)\}]}{C_{X_m} - C_{X_0} \exp[\mu\{\varphi_w (t_L - t_0) + (t_c - t_L)\}]}$$

(5) Stationary Phase

$$\phi_w = \frac{C_{X_D}}{C_X}$$

$$\frac{dC_P}{dt} = \alpha\mu C_{X_D} + \beta(C_{X_f} - C_{X_D})$$

$$\frac{dC_P}{dt} = (\alpha\mu - \beta)\, C_{X_D} + \beta C_{X_f}$$

$$C_{P_f} - C_{P_D} = ((\alpha\mu - \beta)\, C_{X_D} + \beta\, C_{X_f})\,(t_f - t_d)$$

$$C_{P_f} - C_{P_D} = \left((\alpha\mu - \beta)\left(\frac{\mu A}{B} - \left(\frac{\mu A}{B} - C_{X_o} \exp(\mu(\phi_w (t_L - t_o) + (t_C - t_L)))\right)\right) e^{-B(t_d - t_c)}\right)$$
$$+ \beta(1 + \mu(t_f - t_d))\left[\frac{\mu A}{B} - \left(\frac{\mu A}{B} - C_{X_0} \exp\left(\mu(\varphi_w (t_L - t_0) + (t_c - t_L))\right)\right)\right] \exp(-B(t_f - t_d)) \tag{4.86}$$

This is the expression for C_{P_D} given in the last phase, i.e., post exponential phase.

(b) Presence of Dead Cells

Death of cells occurs in a batch culture. Only viable cells (x_v) generate non viable cells (x_d) at a particular rate.

Assumption 1:
First order death rate for the transformation of viable cells to non-viable cells is assumed.

$$x_v \xrightarrow{k} x_d$$

Balance on cell growth:

$$\frac{dC_{x_v}}{dt} = \mu C_{x_v} - kC_{x_v} \tag{4.87}$$

$$\frac{dC_{x_d}}{dt} = kC_{x_v} \tag{4.88}$$

Growth for total cell population is

$$\frac{dC_{x_T}}{dt} = \frac{d(C_{x_v} + C_{x_d})}{dt} = \mu C_{x_v} \tag{4.89}$$

Case 1

μ is constant, which is applicable for exponential growth phase. From Equation (4.87),

$$\ln\frac{C_{x_v}}{C_{x_{v_0}}} = (\mu - k)t$$

or

$$C_{x_v} = C_{x_{v_0}}\exp[(\mu - k)t] \tag{4.90}$$

Substitution of Equation (4.89) in Equation (4.90) follows

$$\frac{dC_{x_T}}{dt} = \mu C_{x_{v_0}}\exp[(\mu - k)t] \tag{4.91}$$

Integration of Equation (4.91)

$$\int_{C_{x_{T_0}}}^{C_{x_T}} dC_{x_T} = \int_0^t \mu C_{x_{v_0}}\exp[(\mu - k)t]\,dt$$

From Equations (4.90) and (4.91)

$$\frac{C_{x_v}}{C_{x_T}} = \frac{C_{x_{v_0}}\exp[(\mu - k)t]}{C_{x_{T0}} + \dfrac{\mu C_{x_{v_0}}}{(\mu - k)}[e^{(\mu - k)t} - 1]}$$

$$\therefore \qquad C_{x_T} = C_{x_{T0}} + \frac{\mu C_{x_{v_0}}}{(\mu - k)}[e^{(\mu - k)t} - 1] \tag{4.92}$$

Therefore,

$$\frac{C_{x_v}}{C_{x_T}} = \frac{\dfrac{C_{x_{v_0}}}{C_{x_{T0}}}\exp[(\mu - k)t]}{1 + \dfrac{\mu C_{x_{v_0}}}{(\mu - k)C_{x_{T0}}}[e^{(\mu - k)t} - 1]} \tag{4.93}$$

Conditions:

(i) If all cells are viable, Equation (4.93) is

$$\frac{C_{x_v}}{C_{x_T}} = \frac{[\mu - k]\exp[\mu - k]t}{[\mu - k] + \mu(\exp[\mu - k]t - 1)} \tag{4.94}$$

(ii) If $k = \mu$, all cells are dead. There is no solution to Equation (4.89).

(iii) If μ is very small, dead cells will dominate. Result will approach the condition (ii).

Case 2

μ is not a constant, i.e., other than exponential phase of growth. Then

$$\mu = \frac{\mu_m C_S}{K_S + C_S}$$

and $$C_X + Y_{X/S}C_S = C_{X_0} + Y_{X/S}C_{S_0} \tag{4.95}$$

$\therefore$ $$C_S = C_{S_0} + \frac{1}{Y_{X/S}}\left(C_{X_0} - C_X\right) \tag{4.96}$$

Substitution of C_S from Equation (4.96) to Monod's equation,

$$\mu = \frac{\mu_m\left[C_{S_0} + \frac{1}{Y_{X/S}}\left(C_{X_0} - C_X\right)\right]}{K_S + C_{S_0} + \frac{1}{Y_{X/S}}\left(C_{X_0} - C_X\right)} \tag{4.97}$$

From Equations (4.87) and (4.97)

$$\frac{dC_{X_V}}{dt} = \left(\frac{\mu_m\left[C_{S_0} + \frac{1}{Y_{X/S}}(C_{X_0} - C_{X_V})\right]}{K_S + \left[C_{S_0} + \frac{1}{Y_{X/S}}(C_{X_0} - C_{X_V})\right]} - k\right)C_{X_V}$$

$$\frac{dC_{X_V}}{dt} = \left(\frac{\mu_m[Y_{X/S}C_{S_0} + C_{X_0} - C_{x_V}] - kY_{X/S}K_S - kY_{X/S}C_{S_0} - kC_{X_0} + kC_{x_V}}{(K_S + C_{S_0})Y_{X/S} + (C_{X_0} - C_{x_V})}\right)C_{x_V} \tag{4.98}$$

(c) Conversion of Productive Cells to Unproductive Cells

This may cause non-ideal behavior to any class of ideal reactors.

Assumptions are the following in this case.

(i) Productive cells produce unproductive cells due to sudden change in reaction conditions. For example, even if the cells are fatigued, this occurs in the bioreactor. If cells require certain component pressure during reaction, in absence of this pressure, productive cells are converted into unproductive cells, viz., cells carrying plasmid.

(ii) The probability for productive cells to produce unproductive cells is p. Suppose N productive cells produce $N(1 - p)$ productive cells and Np unproductive cells after one division.

For
- X_P as productive cells
- X_u as unproductive cells
- μ_P as specific growth rate of productive cells
- μ_u as specific growth rate of unproductive cells

During exponential growth phase, the growth rate of productive cells is

$$\frac{dC_{X_P}}{dt} = (1 - p)\mu_P C_{X_P} \tag{4.99}$$

C_{X_P} is the number of productive cells per unit volume.

If mass of cells is approximately proportional to the number of cells, Equation (4.99) is also valid in this case.

However, growth rate of unproductive cells,

$$\frac{dC_{X_u}}{dt} = \mu_u C_{X_u} + p\mu_p C_{X_P} \qquad (4.100)$$

μ_p and p are constant.

From Equation (4.99) upon integration we get

$$C_{X_P} = C_{X_{P_0}} \exp((1-p)\,\mu_p t \qquad (4.101)$$

From Equations (4.100) and (4.101)

$$\frac{dC_{X_u}}{dt} = \mu_u C_{X_u} + p\mu_p C_{X_{P_0}} \exp((1-p)\mu_p t)$$

$$\text{I.F} = e^{\int -\mu_u dt} = e^{-\mu_u t}$$

$$\therefore \quad C_{X_u} e^{-\mu_u t} = \int e^{-\mu_u t} p\mu_p C_{X_{P_0}} \exp((1-p)\mu_p t)\,dt$$

$$C_{X_u} = \frac{p\mu_p C_{X_{P_0}}}{(1-p)\mu_p - \mu_u}\left[\exp((1-p)\mu_p t) - \exp(\mu_u t)\right] + C_{X_{u_0}} \exp(\mu_u t) \qquad (4.102)$$

The fraction of productive cells in the total population is given by Equation (4.103).

$$\frac{C_{X_p}}{C_{X_p} + C_{X_u}} = F_P \qquad (4.103)$$

Considering Equations (4.101) to (4.103), we get Equation (4.104).

$$F_P = \frac{\exp((1-p)\mu_p t)}{\exp((1-p)\mu_p t) + \dfrac{p\mu_p}{((1-p)\mu_p - \mu_u)}\left[\exp((1-p)\mu_p t) - \exp(\mu_u t)\right] + \dfrac{C_{X_{u_0}}}{C_{X_{P_0}}} \exp(\mu_u t)} \qquad (4.104)$$

F_P is inversely related to C_{X_u}. Hence, the productivity will decrease in the reactor. In this case t_r needs to be calculated from Equation (4.99) after proper modification with F_P from Equation (4.104).

(d) Growth of Cells in the Reactor Walls

For yeast and bacteria such behavior is rare. However, fungal systems do grow on the reactor components during reaction. Some times those cells do not actively take part in the reaction, but they consume reactant. Any approximation of growth calculation based on cell concentration available in reaction fluid is misnomer and it does not reflect the true reaction time. The mass balance equations cannot be written in a proper form. This non-ideality is appropriate for "flow bioreactor" (Rao and Rao, 2004).

(e) Age of Cell Population

Inoculum to the reactor, particularly for batch bioreactors, shows a lot of variation of cell age among the cells (Hartwell and Unger, 1977). The activity is also different for different active cells. One needs to consider the population dynamics to calculate effective t_r owing to this variation.

It is based on population balance model which looks for time dependence of system and state of the cell. The governing equation relates age of the cell, its position in cell cycle, its total mass or volume, mass of cell constituents and other properties (Tsuchiya *et al.*, 1967, Fredrickson, 1992, Fredrickson *et al.*, 1967).

$$\frac{\partial f(t,\vec{\mathbf{m}})}{\partial t} + \frac{\partial(\vec{\mathbf{m}}', f(t,\vec{\mathbf{m}}))}{\partial \vec{\mathbf{m}}} = 2\int p(\vec{\mathbf{m}},\vec{\mathbf{m}}')\, g(\vec{\mathbf{m}}')\, f(t,\vec{\mathbf{m}}')\, d\vec{\mathbf{m}}' - g(\vec{\mathbf{m}})\, f(t,\vec{\mathbf{m}}) \quad (4.105)$$

With initial conditions

$$f(0,\vec{\mathbf{m}}) = f_0(\vec{\mathbf{m}}) \quad (4.106)$$

where $\vec{\mathbf{m}}$ is the vector of cellular state variable

$\vec{\mathbf{m}}'$ is the mean growth rate vector of $\vec{\mathbf{m}}$.

f stands for density function of the distribution of population state.

g is division rate or probability of division of a cell.

p is partitioning function, i.e., a mother cell in state, $\vec{m}'$ will divide to two new cells of state $\vec{m}'$.

$$p(\vec{\mathbf{m}},\vec{\mathbf{m}}') = 0;\ \text{for } \vec{\mathbf{m}} > \vec{\mathbf{m}}' \quad (4.107)$$

$$p(\vec{\mathbf{m}},\vec{\mathbf{m}}') = p(\vec{\mathbf{m}} - \vec{\mathbf{m}},\vec{\mathbf{m}}');\ \text{for } \vec{\mathbf{m}} < \vec{\mathbf{m}}' \quad (4.108)$$

Simplification of balance equation is difficult. However, a simplified age distribution model is given in (4.109) considering population time since birth of the cell, called age of the cell (t_c) (Lion *et al.*, 1997).

$$\frac{\partial f(t,t_c)}{\partial t} + \frac{\partial f(t,t_c)}{\partial t_c} = -\,g(t_c) f(t, t_c) \quad (4.109)$$

With initial condition $f(0, t_c) = f_o(t_c)$ **and boundary condition suggested by (Schügerl and Bellgardt,** 2000).

$$f(t, 0) = 2\int_0^{\infty} g(t_c)\, f(t,t_c)\, dt_c$$

Equation (4.109) is called M'kendrick-von Foerster equation for cell number density (Webb, 1986). The number of cells in the cycle intervals and in total is given in Equation (4.110) to (4.112).

$$N_D,\ (\text{number of daughter cells}) = \int_0^{t_D} f_D(t,t_c)\, dt_c \quad (4.110)$$

$$N_P,\ (\text{number of parent cells}) = \int_0^{t_P} f_P(t,t_c)\, dt_c \quad (4.111)$$

N_B, (number of cells in budding phase)

$$= \int_{t_D}^{t_D+t_B} f_D(t,t_c)\, dt_c + \int_{t_P}^{t_P+t_B} f_P(t,t_c)\, dt_c \quad (4.112)$$

$$\text{Total number of cells} = N_D + N_P + N_B \quad (4.113)$$

This should be included in modified t_r (reaction time) calculation through a proper relation with μ. Hartwell and Unger (1977) suggested the equation like Equation (4.114).

$$\exp(-\mu(t_B + t_p)) + \exp(-\mu(t_B + t_D)) = 1 \quad (4.114)$$

For varying average doubling time $\left[\frac{\ln 2}{\mu}\right]$ the redistribution of cells is more or less same, keeping the relative distribution pattern in the cycling process. In case of decrease in average doubling time, only daughter cells redistribute by an exponential non-uniform manner (Yuan *et al.*, 1993). The average population is calculated by an analogy of Equation (4.113).

$$N_{\text{total}} = \sum_{i=1}^{n_D} d(i) + \sum_{j=1}^{n_p} p(j) + \sum_{k=1}^{n_B} b(k) \tag{4.115}$$

$$C_{X_N} = \frac{\text{total number of cells}}{\text{unit volume}}$$

$$= \frac{N_{\text{tot}}}{V_L} \tag{4.116}$$

where V_L = volume of reaction. This equation suggests a possibility to modify the equations involving C_X described earlier in this chapter.

(f) Derivation of Batch Time Expression Considering Substrate Inhibition

$$\frac{dC_X}{dt} = \frac{\mu_m}{\alpha_1}\frac{C_S C_X}{C_S + K_S} \tag{4.117}$$

It has been observed that reaction time is more in the presence of high concentration of reactant. This effect is also observed for the reduction in μ values. This phenomenon is called substrate inhibition. Monod's equation does not explain such results. So a modified Monod's model having an inhibition term is considered here to calculate growth rate [Equation (4.117)]

where $\alpha_1 = 1 + \frac{C_S}{K_i}$

K_i is inhibition constant

K_s is saturation parameter

One may consider any other suitable model for this purpose.

However, the final expression will be different.

We have to substitute $$C_S = C_{S_0} - \frac{1}{Y_{X/S}}(C_X - C_{X_0}) \tag{4.118}$$

$$\frac{dC_X}{dt} = \frac{\mu_m}{1 + \dfrac{C_{S_0} - \dfrac{1}{Y_{X/S}}(C_X - C_{X_0})}{K_i}} \frac{\left[C_{S_0} - \dfrac{1}{Y_{X/S}}(C_X - C_{X_0})\right]C_X}{\left[C_{S_0} - \dfrac{1}{Y_{X/S}}(C_X - C_{X_0}) + K_S\right]}$$

$$\frac{dC_X}{dt} = \frac{K_i Y_{X/S}\mu_m}{(K_i Y_{X/S} + C_{S_0}Y_{X/S} - C_X + C_{X_0})} \frac{(C_{S_0}Y_{X/S} - C_X + C_{X_0})C_X}{(C_{S_0}Y_{X/S} - C_X + C_{X_0} + K_s Y_{X/S})} \tag{4.119}$$

$$a = Y_{X/S}C_{S_0} + C_{X_0}$$
$$b = Y_{X/S}C_{S_0} + C_{X_0} + K_i Y_{X/S}$$
$$d = Y_{X/S}C_{S_0} + C_{X_0} + K_S Y_{X/S}$$

Now the expression (4.119) reduces to

$$\frac{dC_X}{dt} = \frac{K_i Y_{X/S}\mu_m (a - C_X)C_X}{(b - C_X)(d - C_X)} \tag{4.120}$$

Then $$\int_{C_{X0}}^{C_X} \frac{(b - C_X)(d - C_X)}{(a - C_X)C_X}\, dC_X = \int_{t_0}^{t} K_i Y_{X/S}\mu_m dt \tag{4.121}$$

However, the intergral portion, after simplification, becomes:

$$\frac{(b - C_X)(d - C_X)}{(a - C_X)C_X} = \frac{bd - (b + d)C_X + C_X^{\,2}}{(a - C_X)C_X}$$

$$= -1 + \frac{(a - (b + d))C_X + bd}{(a - C_X)C_X}$$

$$= -1 + \frac{bd}{(a - C_X)C_X} + \frac{aC_X}{(a - C_X)C_X} - \frac{(b + d)C_X}{(a - C_X)C_X}$$

$$= -1 + \frac{bd}{a}\left[\frac{1}{C_X} + \frac{1}{(a - C_X)}\right] + \frac{a - (b + d)}{(a - C_X)}$$

$$\int_{C_{X0}}^{C_X} \left[-1 + \frac{bd}{a}\left[\frac{1}{C_X} + \frac{1}{(a - C_X)}\right] + \frac{a - (b + d)}{(a - C_X)}\right] dC_X = \int_{t_0}^{t} K_i Y_{X/S}\mu_m dt$$

$$-(C_X - C_{X_0}) + \frac{bd}{a}\left[\ln\frac{C_X}{C_{X_0}} - \ln\left(\frac{(a - C_X)}{(a - C_{X_0})}\right)\right] - (a - (b + d))\ln\left(\frac{(a - C_X)}{(a - C_{X_0})}\right) = K_i Y_{X/S}\mu_m (t - t_0) \tag{4.122}$$

We need to express this equation in dimensionless form using the following dimensionless terms. Given that

$$X_0^* = \frac{C_{X_0}}{Y_{X/S}C_{S_0}}$$

$$K_S^* = \frac{K_S}{C_{S_0}}$$

$$S^* = \frac{C_S}{C_{S_0}}$$

$$K_i^* = \frac{K_i}{C_{S_0}}$$

Considering various components of Equation (4.122), we get

$$C_X - C_{X_0} = Y_{X/S}(C_{S_0} - C_S) = Y_{X/S}C_{S_0}(1 - S^*) \tag{4.123}$$

$$\frac{bd}{a} = \frac{(Y_{X/S}C_{S_0} + C_{X_0} + K_i Y_{X/S})(Y_{X/S}C_{S_0} + C_{X_0} + K_S Y_{X/S})}{Y_{X/S}C_{S_0} + C_{X_0}},$$

$$= \frac{Y_{X/S}C_{S_0}(1 + X_0^{\,*} + K_i^{\,*})(1 + X_0^{\,*} + K_S^{\,*})}{(1 + X_0^{\,*})} \tag{4.124}$$

$$\ln\frac{C_X}{C_{X_0}} = \ln\frac{C_{X_0} + Y_{X/S}C_{S_0} - Y_{X/S}C_S}{C_{X_0}},$$

$$= \ln\left(\frac{1 + X_0^* - S^*}{X_0^*}\right) \tag{4.125}$$

$$\ln\left(\frac{a - C_X}{a - C_{X_0}}\right) = \ln S^* \tag{4.126}$$

and

$$a - b - d = -Y_{X/S}C_{S_0} - K_iY_{X/S} - C_{X_0} - K_SY_{X/S}$$

$$= -Y_{X/S}C_{S_0}\,[1 + K_i^* + X_0^* + K_S^*] \tag{4.127}$$

Substituting Equations (4.123) to (4.127) in Equation (4.122), we get

$$-Y_{X/S}C_{S_0}(1 - S^*) + \frac{Y_{X/S}C_{S_0}(1 + K_i^* + X_0^*)(1 + X_0^* + K_S^*)}{(1 + X_0^*)}\left[\ln\left(\frac{1 + X_0^* - S^*}{X_0^*}\right) - \ln S^*\right]$$
$$+Y_{X/S}C_{S_0}\left[1 + K_i^* + X_0^* + K_S^*\right]\ln S^* = \mu_m K_i Y_{X/S}(t - t_0)$$

Dividing the modified Equation (4.122) through out by $Y_{X/S}\,C_{S_0}$, we get Equation (4.128),

$$(S^* - 1) + \frac{(1 + K_i^* + X_0^*)(1 + X_0^* + K_S^*)}{(1 + X_0^*)}\ln\left(\frac{1 + X_0^* - S^*}{X_0^*S^*}\right)$$
$$+\left[1 + K_i^* + X_0^* + K_S^*\right]\ln S^* = \mu_m\frac{K_i}{C_{S_0}}(t - t_0) \tag{4.128}$$

but $\dfrac{K_i}{C_{S_0}} = K_i^*$

$$\left(\frac{S^* - 1}{K_i^*}\right) + \frac{(1 + K_i^* + X_0^*)(1 + X_0^* + K_S^*)}{K_i^*(1 + X_0^*)}\ln\left(\frac{1 + X_0^* - S^*}{X_0^*S^*}\right)$$
$$+\frac{[1 + K_i^* + X_0^* + K_S^*]}{K_i^*}\ln S^* = \mu_m(t - t_0)$$

$$\left(\frac{S^* - 1}{K_i^*}\right) + \frac{(1 + X_0^*)}{K_i^*}\left[\left(1 + \frac{K_i^*}{(1 + X_0^*)}\right)\left(1 + \frac{K_S^*}{(1 + X_0^*)}\right)\right]\ln\left(\frac{1 + X_0^* - S^*}{X_0^*S^*}\right)$$
$$+\frac{[1 + K_i^* + X_0^* + K_S^*]}{K_i^*}\ln S^* = \mu_m(t - t_0) \tag{4.129}$$

Hence, the desired equation is Equation (4.130).

$$\mu_m(t - t_0) = \left(\frac{S^* - 1}{K_i^*}\right) + \left(\frac{K_S^*}{(1 + X_0^*)}\right)\ln\left(\frac{1 + X_0^* - S^*}{X_0^*S^*}\right) + \left(1 + \frac{1 + X_0^* + K_S^*}{K_i^*}\right)\ln\left(\frac{1 + X_0^* - S^*}{X_0^*}\right) \tag{4.130}$$

Equation (4.130) is not the only solution to modify t_r for substrate inhibition. If one considers other inhibition model given in Chapter 1, Equation (4.130) will take different forms.

(g) Physical Factors Causing Non-ideality

Discussion on the non-idealthy caused by physical factors is not only restricted to batch reactor, but one can also apply to other reactors as well.

(i) Flow Patterns and Turbulence

Agitator blade supplies energy for primary motion in the liquid. Energy transferred to the liquid is not totally used for mixing and gas dispersion. The centrifugal acceleration produced by the movement of the fluid causes secondary flow which is composed of radial and axial components. This results in the formation of two large coaxial vortices one above the impeller and another below the impeller plane. The secondary flow is responsible for homogenous mixing in the reactor. If this is not homogeneous, conversion of reactant is not uniform throughout the reaction medium. This will affect reaction time (t_r).

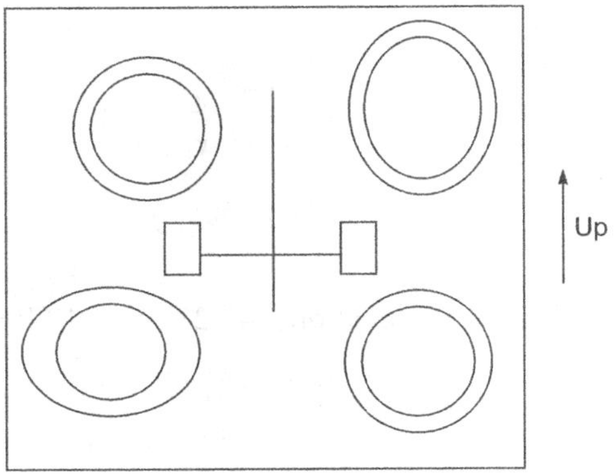

Figure 4.10 Schematic view of secondary flow (based on the concept of Brauer, 1979).

Behind the baffles and around the marginal areas of the reactor, there exist poorly mixed regions. Radial and tangential velocity components are more or less equal around the blade, but the tangential component reduces as the distance increases from the blade.

It appears that gas dispersion is a multi-stage process:

- The agitator collects existing gas bubbles. They coalesce to form gas trail which is attached to the horizontal edges of the turbine blade.
- Gas flows into the vortex cavity and is dispersed.
- Bubbles reach right to the core of the rotating vortex trail and transform into a packed tubular structure.
- Bubbles strip off from the trail by turbulence.

The above process is also followed by the cell present in the reaction. This might cause the less productive cell which influences t_r (reaction time).

Schügerl (1991) has classified gas trails into three categories.

- Small trails – produce low aeration rates
- Medium trails – stick to the agitator blade
- Large trails – cover the whole surface area of the blade caused by high gas throughput rates.

Factors influencing gas-trails

The following factors are important in this context.

- Number of blades: Increase in number of blades reduces gas trails.
- Surfactants: They do not reduce gas trails, but reduce the size of the bubbles.

(ii) Relative Gas Hold-up

This is the function of the properties of the reactants. It is difficult to establish a general relationship for this effect.

As a general rule two relationships appear important for gas holdup (Loiseau *et al.*, 1977).

$$\varepsilon_{Gh} = 0.011 w_{SG}^{0.36} \sigma^{-0.056} \eta^{-0.056} \left(\frac{P_G + P_B}{V_L} \right)^{0.27} \tag{4.131}$$

(Loiseau *et al.*, 1977, for non-foaming systems), where

ε_{Gh} is relative gas hold-up.

w_{SG} is superficial gas velocity (m/s) = q_G/A.

σ is surface tension (Nm^{-1}).

η is dynamic viscosity of liquid (Pa s).

P_G is power input through agitator in aerated system (kg.m^2/s^3).

P_B is power input through gas expansion (kg.m^2/s^3).

$\frac{P_G}{V_L}$ is specific power uptake by agitator (W/m^3).

$\frac{P_B}{V_L}$ is specific power uptake by the system by isothermal gas expansion (W/m^3).

A is cross sectional area of the reactor (m^2).

V_L is reaction volume (m^3).

Correlations for P_G and P_B are given by Schügerl (1990).

$$P_G = 0.83\phi^{0.45}$$

where

$$\phi = \frac{P_R^2 N_R d_R^3}{q_G^{0.56}}$$

and

$$\frac{P_B}{V_L} = \frac{\rho_G q_G RT}{M_G V_L} \ln\left(\frac{p_s}{p_0} \right) \tag{4.132}$$

where

ρ_G being gas density

M_G being molecular mass of gas

R being universal gas constant

T being temperature in K

p_s, p_o being pressure over liquids or aerator

q_G being gas throughput, s^{-1}

N_R being impeller speed, s^{-1}

d_R being impeller diameter, m

P_R being power input through agitater without aeration, kg m^2/s^3

Another Equation (4.133) for foaming system is described by Schügerl (1990).

$$\varepsilon_{Gh} = 0.005 {w_{SG}}^{0.24} \left(\frac{P_G + P_B}{V_L} \right)^{0.57} \tag{4.133}$$

where

$$P_G = 0.69M^{0.45} \text{ for } M < 2 \times 10^3$$
$$P_G = 1.88M^{0.31} \text{ for } M \geq 2 \times 10^3$$

(iii) Recirculation of Old Gas

This is measured by recirculation coefficient (α) defined by Schügerl (1990) in Equation (4.134).

$$\alpha = \frac{q'_G}{q_G} \frac{\text{coalescing, recirculating volumetric gas throughput}}{\text{aeration rate}} \tag{4.134}$$

If $\alpha > 1$, gas trail consists of 'old' recirculated gas.

If $\alpha < 1$, gas is hardly consumed in the reaction. This condition might delay the reaction process. So, the influence on reaction time (t_r) is prominent.

(iv) Mixing

Homogenous reaction is characterized by mixing number ($N_R\ \theta$).

Therefore, the mixing number ($N_R\ \theta$) = mixing time × impeller speed

A few correlations are given in Table 4.4 (Schügerl, 1990).

Table 4.4 *Mixing number correlations in aerated and non-aerated systems*

(1) $N_R\theta \propto N_R^a$, the value of 'a' depends on the type of agitator.
(2) $N_R\theta \propto \left(\frac{D_R}{d_R}\right)^a$, where D_R = reactor diameter and d_R = impelle diameter. 'a' depends on the tye of agitator.
(3) $N_R\theta_{0.95} = 2.23\left(\frac{D_R}{d_R}\right)^{-1.93}\left(\frac{d_R}{h_R}\right)^{-0.47}(\sin\alpha)^{-0.55}{Z_R}^{-0.26}$ (Henzler, 1978) This equation is applicable to pitched blade, propeller turbines and flat blade disc turbine where Z_R = number of impellers, h_R = height of the agitator, a = impeller angle.
(4) $N_R\theta = 6.7\left(\frac{d_R}{D_R}\right)^{-\frac{5}{3}}(Ne_R)^{-1/3}$ This equation applies to turbulent flow with constant mixing quality.
(5) $\frac{\theta_G}{\theta} = 1 + 7.5\eta^{0.27}E_G$, for disc turbine in stirred reactor (Einsele and Finn, 1980) where θ_G = Mixing time in aerated reactor θ = mixing time in non-aerated reactor η = dynamic viscosity (Pa s) E_G = relative gas content This equation is valid over a certain range of D_R, N_R, P_G/V_L, V_L and θ_G.

From the relations of oxygen feed rate, $\frac{P}{V_L}$, W_{SG} and growth rate of the organism are correlated in Equation (4.134a).

Following relations are for semi-quantitative comparison (Schügerl, 1990):

$$Q_{O_2} = \frac{dC_X}{Y_{X/O}} \quad \text{and} \quad Q_{O_2} = \left(\frac{P}{V_L}\right)^b W_{SG}^C \tag{4.134a}$$

where Q_{O_2} is oxygen transfer rate $= k_La(C_{O_{2i}} - C_{\text{critical } O_2})Y_{X/O}$, $C_{O_{2i}}$ is the concentration of oxygen at the interface, $C_{\text{critical } O_2}$ is the critical concentration of oxygen below which organism shows different metabolism, a, b, c are empirical constants depending on the viscosity of reacting fluid and the reactor geometry.

This suggests at different $\frac{P}{V_L}$ and W_{SG} to calculate reaction time (t_r) for a given reactor configuration.

4.3.5 Quantitative Evaluation of Batch Processes

The batch reactor exhibiting concentration time profile is an integral reactor. Integral evaluation method is adopted in shake flask experiments. In general, two techniques are used to evaluate batch processes, viz., integral and differential techniques.

(a) Integral Method

Growth rate is defined as

$$\frac{dC_X}{dt} = \mu C_X \quad \text{or} \quad \mu = \frac{1}{C_X}\frac{dC_X}{dt} \tag{4.135}$$

So the rate equations defining μ must be integrated to determine μ in the time range between t_1 and t_2.

$$\therefore \qquad \mu = \frac{\ln C_{X_2} - \ln C_{X_1}}{t_2 - t_1} \tag{4.136}$$

This means that μ can be expressed as the function of initial reactant concentration by the integral method.

$$\mu = f(C_{S0}) \tag{4.137}$$

(b) Differential Method

This involves differentiation of experimental C_X vs. t plots with the application of graphical, numerical or analytical techniques (Fig. 4.11).

From the slope of the C_X vs. t in batch reactor, μ can be determined from measurements within a certain interval (Δt).

$$\therefore \qquad \mu = \frac{1}{\bar{C}_X}\frac{\Delta C_X}{\Delta t} \tag{4.138}$$

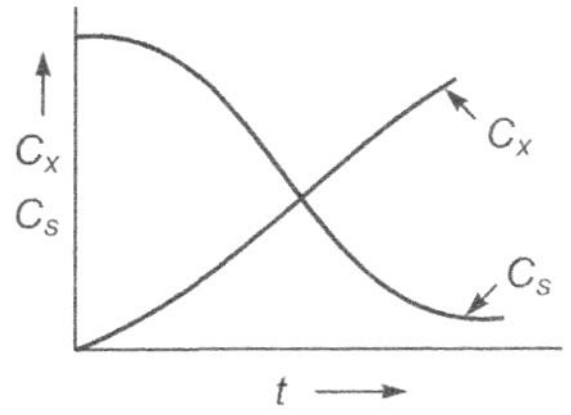

Figure 4.11 Plot of C_X, C_S vs. t.

where $\bar{C}_X$ is mean value of cell concentration in the time interval of Δt.

Similar procedure of integral or differential evaluation is to be followed for the identification of kinetic model. In this regard, two specific cases are highlighted here.

Case 1: Single component with constant catalyst concentration

Specific examples are enzyme reactions and waste water processes with constant sludge input.

The integrated form of the Henri–Michaelis–Menten (H-M-M) equation can be used in the Walker-Schmidt diagram (Walker and Schmidt, 1944) for parameter estimation (Fig. 4.12).

$$r_{\max} t = K_M \log \frac{C_{S_0}}{C_S} + (C_{S_0} - C_S)$$

$$\therefore \quad \frac{(C_{S_0} - C_S)}{t} = -\frac{K_M}{t} \log \frac{C_{S_0}}{C_S} + r_{\max} \tag{4.139}$$

Linearization of this plot is given in Figure 4.12. In the plot, assuming $n = 0$, 1, or 2 for the rate $(r) = k \cdot C^n$. In analogy, such a plot can be used to explain involving cells. However, Equation (4.139) will be modified based on Monod's model.

Case II: Multicomponent systems

For multicomponent systems where both C_s and C_X vary, parameter estimation requires simultaneous solution of both differential equations of r_x and r_S. Integration is possible by elimination with C_s or C_X.

(b) Advantage of Differential Method Over Integral Method

One experimental run at optimal conditions is necessary. However, disadvantage is that the significant period for kinetic model identification is very short.

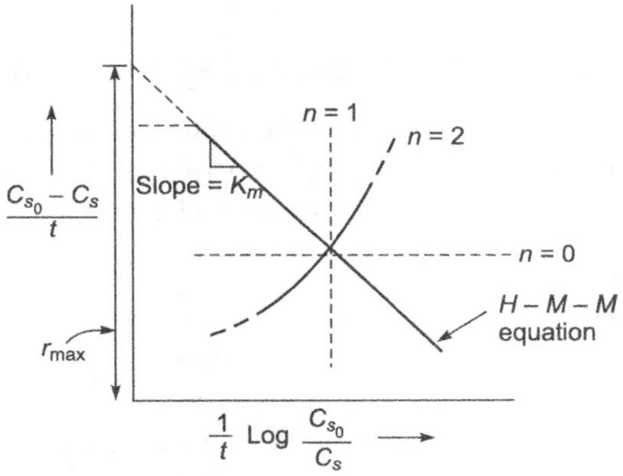

Figure 4.12 Example of Walker–Schmidt diagram.

(c) Application of Batch Reactor Studies

(1) Calculation of $\mu_{\max}$ and K_s

They are not accurate from batch runs as the study always suffers dynamic changes.

(2) Calculation of μ

Preliminary experimental values can be obtained which will allow one to calculate the following

$$\text{doubling time} = t_d = \frac{\ln 2}{\mu} = \frac{1}{\delta}$$

where δ means division rate of cell.

$$\text{Division rate of cells on a number basis} = \frac{d \log_2 C_n}{dt}.$$

If cells are divided N times after time t, total number of cells will be

$$C_n = C_{n_0} \times 2^N \tag{4.140}$$

$$\text{The average division rate} = \delta = \frac{N}{t}$$

$$\therefore \quad N = \log_2 C_n - \log_2 C_{n_0}$$

Since

$$\bar{\delta} = \frac{\log_2 C_n - \log_2 C_{n_0}}{t} \tag{4.141}$$

Division rate is the slope of $\log_2 C_n$ vs. t plot. This is constant during exponential phase of growth. On the other hand, growth rate is calculated from the slope of the plot C_n vs. t.

(3) Characterization of Bio-processes

The nature of production is classified in three ways, viz., growth-associated, non-growth-associated, and mixed growth-associated. Analysis to this problem is partially quantitative. Two approaches are generally used for the characterization of the bioprocesses.

(1) Luedeking and Piret approach

The Luedeking and Piret Equation (2000) contains growth associated term (α) and non-growth associated term (β) in the following form

$$\frac{1}{C_X}\frac{dC_P}{dt} = q_p = \alpha\mu + \beta \tag{4.142}$$

where q_p is specific product formation rate in time^{-1}.
α has no unit whereas β has unit in time^{-1} (h^{-1}).

Caution: We should not compare the values of α and β as they are of different units.

One can consider the magnitude of α and β. If one finds negligible β, the process is called growth associated. Otherwise it is non-growth associated process. Most of the biological processes are neither growth associated nor non-growth associated.

(2) Alternative approach

The analysis of Wang *et al.* (1980) may be applied here to characterize the bioprocesses.

Calculations of $\frac{q_P}{q_{P_{max}}}$ and $\frac{\mu}{\mu_{max}}$ as a function of reaction time lead to the analysis of the processes as per the description given in the Figures 4.13–4.15.

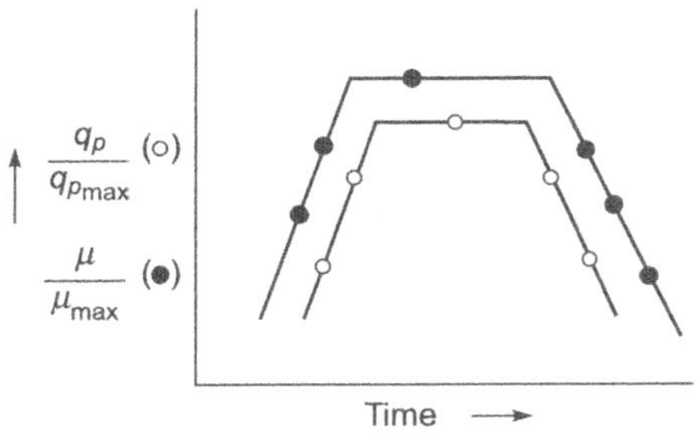

Figure 4.13 Growth associated process.

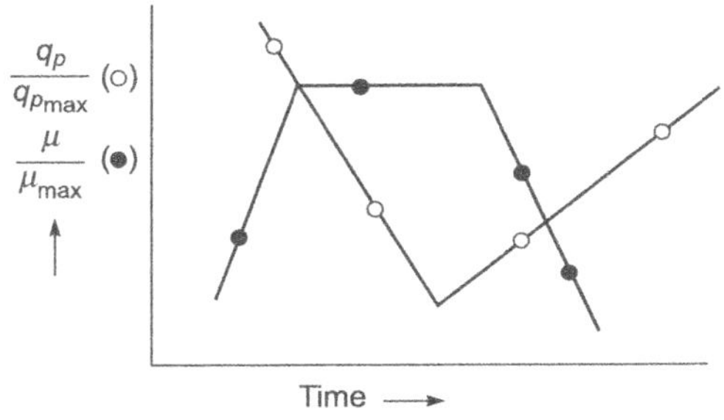

Figure 4.14 Non-growth associated process.

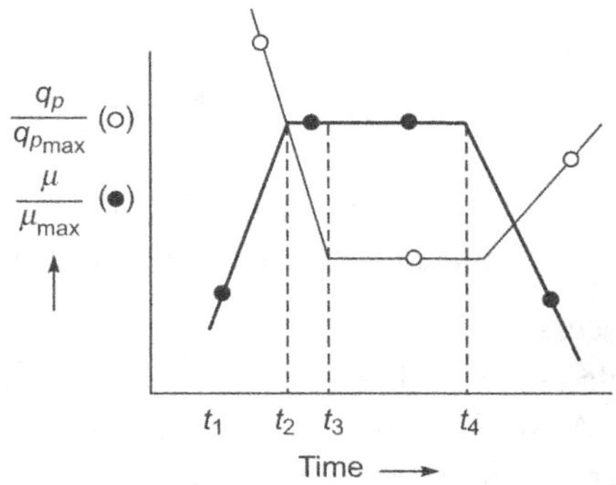

Figure 4.15 Mixed growth associated process. t_1 to t_3 – Non-growth associated, t_3 and t_4 – Growth associated process and later non-growth associated process.

(4) Calculation of lag time (t_{lag})

Let us consider a typical cell mass concentration and time plot (Fig. 4.16). Take slope on the graph. Consider a point where one will get a greatest slope. In this case, let A be such a point. Draw the tangent and extend it to the time axis. It will touch the time axis at point B. $\overline{OB}$ is called lag growth time.

$$\therefore \qquad t_{lag} = \overline{OB}$$

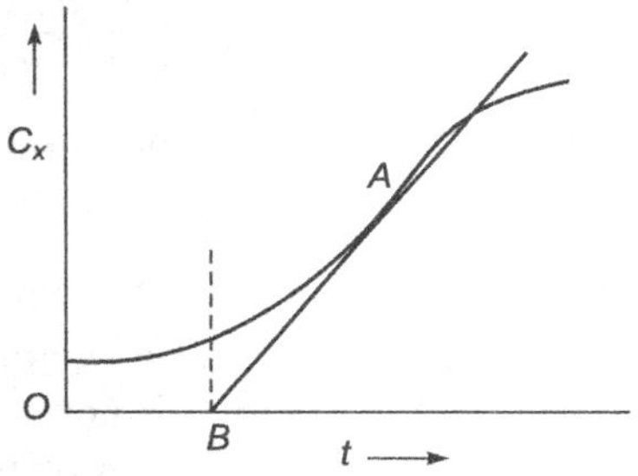

Figure 4.16 C_X vs. t plot.

(5) Studies on various factors

To evaluate critical concentrations of reactant, product, and other bio-chemical components, batch study is more reliable. To find optimal temperature, pH and ionic strength, batch bio-reactor studies are preferred than other mode of reactor operation. For the purpose of replication and repetition batch bioreactor studies are efficient in this context.

(6) Determination of rate constant for a batch data

One can combine the concepts of mass balance, rate law and stoichiometry to derive an expression for rate constant. Initial batch runs will provide information for starting a CFSTBR.

SECTION A: BIOREACTORS FOR SUBMERGED LIQUID FERMENTATION OF MICROBIAL CELLS

Part 2: Continuous Flow Bioreactors

4.4 INTRODUCTION

Ideal continuous flow reactors are of two types—Continuous flow stirred tank bioreactor (CFSTBR) and Continuous plug flow reactor (PFTR).

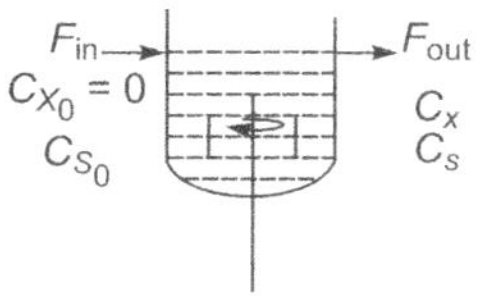

Figure 4.17 CFSTBR.

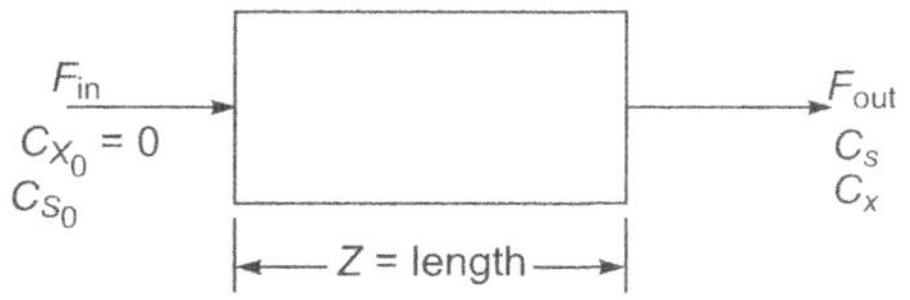

Figure 4.18 PFTR.

Similar type of reactors used for chemical reactions are different having no involvement of cells. For liquid phase chemical reactions, the design equation for the first order reaction is

$$V = \frac{F_O X}{(-r_S)_{\text{exit}}} \tag{4.143}$$

where X is fractional conversion of reactant, and
V is volume necessary to achieve conversion (X).

For constant volume CFSTBR,

$$V = v_o \left(\frac{C_{S_0} - C_S}{-r_S} \right) \tag{4.144}$$

where v_o is feed flow rate or reactant flow rate (volume/time)

$$\therefore \qquad \tau = \frac{V}{v_o} = \frac{C_{S_0} - C_S}{-r_S} \tag{4.145}$$

where τ is space time.

The design equation for tubular reactor is described by Equation (4.146)

$$-r_S = F_{S_0} \frac{dX}{dV} \tag{4.146}$$

Such simple treatment is not possible for a bioreactor even in ideal situation.

4.4.1 Purpose of Continuous Flow Reactors

In batch reactor, environment changes continuously. Growth of organism, product formation, and reactant utilization terminate after a certain time interval. Using continuous flow reactor, reaction is fed with fresh reactant and continuously cells and products are withdrawn from the reactor.

To sustain growth (to maintain the cells in a state of exponential growth phase) and product formation for a longer period of time, continuous flow reactors are used. Continuous culture is an important system to produce desired products under optimal environmental conditions.

Truly speaking, ideal PFTR is not possible to operate in biological system. In this regard, the **discussion in this chapter is restricted to CFSTBR. In CFSTBR, the growth chamber is composed of** sterile medium (reactants). Once growth has been initiated, fresh medium is continuously pumped in from the sterile reactant source. They require a few control devices for the control of parameters, viz., **pH, agitation, dissolved oxygen or redox. CFSTBR can be operated in chemostat (constant chemical** composition in culture) or in turbidostat (constant turbidity in the culture) modes (Pirt, 1975).

Let us consider characteristic features of chemostant and turbidostat.

4.4.2 Differences between Turbidostat and Chemostat Operations

Chemostat	*Turbidostat*
1. Chemical environment in culture is constant.	1. Turbidity in culture is constant.
2. Flow rate of feed is adjusted in terms of growth rate of organisms.	2. Cell concentration in the culture vessel is maintained constant by monitoring the optical density of the culture that controls the feed flow rate.
3. Chemostat is widely used and is simpler to operate.	3. It requires more elaborate arrangement than chemostat as the environment is more dynamic.
4. This is used for all reactants and organisms.	4. Turbidostat is recommended for the following case: • In selecting populations to withstand a desired environmental stress cell concentration is maintained constant. • Selection of variants or mutants with desirable properties. • Bacterial and yeast cell culture with soluble reactants.

4.4.3 Ideal CFSTBR – Chemostat

(a) Possibilities of Chemostat

If one considers the plot of C_X and C_S vs. t (Fig. 4.19), one can easily visualize three different operations designated as 1, 2, and 3.

For (1) Rate of growth of cell < Rate of washout of cell.

- Cell concentration will decrease gradually in the reactor.
- Reactant concentration in the exit stream will approach the feed concentration (C_{S_0}).

For (2) Rate of washout of cell = Rate of growth of cells in the reactor.

- This can be considered as ideal chemostat operation and a situation of steady state. Stability of the steady state needs to be determined.

For (3) Rate of growth of cell in the reactor > Rate of washout of cells.

- Therefore, cell concentration will increase in the reactor.
- Reactant concentration will be depleted in the reactor.

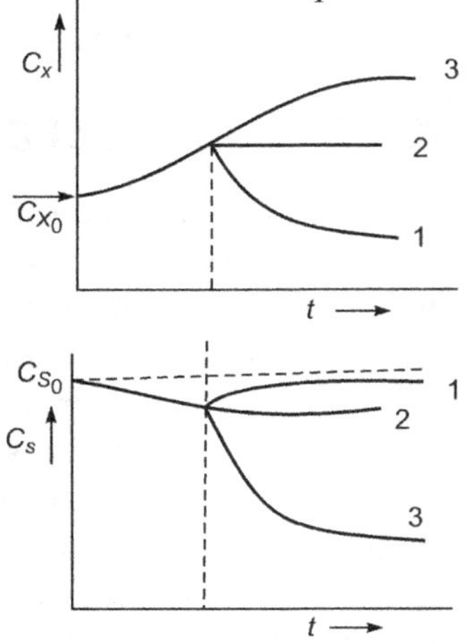

Figure 4.19 C_X and C_S vs t.

In extreme case, situation (3) will mimic batch reactor condition if flow rates to the reactor approach zero. For immediate clarification of the reader, washout of cells means the cells coming out from the reactor through the output stream or exit stream. Washout condition will be dealt in an elaborate way in this chapter.

(b) Mass Balances

For constant volume (V) reaction of CFSTBR, mass balance equations are given below.

For cell

Rate of accumulation of cells = Rate of cells entering – Rate of cells leaving + Net rate of generation of live cells

$$V\frac{dC_X}{dt} = FC_{X_0} - FC_X + (r_g - r_d)V \tag{4.147}$$

where $r_g = \mu_g C_X$ and
$r_d = k_d C_X$ and
$F_{\text{in}} = F_{\text{out}}$

For reactant

Rate of accumulation of reactant = Rate of reactant entering – Rate of reactant leaving + Rate of reactant generation

$$V\frac{dC_S}{dt} = FC_{S_0} - FC_S - r_S V \tag{4.148}$$

where r_s is calculated from the Equation (4.149).

Net rate of reactant consumption = Rate of reactant consumed by cells + Rate of reactant consumed to form product + Rate of reactant consumed for maintenance of cell

$$-r_s = Y_{S/X} r_g + Y_{S/P} r_p + mC_x \tag{4.149}$$

Considering negligible maintenance, Equation (4.149) is modified to Equation (4.150).

$$-r_s = Y_{S/X}\, r_g + Y_{S/P}\, r_p \tag{4.150}$$

Substituting

$$r_g = \mu_g C_X$$

$$r_P = q_P C_X$$

$$Y_{S/X} = \frac{1}{Y_{X/S}}$$

$$Y_{S/P} = \frac{1}{Y_{P/S}}$$

Consider maximum value of cell yield $\left(Y^m_{X/S}\right)$ independent of time and q_P the specific rate of extra cellular product formation.

Equation (4.149) is further modified to Equation (4.151).

$$-r_s = \frac{\mu_g C_X}{\left(Y^m_{X/S}\right)} + \frac{q_P C_X}{Y_{P/S}} \tag{4.151}$$

From Equations (4.148) and (4.151), we get

$$V\frac{dC_S}{dt} = FC_{S_0} - FC_S - \frac{\mu_g C_X}{\left(Y^m_{X/S}\right)}V - \frac{q_P C_X}{Y_{P/S}}V \tag{4.152}$$

Assuming

- Negligible extracellular product formation
- Negligible death rate ($r_d = 0$)
- Sterile feed ($C_{X_0} = 0$)
- Growth rate follows Monod's equation
- Steady state CFSTBR, $\frac{dC_X}{dt} = 0$ and $\frac{dC_S}{dt} = 0$

Equation (4.147) becomes

$$V\frac{dC_X}{dt} = FC_{X_0} - FC_X + (r_g - r_d)V \tag{4.153}$$

$$\overset{0}{\cancel{\frac{dC_X}{dt}}} = \frac{1}{V}\overset{0}{\cancel{FC_{X_0}}} - \frac{F}{V}C_X + (r_g - \overset{0}{\cancel{r_d}})$$

$$-\frac{F}{V}C_x + r_g = 0$$

$$\cancel{-}\frac{F}{V}\cancel{C_X} = \cancel{-}\mu_g\cancel{C_X}$$

$$\mu_g = \frac{F}{V} = D$$

$$\boxed{\mu_g = D} \tag{4.154}$$

D is called dilution rate and has the unit of $(\text{time})^{-1}$.

$$D = \frac{F}{V}$$

For a constant volume reaction $D \equiv f(F)$ where F = flow rate of feed reactant.

If F increases, D will increase. This will produce more cells in exit stream which means that cells are diluted in the reaction volume in the reactor. Similarly if F is low, D is reduced.

For batch reactor, $F = 0$, so $D = 0$.

Again

$$\mu = D = \frac{F}{V} = \frac{1}{\tau} = 1/\,\textbf{(Mean residence time)}$$

$$\therefore \qquad \tau = \frac{V}{F}$$

From Equation (4.147), considering steady state operation if C_{X_0} is non-zero then

$$FC_{X_0} - FC_X + Vr_g = 0$$

$$\frac{V}{F} = \left(\frac{C_X - C_{X_0}}{r_g}\right) = \tau$$

$$\text{Plot} \frac{1}{r_X} \text{vs.} C_X, \text{ for } r_d = 0 \text{ and } r_g = r_x \tag{4.155}$$

Therefore, the required residence time = $(C_X - C_{X_0})$ times $(1/r_g)$. This is equal to the area of the rectangle (ABCD) of width $(C_X - C_{X_0})$ and height $1/r_x$ in the plot of Figure 4.20.

This plot can help to compare how effective is the reactor. This means that shorter the residence time in reaching a cell concentration, the more effective is the reactor.

This deduction refers to one limiting reactant.

So $$\mu_g = D = \frac{\mu_m C'_S}{K_S + C'_S} \tag{4.156}$$

where C'_S is the steady state concentration of limiting reactant.

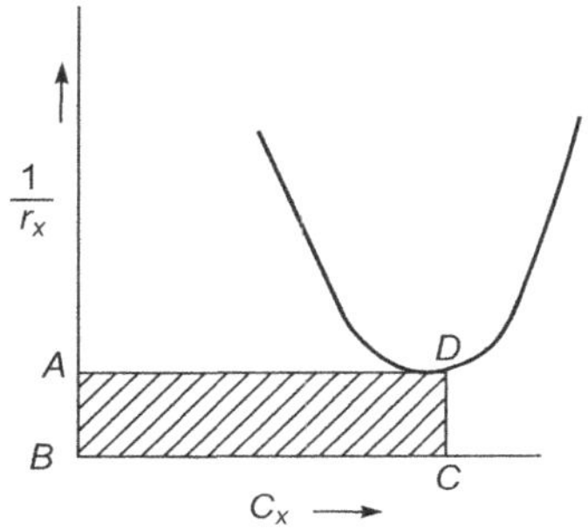

Figure 4.20 Plot of $\frac{1}{r_x}$.

(c) Significance of Equation (4.156)

1. If feed flow rate is very high, D will set a value greater than μ_m, the cell cannot divide quickly enough to maintain itself in the reactor. Hence, cells will come out in the exit stream. This is called wash out phenomenon of cells.
2. Calculation of steady state reactant concentration

   In Equation (4.156) C_S is replaced by C'_S (steady state reactant concentration). Then

$$D = \frac{\mu_m C'_S}{K_S + C'_S}$$

$$C'_S = \frac{K_S D}{\mu_m - D} \tag{4.157}$$

3. Evaluation of Monod's kinetic parameters

   From Equation (4.156), we can make double reciprocal expression,

$$\frac{1}{\mu} = \frac{K_S}{\mu_m}\frac{1}{C'_S} + \frac{1}{\mu_m}$$

At various feed flow rates, steady state reactant concentration is measured which will be used to test various kinetic models.

As $\mu = D$ at steady state, microorganisms following Monod's equation, the plot of $\frac{1}{D}$ vs. $\frac{1}{C'_S}$ will give a straight line plot (Fig. 4.21).

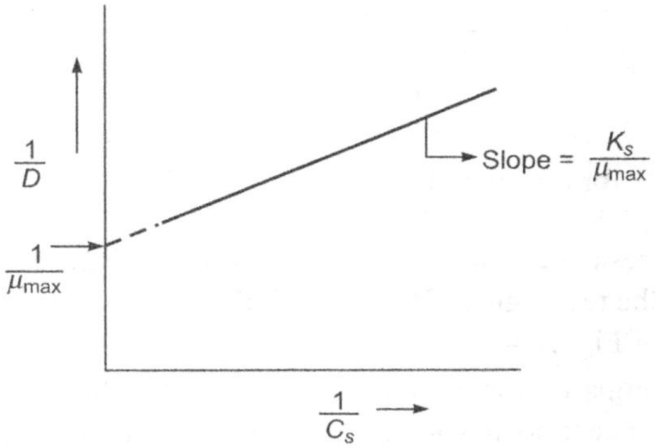

Figure 4.21 Plot of $\frac{1}{D}$ vs. $\frac{1}{C'_S}$.

4.4.4 Application of Single Stage CFSTBR

(a) Estimation of Kinetic Parameters: For Simple Kinetics

The plot of C'_X and C'_S vs. mean residence time (τ) is given in (Fig. 4.22).

The performance equation can be rearranged as

$$\frac{1}{C'_S} = \frac{\mu_m}{K_S}\tau - \frac{1}{K_S}, \text{ where } \tau = \frac{1}{D}$$

A plot of $\frac{1}{C'_S}$ vs. τ will give μ_m and K_S.

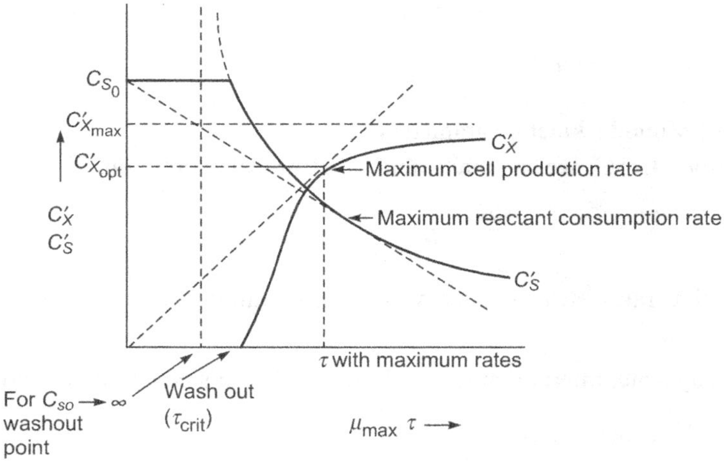

Figure 4.22 Plot of C'_S and C'_X vs. mean residence time (τ).

(b) Calculation of Steady State Cell Concentration

From the mass balance for a single limiting reactant, following considerations are made.

- Steady State, i.e., $\frac{dC_S}{dt} = 0$
- Negligible extra-cellular product formation, i.e, $q_P \approx 0$
- Absence of endogenous metabolism
- Constant $Y_{X|S}$
- $k_d = 0$
- Constant volume of reaction, $F_{\text{in}} = F_{\text{out}}$

Then Equation (4.152) becomes

$$\frac{F}{V}(C_{S_0} - C'_S) = \frac{\mu_g C_X}{Y^m_{X/S}} \tag{4.158}$$

Since Equation (4.154) is $\mu_g = D$, further at steady state using Equations (4.154) and (4.158) gives

$$C_X = Y^m_{X/S}(C_{S_0} - C'_S) \tag{4.159}$$

Using Equations (4.157) and (4.159)

$$C'_X = \text{Steady state cell concentration} = Y^m_{X/S}\left(C_{S_0} - \frac{K_S D}{\mu_m - D}\right) \tag{4.160}$$

4.4.5 Rate of Output of Cell Mass in a Chemostat

Two different solutions exist for this calculation.

They are:

- Analytical solution
- Graphical solution

(a) Analytical Solution

We define rate of output of cell mass/unit volume of culture

$$R = DC_X \tag{4.161}$$

At steady state, $C'_X = Y^m_{X/S}\left(C_{S_0} - \frac{K_S D}{\mu_m - D}\right)$, from Equation (4.160),

$$\therefore \quad R_o = DY^m_{X/S}\left(C_{S_0} - \frac{K_s D}{\mu_m - D}\right) \tag{4.162}$$

Let us visualize the situation in the plot (Fig. 4.23).

Maximum production rate is observed at a particular value of D which is called $D_{\max}$. $D_{\max}$ is obtained by differentiating the rate of output of cell mass with respect to D and equating it to zero.

$$\frac{dR_o}{dD} \approx 0 \tag{4.163}$$

$$\Rightarrow \quad \frac{dR_o}{dD} = Y^m_{X/S}\left(C_{S_0} - \frac{K_S D}{\mu_m - D}\right) + DY^m_{X/S}\left(-\frac{(\mu_m - D)K_S - K_S D(-1)}{(\mu_m - D)^2}\right) = 0$$

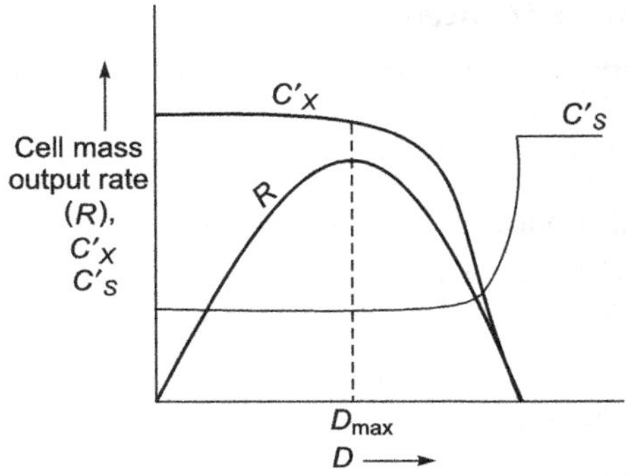

Figure 4.23 Cell mass output rate (R), C'_X, and C'_S as a function of D.

Simplification of the above equation takes the form

$$(\mu_m - D)^2 C_{S_0} - (\mu_m - D) K_S D - \mu_m - DK_S = 0$$

or

$$(C_{S_0} + K_S)D^2 - 2\mu_m(C_{S_0} + K_S)D + \mu_m{}^2 C_{S_0} = 0 \tag{4.164}$$

This is a quadratic equation in D with 2 roots

$$D = \frac{2\mu_m (C_{s_0} + K_s) \pm \sqrt{4\mu_m{}^2 (C_{s_0} + K_s)^2 - 4(C_{s_0} + K_s)\mu_m{}^2 C_{s_0}}}{2(C_{s_0} + K_s)}$$

Considering the positive value of D,

$$\therefore \quad D_{\max} = \frac{\mu_m (C_{s_0} + K_s) \pm \mu_m \sqrt{(C_{s_0} + K_s)^2 - (C_{s_0} + K_s) C_{s_0}}}{(C_{s_0} + K_s)}$$

$$\boxed{D_{\max} = \mu_m \left(1 + \sqrt{\frac{K_s}{K_s + C_{s_0}}}\right)}$$

We know that $\mu_m = D$ and D cannot be greater than μ_m when there is no washout. So, considering the negative value

$$\boxed{D_{\max} = \mu_m \left(1 - \sqrt{\frac{K_s}{K_s + C_{s_0}}}\right)}$$

So that C'_X at this $D_{\max}$ is

$$C'_{X_m} = Y^m_{X/S} \left(C_{S_0} - \frac{K_S D_{\max}}{\mu_m - D_{\max}} \right) \tag{4.166}$$

$$= Y^m_{X/S} \left(\frac{C_{S_0}(\mu_m - D_{\max}) - K_S D_{\max}}{\mu_m - D_{\max}} \right)$$

$$= Y^m_{X/S} \left(\frac{C_{S_0}\mu_m - D_{\max}\left(C_{S_0} - K_S\right)}{\mu_m - D_{\max}} \right)$$

Assuming $C_{S_0} >> K_S$

Then
$$C'_{X_m} = Y^m_{X/S}\left(\frac{C_{S_0}\mu_m - D_{\max}C_{S_0}}{\mu_m - D_{\max}}\right)$$

$$\boxed{C'_{X_m} = Y^m_{X/S}C_{S_0}} \quad (4.167)$$

Alternative solution to this problem is

$$C'_{X_m} = Y^m_{X/S}\left(C_{S_0} - \frac{K_S\mu_m\left(1+\sqrt{\dfrac{K_S}{K_S+C_{S_0}}}\right)}{\mu_m - \mu_m\left(1+\sqrt{\dfrac{K_S}{K_S+C_{S_0}}}\right)}\right)$$

$$= Y^m_{X/S}\left(C_{S_0} - \frac{K_s\left(1+\sqrt{\dfrac{K_S}{K_S+C_{S_0}}}\right)}{1-1-\sqrt{\dfrac{K_S}{K_S+C_{S_0}}}}\right)$$

$$= Y^m_{X/S}\left(C_{S_0} + \frac{K_S\left(1+\sqrt{\dfrac{K_S}{K_S+C_{S_0}}}\right)}{\sqrt{\dfrac{K_S}{K_S+C_{S_0}}}}\right)$$

$$= Y^m_{X/S}\left(C_{S_0} + \frac{K_S\left(\sqrt{K_S+C_{S_0}}+\sqrt{K_S}\right)}{\sqrt{K_S}}\right)$$

$$= Y^m_{X/S}\left(C_{S_0} + \left(\sqrt{K_S(C_{S_0}+K_S)} + K_S\right)\right)$$

$$= Y^m_{X/S}\left(\left(C_{S_0}+K_S\right) + \left(\sqrt{(C_{S_0}+K_S)K_S}\right)\right)$$

$$= Y^m_{X/S}(C_{S_0}+K_S)\left(1+\sqrt{\frac{K_S}{(C_{S_0}+K_S)}}\right)$$

(Assuming $C_{S_0} >> K_S$)

$$C'_{X_m} = Y^m_{X/S}C_{S_0}\text{, which is Equation (4.167).}$$

(b) Graphical Solution

CFSTBR operates with sterile feed ($C_{X_0} = 0$). The Equation (4.161) is

$$R = DC_x$$

or in terms of mean residence time (τ).

$$R = \frac{C_X}{\tau} \tag{4.168}$$

Let us have a plot of C_X vs. τ and C_S vs. τ (Fig. 4.24).

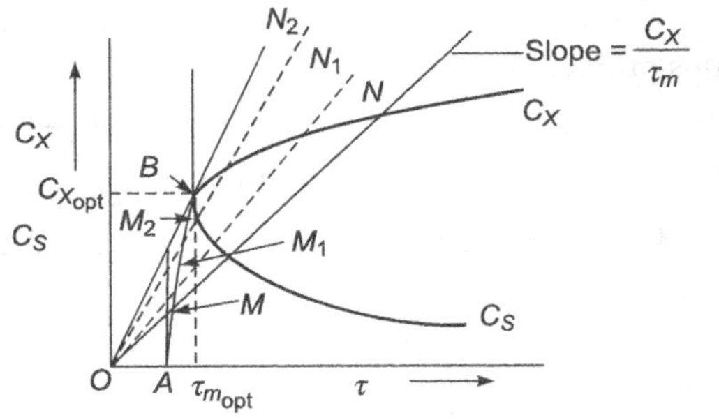

Figure 4.24 C_X and C_S vs. τ

Let us draw a line $\overline{ON}$ through the origin of C_X (or C_S) and τ axes. $\overline{ON}$ touches C_X plot at two points, namely, M and N. The slope of $\overline{ON}$ line is $\frac{C_X}{\tau}$.

The output rate or productivity at points M and N is same.

This is justified in the following way

$$\text{Productivity at } M = \frac{C_{X_1}}{\tau_1}$$

$$\text{Productivity at } N = \frac{C_{X_2}}{\tau_2}$$

$$\frac{C_{X_1}}{\tau_1} = \frac{C_{X_2}}{\tau_2} \text{ is possible.}$$

If
$$C_{X_1} < C_{X_2},$$
$$\tau_1 < \tau_2$$

From the plot of Figure 4.24 we can say that this is true. However, to operate at N is much safer than at M because a slight change in τ at M can drastically change C_x at M. It is likely that the process will reach a value at A having no cell mass. If one rotates ON line towards left, the condition of M at new position (M_1) might improve, but it will be unstable (refer plot 4.24 line ON_1).

To achieve maximum productivity (R_o) is the slope of a line which touches C_x plot at one and only one point, i.e., ON_2 which touches at point B on C_x vs. τ plot.

Let us see the solutions from graphical and analytical approaches.

The cell productivity at steady state with sterile feed

$$R_o = \frac{C_X}{\tau} = D\,C_x$$

$$r_g = \frac{\mu_m C_S C_X}{C_S + K_S} \quad \text{(using Monod's equation)}$$

For maximum productivity, $\frac{dr_g}{dC_X} = 0,$

and substituting, $C_S = C_{S_0} - \frac{C_X}{Y^m_{X/S}}$

The solution will give $C_{X\,\text{optimum}}$.

Assuming $C_{S_0} >> K_S$

Maximum productivity = $\frac{C_{X_{opt}}}{\tau_{opt}}$

4.4.6 Mean Residence Time (τ)

This is defined as the time elapsed by the components with reaction medium in the reactor. In a biological system, major components are cell and reactants (more precisely the limiting reactant).

Mathematically, it is defined in a simpler form

$$\tau = \frac{V}{F} \tag{4.169}$$

For a CFSTBR with sterile feed, single limiting reactant, no endogenous metabolism and negligible death rate constant, τ is expressed by Equation (4.170).

$$\tau = \frac{C_X - C_{X_0}}{r_g} \tag{4.170}$$

where C_{X_0} – initial cell mass concentration
C_X – final cell mass concentration
r_g – rate of growth of cell

For $C_{X_0} = 0$, Equation (4.170) becomes $\tau = \frac{C_X}{r_g}$ (4.171)

Assuming Monod's kinetics, Equation (4.171) is rearranged and differentiated with respect to C_x

$$\frac{C_X}{\tau} = r_g = \frac{\mu_m C_S C_X}{C_S + K_S} \tag{4.172}$$

Equation (4.172) is also called cell productivity.

Upon differentiation of Equation (4.172)

$$\frac{dr_g}{dC_X} = \frac{d}{dC_X}\left(\frac{\mu_m C_S C_X}{C_S + K_S}\right) \tag{4.173}$$

$$\frac{d}{dC_X}\left(\frac{\mu_m C_S C_X}{C_S + K_S}\right) = 0$$

Substituting $$C_S = C_{S_0} - \frac{C_X}{Y^m_{X/S}}$$

$$\frac{d}{dC_X}\left(\frac{\mu_m\left(C_{S_0}-\frac{C_X}{Y^m_{X/S}}\right)C_X}{K_s+\left(C_{S_0}-\frac{C_X}{Y^m_{X/S}}\right)}\right)=0$$

$$\frac{d}{dC_X}\left(\frac{\mu_m\left(C_{S_0}Y^m_{X/S}-C_X\right)C_X}{K_SY^m_{X/S}+C_{S_0}Y^m_{X/S}-C_X}\right)=0$$

$$\frac{(\mu_mC_{S_0}Y^m_{X/S}-2\mu_mC_X)(K_SY^m_{X/S}+C_{S_0}Y^m_{X/S}-C_X)+\mu_mC_X(C_{S_0}Y^m_{X/S}-C_X)}{(K_SY^m_{X/S}+C_{S_0}Y^m_{X/S}-C_X)^2}=0$$

$$(\mu_mC_{S_0}Y^m_{X/S}-2\mu_mC_X)(K_SY^m_{X/S}+C_{S_0}Y^m_{X/S}-C_X)+\mu_mC_X(C_{S_0}Y^m_{X/S}-C_X)=0$$

$$K_SC_{S_0}(Y^m_{X/S})^2-2K_SY^m_{X/S}C_X+(C_{S_0}Y^m_{X/S})^2-2C_{S_0}Y^m_{X/S}C_X-C_{S_0}Y^m_{X/S}C_X+2C_X{}^2+C_{S_0}Y^m_{X/S}C_X-C_X{}^2=0$$

$$C_X{}^2-2Y^m_{X/S}(C_{S_0}+K_S)C_X+C_{S_0}Y^m_{X/S}{}^2(C_{S_0}+K_S)=0$$

$$\therefore \qquad C_{X_{\text{opt}}}=\frac{2Y^m_{X/S}(C_{S_0}+K_S)\pm\sqrt{4Y^m_{X/S}{}^2(C_{S_0}+K_S)^2-4C_{S_0}Y^m_{X/S}{}^2(C_{S_0}+K_S)}}{2}$$

$$=Y^m_{X/S}\left((C_{S_0}+K_S)\pm(C_{S_0}+K_S)\sqrt{1-\frac{C_{S_0}}{(C_{S_0}+K_S)}}\right)$$

Hence, $$C_{X_{\text{opt}}}=Y^m_{X/S}(C_{S_0}+K_S)\left(1\pm\sqrt{\frac{K_S}{(C_{S_0}+K_S)}}\right)$$

Consider $$\alpha=\sqrt{\frac{(C_{S_0}+K_S)}{K_S}}$$

Then $$C_{X_{\text{opt}}}=Y^m_{X/S}(C_{S_0}+K_S)\left(1\pm\frac{1}{\alpha}\right)$$

Considering negative sign in the expression of $C_{X_{\text{opt}}}$,

$$C_{X_{\text{opt}}}=Y^m_{X/S}(C_{S_0}+K_S)\left(1-\frac{1}{\alpha}\right)$$

From α values, $\alpha^2=\dfrac{(C_{S_0}+K_S)}{K_S}$

$$C_{S_0}=K_S(\alpha^2-1);\quad K_S=\frac{C_{S_0}}{(\alpha^2-1)}$$

$$\therefore \qquad C_{X_{\text{opt}}}=Y^m_{X/S}C_{S_0}\left(1+\frac{1}{(\alpha^2-1)}\right)\left(1-\frac{1}{\alpha}\right)=Y^m_{X/S}C_{S_0}\left(\frac{\alpha^2-1+1}{(\alpha^2-1)}\right)\left(\frac{\alpha-1}{\alpha}\right)$$

$$=Y^m_{X/S}C_{S_0}\left(\frac{\alpha}{1+\alpha}\right)$$

$$C_{S_{\text{opt}}} = C_{S_0} - \frac{C_{X_{\text{opt}}}}{Y_{X/S}^m} = C_{S_0} - \frac{Y_{X/S}^m C_{S_0}\left(\dfrac{\alpha}{1+\alpha}\right)}{Y_{X/S}^m}$$

$$C_{S_{\text{opt}}} = C_{S_0}\left(1 - \frac{\alpha}{1+\alpha}\right)$$

So $$C_{S_{\text{opt}}} = \frac{C_{S_0}}{1+\alpha}$$

τ_{opt} from $$\frac{C_{X_{\text{opt}}}}{\tau_{\text{opt}}} = \frac{\mu_m C_{S_{\text{opt}}} C_{X_{\text{opt}}}}{K_S + C_{S_{\text{opt}}}}$$

$\therefore$ $$\tau_{\text{opt}} = \frac{K_S + C_{S_{\text{opt}}}}{\mu_m C_{S_{\text{opt}}}}$$

$$\tau_{\text{opt}} = \frac{K_S + \dfrac{C_{S_0}}{1+\alpha}}{\mu_m \dfrac{C_{S_0}}{1+\alpha}}$$

$$= \frac{K_S(1+\alpha) + C_{S_0}}{\mu_m C_{S_0}}$$

$$= \frac{\dfrac{C_{S_0}}{\alpha^2 - 1}(1+\alpha) + C_{S_0}}{\mu_m C_{S_0}}$$

$$= \frac{\dfrac{1}{\alpha - 1} + 1}{\mu_m}$$

$$= \frac{\alpha}{\mu_m(\alpha - 1)}$$

The best productivity from chemostat, when Monod's equation is applicable, is

$$D_{\text{opt}} C_{X_{\text{opt}}} = \frac{C_{X_{\text{opt}}}}{\tau_{\text{opt}}}$$

$$= \frac{Y_{X/S}^m C_{S_0}\left(\dfrac{\alpha}{1+\alpha}\right)}{\dfrac{\alpha}{\mu_m(\alpha - 1)}}$$

$$= Y_{X/S}^m C_{S_0} \mu_m \left(\frac{\alpha - 1}{\alpha + 1}\right)$$

Application of mean residence time

Following are the applications of τ.

(i) $\tau = \frac{1}{\mu} = \frac{1}{D}$ (4.174)

(ii) Steady state reactant concentration

$$C_S' = \frac{K_S}{\tau\mu_m - 1}, \text{ for } \tau\mu_m > 1 \quad (4.175)$$

If $\tau\mu_m < 1$, growth rate of cells is less than rate of washout of cells. More and more cells will come out from the reactor in the exit stream.

(iii) Steady state: $C_X' = \left(Y_{X/S}^{m}\right)\left(C_{S_0} - \frac{K_s}{\tau\mu_m - 1}\right)$ (4.176)

(iv) To set optimal operation of CFSTBR

$$\tau_{\text{optimal}} = \frac{1}{\mu_m}\left(\frac{\alpha}{\alpha - 1}\right) \quad (4.177)$$

We have gained sufficient information on batch bioreactors and CFSTBR analysis on their design. It is interesting to see some distinctive features of both the bioreactors in the next section.

4.4.7 Comparison of Batch Bioreactor and Single Stage CFSTBR

Batch bioreactor	***Single stage CFSTBR***
(i) For an ideal batch reactor $\frac{1}{r_g}$ vs. C_X plot is 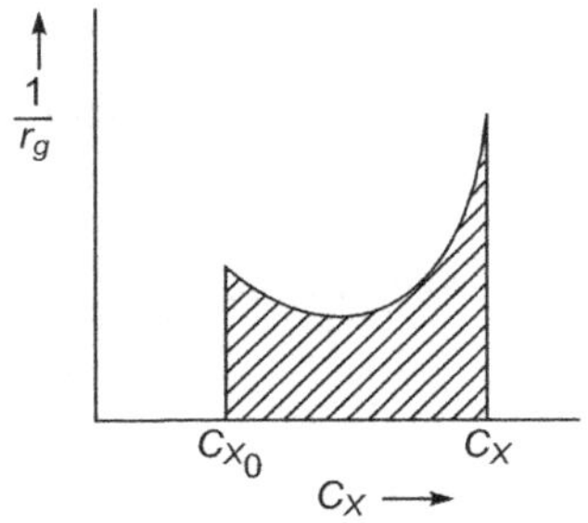 Shaded area $= \tau_b - \tau_{\text{lag}}$ τ_b = batch time τ_{lag} = lag growth phase time	(i) For ideal CFSTBR $\frac{1}{r_g}$ vs. C_X plot is 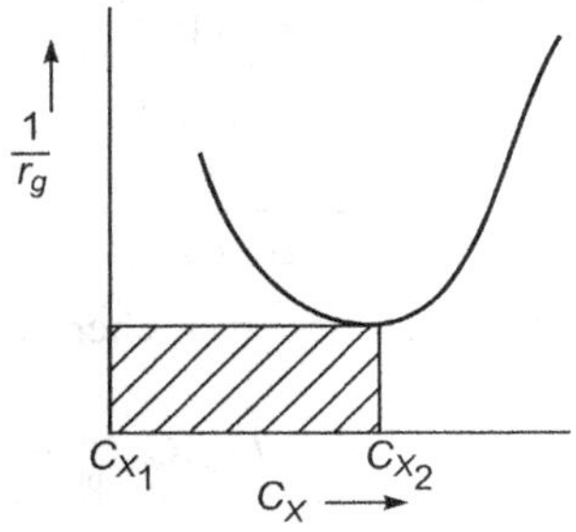 Residence time $= \frac{C_{X_2} - C_{X_1}}{r_g}$

Contd.

Batch bioreactor	*Single stage CFSTBR*
(ii) If the product is synthesized at stationary phase and cell concentration achieved only at stationary phase, batch reactor is a better choice than CFSTBR as the residence time reached is lesser than CFSTBR in this case.	(ii) CFSTBR is most productive when $\frac{1}{r_g}$ is minimum to achieve a particular C_x.
(iii) Cell mass productivity compared under similar conditions, i.e., $C_{S_0} >> K_s$. The problem is simplified by assuming $\mu \rightarrow \mu_{max}$ for exponential phase. Time for exponential phase of growth of cell, $$t_{log\ phase} = \frac{1}{\mu_m} \ln\left(\frac{C_X}{C_{X_0}}\right)$$ We have seen the total batch time except $t_{log\ phase}$ are considered one time $= t_a$ $$\therefore \quad t_{batch} = \frac{1}{\mu_m} \ln\left(\frac{C_X}{C_{X_0}}\right) + t_a$$ C_S is same for batch and chemostat. $\therefore$ The volumetric productivity $$= \frac{\left(Y^m_{X/S}\right)C_{S_0}}{t_{batch}}$$	(iii) Maximum productivity occurs at μ_m and $\mu_m\left(Y^m_{X/S}\right)C_{S_0}$

$$\therefore \quad \frac{\text{Productivity of chemostat}}{\text{Productivity of batch}}$$

$$= \frac{\mu_m\left(Y^m_{X/S}\right)C_{S_0}}{\dfrac{\left(Y^m_{X/S}\right)C_{S_0}}{\dfrac{1}{\mu_m}\ln\left(\dfrac{C_X}{C_{X_0}}\right) + t_a}}$$

$$= \ln\left(\frac{C_X}{C_{X_0}}\right) + \mu_m t_a$$

If we know C_{X_0} and C_X in the form of a relation $C_X = pC_{X_0}$ then the productivity of chemostat is at least $\ln p$ times the productivity of the batch.

4.4.8 Washout Condition

To know the effect of increasing dilution rate, one can combine cell mass balance and the definition of cell growth rate. Assuming $r_d = 0$ with sterile feed conditions,

$$\frac{dC_X}{dt} = 0 - DC_X + \mu C_X$$

or,

$$\frac{dC_X}{dt} = (\mu - D)C_X$$

If $D > \mu$, $\frac{dC_X}{dt}$ will be negative and C_x will continue to decrease till all cells will washout.

$\therefore$ $$C_X = 0$$

The D is, $$D_{\text{washout}} = \frac{\mu_m C_{S_0}}{K_S + C_{S_0}} \tag{4.178}$$

This indicates that there is no cell in the reactor and hence there is no reaction. The input reactant condition (C_{S_0}) is the reactant level in the reactor and in the exit stream.

So far we have considered only the involvement of live cells and reactant in a sterile feed mode of operation of single stage CFSTBR. Following factors will be analyzed on single stage CFSTBR operation:

- Endogenous metabolism
- Maintenance of cells
- Wall growth
- Inhibition
- Product synthesis

The effects of endogenous metabolism, maintenance of cells, wall growth, and inhibition phenomena are discussed under non-ideal behavior in CFSTBR.

Product formation alone is considered in the following discussion.

Mass balance Equations are (4.179) to (4.181).

For cell growth: $\frac{dC_X}{dt} = (\mu - D)C_X$, where $\frac{dC_X}{dt}$ at steady state. (4.179)

For reactant utilization: $$\frac{dC_S}{dt} = D(C_{S_0} - C_S) - \frac{1}{(Y^m_{X/S})}\mu C_X - \frac{1}{(Y^m_{P/S})}(\alpha\mu C_X + \beta C_X) \tag{4.180}$$

where $\frac{dC_S}{dt} = 0$ at steady state.

For product formation: $\frac{dC_P}{dt} = \alpha\mu C_X + \beta C_X - DC_P$, where $\frac{dC_P}{dt} = 0$ at steady state. (4.181)

We need to calculate steady state concentrations of C_X, C_P, and C_S, i.e., C'_X, C'_P, and C'_S, respectively. From reactant mass balance

$$D(C_{S_0} - C'_S) - \frac{1}{(Y^m_{X/S})}\mu C'_X - \frac{1}{(Y^m_{P/S})}(\alpha\mu C'_X + \beta C'_X) = 0 \tag{4.182}$$

As $C'_S = \dfrac{DK_S}{\mu_m - D}$ for single limiting reactant with no endogenous metabolism.

$$D\left(C_{S_0} - \frac{DK_S}{\mu_m - D}\right) - \frac{1}{(Y^m_{X/S})}\mu C'_X - \frac{1}{(Y^m_{P/S})}(\alpha\mu C'_X + \beta C'_X) = 0 \tag{4.183}$$

$$\therefore \quad C'_X\left(\frac{\mu}{(Y^m_{X/S})} + \frac{(\alpha\mu + \beta)}{(Y^m_{P/S})}\right) = D\left(C_{S_0} - \frac{DK_S}{\mu_m - D}\right) \tag{4.184}$$

$$C'_X = \frac{D\left(C_{S_0} - \dfrac{DK_S}{\mu_m - D}\right)}{\left(\dfrac{\mu}{(Y^m_{X/S})} + \dfrac{(\alpha\mu + \beta)}{(Y^m_{P/S})}\right)}$$

$$C'_X = \frac{\left(C_{S_0} - \dfrac{DK_S}{\mu_m - D}\right)(Y^m_{X/S})(Y^m_{P/S})}{\left(Y^m_{P/S} + Y^m_{X/S}\left(\alpha + \dfrac{\beta}{D}\right)\right)} \tag{4.185}$$

$$C'_P = \alpha C'_X + \frac{\beta}{D} C'_X \tag{4.186}$$

Product productivity $= DC_p$. (4.187)

Example 4.4

The production of an enzyme is desired in a chemostat operation. The system operates at steady state. A few related parameters are given below:

$$D = 0.5\ \text{h}^{-1}$$
$$\mu_m = 0.8\ \text{h}^{-1}$$
$$K_S = 0.05\ \text{kg/m}^3$$
$$Y^m_{X/S} = 0.4$$
$$C_{S_0} = 20\ \text{kg/m}^3$$

Calculate reactant and cell concentrations at steady state.

Solution: From Equation (4.157), we know

$$C_S' = \frac{K_S D}{\mu_m - D} = \frac{K_S}{\tau\mu_m - 1} = \frac{0.05}{\frac{0.8}{0.5} - 1}$$

$$= 0.0833 \text{ kg/m}^3$$

$$C_X' = Y_{X/S}^m \left(C_{S_0} - \frac{K_S D}{\mu_m - D} \right)$$

$$= 0.4\left(20 - \frac{0.05 \times 0.5}{0.8 - 0.5} \right)$$

$$= 7.966 \text{ kg/m}^3$$

Example 4.5

We cultivated an organism in a 5 m^3 CFSTBR – chemostat. The flow rate of sterile feed is 1m^3/h. The single limiting reactant level was 10 kg/m^3. Other parameters for the organism are: μ_m = 0.30 h^{-1}, K_s = 0.6 kg/m^3, $Y_{X|S}^m$ = **0.4. What will be the steady** state cell concentration of the outlet stream?

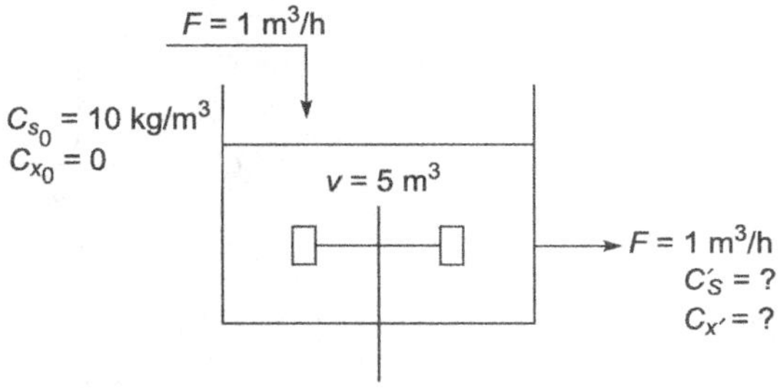

Figure for Example 4.5.

Solution

From the data:

Other available data for the organism:

$$\mu_m = 0.30 \text{ h}^{-1}$$
$$K_S = 0.6 \text{ kg/m}^3$$
$$Y_{X|S}^m = 0.4$$

The material balance for the cell is

$$F\overset{0}{C_{X_0}} - FC_X' + Vr_g = V\frac{d\overset{0}{C_X}}{dt} \tag{4.188}$$

At steady state,

$$\therefore \qquad FC_X' = Vr_g$$

We can assume the Monod's kinetics for growth:

$$r_g = \frac{\mu_m C_S C_X}{K_S + C_S}$$

$\therefore$ Equation (4.188) can be written as follows.

$$FC_X' = V\frac{\mu_m C_S' C_X'}{K_S + C_S'}$$

$$\therefore \qquad F = V\frac{\mu_m C_S'}{K_S + C_S'}$$

Substituting the values,

$$1 = 5 \times \frac{0.3\,C'_S}{0.6 + C'_S}$$

$\therefore$ $$C'_S = 1.2 \text{ kg/m}^3$$

Also we know,

$$C'_X = Y^m_{X/S}\,(C_{S_0} - C'_S), \text{ therefore } C'_X = 3.52 \text{ kg/m}^3.$$

You can assume any other growth model without inhibition, endogenous respiration, maintenance, etc.

> You can also practice this problem with the logistic model for growth
>
> $$r_g = \mu_m \left(1 - \exp\left(-\frac{C_S}{K_S}\right)\right) C_X$$

4.5 PLUG FLOW TUBULAR REACTOR (PFTR)

PFTR is an alternative to mixed operation (batch, CFSTBR). Ideally no mixing occurs in this reactor. There is no interaction with neighboring fluid elements. No concentration or temperature gradient is considered in this reactor. There is only one spatial coordinate along the length of the reactor 'Z'.

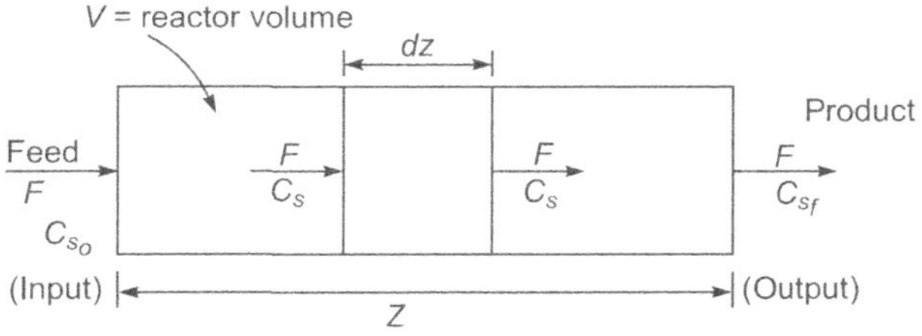

Figure 4.25 Schematic of ideal PFTR operation

Fluid in a PFTR flows at a constant velocity. Therefore, all parts of liquid have identical residence time. As reaction proceeds, reactant concentration and product concentration vary along the length of the reactor.

So the general mass balance for reaction component 'i' in a volume element Adz (A = cross sectional area of the reactor) is

$$\frac{\partial C_i}{\partial t} = -u_z \frac{\partial C_i}{\partial z} + q_i\,(C) \tag{4.189}$$

where, u_z is linear velocity of fluid flow in z direction (m^3/hr).

q_i is the volumetric flow rate

C_i is the concentration of ith chemical component (kg/m^3)

Transient mass balance for PFTR can be solved by the method of Aris and Amundson (1973).

$C_i(z = 0, t = 0)$ must be known for the reactor. If $C_i(z = 0, t)$ is constant in time, a steady state profile $C_i(z)$ will develop for the reactor. If the PFTR is studied at steady state, it operates in a continuous mode. A continuous injection of cells at $z = 0$ is not practical. Inoculation is generally done by placing the PFTR in the exit stream of the chemostat (CFSTBR). The other way is to recycle the cells from the outlet to $z = 0$.

As u_z is constant, the steady state model becomes

$$\frac{d\mathbf{c}}{dt'} = q(\mathbf{c}) \tag{4.190}$$

where,

$$\mathbf{c}(t' = 0)\ \mathbf{c}_0$$

$$t' = \frac{z}{u_z}$$

Assuming maintenance free Monod's equation and inserting appropriate initial values of the variables, following mass balances are obtained, i.e., Equations (4.191) to (4.193).

$$\frac{dC_X}{dt} = \mu C_X,\ C_X(t = 0) = C_{X_0} \tag{4.191}$$

$$\frac{dC_S}{dt} = -Y^m_{X/S}\,\mu C_X\,,\ C_S(t = 0) = C_{S_0} \tag{4.192}$$

$$\frac{dC_P}{dt} = Y^m_{P/S}\mu C_X\,,\ C_P(t = 0) = C_{P_0} \tag{4.193}$$

From the simplest Monod's kinetics

$$C_S = C_{S_0} - Y^m_{X/S}(C_X - C_{X_0}) \tag{4.194}$$

$$C_P = C_{P_0} + Y^m_{X/P}(C_X - C_{X_0}) \tag{4.195}$$

$$\therefore \quad \frac{dC_X}{dt} = \frac{\mu_m C_S}{K_S + C_S} C_X$$

$$Y^m_{X/S} = \frac{C_X - C_{X_0}}{C_{S_0} - C_S}$$

$$\int_{C_{X_0}}^{C_X} \frac{(K_S + C_S)}{\mu_m C_S C_X} dC_X = \int_0^t dt$$

or

$$\int_{C_{X_0}}^{C_X} \frac{(K_S + C_S)}{C_S C_X} dC_X = \mu_m t = \mu_m \frac{V}{F} = \mu_m \tau$$

The solution yields

$$\mu_m \tau = \left(1 + \frac{K_S Y^m_{X/S}}{C_{X_0} + C_{S_0} Y^m_{X/S}}\right) \ln\left(\frac{C_X}{C_{X_0}}\right) + \frac{K_S Y^m_{X/S}}{C_{X_0} + C_{S_0} Y^m_{X/S}} \ln\left(\frac{C_{S_0}}{C_S}\right) \quad (4.196)$$

where τ is the residence time.

This is the design equation of an ideal PFTR.

Example 4.6

The inlet and outlet conditions of the PFTR are

$$C_{S_0} = 0.015 \text{ kg/m}^3$$
$$C_S = 0.003 \text{ kg/m}^3$$
$$C_{X_0} = 0.004 \text{ kg/m}^3$$
$$C_X = 0.052 \text{ kg/m}^3$$

μ_m for the organism $= 1.3 \text{ h}^{-1}$

$K_s = 0.004 \text{ kg/m}^3$ and $Y^m_{X|S} = 0.2$. Calculate the residence time in the reactor.

Solution: From Equation (4.196), substituting the available data

$$1.3\tau = \left[1 + \frac{0.2 \times 0.004}{0.004 + 0.015 \times 0.2}\right] \ln\left(\frac{0.052}{0.004}\right) + \frac{0.004 \times 0.02}{0.004 + 0.015 \times 0.2} \ln\left(\frac{0.015}{0.003}\right)$$

$$= 2.54 \text{ h}$$

4.5.1 Comparison of Ideal Mixed Flow (Batch and CFSTBR) and Plug Flow Tubular Reactors

Semi-continuous reactor is not included for this kind of comparison as the volume in semi-continuous mode of reactor operation is not fully utilized.

1. Kinetic Characteristics

Batch and PFTR are the same.

2. Reactant Concentration

Progressive decrease in the reactant concentration with time of reaction is observed for both batch and PFTR.

For CFSTBR, concentration is same all through for a single well-mixed reactor.

3. Benefits

A particular reactor design or mode of operation depends on the kinetics of the reaction.

For Zero-order Reactions:

No difference exists between these reactors in terms of overall conversion rate.

For 1st Order Reactions:

Batch and PFTR: The rate of reaction decreases as the concentration of reactant decreases. In the beginning of batch reaction or at the inlet of the PFTR, reactant level is high. Subsequently the reaction rate falls during the course of the reaction.

CFSTBR: Reactant entering in the reactor is immediately diluted to the final or steady state concentration. The rate of reaction is comparatively low for the entire reactor. Lower reactant concentration and lower product concentration is achieved in a CFSTBR than the batch and the PFTR of the same volume.

For Michaelis–Menten Reaction Kinetics

The explanation is same like first order rate kinetics.

4. Preference

In practice, batch reactor is preferred over PFTR due to the problem indicated earlier in Chapter 3 and in this chapter.

However, batch reactor has large down time between batches. This needs to be minimized for efficient batch reactor strategy.

5. Auto Catalytic Reactions

Catalyst is produced by the reaction. Therefore, the volumetric rate of reaction increases as the conversion proceeds. Volumetric reaction rate increases till reactant conversion is low.

Batch Culture

Rate of reaction is low as a few cells are present in the reactor initially.

CFSTBR

Reactant immediately after entering into reactor is exposed to high active cells or catalyst present. Rate of conversion is high in CFSTBR chemostat than batch and PFTR. For most biological reactions, CFSTBR theoretically offers more advantage than other modes of reactors.

6. Overall Practice

Despite productivity benefits associated with CFSTBR most of the process is carried out batch wise due to

- lower risk of contamination.
- less problems of control failure.
- risk with genetically unstable cells.
- flexible production system.

7. Commonality

If number of stages in CFSTBR increases, conversion approaches those of mixed batch reactor or PFTR. In the same stage, concentration is uniform, but there is a drop in the next stage of CFSTBR and so on (Fig. 4.26).

We have seen the limitations of a single stage single input CFSTBR. To improve the performance of such reactor a few alternative approaches, viz., recycle reactors and combination of reactors, are considered in the following sections.

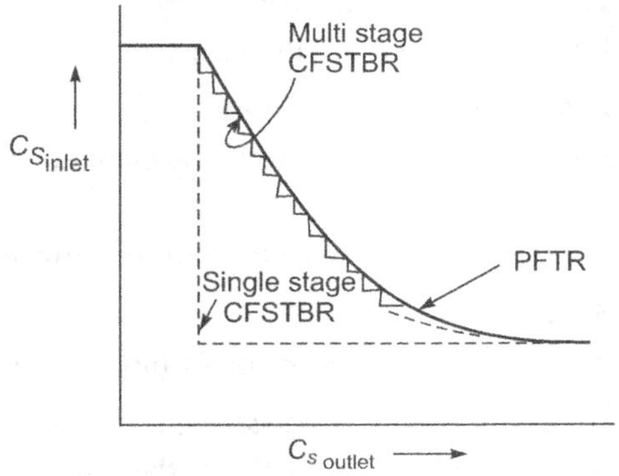

Figure 4.26 Comparison of PFTR and CFSTBR (single and multi stage).

4.6 RECYCLE BIOREACTORS

4.6.1 Objectives

- Cell recycle increases cell density in the reactor, for example: animal cell reactor and waste water treatment reactors.
- Increased cell density assures high rate of reactant utilization and higher product concentration.
- Greater flexibility in operation is obtained by increasing the level of critical dilution rate.
- Reduction in inhibitory components happens in the reaction phase.

4.6.2 Recycling in Biological Reactions

It is done in two ways. Recycling of cell-free culture filtrate containing a number of components is done to stimulate the reactions. For example, penicillin acylase synthesis by *E.coli* is enhanced when part of the cell free culture filtrate is recycled into the reactor (Babu and Panda, 1990). Alternatively, concentrated cells are recycled into the reactor. We deal with the reactor with cells recycled into the reaction phase.

Single stage chemostat–CFSTBR and PFTR are considered here as the examples. These two reactors have common feature like the reaction takes place between reactants entering and products leaving the reactor. A few schematic are given below to understand the process in detail (Figs. 4.27–4.29).

(a) Single stage chemostat-CFSTBR

(b) Perfusion system

Design Criteria

The design criterion of this category of the reactors is the recycle ratio.

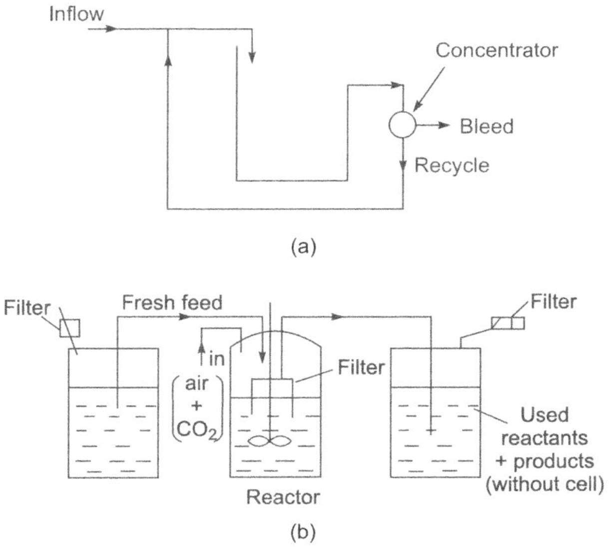

Figure 4.27 **(a) and (b)** Perfusion system to retain cells in the reactor (applicable to animal cell cultivation).

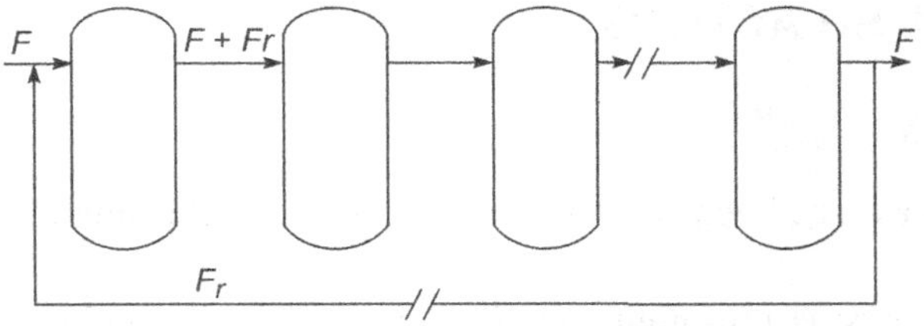

Figure 4.28 Schematic of multistage CFSTBR recycle reactor.

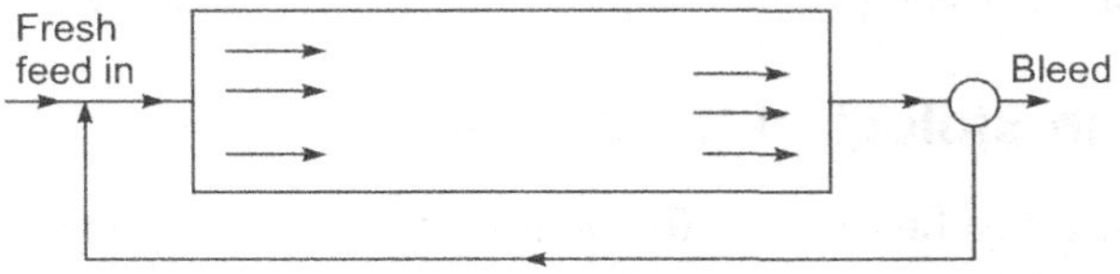

Figure 4.29 Schematic of PFTR recycle reactor.

4.6.3 Analysis of Recycle Reactors

(a) CFSTBR Recycle

We recall the schematic of single stage CFSTBR recycle reactor (Figs. 4.27a and 4.30).

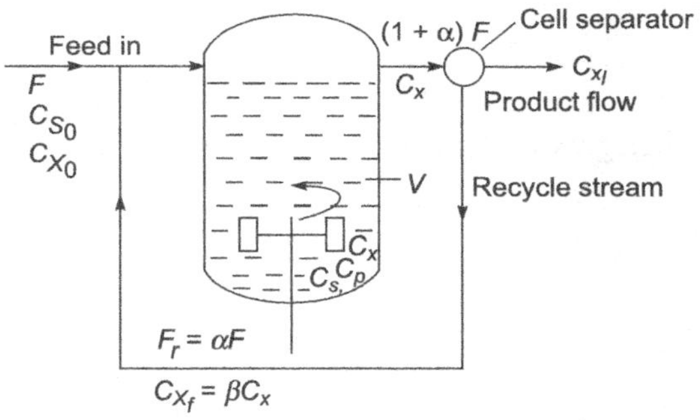

Figure 4.30 Single stage CFSTBR with recycle.

Material Balance on Cell around the Reactor

$$\text{Accumulation} = \text{Input} - \text{Output} + \text{Generation}$$

$$V\frac{dC_X}{dt} = FC_{X_0} + \alpha\beta FC_X - (1+\alpha)FC_X + V\mu_{\text{net}}C_X \tag{4.197}$$

where α is the recycle ratio

β is concentration factor, i.e., cell mass concentration in recycle stream/cell mass concentration in the reactor effluent

C_{X_0} is cell mass concentration in the feed.

C_X is cell mass concentration in the effluent.

C_{X_1} is cell mass concentration in the effluent from the cell separator.

Material Balance on Growth Limiting Reactant around the Reactor

$$V\frac{dC_S}{dt} = FC_{S_0} + \alpha FC_S - (1+\alpha)\, FC_S - V\frac{\mu_g C_X}{Y^m_{X/S}} \tag{4.198}$$

From cell mass balance around the reactor at steady state and with sterile feed $C_{X_0} = 0$, Equation (4.197) is modified to Equation (4.199).

$$\alpha\beta FC_X - (1+\alpha)FC_X + V\mu_{net}\, C_X = 0,\ C_X \neq 0$$

$$\therefore \qquad \mu_{net} = (1+\alpha-\alpha\beta)D \tag{4.199}$$

That means the chemostat with recycle can be operated at a higher dilution rate (D) than the similar capacity chemostat (Fig. 4.31).

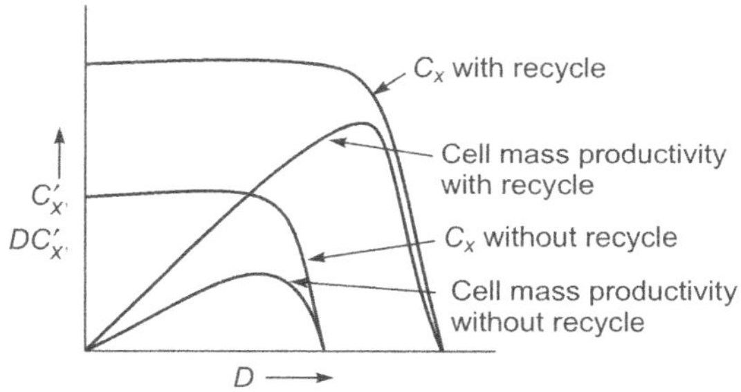

Figure 4.31 Comparison of cell recycle and without recycle CFSTBR.

From reactant mass balance at steady state,

$$FC_{S_0} + \alpha FC'_S - (1+\alpha)FC'_S - V\frac{\mu_g C'_X}{Y^m_{X/S}} = 0 \tag{4.200}$$

or,
$$FC_{S_0} + \alpha \cancel{F}C'_S - FC'_S - \alpha \cancel{F}C'_S - V\frac{\mu_g C'_X}{Y^m_{X/S}} = 0$$

or,
$$\frac{F}{V}(C_{S_0} - C'_S)\frac{Y^m_{X/S}}{\mu_g} = C'_X \tag{4.201}$$

C'_X = steady state cell concentration in CFSTBR recycle reactor (also outflow cell concentration)

$$C'_X = \frac{DY^m_{X/S}}{\mu_g}(C_{S_0} - C'_S) \tag{4.202}$$

where C'_S is the steady state reactant concentration (it is same as the reactant concentration in the outlet of the reactor).

Assuming no endogenous metabolism,

$$\mu_{net} = \mu_g \tag{4.203}$$

$$\therefore \quad C'_X = \frac{\not{D} Y^m_{X/S}}{\not{D}(1+\alpha-\alpha\beta)}(C_{S_0} - C'_S) \tag{4.204}$$

$$= \frac{Y^m_{X|S}}{(1+\alpha-\alpha\beta)}(C_{S_0} - C'_S)$$

Steady state reactant concentration in the reactor is derived from μ_{net} (Equation 4.199) expression and Monod's equation without endogenous metabolism.

$$(1 + \alpha - \alpha\beta)\, D = \frac{\mu_m C'_S}{K_s + C'_S} \tag{4.205}$$

$$\Rightarrow \quad (1 + \alpha - \alpha\beta)\, D(K_S + C'_S) = \mu_m C'_S$$

$$\Rightarrow \quad K_S D(1 + \alpha - \alpha\beta) = (\mu_m - D(1 + \alpha - \alpha\beta))\, C'_S$$

$$\therefore \quad C'_S = \frac{K_S D(1+\alpha-\alpha\beta)}{\mu_m - D(1+\alpha-\alpha\beta)} \tag{4.206}$$

$$\therefore \quad C'_X = \frac{Y^m_{X/S}}{(1+\alpha-\alpha\beta)}\left[C_{S_0} - \frac{K_S D(1+\alpha-\alpha\beta)}{\mu_m - D(1+\alpha-\alpha\beta)}\right] \tag{4.207}$$

In terms of τ, C'_S and C'_X are expressed by Equations (4.208) and (4.209), respectively.

$$C'_S = \frac{K_s(1+\alpha-\alpha\beta)}{\mu_m \tau - (1+\alpha-\alpha\beta)} \tag{4.208}$$

$$C'_X = \frac{Y^m_{X|S}}{(1+\alpha-\alpha\beta)}\left[C_{S_0} - \frac{K_S(1+\alpha-\alpha\beta)}{\mu_m \tau - (1+\alpha-\alpha\beta)}\right] \tag{4.209}$$

Significance of α and β

From Equation (4.199), let us consider different values of α and β.

(a) If $\beta = 1$, $C_{X_r} = C_X$, i.e., total cell recycle.
Cell separator is 100% efficient to reject cells in exit stream of separator.

$$\therefore \quad \mu = D$$

(b) If β is reduced, i.e., less than 1, there is a big jump in C'_X. Cell mass productivity will increase in this case.

(c) If $\alpha \neq 1$, $\beta = 1$

$$C'_S = \frac{K_S D}{\mu_m - D} \tag{4.210}$$

$$C'_X = Y^m_{X/S}\left[C_{S_0} - \frac{K_s D}{\mu_m - D}\right] \tag{4.211}$$

(d) If $\alpha = 1$ and $\beta = 1$, $\mu = D$.
C'_S and C'_X values are same as given in case (c).

(e) If $\beta << 1$ and $\alpha = 1$, $\alpha\beta$ is negligible. $\mu = 2D$

(f) If the separator is not highly efficient, β is $<< 1$ and $\alpha \neq 1$, $\alpha\beta$ is negligible. Then

$$C_S' = \frac{K_s D(1+\alpha)}{\mu_m - D(1+\alpha)} \tag{4.212}$$

$$C_X' = \frac{Y_{X/S}^m}{(1+\alpha)}\left[C_{S_0} - \frac{K_s D(1+\alpha)}{\mu_m - D(1+\alpha)}\right] \tag{4.213}$$

Steady state concentrations depend only on α = recycle ratio.

Let us consider a few modification of the example analyzed so far.

Case I

In the previous example, if we consider overall mass balance for cells and reactant including the reactor and settler, the cell balance is expressed by Equation (4.214).

$$FC_{X_0} + \alpha\beta FC_X - \alpha\beta FC_X - FC_{X_1} + V\mu_{\text{net}}C_X = V\frac{dC_X}{dt} \tag{4.214}$$

Reactant balance is expressed by the Equation (4.215).

In the concentrated cell slurry, reactant concentration is negligible.

$$V\frac{dC_S}{dt} = FC_{S_0} - FC_S - V\frac{\mu_g C_x}{Y_{X/S}^m} \tag{4.215}$$

Similarly calculate C_X' and C_S' from cell mass balance at steady state and with sterile feed ($C_{X_0} = 0$).

$$\alpha\beta FC_X - \alpha\beta FC_X - FC_{X_1} + V\mu_{\text{net}}C_X = 0$$

$$\frac{F}{V}C_{X_1} = \mu_{\text{net}}C_X$$

$$DC_{X_1} = \mu_{\text{net}}C_X$$

$$C_{X_1} = \frac{D(1+\alpha-\alpha\beta)C_X}{D}$$

$$C_{X_1} = (1+\alpha-\alpha\beta)C_X \tag{4.216}$$

Case II

Product formation is considered in this case (Fig. 4.32).

Mass Balance Around the Reactor

Reactant Balance

Change in reactant concentration in the reactor is given by Equation (4.217).

$$V\frac{dC_S}{dt} = FC_{S_0} + \alpha FC_S - (1+\alpha)FC_S - \mu\frac{C_X}{Y_{X/S}^m}V - q_p\frac{C_X}{Y_{P/S}^m}V \tag{4.217}$$

where q_p is specific cellular product formation rate.

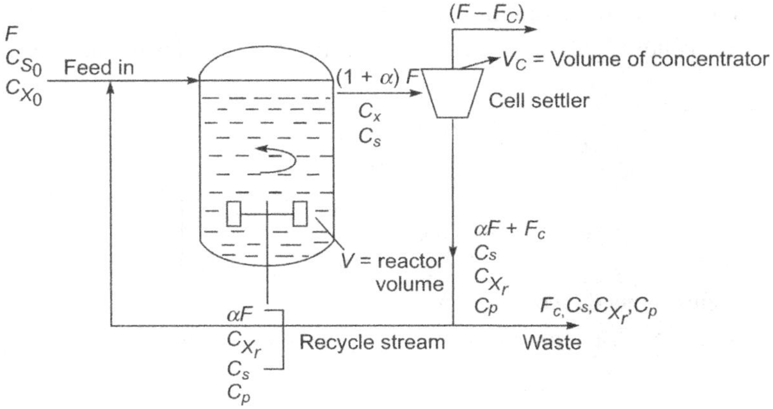

Figure 4.32 CFSTBR with recycle.

Cell Mass Balance (Sterile Feed, $C_{X_o} = 0$)

Equation (4.218) gives the change in cell concentration in the reactor

$$V\frac{dC_X}{dt} = \alpha FC_{X_r} - (1 + \alpha)FC_X + \mu VC_X \tag{4.218}$$

Product Balance

Change in product concentration in the reactor is expressed by Equation (4.219).

$$V\frac{dC_P}{dt} = \alpha FC_P - (1 + \alpha)FC_P + q_P C_X V \tag{4.219}$$

Let us examine the mass balance around concentrator.

Reactant Balance

Change in reactant concentration in the concentrator is given in Equation (4.220).

$$V_c\frac{dC_S}{dt} = (1 + \alpha)FC_S - (F - F_c)C_S - (\alpha F + F_c)C_S \tag{4.220}$$

where V_c is volume of concentrator.

Cell Balance

This is expressed by Equation (4.221).

$$V_c\frac{dC_X}{dt} = (1 + \alpha)FC_X - (F_c + \alpha F)\, C_{X_r} \tag{4.221}$$

No cells in the overflow from the concentrator is assumed.

Product Balance

$$V_c\frac{dC_P}{dt} = (1 + \alpha)FC_P - (\alpha F + F_c)C_P - (F - F_c)C_P \tag{4.222}$$

From Equation (4.217) at steady state, we get

$$FC_{S_0} + \alpha FC_S - (1+\alpha)FC_S - \mu \frac{C_X}{Y_{X/S}^m} V - r_p \frac{C_X}{Y_{P/S}^m} V = 0$$

$$\Rightarrow \quad D\frac{(C_{S_0} - C_S)}{C_X} = \frac{\mu}{Y_{X/S}^m} + \frac{q_P}{Y_{P/S}^m} \tag{4.223}$$

From Equation (4.218) at steady state, we get

$$\alpha FC_{X_r} - (1+\alpha)FC_X + \mu VC_X = 0 \tag{4.224}$$

$$\alpha F\frac{C_{X_r}}{C_X} - (1+\alpha)F + \mu V = 0$$

$$\alpha F\beta - \alpha F - F + \mu V = 0 \left(\because \beta = \frac{C_{X_r}}{C_X}\right)$$

$$\therefore \quad \mu V = F + \alpha F - \alpha F\beta$$

$$\therefore \quad \mu = D(1 + \alpha - \alpha\beta) \tag{4.225}$$

Applying steady state change in product concentration in the reactor [from Equation (4.219)]

$$\alpha FC_P - (1+\alpha)FC_P + q_P C_X V = 0$$

$$\Rightarrow \quad \alpha FC_P - FC_P + \alpha FC_P + q_P C_X V = 0$$

$$-FC_P + q_P C_X V = 0$$

$$\frac{F}{V}C_P = q_P C_X$$

$$DC_P = q_P C_X \tag{4.226}$$

Applying steady state to cell mass balance around concentrator

$$(1+\alpha)F = (F_c + \alpha F)\beta$$

$$\therefore \quad (1 + \alpha - \alpha\beta) = \frac{F_c}{F}\beta$$

Again $\quad D(1 + \alpha - \alpha\beta) = \mu,$

So,

$$\mu = \frac{F_c}{F}\beta D = \frac{F_c}{F}\beta . \frac{1}{\tau_n}$$

$$\mu = \frac{1}{\tau_m} = \frac{1}{\text{mean residence time}}$$

$$\therefore \quad \frac{1}{\tau_m} = \frac{F_c}{F}\beta . \frac{1}{\tau_n}$$

$$\therefore \quad \tau_m = \frac{F}{F_c}\frac{C_X}{C_{X_r}}\tau_n \tag{4.227}$$

where τ_n is hydraulic retention time.

(b) CFSTBR Series with Recycling to the First Stage of the Reactor

Moser (1985) has indicated the complex situation of this system by calculating critical dilution rate.

$$D_{\text{crit}} = \left(\frac{1-\alpha}{\left(1-\alpha\beta^{\frac{1}{N}}\right)^{N}} \right) \mu_m \frac{C_{S_o}}{K_S + C_{S_o}} \tag{4.228}$$

where N is the number of stages.

(c) Continuous Flow-PFTR with Recycling

Let us consider the schematic of this PFTR (Fig. 4.33).

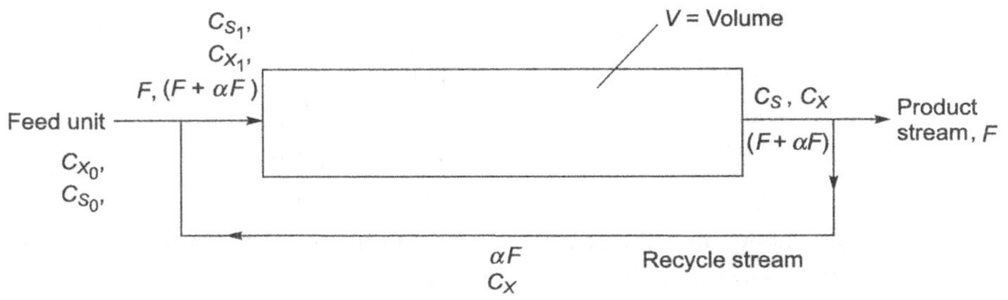

Figure 4.33 PFTR with recycle.

Cell Balance

$$\text{Reactor alone: } V\frac{dC_X}{dt} = (F+\alpha F)(C_{X_1} - C_X) + r_g V \tag{4.229}$$

$$\text{Overall: } V\frac{dC_X}{dt} = FC_{X_0} - FC_X + r_g V$$

$$r_g = \mu_g C_X \tag{4.230}$$

Reactant Balance

$$\text{Reactor alone: } V\frac{dC_S}{dt} = (F+\alpha F)(C_{S_1} - C_{S_1}) - r_g \frac{V}{Y^m_{X/S}} \tag{4.231}$$

$$\text{Overall: } V\frac{dC_S}{dt} = FC_{S_0} - FC_S - r_g \frac{V}{Y^m_{X/S}} \tag{4.232}$$

For PFTR with recycling and without concentration of cell mass system, Levenspiel (1980) proposed the following equation.

$$\mu\tau = (\alpha+1)\left[\frac{K_S}{C_{S_0}} \ln \frac{C_{S_0} + \alpha C_S}{\alpha C_S} + \ln \frac{1+\alpha}{\alpha} \right] \tag{4.233}$$

For processing at some fixed conversion an optimal recycle ratio is to be calculated which is given in detail by Levenspiel (1980). He has suggested $\frac{\partial C_X}{\partial \alpha} = 0$ in the equation of recycle reactors.

$$\frac{K_S}{C_{S_0}} \ln \frac{C_{S_o} + \alpha C_S}{\alpha C_S} + \ln \frac{1+\alpha}{\alpha} = \frac{1+\alpha}{\alpha} \frac{K_S}{C_{S_0} + \alpha C_S} + \frac{1}{\alpha} \tag{4.234}$$

This equation can be solved by trial and error method. Optimal α can be found to be the function of $\left(\frac{K_S}{C_{S_o}}\right)$ and $\left(\frac{C_S}{C_{S_o}}\right)$.

Grieves *et al.*, (1964) have suggested PFTR with recycling of concentrated cell mass.

$$-\mu_m \tau = (\alpha + 1)\left[\frac{(1+\alpha)K_S}{\frac{\alpha\beta(C_{S_0} - C_S)}{1+\alpha-\alpha\beta} - C_{S_0} + \alpha C_S} \ln \frac{C_S \alpha\beta}{C_{S_i} + C_S \alpha} - \ln \frac{1+\alpha}{\alpha\beta}\right] \tag{4.235}$$

Setting a value of $\frac{C_{S_0}}{C_S} = 1$ to this equation, one can calculate D_{crit} by Equation (4.236).

$$D_{\text{crit}} = \frac{1}{\tau_{\text{crit}}} = \frac{\mu_m}{(1+\alpha)\ln\left(\frac{1+\alpha}{\alpha\beta}\right)} \frac{C_s}{K_S + C_s} \tag{4.236}$$

SECTION A: BIOREACTORS FOR SUBMERGED FERMENTATION OF MICROBIAL CELLS

Part 3: Combination of Bioreactors

4.7 COMBINATION OF BIOREACTORS

We begin with a question to understand the use of combination of reactors in biological reactions.

Question ***What are the criteria to choose optimal reactor strategy for maximum productivity?***

Answer: We say

- The shape of $\frac{1}{r_X}$ vs. C_X plot
- Process requirement such as final conversion.

Brief review of ideal reactors on this basis is given here.

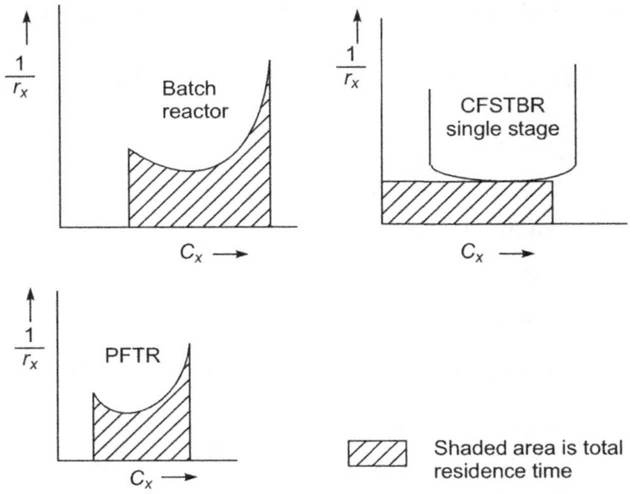

Figure 4.34 Plot of $\frac{1}{r_X}$ vs. C_X for ideal reactors.

(1) If the objective is to obtain $C_{X_{\text{opt}}}$ in lesser time, definitely the productive reactor is the one whose shaded area is small or the one having $\frac{1}{r_X}$ minimum. Clearly it shows the CFSTBR.

(2) If we choose $C_{X_{\text{desired}}} < C_{X_{\text{opt}}}$, single stage CFSTBR is the choice.

(3) If $C_{X_{\text{desired}}} > C_{X_{\text{opt}}}$, case (2) is not the answer. In this case, we need more residence time. Here, the question of multistage reactors comes into play.

(4) In multistage system (combination of continuous flow reactors), the selection depends on combining the following ideal reactors.

- CFSTBR plus CFSTBR plus CFSTBR and so on
- CFSTBR plus PFTR

Certainly, batch reactor has no place in the combination as F=0. If the cell concentration to be reached is in the stationary phase, batch reactor is a better choice than a CFSTBR.

4.7.1 Combination of Continuous Flow Bioreactors

Justification

(1) The physiological role of the biological system cannot be ignored in this case. The physiological behavior is connected to variation in productivity in different phases of cells. In continuous flow single stage reactor, the physiological status of the cell cannot be satisfied as continuous flow mode of operation is controlled by dilution rate (D). D gives an idea of residence time in a reactor. This physiological factor, to some extent, can be satisfied if one increases the residence time, i.e., not by using a single flow reactor. This is done by adding single flow reactor in series.

(2) For obtaining maximum conversion with almost complete utilization of reactants, one can use combination of reactors. However, the combination of CFSTBR plus CFSTBR gives more volume than CFSTBR plus PFTR. Let us examine the $\frac{1}{r_X}$ vs. C_X plot (Fig. 4.35).

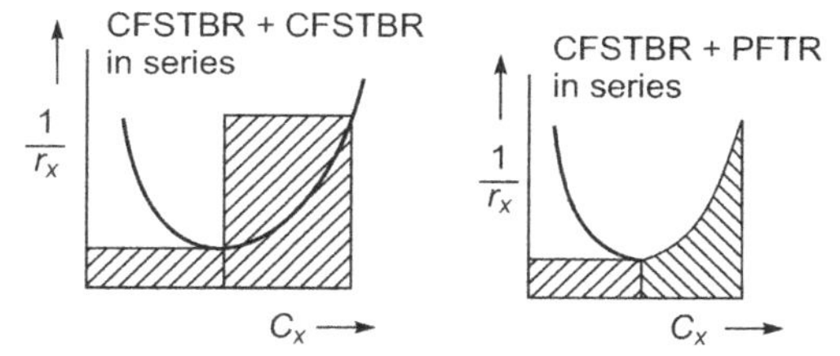

Figure 4.35 $\frac{1}{r_X}$ vs. C_X plots for same reaction using two different combinations of bioreactors in series.

For complete utilization of reactant, PFTR following CFSTBR is the choice.

(3) Growth and production phases have different optimal conditions for reactant concentration, physical and physicoehemical factors (pH, temperature, shear forces). Definitely multistage reactors are required to process them.

(4) If multiple reactants influence production in different phases, multistage reactors is used.

(5) For diauxic or triauxic growth, or for inhibition reaction system, or for production using complex reactants (polysaccharides-chitin, starch, lingo-cellulose and hemi-cellulose materials), or for addition of inducible parameters to improve genetic stability of a plasmid-carrying recombinant DNA, multistage continuous flow reactors are better choice.

4.7.2 Classification of Multistage Bioreactors

Multistage reactors are classified into two major categories (Fig. 4.36).

(a) ***CFSTBR multistage***

(b) ***CFSTBR plus Continuous flow PFTR in Series***

For a combination of CFSTBR followed by PTRR, CFSTBR operates under optimal conditions. PFTR utilizes reactant from the output stream to utilize it completely.

In other case, PFTR followed by CFSTBR can be used but to maintain PFTR with cell it is difficult. PFTR needs continuous addition of inoculums. CFSTBR after PFTR will receive very less concentration of reactant at which to run the system may not be viable.

In this section, we shall deal with the following combinations:

- CFSTBR-single stream
- CFSTBR-multi stream
- CFSTBR-followed by PTFR.

CFSTBR multistage with recycling is discussed under recycle reactors.

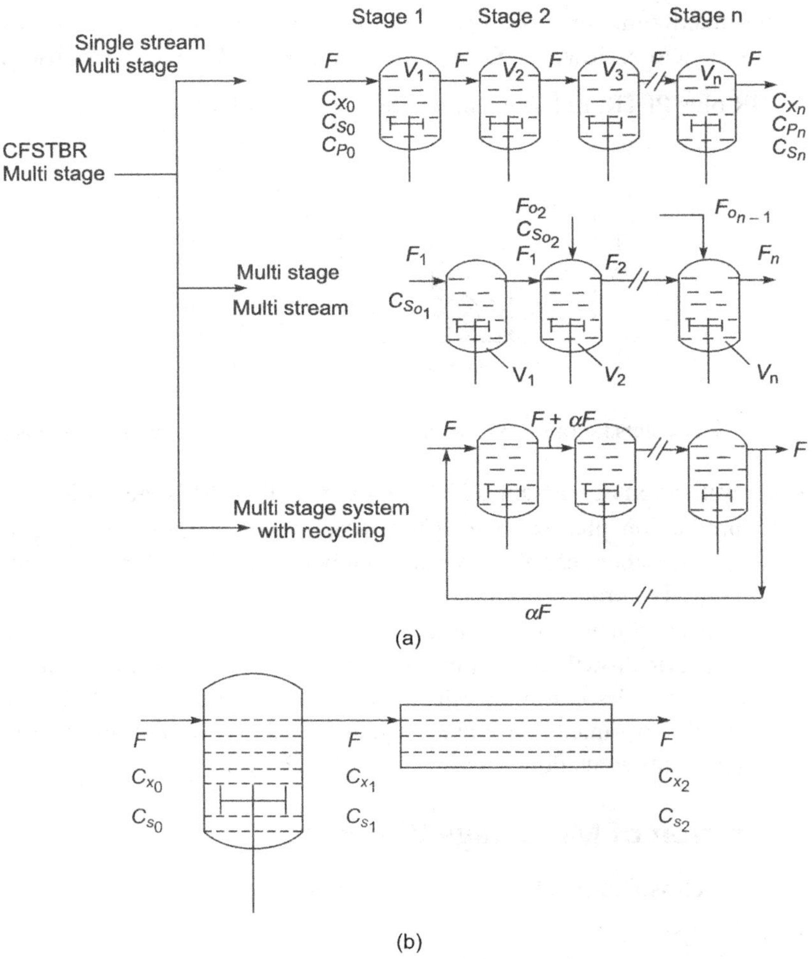

Figure 4.36 Multi stage bioreactors.

4.7.3 Analysis of CFSTBRs in Series with Single Stream

In this approach, the first stage operates at optimal conditions. Assumptions are

(1) Monod's equation is applicable
(2) Sterile feed, $C_{X_0} = 0$
(3) Constant volume

Cell mass balance for the second reactor at steady state condition is

$$FC_{X_1} - FC_{X_2} + V_2 \frac{\mu_m C_{S_2} C_{X_2}}{K_S + C_{S_2}} = 0 \tag{4.237}$$

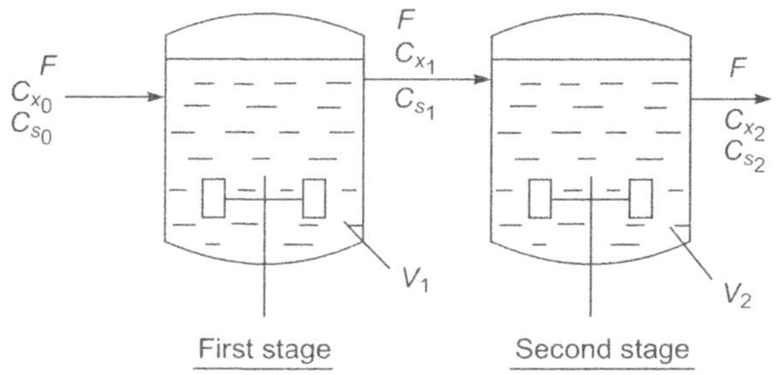

Figure 4.37 Two CFSTBR in series.

This is a simplified approach. Growth in the second stage (in this case, otherwise subsequent stages for cascade systems) is unbalanced growth. Of course, it is in the steady state. Organisms from the previous stage, entering the next stage, is continuously in a state of adapting the new system. Therefore, the unstructured model cannot give all details as per our expectations.

(a) Real Approach

The second stage is a more complicated balance equation (Herbert, 1964).

$$C_{X_2} = Y^{m}_{X/S}(C_{S_0} - C_{S_2}) \tag{4.238}$$

From this C_{X_2} can be calculated if one knows C_{S_2}.

C_{S_2} can be calculated from the Equation (4.239).

$$(\mu_m - D_2)C_{S_2}^2 - \left(\mu_m C_{S_0} - \frac{K_S D_1 D_2}{\mu_m - D_1} + K_S D_2\right)C_{S_2} + \frac{K_S{}^2 D_1 D_2}{\mu_m - D_1} = 0 \tag{4.239}$$

The quadratic equation gives two positive values of C_{S_2}. The smaller value corresponds to C_{S_2} is the required one, whereas other value of C_{S_2} (which is greater than C_{S_1}) is biologically imaginary quantity, but mathematically it is a real value for "reverse growth".

(b) Application of Simpler Approach to 'n' Number of Reactors in Series

The material balance for the microorganisms for n-reactors in series is of the following form (Fig. 4.38).

$$F(C_{X_{N-1}} - C_{X_N}) + V_N r_{gN} = 0 \tag{4.240}$$

where $$r_{gN} = \frac{\mu_m C_{S_N} C_{X_N}}{K_S + C_{S_N}} \tag{4.241}$$

$$Y^{m}_{x/s} = \frac{C_{X_N} - C_{X_{N-1}}}{C_{S_{N-1}} - C_{S_N}}$$

Therefore, the generalization is

$$r_{gN}(C_{X_N}, C_{S_N}) = D_N(C_{X_N} - C_{X_{N-1}}) \tag{4.242}$$

$$\therefore \quad \tau_N = \text{Mean residence time} = \frac{1}{D_N} = \frac{C_{X_N} - C_{X_{N-1}}}{r_{g_N}(C_{X_N}, C_{S_N})} \tag{4.243}$$

$$\therefore \quad r_{S,N} = \frac{1}{Y^{m}_{X/S}} r_{g_N}(C_{X_N}, C_{S_N}) \tag{4.244}$$

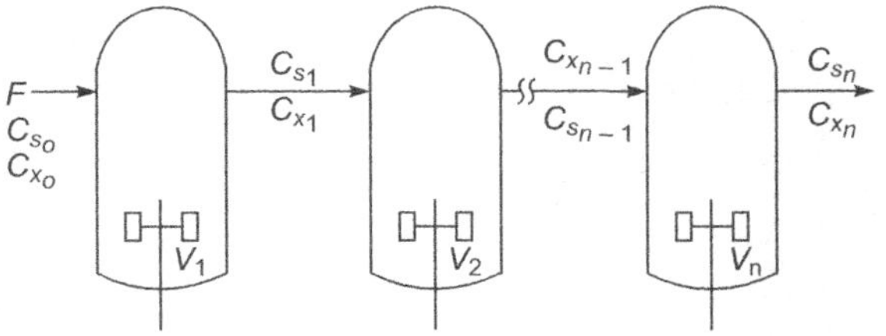

Figure 4.38 Multi stage CFSTBR connected in series.

$$= D_N(C_{S_N} - C_{S_{N-1}})$$

$$\therefore \qquad \tau_N = \frac{1}{D_N} = \frac{C_{S_N} - C_{S_{N-1}}}{r_{S,N}} \tag{4.245}$$

Similarly for product

$$r_{P_N}(C_{X_N}, C_{S_N}, C_i) = D_N(C_{P_N} - C_{P_{N-1}}) \tag{4.246}$$

By solving them simultaneously, we will get dilution rate with known cell mass concentration or vice-versa.

(c) Real Approach Extended to n-reactors in Series

Cell Mass Balance

$$V_N \frac{dC_{X_N}}{dt} = FC_{X_{N-1}} - FC_{X_N} + V_N \mu_N C_{X_N} \tag{4.247}$$

Reactant Balance

(i) V_f product-synthesis is not considered.

$$V_N \frac{dC_{S_N}}{dt} = FC_{S_{N-1}} - FC_{S_N} - \left(\frac{\mu_N}{Y_{X/S}^m} C_{X_N} \right) V_N \tag{4.248}$$

(ii) If product synthesis is considered,

$$V_N \frac{dC_{S_N}}{dt} = FC_{S_{N-1}} - FC_{S_N} - \left(\frac{\mu_N}{Y_{X/S}^m} C_{X_N} \right) V_N - \left(\frac{q_P}{Y_{P/S}^m} C_{X_N} \right) V_N \tag{4.248a}$$

Product Balance

$$V_N \frac{dC_{P_N}}{dt} = FC_{P_{N-1}} - FC_{P_N} + (\mu_N \, C_{X_N} \, Y_{P/X}^m) \, V_N \tag{4.249}$$

Steady state concentrations are given below.

$$C'_{X_N} = \frac{DC'_{X_{N-1}}}{D - \mu_N} \quad \text{for } (N \neq 1)$$

$$C'_{S_N} = C'_{S_{N-1}} - \frac{1}{D} \frac{1}{Y_{X/S}^m} \mu_N C'_{X_N} \quad \text{for no product formation.}$$

$$C'_{S_N} = FC'_{S_{N-1}} - \frac{1}{D} \left(\frac{\mu_N}{Y_{X/S}^m} + \frac{q_P}{Y_{P/S}^m} \right) C'_{X_N}, \text{ for product formation.}$$

$$C'_{P_N} = \frac{1}{D}(DC_{P_{N-1}} + Y^{m}_{P/X}\mu_N C'_{X_N}) \tag{4.250}$$

The estimation of C_X and C_S at a particular stage can be done by knowing the dilution rate. However, the graphical solution suggested by Deindorfer and Humphrey (1961) and Luedeking and Piret (2000) have been applied with the following stipulations.

(1) Reactors of same volume will wash out at D_{critical} which is same for the single stage CFSTBR.

(2) For reactors of unequal volume, the entire system will washout unless there is at least one reactor whose dilution rate is less than D_{critical}. This value may be smaller than the D_{critical} of single stage CFSTBR.

$$\therefore \qquad D_{\text{av}} = \frac{F}{NV}$$

$$\therefore \qquad D_{\text{av}_{\text{critical}}} = \frac{F_C}{NV} = \frac{D_C}{N}$$

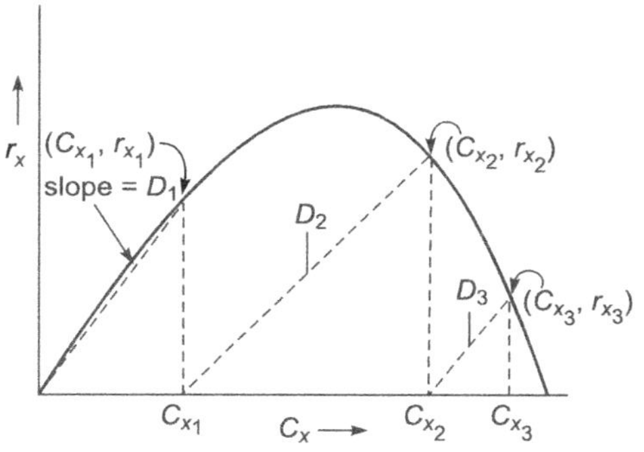

Figure 4.39 Plot of r_X vs. C_X.

Let us describe the graphical solution. A plot of r_X vs. C_X is available from which one needs to find out the number of stages required and other parameters.

Basic concept is Equation (4.251).

$$D_j = \frac{F}{V_j} = \frac{r_{X_j}}{C_{X_j} - C_{X_{j-1}}} \tag{4.251}$$

where, $j = 1, 2, \ldots\ldots n$.

For the first reactor:

$$D_1 = \frac{F}{V_1} = \frac{r_{X_1}}{C_{X_1}}, \text{ where } C_{X_0} = 0 \text{ for sterile feed,} \tag{4.252}$$

The slope of the straight line connecting the origin and (C_{X_1}, r_{X_1}) in the r_x vs. C_X plot is D_1. Similarly, for the second reactor, the slope of the line joining $(C_{X_1}, 0)$ and (C_{X_2}, r_{X_2}) is D_2. This will continue till we get complete number of stages for the r_x vs. C_x plot. One can try with any such plot and calculate the number of stages.

(d) Multi Stream Multistage CFSTBR

Simplest form of the configuration is to feed same reactant to all stages (Fig. 4.40).

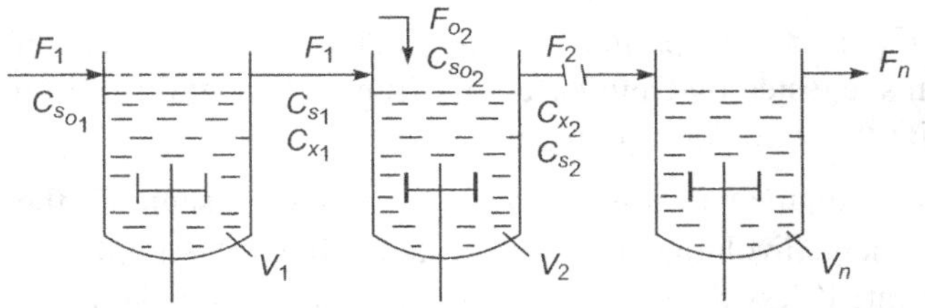

Figure 4.40 Multi stream multi stage CFSTBR.

In this case, for the second stage,

$$D_{02} = \frac{F_{02}}{V_2}, \; D_{12} = \frac{F_1}{V_2}$$

$$\therefore \qquad D_2 = D_{02} + D_{12}$$

As per Herbert (1964), for the second stage under steady state mass balance,

$$C_{X_2} = Y^{m}_{X/S}\left(\frac{D_{12}C_{S_{01}}}{D_2} + \frac{D_{02}C_{S_{02}}}{D_2} - C_{S_2}\right) \tag{4.253}$$

and

$$(\mu_m - D_2)C^2_{S_2} - \left(\frac{\mu_m + D_{12}C_{S_{01}}}{D_2} + \frac{(\mu_m - D_2)D_{02}C_{S_{02}}}{D_2} - (D_{12}C_{S_1} + K_S D_2)C_{S_2}\right) + K_S D_{02} C_{S_{03}} + K_S D_{12} C_{S_1} = 0 \tag{4.254}$$

C_{S_2} can be calculated by incorporating C_{S_1} and steady state behavior for any value of D_1 and D_2 can be calculated, if one knows the values of the parameters.

(e) Simpler Approach

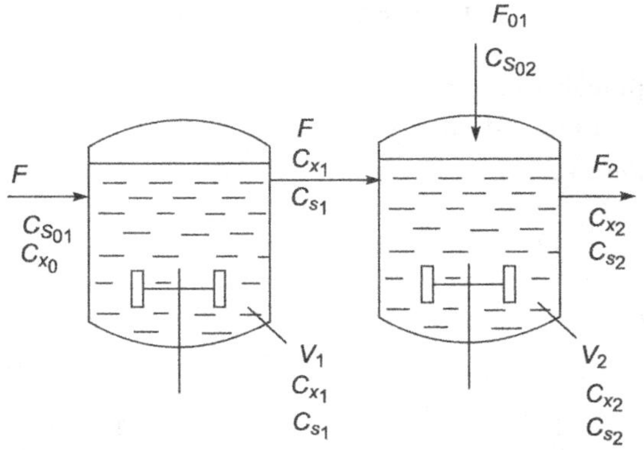

Figure 4.41 Two stage chemostat in series with input of reactant in both the stages.

Assumptions

For the simpler approach, we assume

- No endogenous metabolism

- Monod's kinetics applied to growing organism
- Steady state conditions applied, and
- Sterile feed, $C_{X_0} = 0$.

For the First Stage

$$C_{S_1} = \frac{K_S D_1}{\mu_m - D_1} \tag{4.255}$$

$$C_{X_1} = Y^m_{X/S}(C_{S_{01}} - C_{S_1}) \tag{4.256}$$

For the Second Reactor

Cell mass balance

$$FC_{X_1} - FC_{X_2} + \mu_2 V_2 C_{X_2} = V_2\left(\frac{dC_{X_2}}{dt}\right) \tag{2.257}$$

At steady state

$$\mu_2 = \frac{F}{V_2}\left(1 - \frac{C_{X_1}}{C_{X_2}}\right) \tag{4.258}$$

$$\mu_2 = D_2\left(1 - \frac{C_{X_1}}{C_{X_2}}\right) \tag{4.259}$$

This is valid when $\frac{C_{X_1}}{C_{X_2}} < 1$ and $\mu_2 < D_2$.

The limiting reactant balance is

$$FC_{S_1} - FC_{S_2} - \frac{\mu_2}{Y^m_{X/S}} V_2 C_{X_2} = V_2\left(\frac{dC_{S_2}}{dt}\right) \tag{4.260}$$

(without considering reactant addition in the second stage).

For steady state condition

$$C_{S_2} = C_{S_1} - \frac{\mu_2}{D_2}\frac{C_{X_2}}{Y^m_{X/S}}$$

where $$D_2 = \frac{F}{V_2}$$

and $$\mu_2 = \frac{\mu_m C_{S_2}}{K_S + C_{S_2}}.$$

For additional feed stream in the second stage, if the second feed stream is also sterile, we get cell mass balance as

$$FC_{X_1} - (F + F_{01})C_{X_2} + V_2 \mu_2 C_{X_2} = V_2 \frac{dC_{X_2}}{dt} \tag{4.261}$$

At steady state,

$$\mu_2 = \frac{(F+F_{01})}{V_2} - \frac{F}{V_2}\frac{C_{X_1}}{C_{X_2}} \tag{4.262}$$

$$= D_2' - \frac{F}{V_2}\frac{C_{X_1}}{C_{X_2}}$$

$$= D_2' - D_1 \frac{C_{X_1}}{C_{X_2}}\frac{V_1}{V_2} \tag{4.263}$$

for $$\mu_2 = \frac{\mu_m C_{S_2}}{K_S + C_{S_2}}.$$

Limiting reactant balance for the second stage of CFSTBR is given in Equation (4.264).

$$FC_{S_1} + \mathrm{F}_{0_1}C_{S_{02}} - (F+F_{01})C_{S_2} - \frac{V_2\mu_2 C_{X_2}}{Y^m_{X/S}} = V_2\frac{dC_{S_2}}{dt} \tag{4.264}$$

At steady state,

$$FC_{S_1} + \mathrm{F}_{01}C_{S_{02}} - (F+F_{01})C_{S_2} = \frac{V_2\mu_2 C_{X_2}}{Y^m_{X/S}} \tag{4.265}$$

As we know from the cell mass balance in the second stage, at steady state condition

$$V_2\mu_2 C_{X_2} = (F+F_{01})C_{X_2} - FC_{X_1} \tag{4.266}$$

Substituting in the steady state limiting reactant balance,

$$Y^m_{X/S}(FC_{S_1} + F_{01}C_{S_{02}} - (F+F_{01})C_{S_2}) = (F+F_{01})\,C_{X_2} - FC_{X_1}$$

$$Y^m_{X/S}(FC_{S_1} + F_{01}C_{S_{02}} - (F+F_{01})C_{S_2}) + FC_{X_1} = (F+F_{01})\,C_{X_2}$$

$\because C_{X_1} = Y^m_{X/S}(C_{S_{01}} - C_{S_1})$ in the first stage of CFSTBR under steady state.

Substitution of C_{X_1} in the above equations, we get

$$Y^m_{X/S}(FC_{S_1} + F_{01}C_{S_{02}} - (F+F_{01})C_{S_2}) + FY^m_{X/S}(C_{S_{01}} - C_{S_1}) = (F+F_{01})\,C_{X_2}$$

$$Y^m_{X/S}\,\cancel{F}C_{S_1} - Y^m_{X/S}\,\cancel{F}C_{S_1} + Y^m_{X/S}FC_{S_{01}} + Y^m_{X/S}F_{01}C_{S_{02}} - Y^m_{X/S}(F+F_{01})C_{S_2} = (F+F_{01})\,C_{X_2}$$

$$\therefore \quad Y^m_{X/S}\left(\frac{F}{F+F_{01}}C_{S_{01}} + \frac{F_{01}}{F+F_{01}}C_{S_{02}} - C_{S_2}\right) = C_{X_2} \tag{4.267}$$

$$\therefore \quad C_{X_2} = Y^m_{X/S}\left(\frac{F}{\dfrac{F+F_{01}}{V_2}V_2}C_{S_{01}} + \frac{F_{01}}{\dfrac{F+F_{01}}{V_2}V_2}C_{S_{02}} - C_{S_2}\right) \tag{4.268}$$

So, $$C_{X_2} = Y^m_{X/S}\left(\frac{F}{D_2'V_2}C_{S_{01}} + \frac{F_{01}}{D_2'V_2}C_{S_{02}} - C_{S_2}\right)$$

We assume Monod's equation is also applicable in the second stage.

$$\mu_2 = \frac{\mu_m C_{S_2}}{K_S + C_{S_2}}$$

Substitution of μ_2 and C_{X_2} in steady state reactant mass balance expression gives

$$FC_{S_1} + F_{01}C_{S_{02}} - (F + F_{01})C_{S2} = \frac{V_2}{Y_{X/S}^{m}}\left(\frac{\mu_m C_{S_2}}{K_S + C_{S_2}}\right)Y_{X/S}^{m}\left(\frac{F}{D_2'V_2}C_{S_{01}} + \frac{F_{01}}{D_2'V_2}C_{S_{02}} - C_{S_2}\right)$$

$$\frac{F}{V_2}C_{S_1} + \frac{F_{01}}{V_2}C_{S_{02}} - D_2'C_{S_2} = \frac{\mu_m C_{S_2}}{K_S + C_{S_2}}\left(\frac{F}{D_2'V_2}C_{S_{01}} + \frac{F_{01}}{D_2'V_2}C_{S_{02}} - C_{S_2}\right)$$

$$(K_S + C_{S_2})\left(\frac{F}{V_2}C_{S_1} + \frac{F_{01}}{V_2}C_{S_{02}} - D_2'C_{S_2}\right) = \frac{\mu_m C_{S_2} F}{D_2'V_2}C_{S_{01}} + \frac{F_{01}\mu_m C_{S_2} C_{S_{02}}}{D_2'V_2} - \mu_m C_{S_2}^2$$

or,

$$K_S\frac{F}{V_2}C_{S_1} + K_S\frac{F_{01}}{V_2}C_{S_{02}} - K_S D_2'C_{S_2} + \frac{F}{V_2}C_{S_1}C_{S_2} + \frac{F_{01}}{V_2}C_{S_2}C_{S_{02}} - D_2'C_{S_2}^2$$
$$+ \mu_m C_{S_2}^2 - \frac{\mu_m C_{S_2} F}{D_2'V_2}C_{S_{01}} - \frac{F_{01}\mu_m}{D_2'V_2}C_{S_2}C_{S_{02}} = 0$$

$$\text{or, } (\mu_m - D_2')C_{S_2}^2 - \left(\frac{F_{01}\mu_m}{D_2'V_2}C_{S_{02}} + \frac{F\mu_m}{D_2'V_2}C_{S_{01}} - \frac{F}{V_2}C_{S_1} - \frac{F_{01}}{V_2}C_{S_{02}} + K_S D_2'\right)C_{S_2}$$
$$+ \left(K_S\frac{F}{V_2}C_{S_1} + K_S\frac{F_{01}}{V_2}C_{S_{02}}\right) = 0 \qquad (4.269)$$

This is a quadratic equation in C_{S_2}. We can have two roots for C_{S_2}. The significance of two roots is given in earlier discussion in CFSTBR in series in single stream- real approach.

(f) CFSTBR followed by PFTR

The combination is given in Figure 4.42.

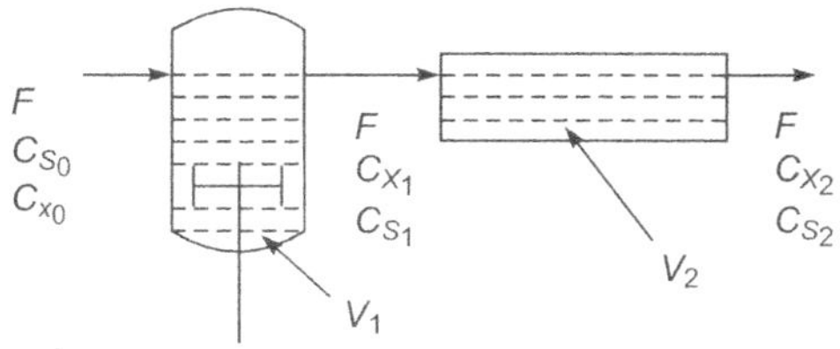

Figure 4.42 Combination of CFSTBR followed by PFTR.

CFSTBR plus PFTR in series

Generally, CFSTBR in this case operates at optimal conditions. So, the values obtained from the mass balances at steady conditions are given below.

$$C_{S_1} = \frac{K_S}{\tau_1\mu_m - 1}$$

$$C_{X_1} = Y^{m}_{X/S}\left(C_{S_0} - \frac{K_S}{\tau_1\mu_m - 1}\right) \quad (4.270)$$

$$C_{P_1} = C_{P_0} + Y^{m}_{P/S}\left(C_{S_0} - \frac{K_S}{\tau_1\mu_m - 1}\right)$$

where τ_1 is the mean residence time in the CFSTBR. This deduction is valid for sterile feed condition (i.e., $C_{P_0} = 0$).

For the second stage of PFTR

∴ Assuming Monod's equation in this case

$$\int_{C_{X_1}}^{C_{X_2}} \frac{dC_X}{r_g} = \int_{C_{X_1}}^{C_{X_2}} \frac{(K_S + C_S)dC_X}{\mu_m C_S C_X}$$

and growth yield $= Y^{m}_{X/S} = \dfrac{C_{X_2} - C_{X_1}}{C_{S_1} - C_{S_2}}$

We can calculate the residence time $= \tau_2$ in this case.

SECTION A: BIOREACTORS FOR SUBMERGED LIQUID FERMENTATION OF MICROBIAL CELLS

Part 4: Semi-Continuous Bioreactors

4.8 SEMI-CONTINUOUS BIOREACTORS

Semi-continuous reactors are unsteady state operations of a reactor. In chemical reactor analysis the initial stage of a CFSTR and semi-batch reactors are the example of unsteady state operation of reactors. Batch reactor is also an example of unsteady state behavior. Figure 4.43 is the example of chemical semi-batch operations.

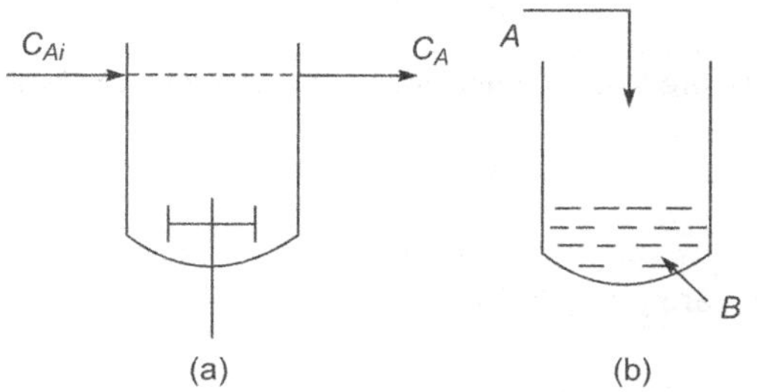

Figure 4.43 (a) Represents the time of reactor operation to reach a steady state. (b) Gradual change in reactant happens in the reactor.

These examples are also considered as transient behavior in reactors (Fogler, 1992). The biological reactions carried out in a reactor do represent similar unsteady behavior. The unsteady state operations of reactors are classified as batch-fed and fed-batch reactors. This is primarily based on flow rate of the components to the reactor. The terminology of semi-batch reactor in chemical engineering may be considered as batch-fed reactor concept in biological reactor.

On the other hand, in fed-batch reactor operation, reactant feeding is controlled and regulated by some microbial parameter which can be easily measured (Suga *et al.*, 1980). With this definition of fed-batch system, the fed-batch results available in the literature are certainly different. One can think of the system which does not conform to this definition of fed-batch (Suga *et al.*, 1980) that might be batch-fed reactors. Fed-Batch reactors defined by Suga *et al.*, (1980) is a variable flow continuous reactor. This variation is to maintain a constant metabolic activity of the cell. They have measured exit CO_2 concentration during reaction. Feed rate to the reactor is regulated on the basis of CO_2 evolution rate.

Moser (1985) suggested the classification of such reactors based on feed rate function into two groups.

1. Exit Flow Rate is Zero

- Input flow rate zero(constant volume)
- Variable input flow rate

Discontinuous feed   Continuous feed

Discontinuous feed and input flow rate equal to zero can be considered as 'semi batch reactor' in chemical reaction system or 'batch-fed' in biological reaction system.

2. Variable-exit Flow Rate

(a) If the exit flow rate = input flow rate, then volume is constant (CFSTBR). This is called transient mode/unsteady state behavior or start-up chemostat or fed-batch operation.

(b) If the exit flow rate ≠ input flow rate, then it is a variable volume reactor. It could be two different modes of operation as given below.

Classification for (b)

Cyclic continuous
Fed-batch   Cyclic discontinuous
Semi-batch or batch-fed

4.8.1 A Few Definitions

It is necessary to clarify some terminologies.

Repeated Fed-batch Culture

The complete medium is fed continuously to a batch culture. The resulting cell mass is adjusted to a preset condition. The process is then repeated. The duration of the periodicity is applied to the culture (Pirt, 1974).

Extended Fed-batch Culture

Environmental sensors measure specific parameters of culture growth and regulate the addition of limiting reactant to the culture. This covers only one cycle of fed-batch (Moser, 1985).

A certain type of extended fed-batch and the cyclic fed-batch for mixed volume systems refer to a periodic withdrawal of a portion of a culture and use of the residual culture to initiate further fed-batch processes.

We shall take the example of

- Semi-Batch
- Start up chemostat (CFSTBR)
- Fed-Batch as per Suga *et al.*, (1980).

for further discussion.

4.8.2 Analysis of Semi-Batch Reactor

The reactor is being filled with the reactant solution which gradually reaches 'V' volume of reaction in a specified time. This can be found from mass balance of the species concerned

Rate in – rate out + rate of generation = rate of accumulation

$$\rho_0 F_0 - 0 + 0 = \frac{d(\rho V)}{dt} \tag{4.271}$$

For constant density systems

$$\rho_0 = \rho$$

$$\therefore \quad \frac{dV}{dt} = F_0 \tag{4.272}$$

Upon integration,

$$V = V_0 + F_0 t$$

$$(V_0 = \text{Volume at } t = 0)$$

or

$$\tau = \tau_0 + t$$

$$\tau = \frac{V}{F_0} \tag{4.273}$$

Assuming for a batch culture

$$C_X = C_{X_0} + Y^m_{X/S}(C_{S_0} - C_S) \tag{4.274}$$

When cell mass concentration reaches maximum, $C_S << C_{S_0}$ and $C_{X_0} << C_X$.

Then C_{X_m} = Maximum cell concentration in the reactor = $Y^m_{X/S} C_{S_0}$ (4.275)

Total amount of cell mass in the reactor = $VC_X = C'_X$

$$C_X = \frac{C'_X}{V} \tag{4.276}$$

Differentiating with time

$$\frac{dC_X}{dt} = \frac{d}{dt}\left(\frac{C'_X}{V}\right) \tag{4.277}$$

$$\Rightarrow \quad \frac{dC_X}{dt} = \frac{V\dfrac{dC'_X}{dt} - C'_X\dfrac{dV}{dt}}{V^2}$$

where $\dfrac{dC'_X}{dt} = \mu_{\text{net}} C'_X$

$$\frac{dV}{dt} = F,$$

$$\Rightarrow \quad \frac{dC_X}{dt} = \frac{V\mu_{\text{net}}C'_X - C'_X F}{V^2}$$

$$= \frac{\cancel{V}(\mu_{\text{net}} - D)C'_X}{\cancel{V}^2}$$

$$= (\mu_{\text{net}} - D)C_X \qquad \left(\because \frac{C'_X}{V} = C_X\right) \tag{4.278}$$

For quasi-steady state approximation when reactant is almost exhausted

$$\frac{dC_X}{dt} = 0 \tag{4.279}$$

$$\therefore \quad \mu_{\text{net}} = D$$

Assuming no maintenance and endogenous metabolism,

$$\mu_{\text{net}} = \mu$$

From Monod's equation approximation,

$$C_S = \frac{K_S D}{\mu_m - D} \equiv \frac{F_0}{V}\frac{K_S}{\mu_m - \dfrac{F_0}{V}}$$

From limiting reactant mass balance:

Rate in – rate out ± rate of generation = rate of accumulation

$$F_0 C_{S_0} - 0 - \frac{\mu_{\text{net}} C'_X}{Y^m_{X/S}} = \frac{dC_S}{dt} \tag{4.280}$$

where C_s = reactant concentration in the reaction medium.

$$\therefore \quad F_0 C_{S_0} - \frac{\mu_{\text{net}} C'_X}{Y^m_{X/S}} = \frac{dC_S}{dt} \tag{4.281}$$

Considering quasi-steady state condition where $C'_X = VC_{X_m}$

$$F_0 C_{S_0} = \frac{\mu_{\text{net}} V C_{X_m}}{Y^m_{X/S}} \tag{4.282}$$

At quasi-steady state with $C_S = 0$

$$\frac{dC_X}{dt} = 0 = \frac{V\left(\frac{dC'_X}{dt}\right) - C'_X\left(\frac{dV}{dt}\right)}{V^2} \tag{4.283}$$

$$V\left(\frac{dC'_X}{dt}\right) = C'_X\left(\frac{dV}{dt}\right)$$

$$\therefore \quad \frac{dC'_X}{dt} = \frac{C'_X}{V}\left(\frac{dV}{dt}\right) = C_X\left(\frac{dV}{dt}\right) \tag{4.284}$$

At quasi-steady state,

$$C_X \to C_{X_m}$$

$$\therefore \quad \frac{dC'_X}{dt} = C_{X_m}\left(\frac{dV}{dt}\right)$$

$$= C_{X_m} F_0$$

$$= F_0 Y^m_{X/S} C_{S_0} \tag{4.285}$$

$$\left(\text{whereas} \quad F_0 C_{S_0} = \frac{D.V}{Y^m_{X/S}} C_{X_m} = \frac{F_0 C_{X_m}}{Y^m_{X/S}}\right)$$

$$\therefore \quad C'_X = C'_{X_0} + F_0 Y^m_{X/S} C_{S_0} t \tag{4.286}$$

Product output at quasi-steady state

$$FP = Y_{P/S} C_{S_0} F \tag{4.287}$$

Specific rate of product formation is

$$\frac{1}{C'_X}\frac{dC'_P}{dt} = q_P$$

$$\therefore \quad \frac{dC'_P}{dt} = q_P C'_X$$

$$\therefore \quad \frac{dC'_P}{dt} = q_p V C_X$$

$$= q_p C_{X_m}(V_0 + F_0 t) \tag{4.288}$$

q_p is constant

$$\therefore \quad C'_P = C'_{P_0} + q_p C_{X_m}\left(V_0 + \frac{F_0 t}{2}\right)t \tag{4.289}$$

In terms of D

$$C'_P = C'_{P_0} + q_p C_{X_m} V\left(\frac{V_0}{V} + \frac{Dt}{2}\right)t \tag{4.290}$$

4.8.3 Fed-Batch Bioreactors

Initially, it is a variable volume with no output so as to develop the cells at active stage. Then it is a continuous flow constant volume reactor. Feed rate depends on metabolic function of the cells.

(a) Advantages over Batch Reactor and CFSTBR

Fed-batch biorteactors have certain advantages over CFSTBR in biological processes.

1. Production of high cell densities due to extension of working time (particularly important in the production of growth associated products)
2. Controlled conditions in the provision of reactants during the fermentation, particularly regarding the concentration of specific reactants, e.g., the carbon source.
3. Control over the production of by-products or catabolite repression effects due to limited provision of substrates solely required for product formation.
4. The mode of operation can overcome and control deviations in the organism's growth pattern as found in batch fermentation.
5. It allows the replacement of water loss by evaporation.
6. Alternative mode of operation for fermentations leading to the generation of toxic components (cells can only metabolize a certain quantity at a time) or the formation of compounds having low solubility.
7. Increase of antibiotic marked plasmid stability by providing the corresponding antibiotic during the time span of the fermentation.
8. No additional special piece of equipment is required compared to the batch fermentation mode of operation.

(b) Disadvantages

Of course, there are some disadvantages associated with the fed batch bioreactors.

1. It requires previous analysis of the microorganism, its requirements and the understanding of its physiology with the productivity.
2. It requires a substantial amount of operator skill for the set up and development of the process.
3. In a cyclic fed batch culture, care should be taken in the design of the process to ensure that toxins do not accumulate to critical inhibitory levels and that nutrients other than those incorporated into

the feed medium become limiting. Operation of many cycles may result in the accumulation of non producing or low producing variants.

4. The quantities of the components to control must be above the detection limits of the available measuring equipment.

(c) Preliminary Knowledge Required to Implement Fed-batch Operation

Before starting a fed-batch process, batch fermentation should be carried out to know the fermentation pattern of the microorganism. From the batch fermentation, the operator should have the knowledge of the following.

1. Best abiotic conditions, viz., temperature, light, agitation, pH, growth medium, etc.
2. Specific needs of precursors, inducers, or other enrichment factors.
3. Understanding of the different growth phases, consumed reactants and produced components (product of interest and by products).
4. The relationship between the cell mass and product formation (growth or non-growth associated product) and the oxygen uptake rates.
5. Limiting reactant for growth and the relationship between the specific growth rate and limiting reactant concentration.
6. Eventual inhibitions from the reactant and/or product.
7. Properly defined objective functions and best parameters to control the reaction.

(d) Parameters to Start and Finish the Feed and Stop the Fed-batch Fermentation

The times at which the feeding should start and finish, as well as the criteria to stop a fed-batch fermentation is very much dependent on the specific cultivation kinetics and the operator's interest. For example, in a reactant limited processes, the feed should start immediately after all reactant is consumed otherwise the process may be difficult to control, namely because of a lag phase due to previous starvation. The common criteria to start the feed are the depletion of reactant which can be measured easily.

The fed batch fermentation should be halted when the production slows down because of cell death, reduced metabolic activity in the culture or accumulation.

(e) Analysis

1. Objective

Feed rate of reactant is adjusted to keep μ (μ = specific growth rate) constant.

From the mass balance

$$\frac{dVC_X}{dt} = \mu VC_X \tag{4.291}$$

$$\frac{dVC_S}{dt} = -\frac{\mu VC_x}{Y} + F(C_{S_0} - C_S)$$

$$\frac{dV}{dt} = F$$

C_S should be constant because $\mu = f(C_S)$.

For quasi-steady state,

$$\frac{d(VC_S)}{dt} = -\frac{\mu VC_X}{Y} + F(C_{S_0} - C_S) = 0$$

$$\therefore \qquad F = \frac{\mu VC_X}{Y(C_{S_0} - C_S)} \qquad (4.292)$$

C_S gives desired μ.

From $$\frac{d(VC_X)}{dt} = \mu VC_X$$

Integrating the equation gives

$$VC_X = (VC_X)_0 \exp(\mu t)$$

$$F = \frac{\mu (VC_X)_0}{Y(C_{S_0} - C_S)} \exp(\mu t) \qquad (4.293)$$

where $(VC_X)_0$ is the total cell mass in the culture in the beginning of fed-batch operation.

The objective of fed-batch bioreactor is to control exit CO_2 from the reactor to manipulate μ. This analysis has been proposed by Suga *et al.*, (1980).

Considering mass balance for cell

$$\frac{d(VC_X)}{dt} = \mu VC_X \qquad (4.294)$$

For limiting reactant

$$\frac{d(VC_S)}{dt} = F(t)(C_{S_0} - C_S) - V\left(mC_X + \frac{1}{Y^m_{X/S}}\mu C_X\right) \qquad (4.295)$$

Change in reaction volume

$$V = V_0 + \int F(t)dt \qquad (4.296)$$

For CO_2

$$F_{CO_2}(C_{CO_2} - C_{CO_2,0}) = \left(m_{CO_2} + \frac{\mu}{Y_{CO_2}}\right)C_X V \qquad (4.297)$$

If μ is a constant,

$$VC_X = V_0 C_{X_0} \exp(\mu t) \qquad (4.298)$$

Assuming

$$C_X = C_{X_0}$$

$$V = V_0 \exp(\mu t)$$

$$C_{CO_2} = \left(m_{CO_2} + \frac{\mu}{Y_{CO_2}}\right)\frac{C_X}{F_{CO_2}} V_0 \exp(\mu t) + C_{CO_2,0} \qquad (4.299)$$

Flow rate of feed is

$$\frac{dV}{dt} = V_0 \mu \exp(\mu t)$$

$$\therefore \quad F = V_0 \mu \exp(\mu t) \tag{4.300}$$

If μ is known, CO_2 in effluent can be controlled to the value as per the above equation.

SECTION B: BIOREACTORS FOR ENZYME REACTIONS AND IMMOBILIZED CELLS

4.9 INTRODUCTION

Like reactors for submerged liquid fermentation using micro organisms, ideal reactors for enzymatic reactions are classified in to three major categories.

(a) Mixed reactors
   (1) Batch reactor
   (2) Semi-batch or semi-continuous reactors.
   (3) Continuous flow reactors

Continuous flow reactors are classified into

- CFSTBR
- Special reactors
   (i) Packed bed reactors
   (ii) Fluidized bed reactors
   (iii) Membrane reactors

(b) Plug flow reactors

The reactors used for enzymatic reactions include free and immobilized enzymes as well as immobilized cell (Prĕnosil *et al.*, 1987; Dunn *et al.*, 2003; Doran, 1995). Immobilized cells and enzyme reactions have almost similar requirements. During continuous flow operations (except plug flow mode), ideally there is no output in terms of enzyme and cell species. In fact, ideal plug flow mode of operation of reactors is possible only with soluble enzyme and soluble reactants.

4.10 INPUT TO KINETIC MODELING OF ENZYME REACTORS

The aim of enzyme reaction engineering is high degree of conversion of substrates with high chemical and enantiomeric selectivity. For enzyme reactor design, one should first develop the rate expression. Then insert this rate expression into the equation for the reactor which we intend to operate and then integrate over the entire conversion. The approach towards the reactor design is given in the following Figure 4.44.

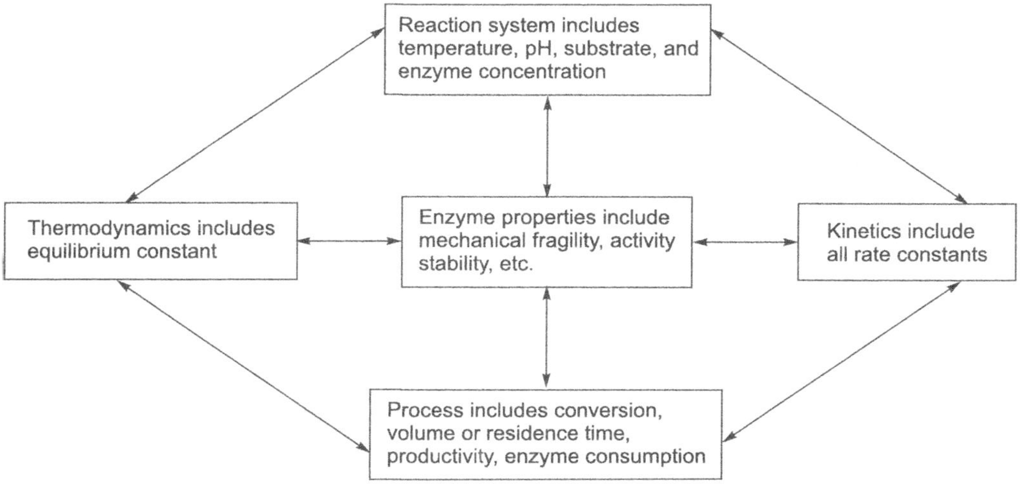

Figure 4.44 General procedures for kinetic modeling of enzyme reactions.

During the initial stage of reactor design, pH and temperature have to be chosen as a function of the properties of the reactants and enzymes. However, most of the enzyme reactions operate in a pH range between 7 and 10 and temperature range between 30°C and 50°C. Then one should select the initial substrate concentration (C_{S_0}) and in the case of two-substrate reactions, the stoichiometric ratio of the two reactants. The enzyme concentration also should be selected, which influences both the achievable space–time-yield and the selectivity in the case of undesired parallel or consecutive side reactions. For multi-enzyme systems, the optimal activity ratio should be found for kinetic analysis. Before the kinetic measurements the activity and stability of all the enzymes involved should be known as a function of the reaction conditions. Enzyme stability is an important aspect of biocatalytic processes. It is usually expressed as an enzyme unit consumption number. It has the dimensions of unit of activity per mass of product. The ideal behavior of these reactors are analysed in the next section.

4.10.1 Ideal Reactors

Three types of ideal reactors are:

(a) Batch reactor
(b) Plug Flow Reactor or PFR
(c) Continuous Flow Stirred Tank reactor or CFSTR.

Analysis of these reactors is considered using Michaelis-Menten kinetics.

(a) Batch Reactor

Objective is to calculate batch reaction time. Michaelis-Menten kinetics is employed for the analysis. In this regard, mass balance equation is given below.

Rate of mass change with time in the reactor = Rate of mass flow into the reactor
– Rate of mass flow out of the reactor
+ Mass rate of generation
– Mass rate of consumption of reactant

The reactant is not generated in the reaction. No inflow and outflow is associated with this reactor. Hence,

Rate of mass flow into the reactor = 0
Rate of mass flow out of the reactor = 0
Mass rate of generation = 0

The mass balance equation is simplified in the form of Equation (4.301).

Rate of mass change with time in the reactor = – Mass rate of consumption of the reactant

$$\frac{d}{dt}\begin{bmatrix}\text{Concentration of reactant}\\ \times\text{Liquid volume}\end{bmatrix} = -\begin{bmatrix}\text{Volumetric rate of reaction}\\ \times\text{Liquid volume}\end{bmatrix}$$

$$\therefore \qquad \frac{d}{dt}(C_S V) = -\left(\frac{r_{\max} C_S}{K_M + C_S}\right)V \tag{4.301}$$

Assuming Michaelis-Menten equation for single reactant – single enzyme – single product system, i.e.,

$$S + E \leftrightarrow ES \rightarrow E + P$$

In batch reactor system, volume is constant

$$\therefore \qquad \frac{d}{dt}(C_S) = -\left(\frac{r_{\max} C_S}{K_M + C_S}\right) \tag{4.302}$$

where C_S is the concentration of reactant
t is the time of reaction
$r_{\max}$ is the maximum reaction rate
K_M is Michaelis-Menten constant having the unit of C_S.

Assuming $r_{\max}$ and K_M are constant in this case.

The result of integration of the Equation (4.302) is the Equation (4.303).

$$-\int_0^t dt = \int_{C_{S_0}}^{C_S} \frac{K_M + C_S}{r_{\max} C_S}\, dC_S$$

$$-t = \int_{C_{S_0}}^{C_S} \left(\frac{K_M}{r_{\max} C_S} + \frac{1}{r_{\max}}\right) dC_S$$

$$= \frac{K_M}{r_{\max}} \ln\left(\frac{C_S}{C_{S_0}}\right) + \frac{(C_S - C_{S_0})}{r_{\max}}$$

$$\therefore \qquad t_{rxn} = \frac{K_M}{r_{\max}} \ln\left(\frac{C_{S_0}}{C_S}\right) + \frac{(C_{S_0} - C_S)}{r_{\max}} \tag{4.303}$$

Typical concentration vs. time plot is given in Figure 4.45.

Figure 4.45 Batch reactor data for C_S, C_P vs. t.

Similar expression can be written in the form of conversion for the reaction $S \rightarrow P$, as $X_S = 1 - C_S/C_{S_0} = C_P/C_{S_0}$.

The design equation of a batch reactor in terms of conversion of substrate is given by equation (4.304).

$$t = C_{S_0} \int \frac{dX_S}{(-r_S)} \tag{4.304}$$

$$r_s = \frac{k_{cat} C_E C_S}{K_M + C_S}$$

$$t_{rxn} = \frac{C_{S_0} X_S}{k_{cat} C_E} - \frac{K_M}{k_{cat} C_E} \ln(1 - X_S) \tag{4.305}$$

(b) Continuous Plug Flow Reactor (Tubular Reactor)

We consider a typical tubular reactor in Figure 4.46.

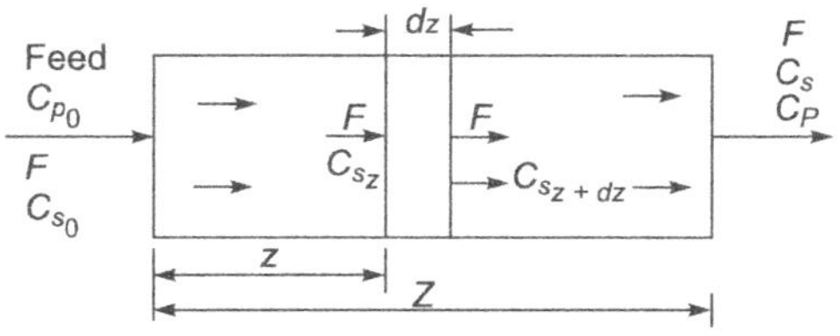

Figure 4.46 A typical tubular reactor for immobilized enzyme.

The concentration of the component varies along the length of the reactor from entrance to exit (Fig. 4.47).

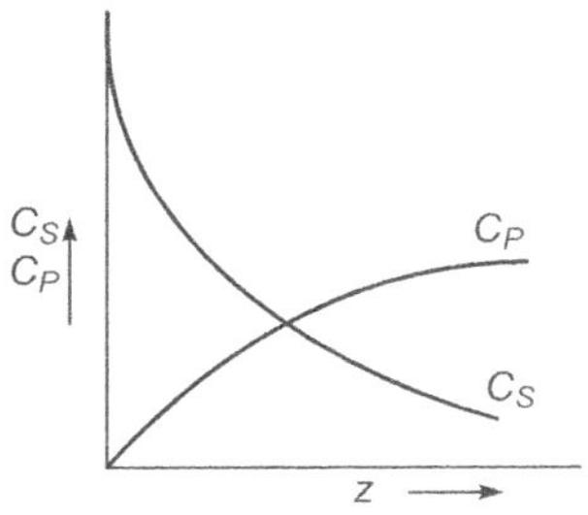

Figure 4.47 Concentration profiles for CFPFTR.

For Free Enzyme

For the analysis, following assumptions are considered:

- Michaelis-Menten equation explains the kinetics.
- No inhibition exists in the system.
- No back mixing appears in the reactor.
- No change in volume of the reactor is observed during the entire period of reaction.

Reactant mass balance is considered in small elemental length of the reactor, *dz*.

Rate of change = Input – Output ± generation + accumulation

At steady state,

$$0 = FC_S\big|_z - FC_S\big|_{z+dz} \pm 0 - \frac{r_{max}C_S}{K_M + C_S}dV$$

As $\quad dV = Adz,$

$$FC_S\big|_z - FC_S\big|_{z+dz} - \frac{r_{max}C_S}{K_M + C_S}Adz = 0$$

Therefore,

$$\frac{FC_S\big|_{z+dz} - FC_S\big|_z}{Adz} = -\frac{r_{max}C_S}{K_M + C_S}$$

Both F and A are constant. Then the expression is valid for entire reactor.
So we can write,

$$\frac{F}{A}\left(\lim_{dz \to 0}\frac{C_S\big|_{z+dz} - C_S\big|_z}{dz}\right) = -\frac{r_{max}C_S}{K_M + C_S}$$

$$\therefore \quad \frac{F}{A}\frac{dC_S}{dz} = -\frac{r_{max}C_S}{K_M + C_S} \tag{4.306}$$

Assumption: $\frac{F}{A}$, K_M and r_{max} are constants.

$$-\frac{F}{A}\int_{C_{S_0}}^{C_S}\frac{K_M + C_S}{r_{max}C_S}dC_S = \int_0^L dz$$

$$-\frac{F}{A}\int_{C_{S_0}}^{C_S}\frac{K_M}{r_{max}}\frac{1}{C_S}dC_S + \int_{C_{S_0}}^{C_S}\frac{1}{r_{max}}dC_S = Z$$

or, $$-\frac{F}{A}\left(\frac{K_M}{r_{max}}[\ln C_S]_{C_{S_0}}^{C_S} + \frac{1}{r_{max}}[C_S]_{C_{S_0}}^{C_S}\right) = Z$$

or, $$-\frac{F}{A}\left(\frac{K_M}{r_{max}}\ln\frac{C_S}{C_{S_0}} + \frac{1}{r_{max}}[C_S - C_{S_0}]\right) = Z$$

$\therefore$ $$\text{Length of the reactor} = \frac{F}{A}\left(\frac{K_M}{r_{max}}\ln\frac{C_{S_0}}{C_S} + \frac{1}{r_{max}}[C_{S_0} - C_S]\right) = Z$$

$$\therefore \quad \frac{Z}{\left(\frac{F}{A}\right)} = \frac{\left(\frac{V}{A}\right)}{\left(\frac{F}{A}\right)} = \frac{V}{F} = \tau = \text{residence time} \tag{4.307}$$

$$\therefore \quad \tau = \frac{K_M}{r_{\max}} \ln \frac{C_{S_0}}{C_S} + \frac{1}{r_{\max}} [C_{S_0} - C_S] \tag{4.308}$$

For a desired conversion of reactant, one can calculate from this expression the length of the reactor (Z) and residence time (τ).

(c) Batch Reactor and PFTR

Following equations are derived based on mass balances in the elemental volume of dV using fractional conversion (X_S) of reactant

$$C_S F_{S_0} - F_{S_0}(C_S + dC_S) - (-r_S)dV = 0$$

$$-F_{S_0}\, dC_S - (-r_S)dV = 0$$

i.e.,

$$\frac{V}{F_{S_0}} = -\int \frac{dC_S}{(-r_S)}$$

$$C_S = C_{S_0}(1 - X_S)$$

$$dC_S = -C_{S_0} X_S \tag{4.309}$$

$$\frac{V}{F_{S_0}} = \tau = C_{S_0} \int \frac{dX_S}{(-r_S)}$$

This is the design equation of PFR. For batch bioreactors, the design equation is equation (4.310).

$$t = C_{S_0} \int \frac{dX_S}{(-r_S)} \tag{4.310}$$

For a simple reaction, $S \rightarrow P$

$$X_S = 1 - \frac{C_S}{C_{S_0}} = \frac{C_P}{C_{S_0}}$$

$$\therefore \quad (-r_S) = \frac{k_{\text{cat}} C_E C_S}{K_M + C_S} \tag{4.311}$$

Substitution of Equation (4.311) in Equation (4.309), gives τ.

$$\tau = C_{S_0} \int \frac{(K_M + C_S)\, dX_S}{k_{\text{cat}} C_E C_S}$$

$$= \frac{C_{S_0}}{k_{\text{cat}} C_E} \int \frac{(K_M + C_S)\, dX_S}{C_S}$$

$$= \frac{C_{S_0}}{k_{\text{cat}} C_E} \int \left(\frac{K_M}{C_S} + 1\right) dX_S$$

and $\quad C_S = C_{S_0}(1 - X_S).$

$$\therefore \qquad \tau = \frac{C_{S_0}}{k_{cat} C_E}\left[X_S - \frac{K_M}{C_{S_0}}\ln(1 - X_S)\right]_0^{X_S}$$

$$\therefore \qquad k_{cat} C_E \tau = C_{S_0} X_S - K_M \ln(1 - X_S) \tag{4.312}$$

(d) CFSTBR: Single Stage with Single Flow

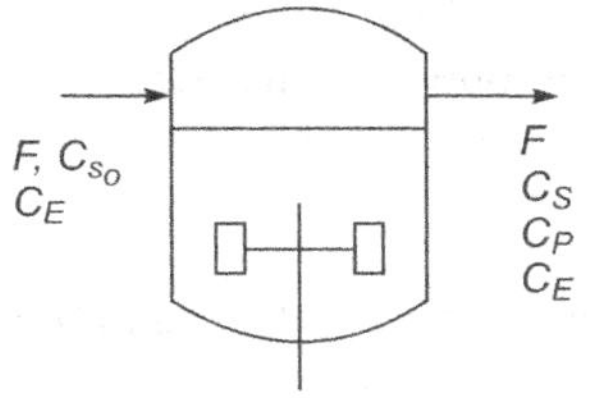

Figure 4.48 CFSTBR for enzymatic reactions.

For Reactions Using Enzymes in Free Form

The mass balance on the reactant at steady state gives

$$FC_{S_0} - FC_S - \frac{r_{max} C_S}{K_m + C_S} V = 0$$

$$D(C_{S_0} - C_S) = \frac{r_{max} C_S}{K_m + C_S}$$

$$D(C_{S_0} - C_S)(K_m + C_S) = r_{max} C_S$$

$$K_m D C_{S_0} - K_m D C_S + D C_{S_0} C_S - D C_S^2 = r_{max} C_S$$

$$D C_S^2 + (K_m D + r_{max} - D C_{S_0}) C_S - K_m D C_{S_0} = 0$$

$$C_S^2 + \left(K_m + \frac{r_{max}}{D} - C_{S_0}\right) C_S - K_m C_{S_0} = 0 \tag{4.313}$$

This is a quadratic equation in C_S, i.e., steady state reactant concentration.

$$C_S' = \frac{-\left(K_m + \frac{r_{max}}{D} - C_{S_0}\right) \pm \sqrt{\left(K_m + \frac{r_{max}}{D} - C_{S_0}\right)^2 - 4K_m C_{S_0}}}{2} \tag{4.314}$$

If we know C_{S_0}, K_M, r_{max} for a given conversion of reactant, we can calculate D(= dilution rate). Actual reactant and product profiles are given in Figure 4.49.

However, enzyme will not be used effectively in this process. One may consider as theoretically possible.

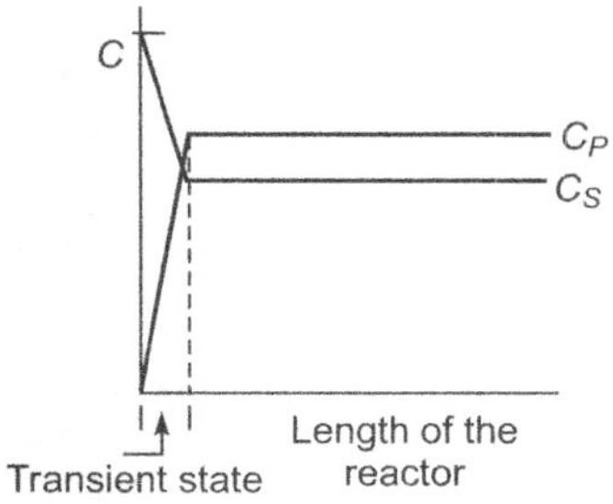

Figure 4.49 C_P and C_S profiles in CFSTBR.

Derivation using Fractional Conversion

Let us take a reaction, $S \xrightarrow{E} P$, having X_S defined by Equation (4.315).

$$X_S = 1 - \frac{C_S}{C_{S_0}} = \frac{C_P}{C_{S_0}} \tag{4.315}$$

The design equation of CFSTBR is expressed by Equation (4.316).

$$\frac{V}{F_{S_0}} = \frac{C_{S_0} X_S}{(-r_S)} \tag{4.316}$$

Assuming Michachis and Menten equation, $r_S = \dfrac{k_{\text{cat}}\, C_E C_S}{K_M + C_S}$

$$\tau = \frac{C_{S_0} X_S (K_M + C_S)}{k_{\text{cat}} C_E C_S}$$

$$C_S = C_{S_0}(1 - X_S)$$

$$\therefore \qquad k_{\text{cat}} C_E \tau = X_S C_{S_0} + \frac{K_M X_S}{(1 - X_S)} \tag{4.317}$$

Let us consider different situations of enzyme reactions in ideal reactors.

4.10.2 Analysis of Ideal Enzyme Reactors: Substrate Inhibition

For uncompetitive inhibition the reaction mechanism is described below.

$$E \underset{}{\overset{+S}{\rightleftarrows}} ES \rightarrow E + P$$

$$+I \; \big\Updownarrow \; K_I$$

$$ESI = ESS$$

The rate law for uncompetitive inhibition is given by Equation (4.318).

$$(-r_S) = \frac{k_{\text{cat}} C_E C_S}{K_M + C_S + \dfrac{C_S^2}{K_{IS}}} \tag{4.318}$$

(a) For Batch Reactor and PFTR

Substituting Equation (4.318) in the design equation of batch [Equation (4.310)] and PFTR [Equation (4.309)], we get

$$\frac{V}{F_{S_0}} = \tau = C_{S_0} \int \frac{\left(K_M + C_S + \dfrac{C_S{}^2}{K_{IS}}\right) dX_S}{k_{\text{cat}} C_E C_S}$$

i.e.,

$$\frac{k_{\text{cat}}\, C_E \tau}{C_{S_0}} = \int \left(1 + \frac{K_M}{C_S} + \frac{C_S}{K_{IS}}\right) dX_S \tag{4.319}$$

Also

$$C_S = C_{S_0}(1 - X_S)$$

On substituting C_S in the above equation and on integrating, we get

$$k_{\text{cat}}C_E\tau = C_{S_0}X_S - K_M \ln(1 - X_S) + \frac{C_{S_0}^2}{2K_{IS}}(2X_S - X_S^2) \tag{4.320}$$

Sintable expression is obtained for batch bioreactor to evaluate reaction time.

(b) For a CFSTBR

The design equation is (4.321).

$$\frac{V}{F_{S_0}} = \tau = \frac{C_{S_0}X_S}{-r_S} \tag{4.321}$$

i.e.,

$$\tau = \frac{C_{S_0}X_S\left(K_M + C_S + \frac{C_S^2}{K_{IS}}\right)}{k_{cat}C_E C_S}$$

$$\therefore \quad k_{\text{cat}}C_E\tau = C_{S_0}X_S\left(\frac{K_M}{C_S} + 1 + \frac{C_S}{K_{IS}}\right)$$

Also

$$C_S = C_{S_0}(1 - X_S)$$

$$\therefore \quad k_{\text{cat}}C_E\tau = C_{S_0}X_S\left(\frac{K_M}{C_{S_0}(1 - X_S)} + 1 + \frac{C_{S_0}(1 - X_S)}{K_{IS}}\right)$$

$$= \frac{X_S K_M}{(1 - X_S)} + C_{S_0}X_S + \frac{X_S C_{S_0}^2(1 - X_S)}{K_{IS}}$$

$$\therefore \quad k_{\text{cat}}C_E\tau = X_S C_{S_0} + \frac{K_M X_S}{(1 - X_S)} + \frac{C_{S_0}^2}{K_{IS}}(X_S - X_S^2) \tag{4.322}$$

4.10.3 Analysis of Ideal Enzyme Reactors: Product Inhibition

We consider competitive inhibition.

$$\begin{array}{l} E \underset{}{\overset{+S}{\rightleftharpoons}} ES \rightarrow E + P \\ \updownarrow \\ EP \end{array}$$

The rate expression is given by Equation (4.323).

$$(-r_S) = r_p = \frac{k_{\text{cat}}C_E C_S}{K_M\left(1 + \frac{C_P}{K_{IP}}\right) + C_S} \tag{4.323}$$

(a) For Batch Reactor and PFTR

$$\frac{V}{F_{S_0}} = \tau = C_{S_0} \int \frac{\left(K_M + C_S + \dfrac{K_M C_P}{K_{IP}}\right) dX_S}{k_{cat} C_E C_S}$$

i.e.,

$$\frac{k_{cat} C_E \tau}{C_{S_0}} = \int \left(1 + \frac{K_M}{C_S} + \frac{K_M C_P}{C_S K_{IP}}\right) dX_S \qquad (4.324)$$

and $C_S = C_{S_0}(1 - X_S)$.

On substituting C_S in the above equation and on integrating, we get

$$k_{cat} C_E \tau = C_{S_0} X_S \left(1 - \frac{K_M}{K_{IP}}\right) - \left(1 + \frac{C_{S_0}}{K_{IP}}\right) K_M \ln(1 - X_S) \qquad (4.325)$$

With proper modification of this equation it can be used for batch reactor.

(b) For a CFSTBR

The design equation is

$$\frac{V}{F_{S_0}} = \tau = \frac{C_{S_0} X_S}{(-r_S)}$$

i.e.,

$$\tau = \frac{C_{S_0} X_S \left(K_M + C_S + \dfrac{K_M C_P}{K_{IP}}\right)}{k_{cat} C_E C_S}$$

$$k_{cat} C_E \tau = C_{S_0} X_S \left(\frac{K_M}{C_S} + 1 + \frac{K_M C_P}{K_{IP} C_S}\right)$$

and $C_S = C_{S_0}(1 - X_S)$.

On substituting C_S in the above equation and on integrating this results Equation (4.326).

$$\therefore \qquad k_{cat} C_E \tau = X_S C_{S_0} + \frac{K_M X_S}{(1 - X_S)} + \frac{K_M X_S^2}{(1 - X_S)} \frac{C_{S_0}}{K_{IP}} \qquad (4.326)$$

4.10.4 Steps for Enzyme Reactor Design

The main aim of the reactor design is to find the volume for a given conversion or to find the conversion for an existing reactor of given volume.

The brief steps for a new reactor design are summarized below.

(i) Specify the conversion, X_S

(ii) Then find C_S by the following equation.

$$C_S = C_{S_0}(1 - X_S)$$

(iii) Find the rate expression for the given condition.

Example: for the simplest classical case, we use Michaelis and Menten equation, i.e.,

$$-r_s = \frac{k_{cat} C_E C_S}{K_M + C_S}$$

(iv) Insert this rate expression into the reactor design equation that we intend to operate.

For batch and PFTR, the design equation is, $t = C_{S_0} \int \frac{dX_S}{(-r_S)}$ and

$$\frac{V}{F_{S_0}} = \tau = C_{S_0} \int \frac{dX_S}{-r_S}, \text{ respectively.}$$

For CFSTR,

$$\frac{V}{F_{S_0}} = \tau = \frac{C_{S_0} X_S}{-r_S}$$

(v) Then find the volume of the reactor and also the residence time, τ.

4.10.5 Immobilized Enzyme Reactions

Analysis of basic reactors is made using immobilized enzyme

(a) Plug Flow Reactor

One can use Equation (4.327) to calculate τ.

$$\frac{F}{A}\frac{dC_S}{dz} = -\eta_{eff} \frac{r_{max} C_S}{K_M + C_S} \tag{4.327}$$

Then the final expression for τ is the Equation (4.328).

$$\tau_{im} = \frac{K_M}{r_{max}\eta_{eff}} \ln\left(\frac{C_{S_0}}{C_S}\right) + \frac{C_{S_0} - C_S}{r_{max}\eta_{eff}} \tag{4.328}$$

This is possible if we know from the separate experiment the value of η_{eff}, being assumed constant all through the length of the reactor.

In practical sense, η_{eff} is not constant which a function of C_S. Again C_S is a function of z.

Example 4.6

For immobilized enzyme in a tubular reactor of length, (L), C_E varies from inlet to outlet of the reactor. The reactor has the void fraction (ε). Calculate outlet concentrations of reactant and product.

Solution Following considerations are made for this analysis.

1. The kinetics is

$$\text{Reactant utilization rate} = -r_S = k_P C_E \frac{C_S}{K_m + C_S + \frac{C_P}{K_i}}$$

2. The mechanism of the reaction is

$$S \rightarrow P$$

Therefore, for reactant

$$\frac{dC_S}{dz} = -\frac{r_S}{u_z}$$

where u_z is the velocity of the reactant moving = F/A. In this case, it is $\frac{F}{A\varepsilon}$, where A is the cross sectional area of the reactor.

For product, $\frac{dC_P}{dz} = \frac{r_P}{u_z}$

From stiochiometry, $r_S = -r_P$

For linear variation of enzyme concentration, $C_E = C_Z + C_{E_i}$

Assumption: $z = 0$; $C_S = C_{S_0}$; $C_P = C_{P_0}$

We integrate the equation numerically to get the concentration profiles of C_S and C_P along the length of the reactor.

(b) For CFSTBR

Effectiveness factor is considered important. Hence,

$$D(C_{S_0} - C_S) = \eta_{\text{eff}} \frac{r_{\max} C_S}{K_M + C_S} \tag{4.329}$$

From the above expression, if η_{eff} is known, one can calculate required dilution rate for a given reactant conversion.

Example 4.7

Let us consider the reaction associated with inhibition. Calculate the outlet concentration of the reactant and product.

Solution We assume the following:

(1) reactor with flow streams as shown in the Figure 4.50.

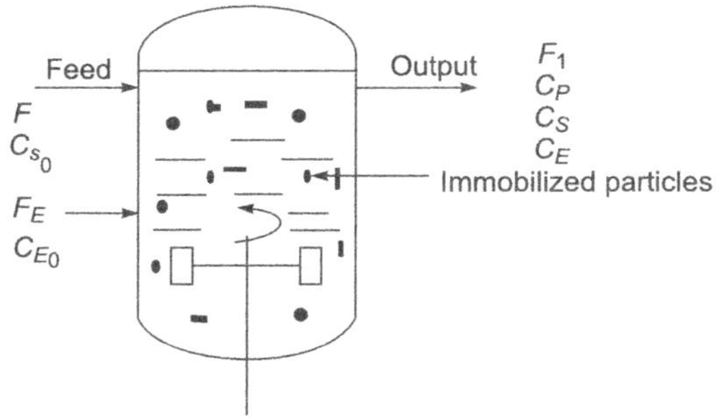

Figure 4.50 CFSTBR using immobilized system.

Considering the mass balances

Enzyme Mass Balance:

$$V\frac{dC_E}{dt} = F_E C_{E_0} - (F_E + F)C_E \tag{4.330}$$

Reactant Mass Balance:

$$V\frac{dC_S}{dt} = FC_{S_0} - (F_E + F)C_S - r_S V \tag{4.331}$$

Product Mass Balance:

$$V\frac{dC_P}{dt} = -(F_E + F)C_P + r_P V \tag{4.332}$$

Flow Rate Balance:

$$\frac{dV}{dt} = (F_E + F) - F_1 \tag{4.333}$$

For constant volume, $(F_E + F) = F_1$ (4.334)

At steady state, Equation (4.330) is given below.

$$F_E C_{E_0} = (F_E + F)C'_E$$

$$\therefore \quad C'_E = \frac{F_E}{(F_E + F)} C_{E_0} \tag{4.335}$$

For Start-up Period:

$$V\int_0^{C_E} \frac{dC_E}{F_E C_{E_0} - (F_E + F)C_E} = \int_0^t dt = t$$

$$V\int_0^{C_E} \frac{dC_E}{F_E C_{E_0} - (F_E + F)C_E} = \int_0^t dt = t$$

$$-\frac{V}{(F_E + F)} \ln\left(\frac{F_E C_{E_0} - (F_E + F)C_E}{FC_{E_0}}\right) = t$$

$$-\frac{(F_E + F)}{F}\frac{C_E}{C_{E_0}} + \frac{F_E}{F} = e^{-\frac{(F_E + F)t}{V}}$$

$$C_E = \frac{FC_{E_0}}{(F_E + F)}\left(\frac{F_E}{F} - e^{-\frac{(F_E + F)t}{V}}\right) \tag{4.336}$$

Steady state concentrations of C_P and C_S calculation are given below.

(1) ideal situation:

For ideal situation, Michaelis and Menten equation is used for the reaction mechanism described by (4.337).

$$S \rightarrow 2P \tag{4.337}$$

From stoichiometry, $r_p = 2(-r_s)$ (4.338)

Considering steady state conditions and initial conditions of C_P to be zero, it is possible to calculate C'_S and C'_P.

At steady state, reactant balance Equation (4.331) is Equation (4.339).

$$FC_{S_0} - (F_E + F)C'_S = r_S V \tag{4.339}$$

$$DC_{S_0} - D'C'_S = \frac{r_{max} C_{S_0}}{K_M + C'_S}$$

where $$D = \frac{F}{V}; D' = \frac{(F_E + F)}{V}$$

$$(DC_{S_0} - D'C'_S)(K_M + C'_S) = r_{max} C'_S$$

$$K_M DC_{S_0} + DC'_S C_{S_0} - D'K_M C'_S - D'C'^2_S - r_{max} C'_S = 0$$

$$DC'^2_S + (D'K_M + r_{max} - DC_{S_0})\, C'_S - K_M DC_{S_0} = 0$$

$$\therefore \quad C'_S = \frac{-(D'K_M + r_{max} - DC_{S_0}) \pm \sqrt{(D'K_M + r_{max} - DC_{S_0})^2 + 4K_M D'DC_{S_0}}}{2D'} \tag{4.340}$$

(2) Introducing inhibition kinetics for the relation,

$$-r_S = k_2 C_E \left(\frac{C_S}{K_m + C_S + \dfrac{C_P}{K_i}} \right) \tag{4.341}$$

For the reaction $S \rightarrow 2P$, the reaction stoichiometry is $r_p = 2(-r_S)$. One can calculate C'_S and C'_P.

(i) Single Stage and Single Flow Chemostat with Immobilized Cells

Ideal chemostat operation with immobilized cells considers no leaching of cells from the immobilized particles. Then the treatment for reactor analysis is same like immobilized enzyme. However, initially in chemostat operation immobilized particles are intact. Later some cells leaked from immobilized particles and they are called as cells in suspension (Fig. 4.51). These cells mix with the product stream. Analysis of this reactor situation should consider both suspended cells and immobilized cells.

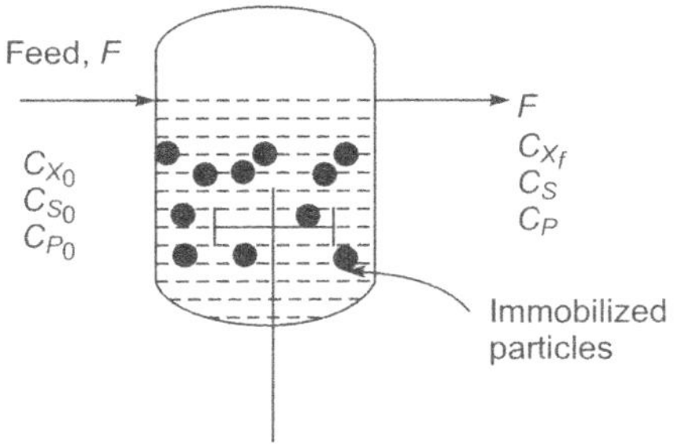

Figure 4.51 CFSTBR single flow with immobilized system.

To analyze this system, assumptions and notations are listed below.

Cells in immobilized state = $C_{X_{im}}$

Cells in suspension = C_{x_f}

Specific growth rate of cells = μ

Cells in inlet = C_{X_0}

Effectiveness factor for immobilized particles = η_{eff}

Considering $C_{X_0} = 0$ and assuming Monod's equation applies for cell growth.

Mass Balance Equation:

For cell:

$$\text{rate of change} = \text{input} - \text{output} \pm \text{generation} + \text{accumulation}$$

At steady state

$$FC_{X_0} - FC_{X_f} + \mu C_{X_f} V + \mu C_{X_{im}} V = 0$$

$$\because \quad C_{X_0} = 0$$

$$-FC_{X_f} + \mu C_{X_f} V + \mu C_{X_{im}} V = 0 \tag{4.342}$$

For immobilized particles diffusional resistances play important role for the growth in immobilized particles. Therefore, the above equation is written as Equation (4.343).

$$-FC_{X_f} + \mu C_{X_f} V + \eta_{\text{eff}} \mu C_{X_{im}} V = 0 \tag{4.343}$$

The dilution rate, $D = \mu\left(1 + \eta_{\text{eff}} \dfrac{C_{X_{im}}}{C_{X_f}}\right)$

For this reason, immobilized system can be run at higher dilution rate than CFSTBR using free cells (cells without immobilization). This is an added advantage for chemostat operation with immobilized particles. Above Equation (4.343) can be reduced to CFSTBR – chemostat for cells if $C_{X_{im}} = 0$ at steady state. This can happen for a poorly immobilized particles which will generate more of C_{X_f} and $\eta_{\text{eff}} = 1$.

For limiting reactant

$$\text{Rate of change} = \text{input} - \text{output} \pm \text{generation} + \text{accumulation}$$

$$V\frac{dC_S}{dt} = FC_{S_0} - FC_S - \frac{\mu C_{X_f} V}{Y^m_{X/S}} - \eta_{\text{eff}} \frac{\mu C_{X_{im}} V}{Y^m_{X/S}}$$

At steady state

$$F(C_{S_0} - C_S) = \frac{\mu V}{Y^m_{X/S}}(C_{X_f} + \eta_{\text{eff}} C_{X_{im}})$$

$$\Rightarrow \quad D(C_{S_0} - C_S) = \frac{\mu}{Y^m_{X/S}}(C_{X_f} + \eta_{\text{eff}} C_{X_{im}})$$

$$\Rightarrow \quad D = \frac{\mu}{Y^m_{X/S}} \frac{(C_{X_f} + \eta_{\text{eff}} C_{X_{im}})}{(C_{S_0} - C_S)} \tag{4.344}$$

If percentage conversion of reactant and C_{X_f} for a particular organism of known specific growth rate (estimated from free cell chemostat operation) are known, it is possible to calculate dilution rate, *D*, for an immobilized cell reactor to operate.

(ii) CFSTBR Series

Understanding of tubular behavior of reactor using immobilized system or soluble enzyme plus insoluble reactant, CFSTBR in series is a better choice. The concentration of reactant decreases from inlet to outlet of a tubular reactor which can be achieved by using large number of CFSTBR in series. In CFSTBR series reactors with single flow, reactant concentration decreases in stepwise manner (Figs. 4.52 and 4.53).

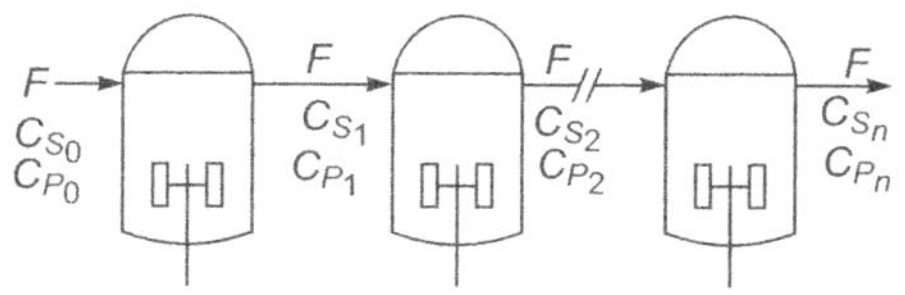

Figure 4.52 CFSTBR in series.

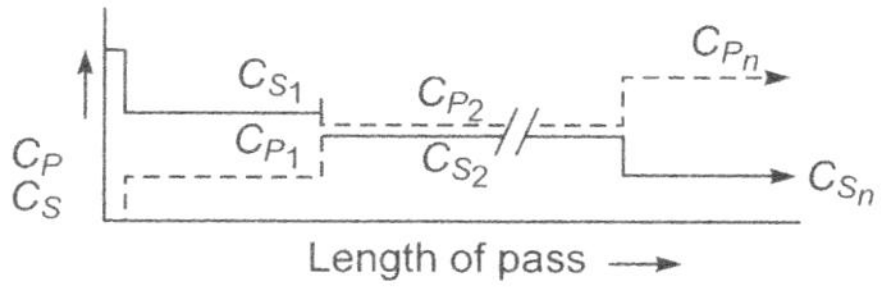

Figure 4.53 C_P and C_S along the length of the reactor.

Mass balances for reactant and product over different stages have been discussed earlier in this chapter (Section 4.7.3).

If we know stoichiometry of the reaction, we can express r_P and r_S easily.

For example, if r_S for soluble enzyme $= \dfrac{r_{max} C_S}{K_M + C_S}$

For the reaction mechanism, S(reactant) $\rightarrow P$(Product),

$$r_S = -r_P \tag{4.345}$$

These equations can be integrated simultaneously to calculate steady state values and also the dynamics of the system.

(c) Semi-continuous Reactor or Semi-batch Reactor

This is popular in submerged liquid fermentation for microorganisms. Let us consider this reactor with the following assumption for enzyme reactions.

Assumption

- Reaction mechanism is, S(reactant) $\rightarrow P$ (product).
- Kinetics of the reaction is expressed by Michaelis and Menten equation.

From the stoichiometry of the reaction, $r_s = -r_P$

From Mass Balances

For reactant

$$\frac{d(VC_S)}{dt} = FC_{S_0} - r_S V \tag{4.346}$$

C_{S_0} is initial concentration of reactant.

For the product

$$\frac{d(VC_P)}{dt} = r_P V \tag{4.347}$$

For the enzyme balance

$$\frac{d(VC_E)}{dt} = 0 \tag{4.348}$$

Change in volume

$$\frac{dV}{dt} = F \tag{4.349}$$

To calculate the effect of flow rate, the value of F would be changed which can be done in successive computer iterations.

EXAMPLE PROBLEMS

4.8 For the production of enzyme by a chemostat, the system is at steady state. A few parameters are available for this production.

$$D = 0.5\ \text{h}^{-1}$$
$$\mu_m = 0.8\ \text{h}^{-1}$$
$$K_S = 0.05\ \text{kg/m}^3$$
$$Y^m_{X/S} = 0.4$$
$$C_{S_0} = 20\ \text{kg/m}^3.$$

Calculate reactant and cell mass concentration at steady state.

Solution

$$C_S' = \frac{K_S}{\mu_m \tau_m - 1}$$

$$\tau_m = \frac{1}{D}$$

$$C_S' = 0.05/(1.6 - 1)$$
$$= 0.0833\ \text{kg/m}^3$$

$$C_X' = Y^m_{X/S}\left(C_{S_0} - \frac{K_S D}{\mu_m - D}\right)$$

$$= 7.9669\ \text{kg/m}^3.$$

4.9 An organism is growing in a CFSTBR with $D = 0.2\ \text{h}^{-1}$. The organism follows reactant inhibition. The model of Luong is used to explain the kinetics of the reaction. Here, $C_{S_0} = 15\text{kg/m}^3$, $\mu_m = 0.08\ \text{h}^{-1}$, $K_S = 0.006\ \text{kg/m}^3$. Assume other suitable data to get optimum values of D, C_S, C_X.

Solution:

For Luong model, we have

$$\mu = \frac{\mu_m C_S}{K_S + C_S}\left(1 - \frac{C_S}{C_S^*}\right)^n$$

$$r_g = \frac{\mu_m C_S}{K_S + C_S}\left(1 - \frac{C_S}{C_S^*}\right)^n C_X$$

Here, $$C_X = Y_{X/S}^m (C_{S_0} - C_S)$$

$$C_S = \left(C_{S_0} - \frac{C_X}{Y_{X/S}}\right)$$

$\therefore$ $$r_g = \frac{\mu_m\left(C_{S_0} - \dfrac{C_X}{Y_{X/S}^m}\right)}{K_S + \left(C_{S_0} - \dfrac{C_X}{Y_{X/S}^m}\right)}\left(1 - \frac{\left(C_{S_0} - \dfrac{C_X}{Y_{X/S}^m}\right)}{C_S^*}\right)^n C_X$$

$$= \frac{\mu_m (C_{S_0} Y_{X/S}^m - C_X) C_X}{Y_{X/S}^m (K_S + C_{S_0}) - C_X}\left(\frac{C_S^* - C_{S_0}}{C_S^*} + \frac{C_X}{C_S^* Y_{X/S}^m}\right)^n$$

It is more easier to write in terms of C_S.

$\therefore$ $$r_g = \frac{\mu_m C_S Y_{X/S}^m}{(K_S + C_S)}(C_{S_0} - C_S)\left(1 - \frac{C_S}{C_S^*}\right)^n$$

$$= \mu_m Y_{X/S}^m \left[\frac{(C_{S_0} C_S - C_S^2)\left(1 - \dfrac{C_S}{C_S^*}\right)^n}{K_S + C_S}\right]$$

For optimal condition, $\dfrac{dr_g}{dC_S} = 0.$

$$(K_S + C_S)(C_{S_0} - 2C_S)\left(1 - \frac{C_S}{C_S^*}\right)^n + (K_S + C_S)(C_{S_0} C_S - C_S^2) n\left(1 - \frac{C_S}{C_S^*}\right)^{n-1}\left[\frac{-1}{C_S^*}\right]^n$$

$$-(C_{S_0} C_S - C_S^2)\left(1 - \frac{C_S}{C_S^*}\right)^n = 0$$

$\Rightarrow$ $$\left(1 - \frac{C_S}{C_S^*}\right)^n \left[K_S (C_{S_0} - 2C_S) - C_S^2 - \frac{(K_S + C_S)(C_{S_0} C_S - C_S^2) n}{(C_S^* - C_S)}\right] = 0$$

Either $\left(1-\frac{C_S}{C_S^*}\right)^n = 0 \Rightarrow C_S^* = C_S$ (not possible)

Or $(K_S C_{S_0} - 2K_S C_S - C_S^2)(C_S^* - C_S) - (K_S + C_S)(C_{S_0} C_S - C_S^2)n = 0$

Here substituting the given values, we will get a cubic equation in C_S.

$$3C_S^3 - 179.76C_S^2 - 2.07C_S + 13.5 = 0$$

By solving the above equation, we get $C_S = 60.002, 0.2688, -0.279$

Since $C_{S_0} = 15 \text{ kg/m}^3;\ C_S = 0.2688 \text{ kg/m}^3$

$\therefore$ $C'_{X_{opt}} = Y_{X/S}(C'_{S_0} - C'_S) = 5.8925 \text{ kg/m}^3$

$$\tau_{opt} = C'_{X_{opt}} \times r_g$$

$$= 2.7073 \text{ h.}$$

$$D_{opt} = \frac{1}{\tau_{opt}} = 0.3694 \text{ h}^{-1}.$$

EXERCISES

4.1 An immobilized enzyme is packed into a 1 m^3 tubular reactor. Following data and kinetic parameters are available:

$$\eta_{neff} = 1$$
$$K_M = 1.5 \text{ kg/m}^3$$
$$r_{max} = 50 \text{ kg/m}^3 \text{ h}$$
$$C_{S_0} \text{ (inlet)} = 12 \text{ kg/m}^3$$

Desired conversion of reactant = 95%

Calculate the flow rate of the feed in the reactor. The reactor operates in plug flow condition.

4.2 An enzyme having r_{max} of 2 m mol/ m^3 s and K_M of 6 mM is used in batch reaction in the presence of a suitable reactant of initial concentration, 15 mM. Graphically, represent batch reaction time as a function of reactant concentration.

4.3 Get the similar plot for the enzyme given in Example 4.2 which experiences thermal decay. $t_{1/2}$ for the enzyme is 6 h.

4.4 The growth of bacteria can be expressed by the rate equation

$$\mu_g = \mu_{max}\left(1 - \frac{C_X}{C_{max}}\right)$$

where $\mu_{max} = 0.8 \text{ h}^{-1}$ and $C_{max} = 15 \text{ kg/m}^3$. Sufficient reactant is supplied to the medium. The cell growth is carried out in a 2m^3 batch reactor. 0.2 gm dry cell equivalent/m^3 of cell is supplied to the reactor as inoculum. Working volume of reaction is 1m^3. Calculate and plot the growth rates and cell concentration as a function of batch reactor time.

4.5 In a batch study for gluconic acid production by *Aspergillus niger* following data are available.

Time(h)	*0*	*10*	*20*	*30*	*40*	*50*	*60*	*70*	*80*
C_X(kg/m^3)	0.3	0.67	1.30	1.70	1.98	3.0	3.6	3.9	4.01
C_S(kg/m^3)	70	65	58	52	47	32	24	15	12
C_P(kg/m^3)	0	1.8	4.0	11.50	21.70	29	29.5	30	30

Calculate K_S, $\mu_{\text{app}_{\max}}$, lag time, total batch time, and cell growth time.

4.6 Use the data from Problem 4.5, show that the production of gluconoic acid is a mixture of growth and non-growth associated systems.

4.7 The growth of *Escherichia coli* used for penicillin acylase production can be expressed by Monod's kinetics with the parameters $\mu_m = 0.93\ \text{h}^{-1}$ and $K_S = 0.705(\text{kg/m}^3)$. The organism was growing on phenylacetic acid as the carbon-source. Assume that the cell yield $Y_{X/S} = 0.6$ kg dry weight of cells/kg reactant if $C_{X_0} = 0.8(\text{kg/m}^3)$ and $C_{S_0} = 9\ (\text{kg/m}^3)$. When the cells started to grow exponentially at $t_0 = 0$, show how ln C_X, C_S, C_X and $\frac{d\ln C_X}{dt}$ change with respect to time.

(Hint: Use Equation (4.39) and from Monod's equation $\frac{d\ln C_X}{dt} = \frac{\mu_m C_S}{C_S + K_S}$ Substitute all the value in Equation (4.50) and prepare a table of ln C_X, C_S, C_X and $\frac{d\ln C_X}{dt}$, t, $\frac{dC_X}{dt}$).

4.8 Develop an expression of specific growth rate related to reactant diffusion. Calculate $K_{S\text{app}}$.

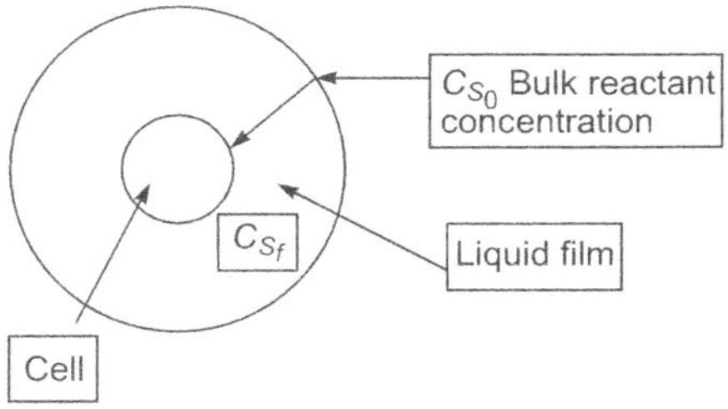

Consider X as cell and S as reactant and ρ_c as cell density, $k_L\,a$ as mass transfer coefficient. Assume steady state condition.

4.9 To calculate batch reaction time by the following expression

$$(t - t_0)\mu_m = \int_{C_{X0}}^{C_X} \left(1 + \frac{K_S Y}{YC_{S_0} + C_{X_0} - C_X}\right)\frac{1}{C_X}\,dC_X$$

Is it accurate?

What are the necessary conditions you need to propose?

4.10 State the possible non-idealities which might influence to calculate the batch reaction time.

4.11 When plotted on an arithmetic paper, the batch growth curve assumes a sigmoidal shape which is predicted by the Monod's equation.

$$\frac{dC_X}{dt} = \frac{\mu_m C_S}{C_S + K_S} C_X$$

There is no endogenous metabolism. Also, $C_X - C_{X_0} = Y_{X/S}(C_{S_0} - C_S)$, where C_{X_0} and C_{S_0} are initial values of cell and reactant concentration and the yield $Y_{X/S}$ is constant in the case. Calculate the time for batch reaction.

4.12 The reactant concentration in the sterile feed to the CFSTBR with recycle is 0.5kg/m^3. The outlet reactant level is 0.005 kg/m^3. The dilution rate is 1.5 h^{-1}. The growth rate of the organism is 0.8 h^{-1}. Fraction of the flow returning to the reactor is 30%. Cell concentration in the concentrate is 0.40 kg/m^3. What is the concentration of cell mass in the reactor?

4.13 A bacterial culture was grown in a batch mode on glucose and the following data were obtained.

Time (h)	*Cell mass concentration (kg/m^3)*	*Glucose concentration (kg/m^3)*
0	0.60	50
10	0.85	49
20	1.30	43
30	2.15	36
40	3.11	23
50	3.29	16

(a) Calculate the growth yields.
(b) What is the value of average growth yield?
(c) q_p and μ are related by a yield term. What is it? Select from $Y_{P/S}$, $Y_{S/P}$, $Y_{P/X}$, $Y_{X/P}$, $Y_{X/S}$, or $Y_{S/X}$.
(d) Suppose $C_{X_{max}} = C_{X_0} + YC_{S_0}$. Which Y is it?

Calculate $C_{X_{max}}$ from the data given above.

4.14 In Problem 4.13, is Y constant? If Y is a constant, assume Monod's equation to calculate batch time of reaction.

4.15 Write down the mass balances around the reactor and cell concentrator separately.

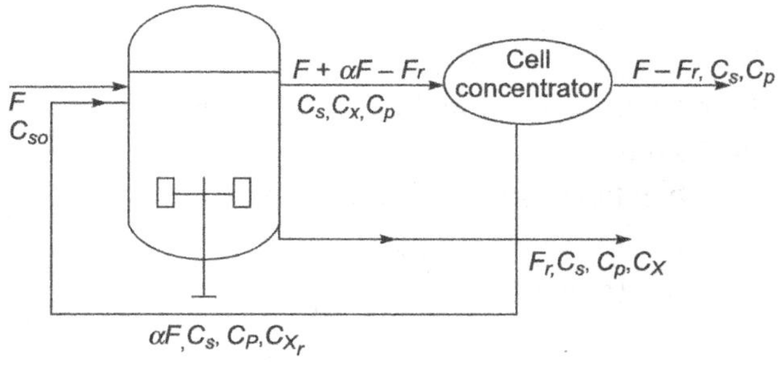

Hint: Consider two flows into the reactor and flows out from the reactor

REFERENCES

Aris R (1969) *Elementary Chemical Reactor Analysis*, Prentice-Hall, Englewood Cliffs, N.J.

Aris R and Armundson NR (Eds) (1973) *Mathemetical methods in chemical engineering 2*, Prentice-Hall, Englewood Cliffs, New Jersey.

Babu PSR (1991) "Studies on biosynthesis of penicillin amidase in *Escherichia coli*, stabilization and immobilization of the enzyme associated with whole cells," Ph.D. Thesis, IIT-Madras.

Brauer H (1979) "Power consumption in aerated stirred tank reactor systems" in: Ghose TK, Fiechter A, and Blakebrough N (Eds) *Advances in Biochemical Engineering*, vol 13, Springer, Berlin, pp 87-119.

Danckwerts PV (1951) "Significance of liquid film coefficients in gas absorption," *Industrial Engineering Chemistry*, **43**, 1460-1467.

Deindorfer FH and Humphrey AE (1961) "Mass transfer from individual gas bubbles". *Industrial Engineering Chemistry*, **53**, 755-759.

Doran PM (Ed) (1995) *Bioprocess Engineering Principles*, Academic Press, London.

Dunn IJ, Heinzle E, Ingham J, and Prĕnosil JE (Eds) (2003) *Biological reaction engineering*, 2nd edn., Wiley-VCH, Weinheim.

Einsele A and Finn RK (1980) *Industrial Engineering Chemistry*, "Process Design and Development" **19**, 600-603.

Fogler HS (Ed) (1992) *Elements of Chemical Reaction Engineering*, 2nd edn., Prentice-Hall, New Delhi, India.

Fredrickson AG (1992) "Growth processes in bioreactors with external sources of biomass: Application of structured continuum models," *AIChE Journal*, **38**, 835-845.

Fredrickson AG, Ramakrishna D, and Tsuchiya HM (1967) *Mathematical Biosciences*, **1**, 327.

Grieves RB, Pipes WO, Milbury WF, and Wood RK (1964) "Piston-flow reactor model for continuous industrial fermentations", *J Applied Chemistry*, **14** (11), 478–486.

Han K and Levenspiel O (1987) "Extended Monod kinetics for substrate, product, and cell inhibition." *Biotechnology and Bioengineering*, **32,** 430-437.

Hartwell LH Unger MW (1977), "Unequal division in *Saccharomyces cerevisiae* and its implications for the control of cell division". *Journal of Cell Biology,* **75**, 422–435.

Herbert D (1964) *Proc. Symp. Continuous Culture of Microorganisms*, Czechoslovak Academy of Sciences, Praha, p. 23.

Henzler HJ (1977) Ph.D. Thesis Technical University, Aachen, Germany.

Knorr D (Ed) (1987) "Biotechnological processes in food production", *Food Biotechnology*, vol 2, Marcel Dekker, New York, Basel.

Kundu S. (1983) "Microbial conversion of cellulose to ethanol", Ph.D. Thesis, IIT-Delhi, India.

Levenspiel O (Ed) (1972) *Chemical Reaction Engineering*, 2nd edn., John Wiley & Sons.

Levenspiel O (1980) "The Monod equation: revisit and a generalization to product inhibition situations". *Biotechnology and Bioengineering*, **22**, 1671–1687.

Liou JJ, Srienc F, and Fredrickson AG (1997) *Chemical Engineering Science*, **52**, 1529.

Loiseau B, Midoux N, and Charpentier J-C (1977), "Some hydrodynamics and power input data in mechanically agitated gas-liquid contactors". *AIChE J*, **23**(6), 931–935.

Ludeking R and Piret EL (2000) A kinetic study of the lactic acid fermentation: Batch process at controlled pH, *Biotechnology and Bioengineering*, **67**, 393-401.

Moser A (1985), in *Biotechnology: Fundamentals of Biochemical Engineering*, Vol. 2, Rehm HJ and Reed G (Eds), VCH, Weinheim, Germany.

Pirt SJ (Ed) (1975), *Principles of microbe and cell cultivation*. Blackwell Scientific, London, 214-215.

Prĕnosil JE, Dunn IJ, and Heinzle E (1987), "Biocatalyst reaction engineering". In *Biotechnology*, Kennedy JF (Ed), VCH, Weinheim, Germany, Vol 7a, 489-545.

Rao VSH and Rao PRS (2004) Global stability in chemostat models involving time delays, wall growth, nonlinear analysis, *Real world applications*, **5**, 141-158.

Suga K, Waki T, Kumano M, Chimange P, Shin SB, and Ichikawa K (1980) Production of cellulose in fed-batch system, pp 371-392, *Proceedings of Bioconversion and Biochemical engineering*, Symposium 2, vol II, Ghose TK (Ed) IITDelhi, New Delhi.

Schügerl K and Bellgardt KH (Eds) (2000) *Bioreaction Engineering: Modeling and Control*, Springer-Verlag, Berlin, Heidelberg.

Schügerl K (Ed) (1991) *Bioreaction Engineering*: *Characteristics features of bioreactors*, vol 2, John Wiley & Sons, New York.

Schügerl K (Ed) (1985) *Bioreaction engineerimg*, vol 1, Wiley, Chichester.

Schmauder HP, Schweizer M and Schewizer LM (2003) *Methods in Biotechnology*, Taylor & Francis, London.

Thilakavathi M, Panda T and Basak T (2007) Modeling of enzyme production kinetics, *Applied Microbiology and Biotechnology*, **73**, 991–1007.

Tsuchiya HM, Fredrickson AG, and Aris R (1967) (Eds) "Dynamics of microbial cell population", In: *Advances in Chemical Engineering*, vol. 6, Drew TE, Hoopes JW, Vermeulen T, New York, NY, Academic Press, pp. 125-206.

Wang DIC, Cooney CL, Demain AL, Dunnill P, Humphrey AE and Lilly MD (Eds) (1979) *Fermentation and Enzyme Technology*, John Wiley & Sons Inc., New York.

Walker AC and Schimdt CLA (1944) *Archives of Biochemistry*, **5**, 445.

Webb GF (1986), A model of proliferating cell populations with inherited cycle length, *J Math Biol*, **23**, 269–282.

Wen CY and Fan LT (Eds) (1975), In: *Models for Flow Systems and Chemical Reactors*, Marcel Dekker, New York.

Wiseman A (Ed) (1995) *Handbook of enzyme biotechnology*, 3rd ed, Ellis Horwood, London, Singapore.

Yuan JQ, Bellgardt KH, Deckwer W-D and Jiang W-S (1993), Modification and verification of the dynamic cell cycling model for baker's yeast, *Bioprocess Eng*, **9**, 173–182.

CHAPTER 5

Analysis of Non-Ideal Behavior in Bioreactors

OBJECTIVES

- To understand the design of real bioreactors, role of non-ideal parameters is necessary to visualize quantitatively in biological reactions. The analysis of those non-ideal parameters is dealt herewith.
- To understand the dynamic behavior of bioreactors and their stability analysis.

5.1 INTRODUCTION

We have discussed the ideal reactors in two categories, viz., perfect mixing and plug flow reactors. Under perfect mixing, batch stirred tank bioreactor (BSTBR) and single stage continuous flow stirred tank bioreactor (CFSTBR) are discussed where "batch cycle time" and "volume of the reactor" are the design criteria for BSTBR and CFSTBR, respectively. One can imagine that bioreactor design ends here. However, the discussion so far is just an overview of the real problem. The ideal concept of bioreactor does not match the real operation of the bioreactor. This is called non-ideality in bioreactor where ideal flows in bioreactor are affected by a number of parameters.

5.2 NON-IDEAL PARAMETERS

Non-ideal parameters are of biological, chemical, and of physical origins.
Biological factors are:

- Maintenance of cells
- Endogenous metabolism
- Wall growth
- Combination thereof and/with physical and chemical factors.

Chemical factors are:

- Inhibition
- Activation.

Physical factors are:

- Dead zones
- Bypassing or short-circuiting
- Channelling
- Axial mixing
- Laminar flow
- Recirculation zones/eddies
- Radial mixing.

The performance equation of the bioreactor may be defined in the following manner:

Output as a function of
- Input
- Organism behavior
  - Sometimes physiological and biochemical changes are difficult to calculate
- Material and energy balances
- Equilibrium and kinetics
  - It includes mass and heat transfer for heterogeneous systems
- Flow and contacting pattern
  - Early or late mixing
- State of aggregation of phases
  - Macro or micro fluid

So, the bioreactor design should include all the parameters effectively. Following diagram tries to consolidate the non-ideality (Fig. 5.1).

Therefore, the performance equation is described here.

Output: f {input, material and energy balances, equilibrium and kinetics, flow and contacting pattern, state of aggregation of the phases, biological parameters}

Equilibrium and kinetics include mass and heat transfer for heterogeneous systems, whereas flow and contacting pattern include RTD, early or late mixing, PFTR and CFSTBR extremes. State of aggregation of the phases says whether the fluid is micro or macro nature.

Let us consider the biological factors responsible for non-ideality.

5.2.1 In CFSTBR

(a) Effect of Endogenous Metabolism

We recall the mass balance of single limiting reactant with no endogenous metabolism (Equation 4.152)

$$FC_{S_0} - FC_S - \left(\frac{V\mu_g C_X}{Y_{X/S}^m}\right) - \left(\frac{Vq_p C_X}{Y_{P/S}^m}\right) = V\frac{dC_S}{dt} \tag{5.1}$$

where q_p is the specific rate of extracellular product formation.

Figure 5.1 Stages of understanding complexity in a real bioreactor (Proper modified form of the lecture note on "Advanced bioreactor theory: Fundamentals of bioreactor design" pp. 179–192, in short term course on 'Microbial Engineering', coordinators: Baradarajan A and Panda T, 1988. Reproduced with kind permission from Professor K. Krishnaiah, Department of Chemical Engineering, IIT Madras).

Figure 5.1 Stages of understanding complexity in a real bioreactor (Proper modified form of the lecture note on "Advanced bioreactor theory: Fundamentals of bioreactor design" pp. 179–192, in short term course on 'Microbial Engineering', coordinators: Baradarajan A and Panda T, 1988. Reproduced with kind permission from Professor K. Krishnaiah, Department of Chemical Engineering, IIT Madras).

For endogenous metabolism, the cell mass balance is given below.

$$\frac{dC_X}{dt} = \mu C_X - DC_X - k_d C_X \tag{5.2}$$

k_d is endogenous rate constant.

At steady state

$$D = \mu_g - k_d = \mu_{\text{net}}$$

or

$$\mu_g = D + k_d \tag{5.3}$$

Assuming no extracellular product formation, then

$$D(C_{S_0} - C_S) - \left(\frac{D + k_d}{Y^m_{X/S}}\right) C_X = 0 \tag{5.4}$$

$Y^m_{X/S}$ is a single constant value independent of growth rate.

$$\therefore \qquad \frac{D(C_{S_0} - C_S)}{C_X} - \left(\frac{D + k_d}{Y^m_{X/S}}\right) = 0$$

or

$$D\left(\frac{1}{Y^{\text{app}}_{X/S}}\right) - D\left(\frac{1}{Y^m_{X/S}}\right) - \frac{k_d}{Y^m_{X/S}} = 0$$

or

$$\left(\frac{1}{Y^{\text{app}}_{X/S}}\right) = \left(\frac{1}{Y^m_{X/S}}\right) + \frac{k_d}{Y^m_{X/S}}\frac{1}{D}$$

or

$$\left(\frac{1}{Y^{\text{app}}_{X/S}}\right) = \left(\frac{1}{Y^m_{X/S}}\right) + \frac{m_S}{D} \tag{5.5}$$

where m_s is the maintenance coefficient $\left(\frac{k_d}{Y^m_{X/S}}\right)$ based on reactant and $Y^{\text{app}}_{X/S}$ is the apparent yield $= \left(\frac{C_X}{C_{S_0} - C_S}\right)$.

We can evaluate $Y^m_{X/S}$ and m_s from the CFSTBR-chemostat study by varying D (Fig. 5.2).

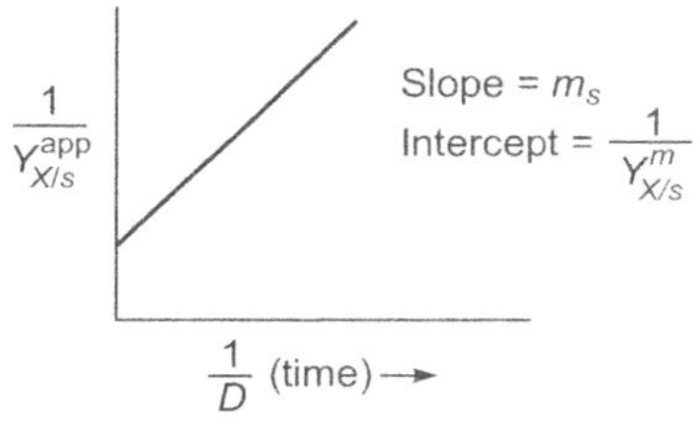

Figure 5.2 Estimation of m_s and $\frac{1}{Y^m_{X/S}}$.

From the above Equations (5.1) and (5.3) assuming Monod's kinetics

$$\mu_{\text{net}} = \mu_g - k_d$$

$$\mu_{\text{net}} = \frac{\mu_m C_S}{K_S + C_S} - k_d$$

Again $\mu_{\text{net}} = D$. Therefore,

$$C'_S = \frac{K_S(D + k_d)}{\mu_m - D - k_d} \quad \text{and} \quad C'_X = \frac{Y^m_{X/S}(C_{S_0} - C'_S)D}{D + k_d}$$

Case 1 Endogenous metabolism with the production of extracellular products

Growth associated product is considered in this example. From the mass balance Equation (5.1)

$$D(C_{S_0} - C'_S) = \frac{1}{Y^m_{X/S}}(D + k_d)\, C'_X + \frac{1}{Y^m_{P/S}}\, q_p C'_X$$

Therefore,

$$C'_S = \frac{K_S\,(D + k_d)}{\mu_m - D - k_d} \tag{5.6}$$

whereas,

$$C'_X = Y^m_{X/S}(C_{S_0} - C'_S)\left(\frac{D}{D + k_d + \dfrac{Y^m_{X/S}}{Y^m_{P/S}} q_P}\right) \tag{5.7}$$

The productivity of the extracellular product is given by DC_P and for cells, it is DC_X. The dilution rate that maximizes the productivity can be found by differentiating product productivity and cell productivity with respect to dilution rate and equating each term to zero.

i.e., $$\frac{d(DC_P)}{dD} = 0 \quad \text{and} \quad \frac{d(DC_X)}{dD} = 0.$$

The optimal value of D (i.e., D_{opt}) will depend on the fact that whether endogenous metabolism alone or endogenous metabolism associated with product formation is being considered.

Case 2 Effect of endogenous metabolism and maintenance

Considering the mass balances of the reactant

$$\frac{dC_S}{dt} = D(C_{S_0} - C_S) - \frac{1}{Y^m_{X/S}}\mu C_X - mC_X$$

where 'm' accounts for the maintenance of cell. Assuming Monod's equation and steady state conditions

$$C'_S = \frac{DK_S}{\mu_m - D}$$

$$C'_X = \frac{D(C_{S_0} - C'_S)}{\left(m + \dfrac{D}{Y^m_{X/S}}\right)} \tag{5.8}$$

where the superscript '′' indicates steady-state conditions.

On the other hand, from the cell mass balance

$$\frac{dC_X}{dt} = \mu_g C_X - DC_X - k_d C_X$$

where k_d is the endogenous metabolism effect. So

$$C'_S = \frac{(D + k_d)K_S}{\mu_m - D - k_d} \tag{5.9}$$

From Equations (5.8) and (5.9), we get a combined effect of maintenance and endogenous metabolism.

(b) Effect of Wall Growth

Wall growth is a serious problem in the bioreactor operation. This mainly happens during mold cultivation. Part of the cells will be attached to the reactor components and consume reactants. Effectively, they do not contribute to product synthesis. In chemostat operation, wall growth interferes the setting of proper dilution rate. In extreme case, one may not achieve true wash out condition if wall growth occurs in the reactor. Quantitation of cell mass in attached growth is really difficult as there is no proper sampling strategy available during the course of reaction. Growth is also non-uniform. In this context, the analysis is based on the models of Howell *et al.*, (1972) and Schügerl (1985). Cell growth and reactant consumption are considered by the cells present both in attached form and in reaction fluid.

Mass Balances

On cells

$$\frac{dC_X}{dt} = D(C_{X_0} - C_X) + \mu_g C_X + \mu_{gw} C_X \tag{5.10}$$

where $\mu_g C_X$ is the growth in reaction fluid and $\mu_{gw} C_X$ is the growth in attached form.

On reactant

$$\frac{dC_S}{dt} = D(C_{S_0} - C_S) - \frac{\mu_g C_X}{Y^m_{X/S}} - \frac{\mu_{gw} C_X}{Y^m_{X/S}} \tag{5.11}$$

Howell *et al.* (1972) assumed

$$r_g = \frac{\mu_m C_S}{K_S + C_S + \dfrac{C_S^2}{K_I}} \tag{5.12}$$

Schügerl (1985) suggested

$$r_{gw} = \frac{\mu_m C_S X_w A}{K_S + C_S + \dfrac{C_S^2}{K_I}} \tag{5.13}$$

where A is the area of the wall covered by the cells and X_w is the wall coating density.

$$X_w = \eta \frac{d(1-\varepsilon)}{v}$$

where η is the effectiveness factor for attached growth influenced by mass transfer limitation, ε is voidage of wall growth layers, d is the thickness of cell layer and v is the volume of a single micro-organism.

For mycelial growth, it is difficult to estimate v. X_w has the dimension of (length)$^{-2}$. It is observed that in the presence of wall growth there is no washout. This will be discussed later in detail in this chapter.

(c) Effect of Reactant Inhibition

Let us consider the inhibition model expressed by Howell *et al.* (1972).

$$\mu = \frac{\mu_m C_S}{K_S + C_S + \dfrac{C_S^2}{K_I}}$$

Cell Mass Balance

$$\frac{dC_X}{dt} = \left(\frac{\mu_m C_S}{K_S + C_S + \frac{C_S{}^2}{K_I}} - D \right) C_X \tag{5.14}$$

Reactant Mass Balance

$$\frac{dC_S}{dt} = D(C_{S_0} - C_S) - \left(\frac{\mu_m C_S}{K_S + C_S + \frac{C_S^2}{K_I}} \right) \frac{C_X}{Y_{X/S}^m} \tag{5.15}$$

We can apply steady state condition to the above equations to calculate C_X' and C_S'.

Biological factors along with physical and chemical parameters cause resultant fluid flow into and out of the reactor. To estimate the non-ideal parameters like axial mixing, channelling, by-pass and dead space in reactors, Danckwerts proposed Residence Time Distribution (RTD) studies. Some of the interrelated factors make up the contacting pattern along with RTD are state of aggregation and earliness and lateness of mixing.

5.3 RESIDENCE TIME DISTRIBUTION—SOME ASPECTS OF MACRO MIXING

The time spent by a packet of fluid right from the entry to the exit of the bioreactor is called residence time for that fluid packet. It is expressed in terms of τ in hour or minutes, etc. Another way of expressing residence time is by dimensionless time, $\theta = \frac{t}{\tau}$.

For different reactors, the requirement for residence time of a fluid is demanded by the process. For example, in bioreactor, fluid containing reactant must stay for sufficient time in the reactor for the maximum conversion of reactant by the organism to product(s). This stay (i.e., residence time) should be minimum so that it does not cause reactant limitation or product inhibition. For reactors used in waste treatment processes, the minimum time should be sufficient enough to settle the solids.

Then one might find that the packet of entering fluid, after reaction, is leaving the reactor at the same time. **It does not happen in the real reactor**. This is called non-ideality in the system. The non-ideal behavior is described critically by Levenspiel (1972) and Fogler (1992) with reference to chemical reactions in the reactor.

Let us examine this situation. The packet of fluid after entering into the bioreactor will be disintegrated into smaller components which will separate and disperse throughout the vessel. Thus, some fraction of this fluid element will rapidly find its way to the exit stream, while other fractions will wander inside the reactor for varying times before leaving the system. Thus the exit stream contains fluid elements of different residence time. The determination of the distribution of these residence times in the exit stream is an important indicator of the mixing and flow patterns within the reactor.

For the estimation of RTD of the fluid in a bioreactor, one needs to perform a stimulus response experiment (Fig. 5.3) in which the input is changed in a desired way and the output is analysed as a response in the exit stream. Input signals are of step and pulse types. In chemical reactor, stimulus response experiment is carried out by an inert component which is known as tracer.

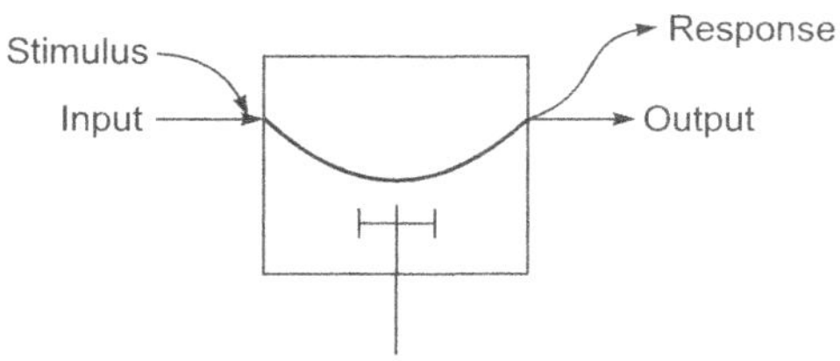

Figure 5.3 Experiments on stimulus response in a bioreactor.

The properties of tracer are the following:

(1) It should be easily identifiable and measurable component.
(2) Physical properties of the main fluid and the tracer must be the same.

One should be careful to take the right amount of tracer without disturbing the hydrodynamic behavior of the fluid.

In biological systems, it is difficult to find such tracer for RTD analysis. In presence of cells or enzymes the tracer may be adsorbed on them which finally will hinder the estimation of RTD.

5.3.1 Ways to Characterize RTD

There are three ways to characterise RTD.

- E-function
- C-function
- F-function

(a) The E-Function (Exit Age Distribution)

The fraction of the fluid in exit stream which has been in the vessel for time between t and $(t + dt)$ is $E(t)dt$.

How can one determine E(t)?

Take a sample from the exit stream of the arrangement shown in Figure 5.3 and determine the fraction of stimulus or tracer in it with a residence time between t and $(t + dt)$ and continue for different times.

$\int_t^{t_1} E(t)dt$ = volume fraction of the sample of the outgoing flow with the residence time between t and t_1 and $\int_0^{\infty} E(t)dt = 1$. A residence time of infinity means that the element never leaves the reactor. To understand this let us examine Figure 5.4.

Less residence time implies less mixing whereas longer residence time is observed for more mixing.

(b) The C-Function

One can estimate the concentration of the tracer in the exit stream and can express the RTD. The plot of the concentration of the tracer in the exit stream as a function of time is called the C-function. Let us observe from Figure 5.5 about the C-function.

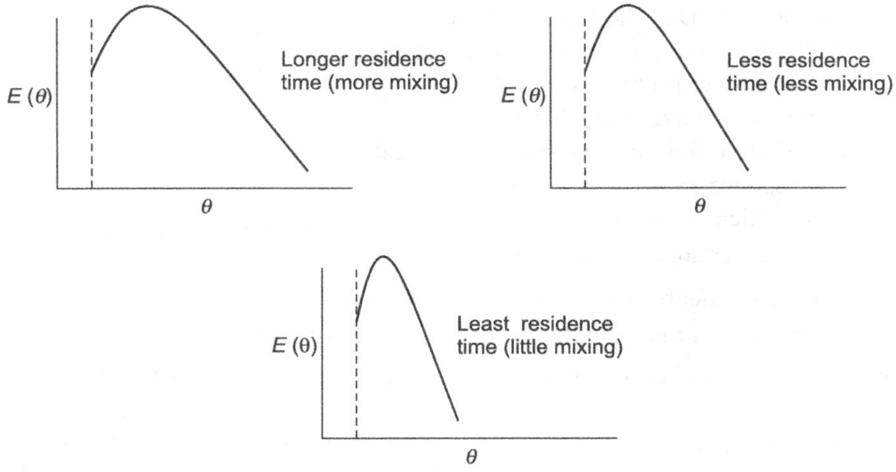

Figure 5.4 $E(\theta)$ vs. θ at different mixing conditions in the bioreactor.

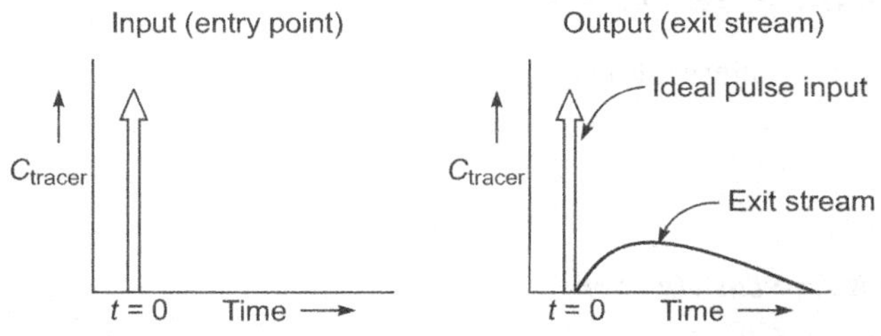

Figure 5.5 C_{tracer} as a function of time.

Hence, total amount of tracer injected in a reactor of volume, V, and a concentration of $C_{Ti} = VC_{Ti}$. The C-function is described by Equation (5.16).

$$C(t) = \frac{C_{T,\,\text{exit}}(t)}{C_{Ti}} \tag{5.16}$$

or in terms of dimensionless time

$$C(\theta) = \frac{C_{T,\,\text{exit}}(\theta)}{C_{Ti}} \tag{5.17}$$

Volumetric mass balance for tracer is described by Equation (5.18).

$$\frac{dM_T}{dt} = f_m C_{Ti} - f_m C_{T,\,\text{exit}} \tag{5.18}$$

where f_m is the mass flow rate.

Integrating the above expression

$$\int_0^\infty dM_T = -M_{T,\,\text{total}} = -VC_{Ti} \tag{5.19}$$

$$= -\int_0^\infty f_m C_{T,\,\text{exit}}\, dt$$

$$\therefore \quad 1 = \int_0^\infty \frac{f_m C_{T,\,\text{exit}}}{VC_{Ti}}\, dt$$

$$= \int_0^\infty \left(\frac{f_m}{V}\right)\left(\frac{C_{T,\,\text{exit}}}{C_{Ti}}\right) dt$$

$$= \int_0^\infty \left(\frac{1}{\tau}\right) C(\theta)\, dt$$

$$= \int_0^\infty C(\theta)\, d\theta \tag{5.20}$$

For a steady-state system $C(\theta)$ and $E(\theta)$ are equal.

(c) The F-Function

Initially there is no tracer anywhere in the reactor. A step-input of tracer is introduced in the inlet fluid stream. Then the tracer is measured in the exit stream as a function of time (Fig. 5.6).

$$F(\theta) = \frac{C_{T,\,\text{exit}}(\theta)}{C_{Ti}} \tag{5.21}$$

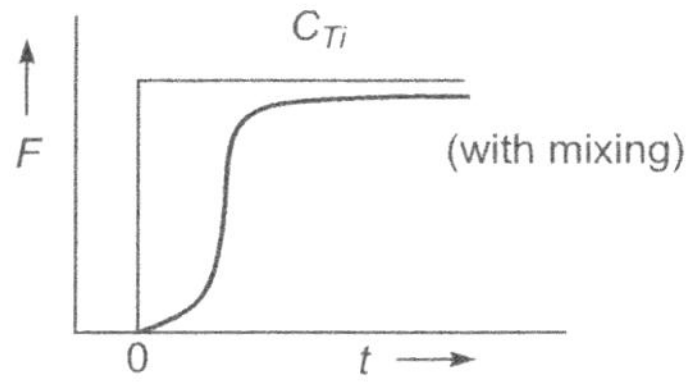

Figure 5.6 F-function plot for a single reactor in mixed flow.

(d) The Physical Significance of F(θ)

The volume fraction in the outgoing flow has a τ (residence time) less than $\theta\left(=\frac{t}{\tau}\right)$.

(e) The Relation between E-, C-, and F-functions

From the definitions of $C(\theta)$ and $E(\theta)$, $C(\theta)\,d\theta$ is defined as the volume fraction in the exit flow with residence time between θ and $d\theta$. C-function gives directly the exit age distribution $E(\theta)$.

For a steady state

$$F(\theta) = \int_0^\theta C(\theta)\, d\theta \tag{5.22}$$

or in general sense

$$F = \int_0^t E dt$$

$$\therefore \quad \frac{dF}{dt} = E \tag{5.23}$$

$$\therefore \quad E = C = \frac{dF}{dt} \tag{5.24}$$

The mean holding time or mean residence time (τ) can be calculated from the tracer concentration (Levenspiel, 1972).

$$\tau = \frac{\int_0^\infty tCdt}{\int_0^\infty Cdt} \approx \frac{\Sigma t_i C_i \Delta t_i}{\Sigma C_i \Delta t_i} \tag{5.25}$$

(f) Examples Related to E-, C-, and F- functions using Ideal Input Signals of Tracer in Bioreactors

General response obtained with ideal input signals.

(i) Ideal Pulse gives exit age distribution function $E(t)$. $E(t)\,dt$ gives the fraction of material that has spent a time between t and $(t + dt)$ in the bioreactor. From the experiment, C vs. t data are collected and plotted in Figure 5.7 (Krishnaiah, 1988).

Area under C vs. t curve is Q.

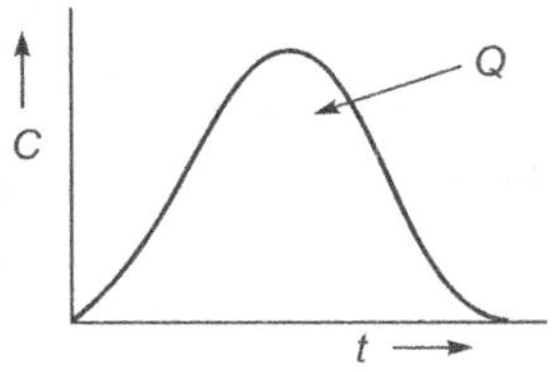

Figure 5.7 *C* vs. *t* plot.

$$Q = \int_0^\infty Cdt$$

To get $E(t)$, it is required to calculate $\left(\frac{C}{Q}\right)$. The plot $E(t)$ vs. t (Fig. 5.8) is the required RTD.

Therefore,

$$\int_0^\infty E(t)dt = 1 \tag{5.26}$$

(ii) Step input: This gives cumulative age distribution function $F(t)$. $F(t)$ is the fraction of material that has spent a time t in the bioreactor vessel. If C_{Ti} is the concentration of the tracer at the inlet for step input, τ, where C_{exit} is the concentration of tracer at any time in the outlet of the bioreactor.

To get $F(t)$, Figure 5.9 is the plot necessary in this regard.

$$\therefore \qquad \tau = \int_0^1 t dF \tag{5.27}$$

The $E(t)$ vs. t plot can be obtained by measuring slope of $F(t)$ vs. t diagram for the step input at various t values (Fig. 5.10).

$$F(t) = \int_0^t E(t)dt$$

$$\frac{dF(t)}{dt} = E(t) \tag{5.28}$$

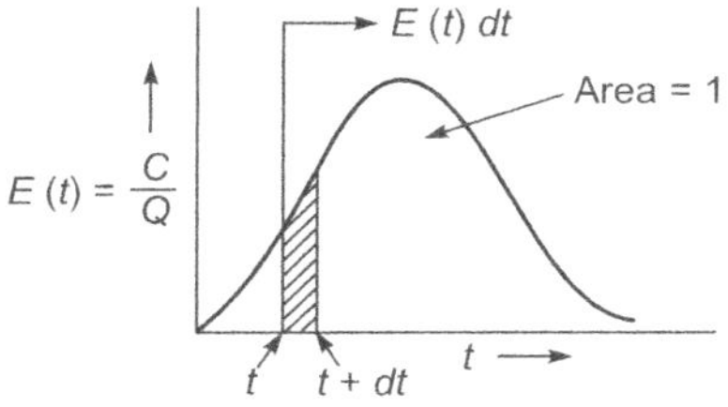

Figure 5.8 $E(t)$ vs. t plot.

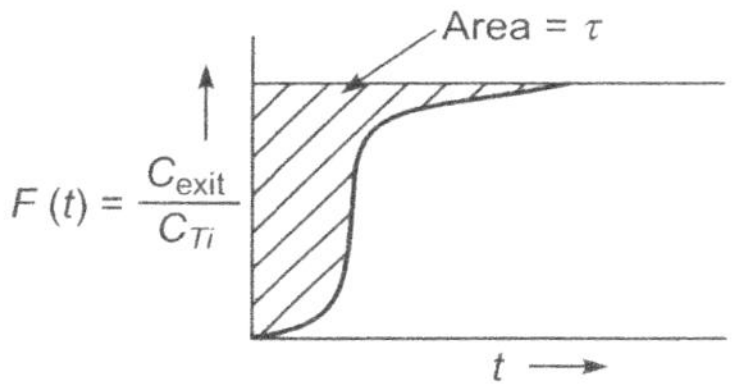

Figure 5.9 $F(t)$ vs. t plot.

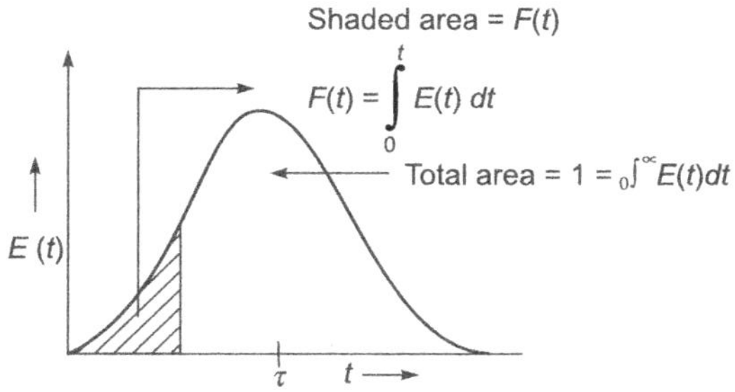

Figure 5.10 $E(t)$ vs. t plot.

5.4 SOME EXERCISE FOR RTD OF IDEAL SYSTEMS (IDEAL BIOREACTORS)

(i) For Plug Flow systems, the $F(t)$ vs. t plots are given in Figure 5.11.

(ii) For Mixed Flow/Perfect Mixed Flow Reactor: $E(t)$ and $F(t)$ vs. t plots are given in Figure 5.12.

(iii) For real reactor, the situation is described below in Figure 5.13.

To model and to understand the real reactor one requires to use non-ideal models like axial dispersion, tanks in series, CFSTBR with bypass and dead zones, etc.

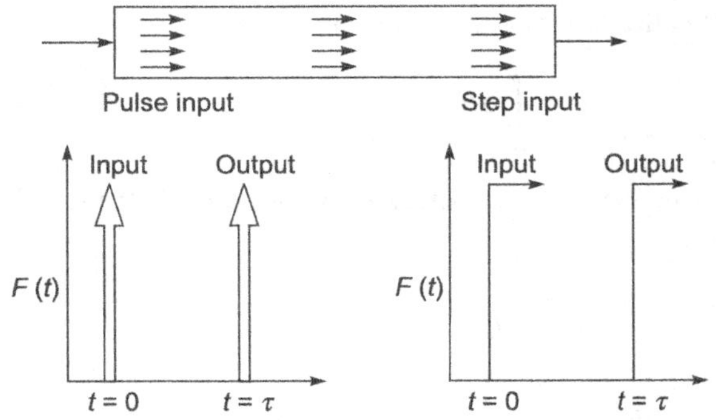

Figure 5.11 $F(t)$ vs. t plot for PFR. For pulse input, $F(t) = \delta(t - \tau)$ whereas for step input, $F(t) = 0$, $t \leq \tau$ and $F(t) = 1$, $t > \tau$.

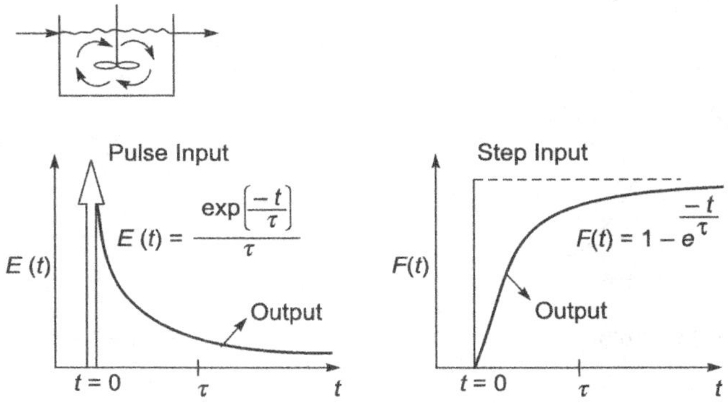

Figure 5.12 $E(t)$ and $F(t)$ vs. t plots.

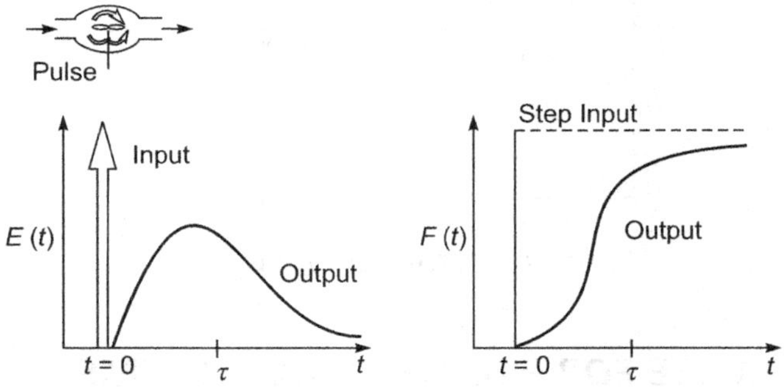

Figure 5.13 $E(t)$ vs. t plot for real reactor.

It is easier to compare the RTDs by using their moments rather than the comparison of entire distribution (Wen and Fan, 1975).

5.5 MOMENTS OF THE DISTRIBUTION

For the consideration of distribution functions like $E(t)$, it is better to calculate the moments of the distribution. For example, the kth moment of $E(t)$ is defined by Equation (5.29).

$$m_k = \int_0^{\infty} t^k E(t)dt \,,\; k = 0, 1, 2, \ldots \tag{5.29}$$

Since a unit amount of tracer is introduced into a bioreactor to observe its E-curve and all tracer eventually must leave the bioreactor vessel, the zeroth moment is $m_0 = 1$.

The first moment (m_1) is the mean of RTD or the mean residence time (τ), under no back diffusion. For a single phase fluid in an arbitrary reactor, the mean residence time (τ) is

$$\tau = m_1 = \frac{V_R}{F} \tag{5.30}$$

where, V_R is the reaction volume and F is flow rate of the feed.

The second moment (m_2) is expressed in terms of variance of distributions. This is the average of the squares of deviations from the mean residence time, i.e.,

$$m_2 = \sigma^2 - m_1^2 \tag{5.31}$$

where σ^2 is the variance for the quantifying spread of the distribution. σ^2 is calculated from the following Equation (5.32).

$$\sigma^2 = \frac{\int_0^{\infty} (t-\tau)^2 C dt}{\int_0^{\infty} C dt} \tag{5.32}$$

or

$$\sigma^2 = \int_0^{\infty} (t-\tau)^2 E(t) dt$$

$$m_2 = \frac{\int_0^{\infty} t^2 C dt}{\int_0^{\infty} C dt} - \tau^2$$

$$\therefore \qquad \sigma^2 \approx \frac{\Sigma t_i^2 C_i \Delta t_i}{\Sigma C_i \Delta t_i} - \tau^2$$

This magnitude is an indication of the spread of the distribution. The third moment is taken about the mean and is related to skewness. The skewness is defined by Fogler (1992).

$$m_3 = s^3 = \frac{1}{\sigma^{3/2}} \int_0^{\infty} (t-\tau)^3 E(t)\, dt \tag{5.33}$$

The magnitude signifies the degree of skewness in the distribution.

5.6 *E*(*t*) OR *F*(*t*) AND THE BIOREACTOR DESIGN

To obtain experimental $E(t)$ or $F(t)$ and the necessary information for the bioreactor design is given in Figure 5.14. F is the flow rate and C_A is the reactant concentration.

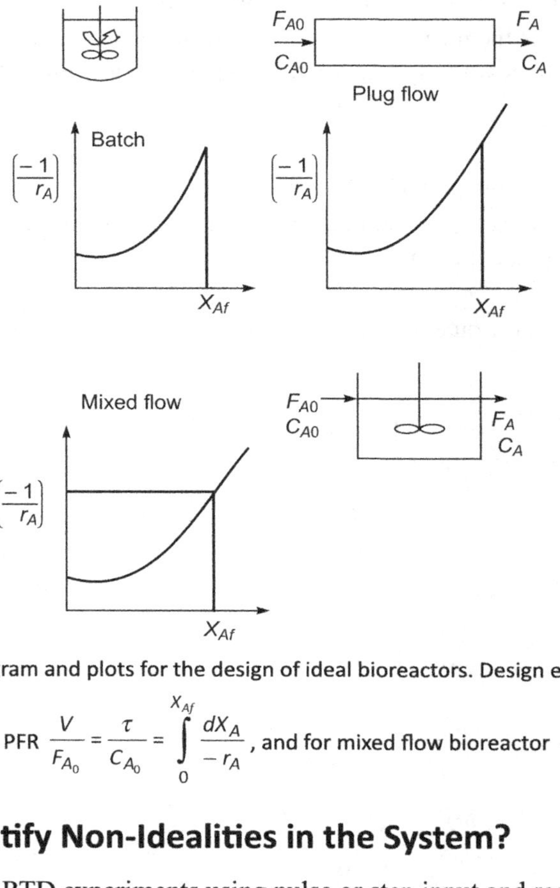

Figure 5.14 Schematic diagram and plots for the design of ideal bioreactors. Design equations for batch reactors

$$t = C_{A_0} \int_0^{X_{Af}} \frac{dX_A}{-r_A}, \text{ for PFR } \frac{V}{F_{A_0}} = \frac{\tau}{C_{A_0}} = \int_0^{X_{Af}} \frac{dX_A}{-r_A}, \text{ and for mixed flow bioreactor } \frac{V}{F_{A_0}} = \frac{\tau}{C_{A_0}} = \frac{X_{Af}}{-r_A}.$$

5.6.1 How to Identify Non-Idealities in the System?

It is defined by conducting RTD experiments using pulse or step input and recording the corresponding outputs. From Figure 5.15, we can have an idea of the type of non-ideality existing in the reactor for pulse input.

5.6.2 Assessment of Non-Ideality

For bioreactor design, it is necessary to know, as far as possible, the real kinetic parameters and quantitative information of the factors influencing those parameters and hence the bioreactor performance. In order to estimate the possible quantitative information for the design of real reactor, the real bioreactor is assumed to be the combination of phenomena occuring in ideal reactors. Those information are compared with the ideal analytical tools. Some of them depict the ideal behavior. Others show the modification of basic ideal tools or systems. For example, to analyze the plug flow tubular reactor it is difficult to find

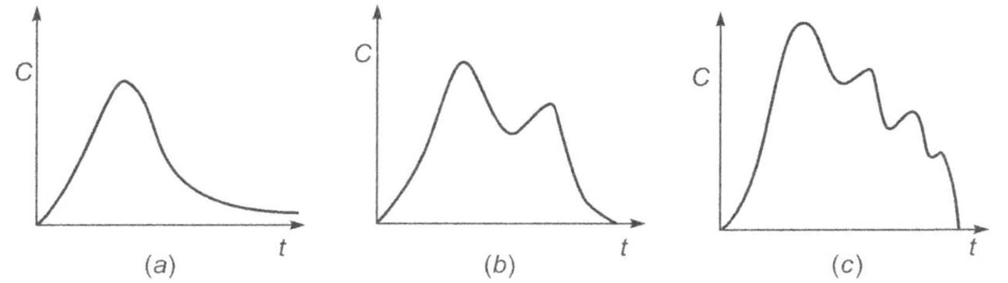

Figure 5.15 Nature of non-ideality in the bioreactor detected by the output of the response. (a) Usually long tail represents dead space. (b) Double peaks could be due to bypassing or channelling. (c) Multiple peaks could be due to internal recirculation.

a real solution in terms of the design equation available for ideal plug flow reactor. In this case, a large number of CFSTBRs are assumed to be connected in series. Figure 5.16, mentioned earlier for single input multistage CFSTBR, represents similar concept.

Therefore, the analysis of complex real reactor can be made by the proper combination of ideal reactor behavior (called "model"). The modelling of non-ideal reactor is discussed here on the basis of RTD, bypass, axial dispersion and micro mixing. However, there are other basis too. For example, the models for fluidized bed reactors include

- Dispersion and tanks in series model
- RTD model
- Contact time distribution model
- Two region model
- Hydrodynamic flow model (Levenspiel, 1972)

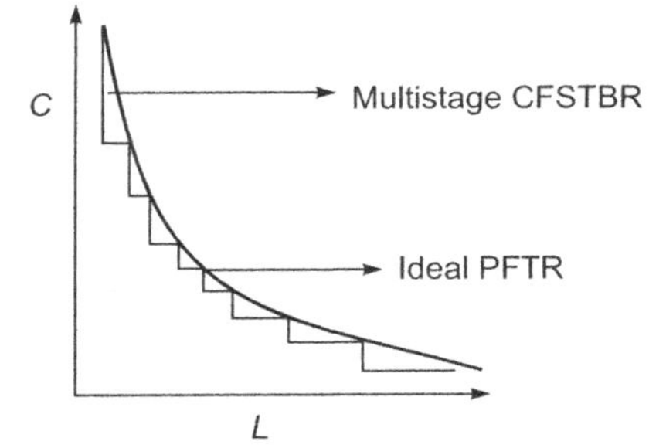

Figure 5.16 Comparison of concentration along the axial length of the plug flow tubular reactor with single flow multistage CFSTBR.

5.7 MODELS FOR NON-IDEAL FLOW

Two major groups of models are single parameter models (dispersion and tanks in series model) which effectively represent packed bed or tubular reactors and multi parameter models for mixed flow bioreactors.

5.7.1 Single Parameter Models

A single parameter is used to analyse the non-ideality of the reactor. This parameter is usually calculated by analysing the RTD determined from the tracer test.

Examples

Models for non-ideal tubular reactors:

Model	*Parameter*
Tanks-in-series model	Number of tanks
Dispersion model	Dispersion coefficient

Models for non-ideal CFSTBRs

Model	*Parameter*
Reactor dead volume model	Dead volume(where there is no reaction)
By-pass model	The fraction of fluid by-pass to the reactor exit which is unreacted.

5.7.2 The Purpose of the Parameters

After calculating the parameters, conversion of a component can be determined at the exit stream of the reactor. Assuming non-ideal tubular reactor theory, the reactor can be empty or packed with reactant or immobilized catalysts or the reaction medium. There is no question of flat velocity profile which is assumed for PFR. There exists axial mixing. To analyse the non-ideal situation in the tubular reactor, two different approaches are:

1. A non-ideal tubular reactor, equivalent to a series of identically sized CFSTBRs and
2. A non-ideal tubular reactor, equivalent to a PFR with axial mixing.

5.7.3 Models for Non-Ideal Tubular Reactors

1. One Parameter Model

(a) Tanks-in-series Model

A real tubular reactor can be expressed as a number of equally sized tanks-in-series. Each tank behaves as an ideal CFSTBR. The number of reactors needed, n (the single parameter), is determined from the $E(t)$ curve.

For n reactors in series, $E(t)$ is given by Equation (5.34).

$$E(t) = \frac{t^{n-1} e^{\frac{-t}{\tau_i}}}{(n-1)!\,\tau_i^n} \tag{5.34}$$

where $\tau_i = \dfrac{\tau}{n}$.

The residence time is given below

$$t_m = \tau = n\tau_i$$

In dimensionless form, $\theta = \dfrac{t}{\tau} = \dfrac{t}{n\tau_i}$ (5.35)

$$n\theta = \frac{t}{\tau_i}$$

$$E(\theta) = \tau E(t) = \frac{n(n\theta)^{n-1} e^{-n\theta}}{(n-1)!} \tag{5.36}$$

where V_i is the volume of ith reactor
F is the flow rate of the tracer
τ_i is the residence time of fluid in i^{th} reactor
N_0 is number of moles
θ is dimensionless time

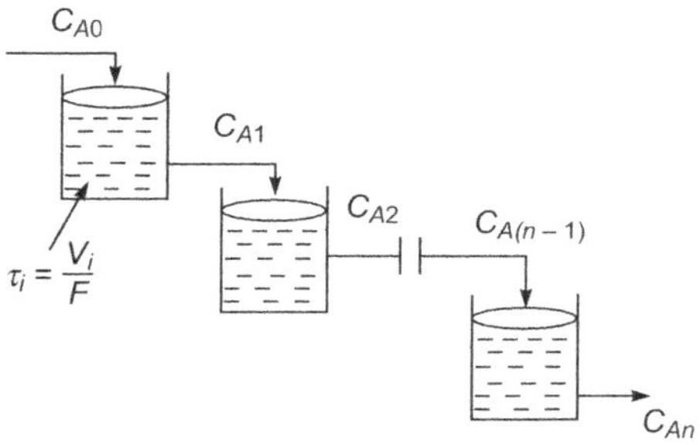

Figure 5.17 Tanks-in-series.

Calculation of number of tanks-in-series (Fig. 5.17)

σ_θ^2 (dimensionless variance) is calculated from the tracer experiment in this case.

$$\sigma_\theta^2 = \frac{\sigma^2}{\tau^2} = \frac{\int_0^\infty (t-\tau)^2 E(t)dt}{\tau^2} \tag{5.37}$$

$$\sigma_\theta^2 = \frac{\sigma^2}{\tau^2} = \int_0^\infty (\theta-1)^2 E(\theta)d\theta$$

Carrying out the integration for n tanks-in-series

$$\sigma_\theta^2 = \frac{\sigma^2}{\tau^2} = \frac{1}{n}$$

$$n = \frac{\tau^2}{\sigma^2} = \frac{1}{\sigma_\theta^2}$$

Calculation of conversion

(i) For first order reactions

$$X = 1 - \frac{1}{(1+\tau_i k)^n}, \quad \tau_i = \frac{\tau}{n} \tag{5.38}$$

(ii) For multiple reactions, the sequential equations must be solved.

$$V_i = \frac{V}{n}$$

$$V_1 = F\frac{(C_{A0} - C_{A1})}{-r_{A1}}$$

$$V_2 = F\frac{(C_{A1} - C_{A2})}{-r_{A2}}$$

....

$$V_n = F\frac{(C_{A(n-1)} - C_{An})}{-r_{An}} \tag{5.39}$$

(b) The Dispersion Model

Dispersion coefficient (De) is the single parameter in the dispersion model which is used most often for non-ideal tubular reactor. To understand the effect of dispersion on concentration in a tubular reactor,

Figure 5.18 shows the broadening of pulse along the length of the reactor.

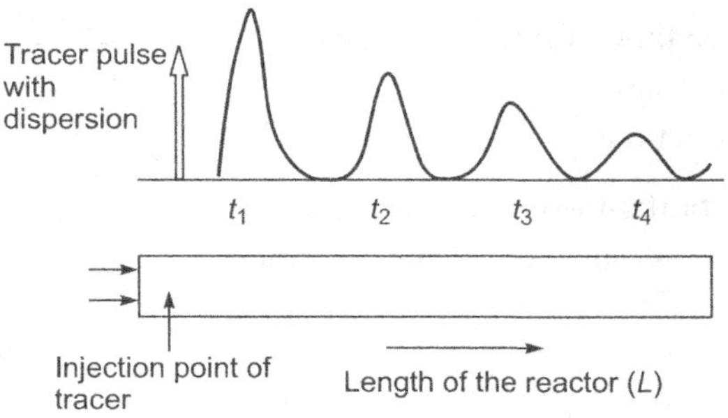

Figure 5.18 Dispersion in a tubular reactor.

This dispersion is governed by an analogy to Fick's law of diffusion associated with the flow. In this, the flow of tracer has two distinct components. Convection (caused by the bulk flow of the fluid) and the dispersion (resulted from molecular and turbulent diffusion).

The dispersion coefficient can be found by a pulse tracer experiment

$$De\frac{\partial^2 C_T}{\partial l^2} - U\frac{\partial C_T}{\partial l} = \frac{\partial C_T}{\partial t} \tag{5.40}$$

where C_T is tracer concentration
l is the length of the reactor
U is superficial velocity
De is dispersion coefficient (m^2/s)

Equation (5.40) is the unsteady state tracer balance equation (Fogler, 1992). Equation (5.41) is the dimensionless form of Equation (5.40).

$$\frac{1}{Pe}\frac{\partial^2 \Phi}{\partial \overline{L}^2} - \frac{\partial \Phi}{\partial \overline{L}} = \frac{\partial \Phi}{\partial \theta} \tag{5.41}$$

Pe = Peclet Number
= rate of transport by convection/rate of transport by dispersion
$= \frac{UL}{D_e}$

$\Phi = \frac{C_T}{C_{T0}}$ is the dimensionless concentration of tracer,

$\overline{L} = \frac{l}{L}$ is the dimensionless length and

$\theta = \frac{Ut}{L}$ is the dimensionless time.

For the closed-closed boundary condition the solution to the tracer balance at $\overline{L} = 1$(exit) at any time (θ) is $\Phi(1, \theta)$ which is expressed by Equation (5.42) (Fogler and Gürman, 2008).

$$E(\theta) = \frac{\Phi(1,\theta)}{\int_0^{\infty} \Phi(1,\theta)\,d\theta} \tag{5.42}$$

where

$$E(\theta) = \exp\left(\frac{Pe}{2}\right)\sum_{i=1}^{\infty}(-1)^{i+1}\frac{8\alpha_i^2}{4\alpha_i^2 + 4Pe + Pe^2}\exp\left(\frac{-\theta(Pe^2 + \alpha_i^2)}{4Pe}\right)$$

$\sigma_\theta^2 = \int_0^{\infty} \theta^2 E(\theta)d\theta - 1$ and α_i is called Eigen values. α_i is calculated from $\alpha_i = \tan^{-1}\dfrac{4Pe\alpha_i}{4\alpha_i - Pe^2}$

(Froment and Bischoff, 1990).

Integrating the above equation and if α_i is large, the exponential term decays. This results in Equation (5.43) (Fogler, 1992).

$$\frac{\sigma^2}{t_m^2} = \frac{2}{Pe}\left[1 - \frac{1}{Pe}(1 - e^{-Pe})\right] \tag{5.43}$$

If σ^2 and t_m are calculated from the RTD data, Pe can be evaluated from the above equation. This is applicable for closed-closed vessel boundary conditions. Fogler (1992) has described Equation (5.44) for open-open boundary condition.

$$\frac{\sigma^2}{t_m^2} = \frac{2}{Pe} + \frac{8}{Pe^2} \tag{5.44}$$

According to Aris (1959), this equation is modified for sloppy tracer input (i.e., not a clean injection of tracer).

$$\frac{\Delta\sigma^2}{t_m^2} = \frac{2}{Pe} \tag{5.45}$$

where $\Delta\sigma^2 = \sigma_{\text{inlet}}^2 - \sigma_{\text{exit}}^2$.

Flow and dispersion with reaction

Equation (5.46) is the guiding equation.

$$\frac{De}{U}\frac{d^2C_A}{dl^2} - \frac{dC_A}{dl} + \frac{r_A}{U} = 0 \tag{5.46}$$

For first order reaction this equation is written as

$$\frac{De}{U}\frac{d^2C_A}{dl^2} - \frac{dC_A}{dl} + \frac{kC_A}{U} = 0 \tag{5.47}$$

where k is the first order reaction rate constant. This equation can be solved analytically. In dimensionless form, the above equation can be written as

$$\frac{1}{Pe}\frac{d^2\Phi}{d\bar{L}^2} - \frac{d\Phi}{d\bar{L}} - Da\Phi = 0 \tag{5.48}$$

where Da is Damköhler number = rate of consumption of reactant by reaction/rate of transport of reactant by convection.

$$Da = \frac{kC_{A0}^{n-1}L}{U}$$

$$\therefore \quad Da = kC_{A0}^{n-1}\tau$$

Using this form and considering Danckwerts' boundary conditions, conversion can be calculated and expressed by Equation (5.49) for closed-closed system for a first order reaction (Fogler, 1992).

$$X = 1 - \Phi_L$$

$$X = 1 - \frac{4q\exp\left(\frac{Pe}{2}\right)}{(1+q)^2\exp\left(\frac{Peq}{2}\right) - (1-q)^2\exp\left(\frac{-Peq}{2}\right)} \tag{5.49}$$

where

$$q = \sqrt{1 + \frac{4Da}{Pe}} \tag{5.50}$$

Example 5.1

Consider an enzyme reaction which is performed in a reactor of length L and fluid velocity U. Axial dispersion is predominant in the reactor. Evaluate the performance of the reactor. Assume S is the reactant for this enzyme.

Solution Steady state condition is assumed in the reactor which carries out reaction associated with dispersion and flow. It is assumed that Equation (5.46) is applicable here.

$$\therefore \quad De\frac{d^2C_S}{dl^2} - U\frac{dC_S}{dl} - (-r(S,P)) = 0$$

where l is the longitudinal direction of flow. The boundary conditions of Danckwerts' (1953) at steady state require continuity of flux at both ends of the reactor.

$$\text{At } l = 0, \quad C_S = C_{S_0} + \frac{De}{U}\frac{dC_S}{dl}$$

$$\text{At } l = L, \quad \frac{dC_S}{dl} = 0$$

It is assumed that this reaction follows Michaelis-Menten kinetics with competitive product inhibition and without any deactivation.

$$\therefore \qquad -r(S, P) = -\frac{dC_S}{dt} = \frac{r_{\max} C_S}{K_m\left(1 + \frac{C_P}{K_I}\right) + C_S}$$

This expression depends on the type of inhibition mechanism.

where $r_{\max}$ is the maximum reaction rate and K_m is the Michaelis-Menten constant (mol/l), K_I is the inhibition constant (mol/l) and C_P is the product concentration.

Assuming no product in the feed, i.e., $C_{P_0} = 0$.

∴ Reaction stoichiometry yields,

$C_{S_0} = C_P + C_S$, assuming no intermediates of by-products of this reaction.

In analogy to Equation (5.48), the dimensionless form of the equation is:

$$\frac{1}{Pe}\frac{d^2\Phi}{d\bar{L}^2} - \frac{d\Phi}{d\bar{L}} - \theta\frac{\Phi}{K + \Phi(1 - K') + K'} = 0 \qquad (5.51)$$

where $\Phi = \frac{C_S}{C_{S_0}}$, $Pe = \frac{UL}{De}$, $\bar{L} = \frac{l}{L}$, $K = \frac{K_m}{C_{S_0}}$, $K' = \frac{K_m}{K_I}$, $\theta = \frac{r_{\max}\tau}{C_{S_0}}$

The boundary conditions are

At $\bar{L} = 0$, $\Phi = 1 + \frac{1}{Pe}\frac{d\Phi}{d\bar{L}}$

At $\bar{L} = 1$, $\frac{d\Phi}{d\bar{L}} = 0$

Solution of this equation requires numerical integration of a non-linear second order differential equation to calculate conversion in the reactor. Other methods are reverse shooting method (Nauman, 1987) and double precision IMSL library subroutine with the boundary conditions given above (IMSL software, 1987).

If K_m and K_I are equal, i.e., $K' = 1$, the above dimensionless Equation (5.51) becomes

$$\frac{1}{Pe}\frac{d^2\Phi}{d\bar{L}^2} - \frac{d\Phi}{d\bar{L}} - \frac{\theta\Phi}{K + 1} = 0 \qquad (5.52)$$

The boundary conditions are at $\bar{L} = 0$, $\Phi = 1 + \frac{1}{Pe}\frac{d\Phi}{d\bar{L}}$

and at $\bar{L} = 1$, $\frac{d\Phi}{d\bar{L}} = 0$

The analytical solution is similar to Φ_L described in Equation (5.49) where q can be described by Equation (5.53).

$$q = \sqrt{1 + \frac{4\theta}{(1 + K)Pe}} \qquad (5.53)$$

This is similar to the solution obtained by Danckwerts (1953).

Reactant conversion

As defined earlier by Fogler (1992),

$$X = 1 - \Phi_L \qquad (5.54)$$

Yield

Yield can be defined in two ways. It is the ratio of the moles of desired product formed to the moles of limiting reactant reacted. Another definition is that it is the ratio of the moles of desired product formed to the moles of limiting reactant fed to the reactor.

Note that yield is always with respect to the product whereas conversion is always with respect to the reactant.

However, Lortie (1994) and Abu-Reesh (1997) defined yield as the ratio of the amount of reactant converted to the maximum amount converted during one residence time.

$$Y = \frac{C_{S_0} - C_S|_{l=L}}{r_{max}\frac{L}{U}} = \frac{1-\Phi_L}{\theta} = \frac{X}{\theta} \tag{5.55}$$

Example 5.2 Solve Example 5.1 using the concept of the tanks-in-series model.

Solution Considering n number of CFSTBRs of equal size in series, the reactant balance over any stage at steady state gives

$$\tau_i = \text{residence time} = \frac{C_{S_{i-1}} - C_{S_i}}{-r(S, P)}$$

$$= \frac{(C_{S_{i-1}} - C_{S_i})}{\left(\dfrac{r_{max} C_{S_i}}{K_m\left(1 + \dfrac{C_{P_i}}{K_I}\right) + C_{S_i}}\right)} \tag{5.56}$$

where $i = 1, 2, 3 \ldots n$

This is a type of enzymatic reaction involving one reactant and one product, i.e., $S \rightarrow P$.

Reactor performance

The conversion is

$$X = 1 - \Phi_n$$

For $$K_m = K_I \text{ and } K' = 1$$

$$X = 1 - \frac{1}{\left(\dfrac{\theta_{n,\text{ total}}}{n(K+1)}\right)^n}$$

The yield is given by

$$Y = \frac{C_{S_0} - C_{S_n}}{r_{max}\tau_{n,\text{ total}}} = \frac{1-\Phi_n}{\theta_{n,\text{ total}}} = \frac{X}{\theta_{n,\text{ total}}} \tag{5.57}$$

Correlation between *Pe* and *n*

It is possible to get a relation between *Pe* and *n* by equating the variance of the axial dispersion and tanks-in-series models. Using the closed-closed boundary conditions (Danckwerts, 1953) the following relation is obtained.

$$n = \frac{Pe^2}{2(Pe - 1 + e^{-Pe})} \tag{5.58}$$

For very large Pe with deviation from PTFR, $Pe = 2n$.

According to Elgeti (1996), the relation between Pe and n covering whole range of dispersion is

$$Pe = 2(n - 1)$$

When $Pe = 0$, $n = 1$ (for CFSTBR).

When $Pe \rightarrow \infty$, $n = \infty$, i.e., PTFR.

5.7.4 Models for Non-Ideal CFSTBRs

(1) Single Parameter Models

(a) Reactor Dead Volume as Non-ideality

This is a common phenomenon for all bioreactors where mixing of cells and reactants is difficult near to the agitator. A schematic representation is described in Figure 5.19.

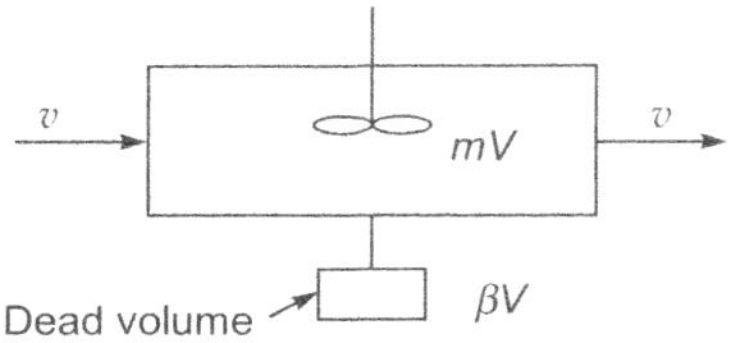

Figure 5.19 CFSTBR with a dead volume.

where v is the flow rate

m is a fraction of active volume

V is volume

β is a fraction of dead volume

N_o is the initial number of moles

ΔN is change in moles

τ is the dimensionless time

C_{Ao} is the initial concentration of A (any component)

$\theta = \frac{t}{\tau}$ is the dimensionless time.

Mass balance in CFSTBR:

$$mV\frac{dC}{dt} = vC_i - vC$$

Integrating the above equation

$$m\tau\frac{dC}{dt} = C_i - C$$

where,

$$\frac{V}{v} = \tau$$

$$\frac{dC}{dt} + \frac{C}{m\tau} = \frac{C_i}{m\tau} \Rightarrow \frac{dy}{dx} + Py = Q$$

$$y = C,$$
$$x = t,$$
$$P = \frac{1}{m\tau}$$
$$Q = \frac{C_i}{m\tau}$$
$$I.F. = e^{\int P dx} = e^{\int \frac{1}{m\tau} dt} = e^{\frac{t}{m\tau}}$$

The solution is

$$y \times I.F. = \int Q \times I.F. dx + constant(C_k)$$
$$C \times e^{\frac{t}{m\tau}} = \int \frac{C_i}{m\tau} e^{\frac{t}{m\tau}} dt + C_k$$
$$= \frac{C_i}{m\tau} e^{\frac{t}{m\tau}} m\tau + C_k$$

for $t = 0, C = 0 \Rightarrow 0 = C_i + C_k \quad \Rightarrow \quad C_k = -C_i$

$\therefore$
$$C \times e^{\frac{t}{m\tau}} = C_i e^{\frac{t}{m\tau}} - C_i$$

$\Rightarrow$
$$\frac{C}{C_i} = 1 - e^{-\frac{t}{m\tau}}$$

$\Rightarrow$
$$\frac{C}{C_i} = F(t) = 1 - e^{-\frac{t}{m\tau}}$$
$$E(t) = \frac{dF}{dt} = \frac{1}{m\tau} e^{-\frac{t}{m\tau}}$$

$\Rightarrow$
$$\tau E(t) = \frac{1}{m} e^{-\frac{t}{m\tau}}$$

The final solution is

$$E(\theta) = \frac{1}{m} \exp\left(-\frac{\theta}{m}\right) \tag{5.59}$$

(b) Reactor with Bypass

Figure 5.20 is an example of CFSTBR at a particular feed flow rate.

where $(1 - \lambda)$ is the fraction of feed stream bypassed
δ is Dirac-delta function
C_a is the tracer concentration coming out of CFSTBR
C_i is inlet concentration of the tracer.

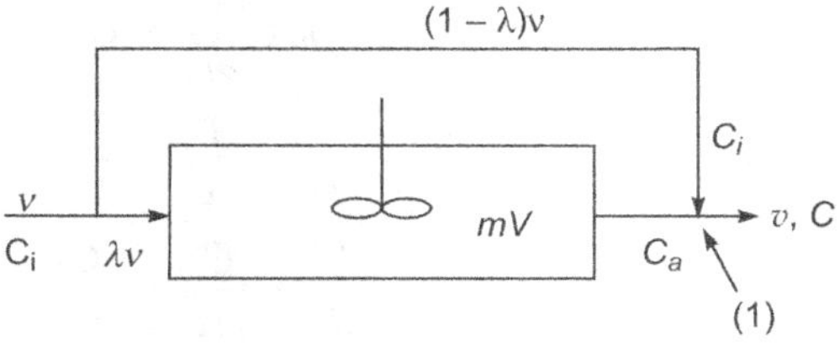

Figure 5.20 CFSTBR with flow bypassing.

Mass balance around the CFSTBR:

$$\frac{dC_a}{dt} + \frac{\lambda}{m\tau} C_a = \frac{\lambda C_i}{m\tau} \tag{5.60}$$

The above equation is in Leibenitz's form $\left[\text{i.e., } \frac{dy}{dx} + Py = Q\right]$.

The solution is given by

$$C_a e^{\frac{\lambda t}{m\tau}} = \int \frac{C_i \lambda}{m\tau} e^{\frac{\lambda t}{m\tau}} dt + C_k.$$

where C_k is a constant.

$$C_a e^{\frac{\lambda t}{m\tau}} = C_i e^{\frac{\lambda t}{m\tau}} + C_k.$$

At $\quad t = 0,\ C_a = 0.$

So $\quad C_k = -C_i$

$$C_a e^{\frac{\lambda t}{m\tau}} = C_i (e^{\frac{\lambda t}{m\tau}} - 1)$$

$$C_a = C_i (1 - e^{\frac{-\lambda t}{m\tau}}) \tag{5.61}$$

C_a **is difficult to measure. In order to write** C_a in terms of measurable quantity(C), consider material **balance around point (1) as shown in Figure 5.20.**

$$C = \lambda C_a + (1 - \lambda) C_i$$

where C is the concentration of the tracer at the exit of the entire system shown above.

Replacing C_a from Equation (5.61) results Equation (5.61a)

$$\frac{C}{\lambda} = C_i \left(\frac{1}{\lambda} - e^{\frac{-\lambda t}{m\tau}} \right)$$

$$\frac{C}{C_i} = (1 - \lambda e^{\frac{-\lambda t}{m\tau}}) \tag{5.61a}$$

i.e.,

$$F(t) = \frac{C}{C_i} = (1 - \lambda e^{\frac{-\lambda t}{m\tau}})$$

$$E(t) = \frac{dF}{dt} = \frac{\lambda^2}{m\tau} e^{\frac{-\lambda t}{m\tau}} \tag{5.62}$$

$$E(\theta) = \tau E(t) = \frac{\lambda^2}{m} e^{\frac{-\lambda \theta}{m}} \tag{5.63}$$

Here, $\quad m = 1$, so

$$E(\theta) = \tau E(t) = \lambda^2 e^{-\lambda\theta}$$

Let us consider the bypass stream. The bypass is like a pulse that is coming out at time $t = 0$. It can be represented by a Dirac-delta function, $\frac{C}{C_i} = (1 - \lambda)\delta(t)$.

$$E(t) = (1 - \lambda)\delta(t)$$
$$E(\theta) = (1 - \lambda)\delta(\theta)$$

So the exit age distribution function for the whole system is

$$E(\theta) = \tau E(t) = \lambda^2 e^{-\lambda\theta} + (1 - \lambda)\delta(\theta) \tag{5.64}$$

5.8 MULTI PARAMETER MODELS

These models require more than one parameter to understand the real behavior in bioreactor (as in fluidized bed reactor). In multi-parameter models, consider the real reactor which consists of different regions, viz., plug, dispersed plug, mixed, dead fluid, interconnected by-pass, recycle, cross flow, etc. They are discussed here for better understanding of the system.

(a) Bioreactor with Bypass and Dead Space

In this system (Fig. 5.21), m is not equal to 1. Therefore, the exit age distribution function is

$$E(\theta) = \tau E(t) = \frac{\lambda^2}{m} e^{\frac{-\lambda}{m}\theta} + (1 - \lambda)\delta(\theta = 0) \tag{5.65}$$

(b) CFSTBR with Dead Space and a PFTR Connected in Parallel

Material balance around the CFSTBR gives

$$(\lambda v)C_i - (\lambda v)C_a = \frac{mV}{v}\frac{dC_a}{dt} \tag{5.66}$$

where C_a is the concentration of the tracer at the exit of CFSTBR.

$$\frac{dC_a}{dt} + \frac{\lambda}{m\tau}C_a = C_i\frac{\lambda}{m\tau}$$

$$C_a e^{\frac{\lambda t}{m\tau}} = C_i e^{\frac{\lambda t}{m\tau}} + C_k. \tag{5.67}$$

where C_k is a constant.

At $\quad t = 0,\ C_a = 0.$

So $\quad C_k = -C_i$

$$C_a = C_i\left(1 - e^{\frac{-\lambda t}{m\tau}}\right)$$

In order to write C_a in terms of measurable quantity(C), consider material balance around point (1) as shown in Figure 5.22.

$$C = \lambda C_a + (1 - \lambda)C_i$$

where C is the concentration of the tracer at the exit of the entire system (Fig. 5.22).

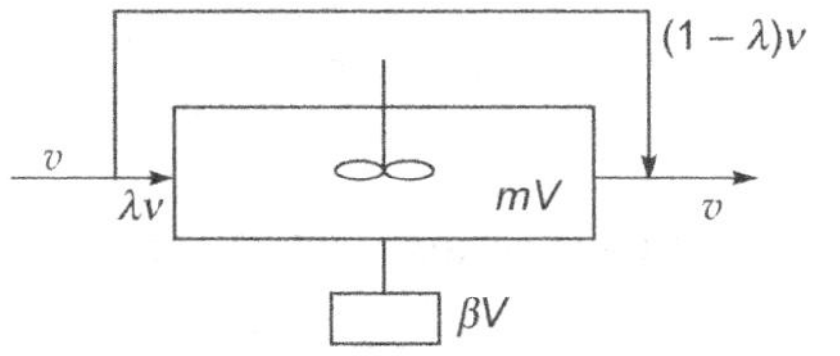

Figure 5.21 CFSTBR with bypass and dead space.

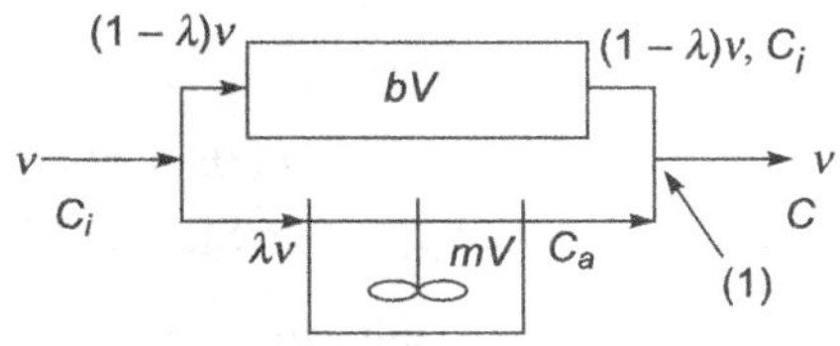

Figure 5.22 CFSTBR and PFR in parallel connection.

From the above relation

$$C_a = \frac{C}{\lambda} - \frac{(1-\lambda)}{\lambda}C_i$$

$$\frac{C}{\lambda} - \frac{(1-\lambda)}{\lambda}C_i = C_i\left(1 - e^{\frac{-\lambda t}{m\tau}}\right)$$

i.e.,

$$F(t) = \frac{C}{C_i} = (1 - \lambda e^{\frac{-\lambda t}{m\tau}})$$

$$E(t) = \frac{dF}{dt} = \frac{\lambda^2}{m\tau}e^{\frac{-\lambda t}{m\tau}}$$

$$E(\theta) = \tau E(t) = \frac{\lambda^2}{m}e^{\frac{-\lambda\theta}{m}}$$

For the PFR

$$E(t) = (1-\lambda)\delta\left(t - \frac{bV}{(1-\lambda)v}\right)$$

$$E(t) = (1-\lambda)\delta\left(t - \frac{b\tau}{(1-\lambda)}\right)$$

$$E(\theta) = (1-\lambda)\delta\left(\theta - \frac{b}{(1-\lambda)}\right) \tag{5.68}$$

The $E(\theta)$ for the whole system is given by Equation (5.69).

$$E(\theta) = \tau E(t) = \frac{\lambda^2}{m}e^{\frac{-\lambda}{m}\theta} + (1-\lambda)\delta\left(\theta - \frac{b}{(1-\lambda)}\right) \tag{5.69}$$

5.9 APPLICATION OF RTD BASED MODELS TO NON-IDEAL BIOREACTORS

Example 5.3

Chemostat is a non-ideal CFSTBR. This can be modelled as a CFSTBR with a dead zone and a by-pass (Fig. 5.23). For a chemical reaction, similar approach is described by Fogler (1992).

The terms related to Figure 5.23 are:

C_{S_0} — Initial reactant concentration, $\frac{\text{kg}}{\text{m}^3}$

C_{S_S} — Exit concentration of the reactant from the well mixed portion of the bioreactor, $\frac{\text{kg}}{\text{m}^3}$

C_S — Exit concentration of reactant from the bioreactor, $\frac{\text{kg}}{\text{m}^3}$

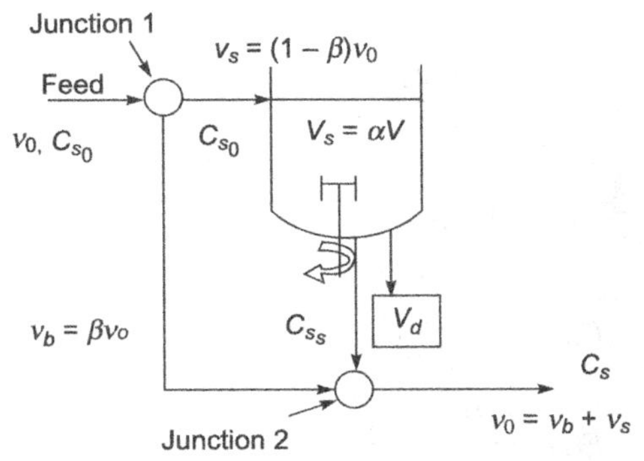

Figure 5.23 Chemostat with bypass and dead zone.

v_0 — Total volumetric flow rate, $\frac{m^3}{h}$

v_s — Volumetric flow rate of the well mixed portion of the bioreactor, $\frac{m^3}{h}$

v_b — Bypass flow rate, $\frac{m^3}{h}$

V — Total volume, m^3

V_d — Dead volume, m^3

V_S — Volume of the well mixed portion of the reactor, m^3

α — Volume fraction of well mixed portion in the reactor

β — Fraction of unreacted bypass stream

C_{Ti} — Initial tracer concentration, $\frac{kg}{m^3}$

C_T — Effluent tracer concentration, $\frac{kg}{m^3}$

C_{TS} — Output tracer concentration from the well mixed zone, $\frac{kg}{m^3}$

The tracer experiment can be done using a positive step input to the well mixed volume (V_s) in Figure 5.23.

The tracer balance is given in Equation (5.70).

Input rate – Output rate = Rate of accumulation

$$v_s C_{Ti} - v_s C_{TS} = V_S \frac{dC_{TS}}{dt} \tag{5.70}$$

The conditions for the step input are

At $t < 0$, $C_T = 0$

At $t \geq 0$, $C_T = C_{Ti}$

Tracer balance around junction 2 in the chemostat is

$$C_T = \left(\frac{v_b C_{Ti} + v_s C_{TS}}{v_0}\right) \tag{5.71}$$

Integration of the above equation yields

$$\frac{C_{TS}}{C_{Ti}} = 1 - \exp\left[-\left(\frac{1-\beta}{\alpha}\right)Dt\right]$$

$$\frac{C_T}{C_{Ti}} = 1 - (1-\beta)\exp\left[-\left(\frac{1-\beta}{\alpha}\right)Dt\right] \tag{5.72}$$

On rearrangement

$$\ln\left(\frac{C_{Ti}}{C_{Ti} - C_T}\right) = \ln\left(\frac{1}{1-\beta}\right) + \left(\frac{1-\beta}{\alpha}\right)Dt \tag{5.73}$$

From the plot of $\ln\left(\frac{C_{Ti}}{C_{Ti} - C_T}\right)$ vs. time (t), α and β values are calculated. These values are important to calculate optimal dilution rate which is a function of C_{S_0}. With reference to Figure 5.23, let us see how the parameters are evaluated to get reactant and cell mass balances.

At the junction 2,

Rate of reactant entering = Rate of reactant leaving

$$C_{S_0} v_b + v_s C_{S_S} = v_0 C_S, \tag{5.74}$$

$$\therefore \quad C_S = \left(\frac{v_b C_{S_0} + v_S C_{S_S}}{v_0}\right)$$

Let

$$\alpha = \frac{V_S}{V}$$

$$\beta = \frac{v_b}{v_0}$$

$$\therefore \quad C_S = \beta C_{S_0} + \frac{v_S}{v_0} C_{S_S}$$

$$= \beta C_{S_0} + \left(\frac{v_0 - v_b}{v_0}\right) C_{S_S}$$

$$= \beta C_{S_0} + (1-\beta) C_{S_S}$$

$$\therefore \quad \frac{dC_S}{dt} = (1-\beta)\frac{dC_{S_S}}{dt} \tag{5.75}$$

Similarly for cells

$$\frac{dC_X}{dt} = (1-\beta)\frac{dC_{X_S}}{dt}$$

Therefore, from unsteady state mass balances it is possible to get various relations for reactant and cells.

Reactant balance is given by Equation (5.76).

Rate of reactant accumulation = rate of reactant entering – rate of reactant leaving + rate of reactant utilized

$$V_S \frac{dC_{S_S}}{dt} = v_S C_{S_0} - v_S C_{S_S} - rV_S \tag{5.76}$$

$$\frac{dC_{S_S}}{dt} = \frac{v_S}{V_S}(C_{S_0} - C_{S_S}) - r$$

$$\frac{dC_{S_S}}{dt} = \left[\frac{(1-\beta)D}{\alpha}\right](C_{S_0} - C_{S_S}) - r \tag{5.77}$$

Cell mass balance (sterile feed)

$$V_S \frac{dC_{X_S}}{dt} = 0 - v_S C_{X_S} + r_X V_S \tag{5.78}$$

$$\frac{dC_{X_S}}{dt} = r_X - \frac{v_S}{V_S} C_{X_S}$$

$$= r_X - \left[\frac{(1-\beta)v_0}{\alpha V_S}\right] C_{X_S}$$

$$= r_X - \left[\frac{(1-\beta)D}{\alpha}\right] C_{X_S} \tag{5.79}$$

These approaches are used for desulfurization of diesel and organosulfur compounds in a chemostat operation (Chowdhury and Dutta Gupta, 2005) and SCP production in a cylindrical and tapered reactor (Gowthaman *et al.*, 1991).

Example 5.4 Calculation of conversion based on the model in Example 5.3.

RTD information alone is insufficient for predicting conversions of reactions with non-linear rates. One needs information on degrees of segregation (Levenspiel, 1972), earliness and lateness of mixing, etc. In most of the cases, the performance equations are mathematically tedious. In this regard, RTD information can be used for predicting conversions provided the model developed will approximate to the real vessel (with non-idealities) reasonably. In this case, the non-ideal parameters are evaluated from RTD data using Cholette and Cloutier model (1959). Those parameters are used to calculate conversions of the reactant with known kinetic expressions for cellular growth. The model is given in Example 5.3. The mass balance of the reactant for perfectly mixed region is

$$v_S C_{S_0} = v_S C_{S_S} + V_S(-r_{S_S}) \tag{5.80}$$

Assuming

$$-r_{S_S} = \frac{\mu_m (C_{S_0} - C_{S_S}) C_{S_S}}{K_S + C_{S_S}} \tag{5.81}$$

Substituting the r_{S_S} value in the mass balance equation gives

$$(v_S - V_S\mu_m)C_{S_S}^2 + (v_S K_S + V_S \mu_m C_{S_0} - v_S C_{S_0})C_{S_S} - v_S K_S C_{S_0} = 0$$

This is a quadratic equation in C_{S_S}. Therefore, the possible relation considered here is

$$C_{S_S} = \frac{K_S}{\left[\left(\frac{1-\alpha}{1-\beta}\right)\mu_m \tau - 1\right]} \tag{5.82}$$

C_{S_S} is unknown and it is expressed in measurable value, i.e., C_S.
From the material balance of the reactant on exit stream, we get

$$v_0 C_S = v_S C_{S_S} + v_b C_{S_0}$$

Also,

$$V = V_S + V_d$$

$$v_0 = v_b + v_S$$

$$\beta = \frac{v_b}{v_0}$$

$$\alpha = \frac{v_S}{V}$$

$$\therefore \quad \frac{C_S}{C_{S_0}} = \frac{\left(\frac{K_S}{C_{S_0}}\right)(1-\beta)^2}{[(1-\alpha)\mu_m \tau - (1-\beta)]} + \beta \tag{5.83}$$

Equation (5.83) can be used to compare the experimental reactant concentration, provided that α and β from RTD studies and K_s and μ_m from batch reactor studies are available. A typical plot can be considered here (Fig. 5.24).

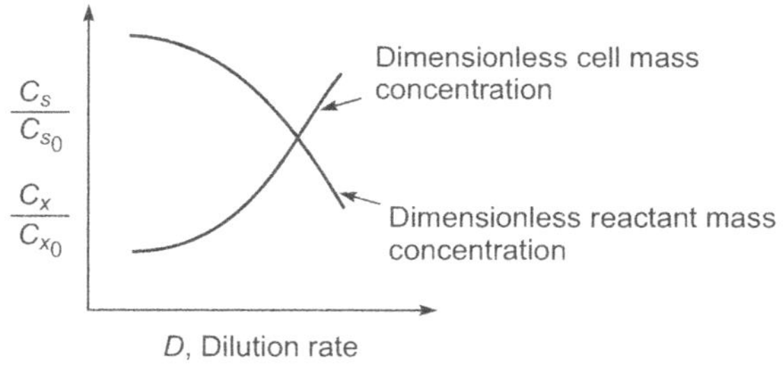

Figure 5.24 Plot of $\frac{C_S}{C_{S_0}}$ and $\frac{C_X}{C_{X_0}}$ vs. dilution rate.

One can use this model to compare the different reactor configurations.

Example 5.5 Calculate α and β values from $E(\theta)$ values.

Solution Assuming the model given in Example 5.3,

$E(\theta)$ = Dimensionless exit age distribution function

$$= \frac{(1-\beta)^2}{(1-\alpha)} \exp\left[-\frac{(1-\beta)}{(1-\alpha)}\theta\right] + \alpha\delta(\theta = 0) \tag{5.84}$$

where α is the fraction of stagnant zone and β is the bypass liquid. Above $E(\theta)$ expression is given by

Cholette and Cloutier (1959). δ is the Dirac delta function which is valid only at $\theta = 0$, where θ is the dimensionless residence time.

From $E(\theta)$ expression we plot ln $E(\theta)$ vs. θ (Fig. 5.25). From this plot, α and β are calculated.

Note: Typical $E(\theta)$ vs. θ plot is given in Figure 5.26.

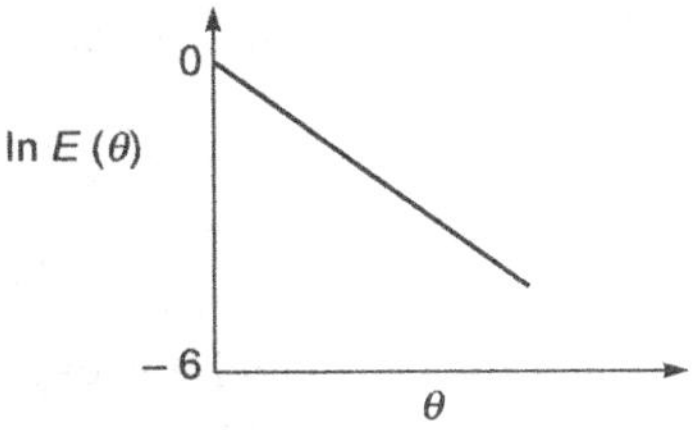

Figure 5.25 A plot of ln $E(\theta)$ vs. θ.

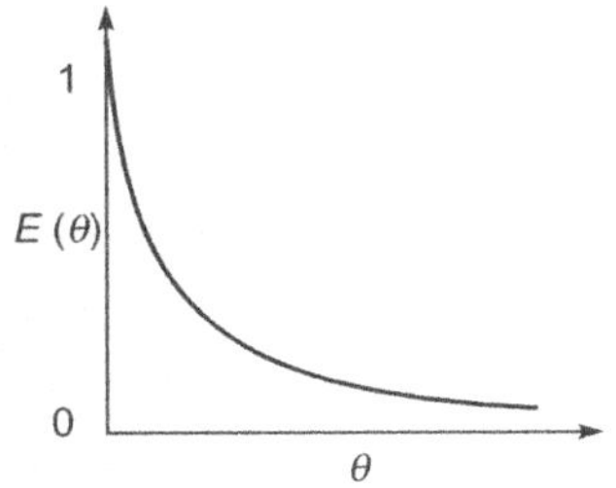

Figure 5.26 RTD of a typical CFSTBR.

RTD does not characterize all aspects of mixing. It indicates how long various elements of effluent fluid have been in the reactor. It does not tell us about the mixing of fluid elements of different ages. To understand this, let us consider two limiting cases.

1. Maximum mixedness The input feed material comes into intimate contact with other fluid elements of all stages.

2. Complete segregation Fluid elements of different ages do not intermix at all while in the reactor and come together only when they are withdrawn in the effluent streams.

Real situations fall in between the above two stages. The RTD of the reactor is completely independent of its micro mixing characteristics. Micro mixing is discussed later in this chapter. In a reactor with complete segregation, each independent fluid element behaves like a small batch reactor. The exit fluid is a blend of the products of these batch reactors which stayed in the reactor for different lengths of time. Therefore, the exit concentration C_a from the reactor when the fluid is completely segregated, is expressed by Equation (5.85).

$$C_a = \int_0^{\infty} C_{a,\text{batch}}(t)E(t)dt \tag{5.85}$$

where $C_{a,\text{batch}}$ - concentration of the component i in a batch reactor after an elapsed time t.
$E(t)dt$ - fraction of the reactor effluent containing fluid elements with residence time near t.

For the above equation reaction conditions are assumed uniform, like temperature, pH, and dissolved oxygen. In a cellular reaction system, cells appearing in the reactor are segregated. It is also assumed that the environment of the cell remains segregated in the flow system.

5.10 DRAWBACKS OF CLASSICAL RTD MEASUREMENTS

In biological reactor for classical RTD measurement, the flow rate is very low compared to the volume of the reactor. This combines with high stirring rate in the reactor which results in almost perfect tracer response showing a very little deviation from ideal situation.

5.10.1 Macro-Mixing

This describes the characteristics of bulk mixing in bioreactor, which is generated by the impeller or its equivalent to change flow pattern in the reacting fluid.

5.10.2 Mixing Time

Mixing time is a useful parameter for assessing mixing efficiency. This is the time required to get a given degree of homogeneity starting from a complete segregated state. For example, in an agitated reactor the time is required for the reactor composition to achieve a specified level of homogeneity following the addition of a tracer at a single point in the reactor. Following figures show the internals which give rise to different circulation and mixing time characteristics (Fig. 5.27).

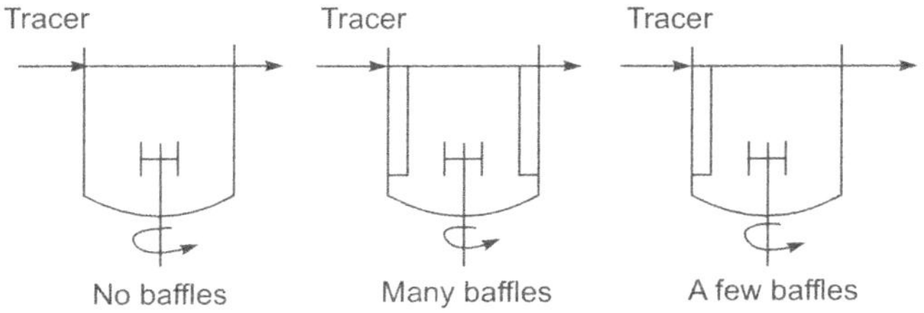

Figure 5.27 Different internals in agitated vessels.

Tracers are acids, bases, and concentrated salt solutions and the corresponding detectors are pH probes and conductivity meters. Definition of mixing time includes the degree of homogeneity required. At t_m, the tracer concentration is relatively steady and the fluid composition is uniform.

For a single phase fluid in a stirred tank provided with baffles and impeller, t_m is defined as

$$t_m = 4t_c \tag{5.86}$$

where t_c is circulation time (Doran, 1995).

The mathematical expression for mixing time depends on the number of baffles, type of impeller, geometry of the impeller, and fluid properties.

For Rushton turbine, the mixing time is

$$N_R t_m = \frac{1.54V}{D_i^3} \text{ at high } \mathrm{Re}_{\text{impeller}} \tag{5.87}$$

where N_R = Rotational speed of the stirrer

$N_R t_m$ = Number of stirrer rotations required to homogenise the liquid

V = Liquid volume

D_i = Impeller diameter

$\mathrm{Re}_{\text{impeller}}$ = Impeller Reynolds number $= \dfrac{N_R D_i \rho}{\mu}$

ρ = Fluid density

μ = Fluid viscosity

The circulation time distribution is defined by Bailey and Ollis (1986) [Equation (5.88)].

$$F_C(t) = \frac{1}{\sigma\sqrt{2\pi}} \exp\left(-\frac{(\ln t - t_l)^2}{2\sigma_l^{\,2}}\right) \tag{5.88}$$

where t_l is log mean circulation time

σ_l is standard deviation of log mean circulation time (Bryant, 1977)

To study macro mixing following tracers can be used (Dunlop, 1990).

- Coloured dyes, whose concentration followed spectrophotometrically in the reactor.
- Acid or base pulse input followed by pH probe.
- Salt pulses measured by conductivity probe whose response is linear.
- Power input to the reactor.

The important dimensionless groups are

$$N_p = \text{Power number} = \frac{P}{\rho N^3 D^5}$$

$$\text{Re} = \text{Reynolds number} = \frac{ND^2}{\nu}$$

$$\text{Aeration number} = \frac{Fg}{ND^3}$$

where P is the power input to the reactor

ρ is the liquid density

N is the rotational speed

D is the impeller diameter (m)

ν is kinematic viscosity (m^2/s)

For fully turbulent regime, where bioreactors are operated, N_p is constant.

5.10.3 Micro Mixing—Another Factor for Non-ideality in the Bioreactor

Micro mixing is a serious problem to be understood well in the hydrodynamics of bioreactor.

How should we measure?

Following methods are used commonly.

1. Hot film anemometer and laser Doppler scattering from small particles, i.e., gas bubbles and cells.
2. Extent of reaction is predicted from models of complete segregation and completely micro-mixed states – This depends on models.
3. Competing reactions – One is kinetically controlled whereas the other is diffusion controlled (Bourne *et al.*, 1977; Bourne, 1984).

The distribution of eddy sizes in a bioreactor is non-uniform. Nagata (1975) measured the distribution of eddy sizes in an agitated water system using hot film anemometer. This gives certain clue where one can add reactant to determine the local micro scale of turbulence and idea for scale-up of bioreactor. Biological response of micro scale is not clearly described in the available literature. However, some attempt has been made in this direction, viz,

(a) Flip-flop Experiment

Reactant is injected in two sites, specific for two different micro-scales eddies.

(b) Dilution Rate Studies

Large change in the yield of cell is observed by changing local micro scale at the injection point.

(c) Micro Mixing Models

Two miscible solutions containing two different chemical species are mixed by turbulence and simultaneously react with each other. The two solutions are divided into several lumps of a small scale. These lumps are changed to much smaller lumps and finally converted to a molecular scale. At this point, they react in the bioreactor. A schematic of micro-mixing model adapted from Liou and Chien (1995) is given in Figure 5.28.

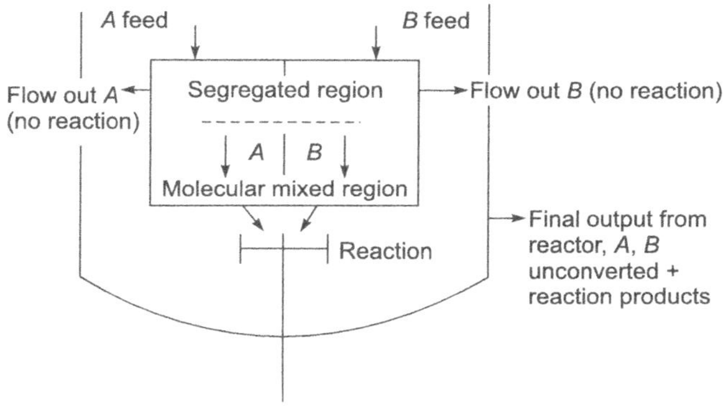

Figure 5.28 A model for micro-mixing.

Molecular mixed regions are homogeneous. The degree of micro-mixing (α_m) is defined by the relative volume ratio of the summation of chemical species at molecular level multiplied by its volume of the total mixing system (Liou and Chien (1995)).

In this example,

$$\alpha_m = \frac{A_{\text{molecular}-\text{level}} V_A + B_{\text{molecular}-\text{level}} V_B}{V} \tag{5.89}$$

This value is constant throughout the vessel, but it is a function of time.

The design equations of the micro-mixing for CFSTBR are described by from Liou and Chien (1995).

5.11 TRANSIENT BEHAVIOR IN BIOREACTORS

Transient behavior indicates the change in properties with respect to time. Our aim is to study the unsteady state or transient behavior exhibited by a bioreactor. Transient reactor operation includes all periods in which environmental conditions change other than lag phases.

5.11.1 Classification of Continuous Change in Environmental Factors

Four broad classes of periodic operations may be considered here (Bailey, 1973).

- **Process life cycle:** The cycle is determined by poisoning of the catalyst.

- **Relaxed steady state:** The system is not capable to follow rapid cycling. This enters to a time invariant value or relaxed steady state.
- **Quasi-steady state periodic operations:** This is opposite to relaxed steady state operation and is a process under time invariant conditions. In such a situation, the period of the input oscillations is large compared to the systems response time. The system results in a steady state relationship with input at any time.
- **Intermediate periodic operations:** It is the intermediate between "relaxed" and quasi-steady state approach. The system response time is of the same order as the imposed function cycle time. This causes resonance. Examples are semi batch-periodic operation, selectivity in periodically forced CFSTBRs and cycled reactors with heterogeneous catalysts.

Lag, overshoot, and oscillations can happen in biological reactors. Detailed transient reactor operations are summarized by Pickett (1982).

In Chapter 4, various unsteady state and transient state approaches in reactor operations have been described using microbial systems. Let us take up another example of multicellular system, i.e., fungal system. Fungal species, in general, cause non-ideality in reactor operation. One of the typical problems is wall growth. In fact, fungal cell development passes through different stages of growth. On that basis, the simple growth kinetics cannot be implemented as used in Chapter 4. It is wise to use population based morphologically structured model.

5.11.2 Characterization of Transient State

To characterize the transient state behavior, the following material balance is the starting point:

$$\frac{dC_i}{dt} = D(C_{i_f} - C_i) + r_i \tag{5.90}$$

This equation applies to each component in the bioreactor model. The dynamic reactor model consists of a set of equations of the above type [Equation (5.90)] which are usually coupled through the rate of formation terms. In general, the rate of formation of component, i, may depend on the concentration of all the other components in the reactor. It will be convenient to use vector-matrix to simplify large sets of equations.

Equation (5.90) can be written as

$$\frac{d\mathbf{C}}{dt} = \mathbf{f}(\mathbf{C}, \mathbf{p}) \tag{5.91}$$

where $\mathbf{C}$ is vector of concentration of m elements or components, i.e., m-vector

$\mathbf{p}$ is q-vector of model parameter, i.e., feed concentration, D, kinetic parameters, etc.

$\mathbf{f}$ is i^{th} component of vector-valued function which is equal to the right hand side function of Equation (5.91).

Equation (5.91) is non-linear. We are interested in dynamic behavior of the reactor near a particular condition which is easy to evaluate the required parameters. This condition is in steady state.

Therefore, Equation (5.91) at steady state is

$$f(C_S, p) = 0 \tag{5.92}$$

To determine the behavior near $\mathbf{C_S}$, $\mathbf{x}$ is introduced which represents the deviations from the steady state.

$$\mathbf{x} = \mathbf{C} - \mathbf{C_S} \tag{5.93}$$

So,

$$\frac{d\mathbf{x}}{dt} = \mathbf{f}(\mathbf{C_S} + \mathbf{x}, \mathbf{p}) \tag{5.94}$$

The right hand side of Equation (5.94) is expanded in Taylor's series

$$\frac{d\mathbf{x}}{dt} = \mathbf{f}(\mathbf{C_S}, \mathbf{p}) + \underline{\underline{\mathbf{A}}}\mathbf{x} + \text{higher order terms} \tag{5.95}$$

Neglecting all second and higher order terms as they are small, Equations (5.92) and (5.95) result in Equation (5.96).

$$\frac{d\mathbf{x}}{dt} = \underline{\underline{\mathbf{A}}}\mathbf{x} \tag{5.96}$$

This is the linear approximation of the system. $\underline{\underline{\mathbf{A}}}$ is a matrix corresponding to a particular steady state. The elements of this matrix are evaluated at each steady state. The element a_{ij} of the ith column of the matrix $\underline{\underline{\mathbf{A}}}$ is defined by Equation (5.97).

$$a_{ij} = \frac{\partial f_i(\mathbf{C}_S, \mathbf{p})}{\partial C_j} \tag{5.97}$$

The solution of Equation (5.95) is usually the Equation (5.98)

$$x(t) = \sum_{i=1}^{m} \alpha_i \beta_i e^{\lambda i t} \tag{5.98}$$

where m is the number of species in the system indicated earlier. The quantities $\boldsymbol{\beta}_i$ and λ_i are corresponding pairs of Eigen vectors and Eigen values of $\underline{\underline{\mathbf{A}}}$, respectively.

As $\lambda = \lambda_i$ satisfies the characteristic equation

$$\det\left(\underline{\underline{\mathbf{A}}} - \lambda\underline{\underline{\mathbf{I}}}\right) = 0 \tag{5.99}$$

$\underline{\underline{\mathbf{I}}} = m \times m$ identity matrix.

The Eigen vector $\boldsymbol{\beta}_i$ satisfies

$$\left(\underline{\underline{\mathbf{A}}} - \lambda\underline{\underline{\mathbf{I}}}\right)\boldsymbol{\beta}_i = 0, \quad i = 1, 2, \ldots m \tag{5.100}$$

The values of α_i are constants to be chosen to fulfil the specified initial conditions $\mathbf{x}(0)$, and hence

$$\sum_{i=1}^{m} \alpha_i \boldsymbol{\beta}_i = \mathbf{x}(0) \tag{5.101}$$

The linearized dynamic model of the reactor provides a strong base for identification of characteristic response time of the system. From Equation (5.98), the time scale for the decay of disturbances from reference steady state $\mathbf{C_S}$ is characterised by λ_i of the matrix $\underline{\underline{\mathbf{A}}}$.

Therefore, the characteristic time of the reactor

$$\mathbf{t_i} = |\lambda_i|^{-1}, \quad i = 1, 2, \ldots m \tag{5.102}$$

Significance of $\mathbf{t_i}$

This is used to examine the relative magnitudes of reactor time scale, with time scales for input variations.

λ_i (Eigen values) are functions of all entries of $\underline{\underline{\mathbf{A}}}$ and depend on $\mathbf{C_S}$ and $\mathbf{p}$. It is difficult to compare the time scales of mixing, reaction, transfer, etc. In general, Eigen values take the form

$$\lambda_i = a_i + b_i i \tag{5.103}$$

General application of λ_i is to determine the stability of the system.

5.11.3 Stability and Dynamic Behavior of Bioreactor

A system is considered at some equilibrium point (or at steady state). The equilibrium point will be stable if the system deviates infinitesimally small around the equilibrium point. It will be unstable if the variables of the process move away from the equilibrium point.

For testing the stability of a system, Ljapunov's theorems is extended to the stability of the non-linear systems. Ljapunov's theorem (Schügerl, 1985) states the following.

Theorem 1:

"If all Eigen values of the characteristic linearized system equations have negative real parts, the non-disturbed motion of the system is stable, irrespective of the effects of higher order terms in the earlier equation."

Theorem 2:

"If at least one of the Eigen values of characteristic linearized system equation has a positive real part, the non-disturbed motion of the system is unstable, irrespective of the effects of the higher order terms in the disturbed motion equation."

Theorem 3:

"If the characteristic linearized system equation has no Eigen values with positive real parts, but does have some vanishing real parts, then the higher order term of the disturbed motion equation should be considered to give a state of stability or unstability as required."

In fact, all Eigen values are required to be calculated to test the system. It is better to expand Equation (5.99) to get the mth order algebraic equation.

$$\lambda^m + a_1\lambda^{m-1} + \ldots + a_{m-1}\lambda + a_m = 0 \tag{5.104}$$

Applying Hurwitz's criterion which suggests all roots of Equation (5.104) have negative real parts if and only if the following conditions are met.

$$a > 0$$

$$\begin{vmatrix} a_1 & a_3 \\ 1 & a_2 \end{vmatrix} > 0$$

$$\cdots\cdots\cdots\cdots\cdots\cdots$$

$$\begin{vmatrix} a_1 & a_3 & a_5 & \cdot & \cdot & \cdot & 0 \\ 1 & a_2 & a_4 & \cdot & \cdot & \cdot & 0 \\ 0 & a_1 & a_3 & \cdot & \cdot & \cdot & 0 \\ 0 & 1 & a_2 & \cdot & \cdot & \cdot & 0 \\ \cdot & \cdot & \cdot & \cdot & \cdot & \cdot & \cdot \\ \cdot & \cdot & \cdot & \cdot & \cdot & \cdot & a_m \end{vmatrix} > 0$$

5.11.4 Stability and Eigen Values

1. If all Eigen values have negative real parts and no imaginary parts, the steady state is said to be locally stable. The plot of the variable as a function of time shows exponential decay of the variable.

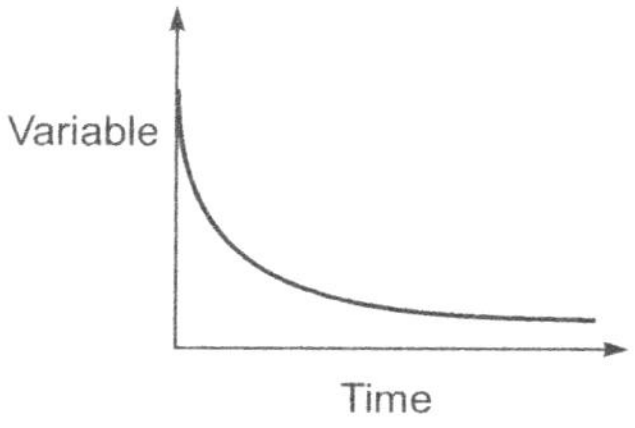

Figure 5.29 Plot of exponential decay of a variable as a function of time.

2. Conversely, if the Eigen value is positive, the steady state is unstable and the variable vs. time shows the following behavior.

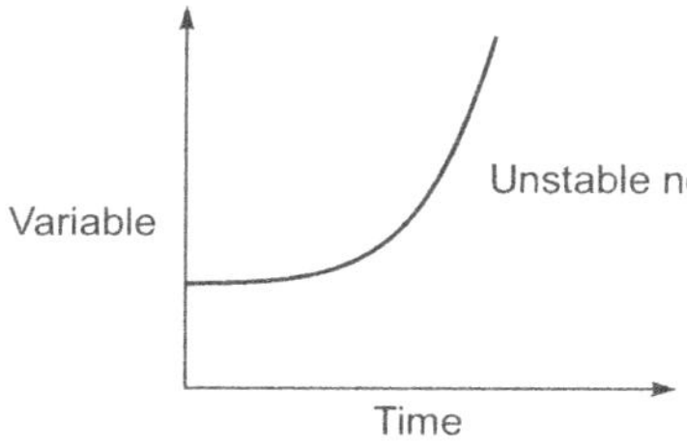

Figure 5.30 Typical plot of unstable mode.

3. If the Eigen values have only imaginary parts and zero real parts, the stability is determined by the higher order terms.

5.11.5 Examples

For studying transient behavior in reactors, cellular reactions are analyzed for stability using the following strategy:

- Cellular growth follows Monod's kinetics (ideal systems)
- Cellular growth with substrate inhibition using modified Monod's kinetics and Han and Levenspiel model.
- Cell growth accomplished by wall growth
- Mixed culture system.

Transient behavior of chemostat

Example 5.6 Cellular growth follows Monod's kinetics (ideal systems)

The governing mass balances are given below.

Cell mass balance is given by

$$\frac{dC_X}{dt} = -DC_X + \mu(C_S)C_X \tag{5.105}$$

Substrate mass balance is given by

$$\frac{dC_S}{dt} = D(C_{S_0} - C_S) - \frac{\mu(C_S)}{Y_{X/S}} C_X \tag{5.106}$$

From the mass balance, the matrix $\underline{\underline{\mathbf{A}}}$ components are given below.

$$a_{11} = \frac{\partial\left(\frac{dC_X}{dt}\right)_s}{\partial C_X} = \mu(C'_S) - D$$

$$a_{12} = \frac{\partial\left(\frac{dC_X}{dt}\right)_s}{\partial C_S} = C'_X\left(\frac{d\mu}{dC_S}\right)_s$$

$$a_{21} = \frac{\partial\left(\frac{dC_S}{dt}\right)_s}{\partial C_X} = -\frac{\mu(C'_S)}{Y_{X/S}}$$

$$a_{22} = \frac{\partial\left(\frac{dC_S}{dt}\right)_s}{\partial C_S} = -\left[\frac{C'_X}{Y_{X/S}}\left(\frac{d\mu}{dC_S}\right)_s + D\right]$$

where superscript '′' indicates steady state values. Subscript 's' in the differential function means at steady state.

For Monod's model

$$\mu(C'_S) = \frac{\mu_m C'_S}{K_S + C'_S} = D \tag{5.107}$$

$\therefore$

$$a_{11} = D - D = 0, \text{ and}$$

$$a_{21} = -\frac{D}{Y_{X/S}}$$

$\therefore$

$$\underline{\underline{\mathbf{A}}} = \begin{pmatrix} a_{11} & a_{12} \\ a_{21} & a_{22} \end{pmatrix} = \begin{pmatrix} 0 & C'_X\left(\frac{d\mu}{dC_S}\right)_S \\ -\frac{D}{Y_{X/S}} & -\left[\frac{C'_X}{Y_{X/S}}\left(\frac{d\mu}{dC_S}\right)_S + D\right] \end{pmatrix} \tag{5.108}$$

The matrix $\left(\underline{\underline{\mathbf{A}}} - \lambda \underline{\underline{\mathbf{I}}}\right)$ has elements

$$\left(\underline{\underline{\mathbf{A}}} - \lambda \underline{\underline{\mathbf{I}}}\right) = \begin{pmatrix} a_{11} - \lambda & a_{12} \\ a_{21} & a_{22} - \lambda \end{pmatrix} \tag{5.109}$$

To get the Eigen values, solve

$$\det\left(\underline{\underline{\mathbf{A}}} - \lambda \underline{\underline{\mathbf{I}}}\right) = 0$$

$$\left|\underline{\underline{\mathbf{A}}} - \lambda \underline{\underline{\mathbf{I}}}\right| = 0$$

$$\lambda^2 - (a_{11} + a_{22})\lambda + (a_{11}a_{22} - a_{12}a_{21}) = 0$$

$$\therefore \quad \lambda^2 + \left[\frac{C'_X}{Y_{X/S}}\left(\frac{d\mu}{dC_S}\right)_S + D\right]\lambda + \frac{D}{Y_{X/S}} C'_X \left(\frac{d\mu}{dC_S}\right)_S = 0 \tag{5.110}$$

Solving the quadratic expression in λ

$$\lambda = \frac{-\left[\frac{C'_X}{Y_{X/S}}\left(\frac{d\mu}{dC_S}\right)_S + D\right] \pm \sqrt{\left[\frac{C'_X}{Y_{X/S}}\left(\frac{d\mu}{dC_S}\right)_S + D\right]^2 - 4\frac{D}{Y_{X/S}} C'_X \left(\frac{d\mu}{dC_S}\right)_S}}{2}$$

$$= \frac{-\left[\frac{C'_X}{Y_{X/S}}\left(\frac{d\mu}{dC_S}\right)_S + D\right] \pm \left[\frac{C'_X}{Y_{X/S}}\left(\frac{d\mu}{dC_S}\right)_S - D\right]}{2} \tag{5.111}$$

The roots are

$$\lambda_1 = \frac{-2D}{2} = -D \tag{5.112}$$

$$\lambda_2 = \frac{-2\frac{C'_X}{Y_{X/S}}\left(\frac{d\mu}{dC_S}\right)_S}{2} = -\frac{C'_X}{Y_{X/S}}\left(\frac{d\mu}{dC_S}\right)_S \tag{5.113}$$

Also,

$$\mu(C'_S) = \frac{\mu_{\max} C'_S}{K_S + C'_S} = D$$

$$\therefore \quad C'_S = \frac{DK_S}{\mu_{\max} - D} \quad \text{and}$$

$$\left(\frac{d\mu}{dC_S}\right)_S = \frac{\mu_{\max} K_S}{(K_S + C'_S)^2}$$

$$\therefore \quad \lambda_2 = -\frac{C'_X}{Y_{X/S}} \frac{\mu_{\max} K_S}{(K_S + C'_S)^2} \quad \text{and}$$

$$\frac{C'_X}{Y_{X/S}} = (C_{S_0} - C'_S) = \frac{C_{S_0}(\mu_{\max} - D) - DK_S}{(\mu_{\max} - D)}$$

$$(K_S + C'_S)^2 = \left[K_S + \frac{DK_S}{\mu_{\max} - D}\right]^2$$

$$= \frac{\mu_{\max}{}^2 K_S^2}{(\mu_{\max} - D)^2}$$

Substituting $(K_S + C'_S)^2$ and $\dfrac{C'_X}{Y_{X/S}}$ in λ_2 results,

$$\lambda_2 = \frac{(\mu_{max} - D)[C_{S_0}(\mu_{max} - D) - DK_S]}{\mu_{max} K_S}$$

Therefore, the Eigen values are

$$\lambda_1 = -D$$

and

$$\lambda_2 = \frac{(\mu_{max} - D)[C_{S_0}(\mu_{max} - D) - DK_S]}{\mu_{max} K_S} \tag{5.114}$$

Example 5.7 Cellular growth with substrate inhibition

Let us assume the inhibition model $\mu = \dfrac{\mu_m C_S}{K_S + C_S + \dfrac{C_S{}^2}{K_I}}$ (5.115)

The governing mass balance equations are Equations (5.116) and (5.117).

Cell mass balance:

$$\frac{dC_X}{dt} = -DC_X + \mu(C_S)C_X \tag{5.116}$$

Reactant mass balance:

$$\frac{dC_S}{dt} = D(C_{S_0} - C_S) - \frac{\mu(C_S)}{Y_{X/S}} C_X \tag{5.117}$$

From the mass balances, the $\underline{\underline{\mathbf{A}}}$ matrix is

$$\underline{\underline{\mathbf{A}}} = \begin{pmatrix} a_{11} & a_{12} \\ a_{21} & a_{22} \end{pmatrix}$$

The elements of $\underline{\underline{\mathbf{A}}}$ are

$$a_{11} = \frac{\partial}{\partial C_X}\left(\frac{dC_X}{dt}\right)_S = \mu(C'_S) - D$$

$$a_{12} = \frac{\partial}{\partial C_S}\left(\frac{dC_X}{dt}\right)_S = C'_X\left(\frac{d\mu}{dC_S}\right)_S$$

$$a_{21} = \frac{\partial}{\partial C_X}\left(\frac{dC_S}{dt}\right)_S = -\frac{\mu(C'_S)}{Y_{X/S}}$$

$$a_{22} = \frac{\partial}{\partial C_S}\left(\frac{dC_S}{dt}\right)_s = -\left(\frac{C'_X}{Y_{X/S}}\left[\frac{d\mu}{dC_S}\right]_S + D\right)$$

At steady state

$$\mu(C'_S) = \frac{\mu_{\max} C'_S}{K_S + C'_S + \frac{C'^2_S}{K_I}} = D$$

So,

$$a_{11} = 0.$$

$$a_{21} = -\frac{\mu(C'_S)}{Y_{X/S}} = -\frac{D}{Y_{X/S}}$$

$\underline{\underline{\mathbf{A}}}$ matrix is

$$\begin{pmatrix} 0 & C'_X \left[\frac{d\mu}{dC_S}\right]_S \\ -\frac{D}{Y_{X/S}} & -\left(\frac{C'_X}{Y_{X/S}}\left[\frac{d\mu}{dC_S}\right]_S + D\right) \end{pmatrix}$$

The $\left(\underline{\underline{\mathbf{A}}} - \lambda \underline{\underline{\mathbf{I}}}\right)$ matrix is

$$\begin{pmatrix} a_{11} - \lambda & a_{12} \\ a_{21} & a_{22} - \lambda \end{pmatrix}$$

Solving det $\left(\underline{\underline{\mathbf{A}}} - \lambda \underline{\underline{\mathbf{I}}}\right)$ to get the Eigen values as per Equation (5.118) is:

$$\lambda^2 - (a_{11} + a_{22})\lambda + (a_{11}a_{22} - a_{12}a_{21}) = 0 \tag{5.118}$$

$$a_{11} + a_{22} = -\left(\frac{C'_X}{Y_{X/S}}\left[\frac{d\mu}{dC_S}\right]_S + D\right)$$

$$a_{11}a_{22} - a_{12}a_{21} = \frac{DC'_X}{Y_{X/S}}\left[\frac{d\mu}{dC_S}\right]_S$$

Therefore,

$$\lambda = \frac{-\left(\frac{C'_X}{Y_{X/S}}\left[\frac{d\mu}{dC_S}\right]_S + D\right) \pm \sqrt{\left(\frac{C'_X}{Y_{X/S}}\left[\frac{d\mu}{dC_S}\right]_S + D\right)^2 - 4\frac{DC'_X}{Y_{X/S}}\left[\frac{d\mu}{dC_S}\right]_S}}{2}$$

$$= \frac{-\left(\frac{C'_X}{Y_{X/S}}\left[\frac{d\mu}{dC_S}\right]_S + D\right) \pm \sqrt{\left(\frac{C'_X}{Y_{X/S}}\left[\frac{d\mu}{dC_S}\right]_S\right)^2 + D^2 - 2\frac{DC'_X}{Y_{X/S}}\left[\frac{d\mu}{dC_S}\right]_S}}{2} \tag{5.119}$$

$$\lambda_1 = -D \tag{5.120}$$

$$\lambda_2 = -\left(\frac{C'_X}{Y_{X/S}}\left[\frac{d\mu}{dC_S}\right]_S\right) \tag{5.121}$$

$$\left(\frac{d\mu}{dC_S}\right)_S = \frac{\mu_{\max}(K_S K_I - C_S'^2)}{\left[K_S + C_S' + \frac{C_S'^2}{K_I}\right]^2} \tag{5.122}$$

$$C_X' = Y_{X/S}(C_{S_0} - C_S') \tag{5.123}$$

Example 5.8 Cellular growth in addition to wall growth in a bioreactor
Specific wall growth rate suggested by Schügerl (1985) is given in Equation (5.124).

$$\mu_w = \frac{\mu_{\max} C_S X_w}{K_S + C_S + \frac{C_S^2}{K_I}} \tag{5.124}$$

Substrate mass balance is

$$\frac{dC_S}{dt} = D(C_{S_0} - C_S) - \frac{1}{Y_{X/S}} \frac{\mu_{\max} C_S C_X}{K_S + C_S + \frac{C_S^2}{K_I}} - \frac{1}{Y_{X/S}} \frac{\mu_{\max} C_S X_w A}{K_S + C_S + \frac{C_S^2}{K_I}}$$

where X_w is the wall coating density (Schügerl, 1985) and A is the area covered by cells.

Cell mass balance is

$$\frac{dC_X}{dt} = D(C_{X_0} - C_X) - \frac{\mu_{\max} C_S X_w A}{K_S + C_S + \frac{C_S^2}{K_I}} + \frac{\mu_{\max} C_S C_X}{K_S + C_S + \frac{C_S^2}{K_I}}$$

From the mass balances, the $\underline{\underline{\mathbf{A}}}$ matrix is

$$\underline{\underline{\mathbf{A}}} = \begin{pmatrix} a_{11} & a_{12} \\ a_{21} & a_{22} \end{pmatrix}$$

The components of $\underline{\underline{\mathbf{A}}}$ are

$$a_{11} = \frac{\partial}{\partial C_X}\left[\frac{dC_X}{dt}\right]_S = -D + \frac{\mu_w C_S}{K_S + C_S + \frac{C_S^2}{K_I}}$$

$$a_{12} = \frac{\partial}{\partial C_S}\left[\frac{dC_X}{dt}\right]_S = C_X'\left(\frac{d\mu}{dC_S}\right)_S$$

$$a_{21} = \frac{\partial}{\partial C_X}\left[\frac{dC_S}{dt}\right]_S = -\frac{\mu_{\max} C_S}{Y_{X/S}\left(K_S + C_S + \frac{C_S^2}{K_I}\right)}$$

$$a_{22} = \frac{\partial}{\partial C_S}\left[\frac{dC_S}{dt}\right]_S = -\left[D + \frac{C_X'}{Y_{X/S}}\left[\frac{dC_S}{dt}\right]_S\right]$$

Solving det $\left(\underline{\underline{\mathbf{A}}} - \lambda\underline{\underline{\mathbf{I}}}\right)$ to get the Eigen values, we obtain,
$\lambda^2 - (a_{11} + a_{22})\lambda + (a_{11}a_{22} - a_{12}a_{21}) = 0$

Example 5.9 Mixed culture system using Monod's model

In the mixed culture, at least two different organisms are used for the reaction in the continuous flow bioreactor. The final rate constant for the product formation will be sum of individual rate constants. Only difference is that the number of parameters will double for the mixed culture using two organisms. So the plot can be visualized as the net effect of individual effects, but the plot will be tilted to the behavior of organism where concentration is more.

Now let us see the rate equation for the two organisms used for the reaction.

According to Monod's kinetics, the specific growth rate of the cell can be given as follows.

For 1st organism:

$$\mu_1 = \mu_{\max_1} \frac{C_S}{K_{S_1} + C_S} \tag{5.125}$$

For 2nd organism:

$$\mu_2 = \mu_{\max_2} \frac{C_S}{K_{S_2} + C_S} \tag{5.126}$$

$$\mu = \mu_1 + \mu_2 \tag{5.127}$$

The substrate and cell balances are

$$\frac{dC_S}{dt} = D(C_{S_0} - C_S) - \frac{\mu C_X}{Y_{X/S}} \tag{5.128}$$

$$\frac{dC_X}{dt} = (\mu - D)C_X \tag{5.129}$$

$$\frac{dC_S}{dt} = D(C_{S_0} - C_S) - \mu_{\max_1} \frac{C_S}{K_{S_1} + C_S} \frac{C_X}{Y_{(X/S)_1}} - \mu_{\max_2} \frac{C_S}{K_{S_2} + C_S} \frac{C_X}{Y_{(X/S)_2}} \tag{5.130}$$

$$\frac{dC_X}{dt} = \left(\mu_{\max_1} \frac{C_S}{K_{S_1} + C_S} + \mu_{\max_2} \frac{C_S}{K_{S_2} + C_S} - D\right)C_X \tag{5.131}$$

In order to find the actual variation of substrate concentration for the mixed culture, one needs to integrate the above equations assuming some reasonable values for the parameters.

Steps in brief to do stability analysis of mixed cultures

(1) Find the steady state values for both the cell cultures.

(2) Find the Jacobian matrix for the given system.

(3) Apply two conditions—one for wash out and the other for non trivial solution for steady state.

Example 5.10 Multiple growth limiting reactant in a mixed culture system

Let us consider S_1 and S_2 as two reactant species in a CFSTBR with two known organisms. The mass balances are Equations (5.132) to (5.135).

$$\frac{dC_{X_1}}{dt} = -DC_{X_1} + \mu_1 C_{X_1} \tag{5.132}$$

$$\frac{dC_{X_2}}{dt} = -DC_{X_2} + \mu_2 C_{X_2} \tag{5.133}$$

$$\frac{dC_{S_1}}{dt} = D(C_{S_{10}} - C_{S_1}) - \frac{\mu_1 C_{X_1}}{Y_{11}} - \frac{\mu_2 C_{X_2}}{Y_{21}} \tag{5.134}$$

$$\frac{dC_{S_2}}{dt} = D(C_{S_{20}} - C_{S_2}) - \frac{\mu_1 C_{X_1}}{Y_{12}} - \frac{\mu_2 C_{X_2}}{Y_{22}} \tag{5.135}$$

where

$$\mu_1 = \mu_1(C_{S_1}, C_{S_2}) \tag{5.136}$$

$$\mu_2 = \mu_2(C_{S_1}, C_{S_2}) \tag{5.137}$$

At steady state, first two equations show

$$\mu_1(C_{S_1}, C_{S_2}) = \mu_2(C_{S_1}, C_{S_2}) = D \tag{5.138}$$

Single value of D cannot show steady state which might represent coexistence of steady state with one reactant.

There could be four possible steady states in the system.

Steady state 1 – Both organisms coexist.

Steady states 2 and 3 – Any one organism is present at steady state.

Steady state 4 – Both organisms washed out.

If functional forms of $\mu_1(C_{S_1}, C_{S_2})$ and $\mu_2(C_{S_1}, C_{S_2})$ are known, contour plots describing constant values of μ_1 and μ_2 can be drawn on a graph of C_{S_1} vs. C_{S_2}. Possible steady states are the intersection of these contour plots of μ_1 and μ_2.

One needs to use extended Monod's Model for two reactants.

$$\mu_1 = \frac{\mu_{m11} C_{S_1}}{K_{11} + C_{S_1} + a_{11} C_{S_2}} + \frac{\mu_{m12} C_{S_2}}{K_{21} + C_{S_2} + a_{21} C_{S_1}} \tag{5.139}$$

$$\mu_2 = \frac{\mu_{m11} C_{S_1}}{K_{12} + C_{S_1} + a_{12} C_{S_2}} + \frac{\mu_{m22} C_{S_2}}{K_{22} + C_{S_2} + a_{22} C_{S_1}} \tag{5.140}$$

Example 5.11 Competition of a mixed culture components in a chemostat

We consider Monod's kinetics model using single reactant for the interaction of two organisms, growing in a mixed culture. The possibilities of μ-C_S diagram can be represented by Figure 5.31.

For cases (i) and (ii), the mass balance equations are

$$\frac{dC_{X_1}}{dt} = -DC_{X_1} + \mu_1 C_{X_1} \tag{5.141}$$

$$\frac{dC_{X_2}}{dt} = -DC_{X_2} + \mu_2 C_{X_2} \tag{5.142}$$

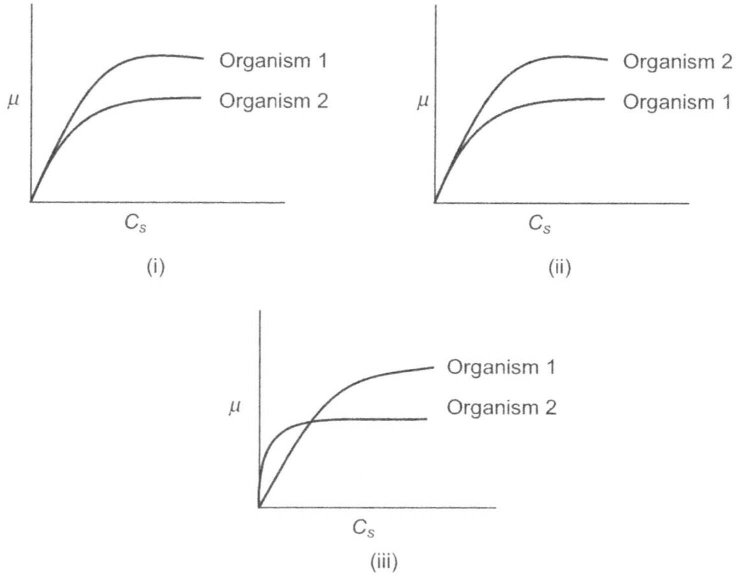

Figure 5.31 (i – iii) Interaction of two organisms

If steady state population of both C_{X_1} and C_{X_2} are to coexist, then

$$D = \mu_1 = \mu_2. \tag{5.143}$$

where co-existence occurs at $D = D_c$.

If $D > D_c$, organism 2 will be washed out for Figures 5.31(i) and (iii).

If $D < D_c$, organism 2 will dominate (Fig. 5.31ii).

Considering a set of mass balances for both organisms and reactant, Equations (5.144) to (5.146) are considered in this case.

$$\frac{dC_{X_1}}{dt} = -DC_{X_1} + \mu_1 C_{X_1} \tag{5.144}$$

$$\frac{dC_{X_2}}{dt} = -DC_{X_2} + \mu_2 C_{X_2} \tag{5.145}$$

$$\frac{dC_S}{dt} = D(C_{S_0} - C_S) - \frac{\mu_1 C_{X_1}}{Y_1} - \frac{\mu_2 C_{X_2}}{Y_2} \tag{5.146}$$

Coexistence can occur at $D = D_c$.

$$D(C_{S_0} - C_S) - \frac{\mu_1 C_{X_1}}{Y_1} - \frac{\mu_2 C_{X_2}}{Y_2} = 0$$

$$(C_{S_0} - C_S) = \frac{C_{X_1}}{Y_1} + \frac{C_{X_2}}{Y_2}$$

If C_S in the feed is less than C_S in solution, then

$$\mu_1(C_S) = \mu_2(C_S) = D_C \tag{5.147}$$

Both organisms will washout, if Monod's expression is assumed,

$$\mu_1 = \frac{\mu_{m1} C_S}{K_1 + C_S}$$

$$\mu_2 = \frac{\mu_{m2} C_S}{K_2 + C_S}$$

Then one can calculate C'_S and D_c. It is possible to evaluate the conditions for positive values of D_c and C'_S.

Example 5.12 Stability analysis of chemostat for substrate inhibition expressed using Han and Levenspiel kinetics

The expression by Han and Levenspiel (1987) that accounts for the substrate inhibition is given by Equation (5.148).

$$\mu = \mu_{\max} \frac{\left(1 - \frac{C_S}{C_S^*}\right)^n C_S}{K_S \left(1 - \frac{C_S}{C_S^*}\right)^m + C_S} \tag{5.148}$$

where

μ is the specific growth rate
C_S is the substrate concentration
$\mu_{\max}$ is the maximum specific growth rate
K_S is the saturation constant for growth limiting substrate
m and n are constants
C_S^* is the critical reactant concentration.

The Han-Levenspiel equation is used for overall cell and substrate mass balances.

The cell balance for the continuous flow bioreactor with substrate inhibition can be given by the following Equation (5.149).

$$\frac{dC_X}{dt} = \left(\mu_{\max} \frac{\left(1 - \frac{C_S}{C_S^*}\right)^n C_S}{K_S \left(1 - \frac{C_S}{C_S^*}\right)^m + C_S} - D \right) C_X \tag{5.149}$$

The substrate balance equation can be written as

$$\frac{dC_S}{dt} = D(C_{S_0} - C_S) - \mu_{\max} \frac{\left(1 - \frac{C_S}{C_S^*}\right)^n C_S}{K_S \left(1 - \frac{C_S}{C_S^*}\right)^m + C_S} \frac{C_X}{Y_{X/S}} \tag{5.150}$$

Concentrations of cell and substrate are evaluated at steady state.

Hence,

$$\frac{dC_X}{dt} = 0 \text{ and } \frac{dC_S}{dt} = 0$$

$$\therefore \quad \left(\mu_{\max} \frac{\left(1 - \frac{C'_S}{C^*_S}\right)^n C'_S}{K_S \left(1 - \frac{C'_S}{C^*_S}\right)^m + C'_S} - D \right) C'_X = 0$$

$$\mu_{\max} \frac{\left(1 - \frac{C'_S}{C^*_S}\right)^n C'_S}{K_S \left(1 - \frac{C'_S}{C^*_S}\right)^m + C'_S} = D \tag{5.151}$$

Equation (5.150) becomes

$$D(C_{S_0} - C'_S) - \mu_{\max} \frac{\left(1 - \frac{C'_S}{C^*_S}\right)^n C'_S}{K_S \left(1 - \frac{C'_S}{C^*_S}\right)^m + C'_S} \frac{C'_X}{Y_{X/S}} = 0 \tag{5.152}$$

Now let us find the cell concentration and substrate concentration with a substrate inhibition at the steady state condition.

$$C'_S = C_{S_0} - \left\{ \mu_{\max} \frac{\left(1 - \frac{C'_S}{C^*_S}\right)^n C'_S}{K_S \left(1 - \frac{C'_S}{C^*_S}\right)^m + C'_S} \right\} \frac{C'_X}{Y_{X/S} D} \tag{5.153}$$

This is the substrate concentration at the steady state. We can write it as

$$C'_S = C_{S_0} - \mu \frac{C'_X}{Y_{X/S} D} \tag{5.154}$$

The cell concentration can be calculated as

$$C'_X = Y_{X/S}(C_{S_0} - C'_S) \tag{5.155}$$

Equations (5.154) and (5.155) give the cell and substrate concentration at steady state.

Figure 5.32 represents steady state substrate concentration vs. dilution rate.

5.12 STABILITY ANALYSIS FOR CONTINUOUS FLOW BIOREACTOR WITH SUBSTRATE INHIBITION

A steady state is stable if, for initial conditions near the steady state, all transients converge to it. If the transients diverge, steady state is called unstable. The diverging transients always end at some other stable state. Stability analysis of a steady state would involve whether the steady state under consideration is stable or not and the information about state-to-state transitions in case of unstable steady states.

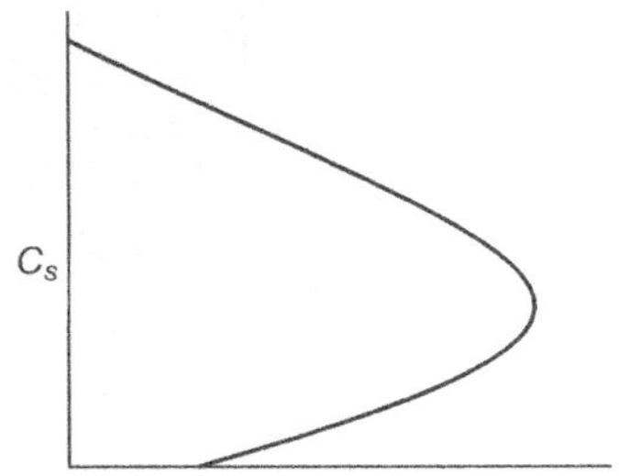

Figure 5.32 Steady state substrate concentrations as a function of dilution rate.

The information about the stability and local dynamics of the steady states is accomplished through linear stability analysis. It should be borne in mind that the results of the linear stability analysis are good only near the steady state. For general (non local) behavior and information about state-to-state transitions, generation of phase plane is suitable.

Let us recall the basic equation for defining stability.

$$\frac{dC_X}{dt} = \left(\mu_{\max} \frac{\left(1 - \frac{C_S}{C_S^*}\right)^n C_S}{K_S \left(1 - \frac{C_S}{C_S^*}\right)^m + C_S} - D \right) C_X = (\mu - D) C_X \tag{5.156}$$

$$\frac{dC_S}{dt} = D(C_{S_0} - C_S) - \mu_{\max} \frac{\left(1 - \frac{C_S}{C_S^*}\right)^n C_S}{K_S \left(1 - \frac{C_S}{C_S^*}\right)^m + C_S} \frac{C_X}{Y_{X/S}}$$

$$= D(C_{S_0} - C_S) - \frac{\mu C_X}{Y_{X/S}} \tag{5.157}$$

The Jacobian matrix for the above system can be written as

$$J = \begin{vmatrix} a_{11} & a_{12} \\ a_{21} & a_{22} \end{vmatrix} = \begin{vmatrix} \dfrac{\partial\left(\dfrac{dC_X}{dt}\right)}{\partial C_X} & \dfrac{\partial\left(\dfrac{dC_X}{dt}\right)}{\partial C_S} \\ \dfrac{\partial\left(\dfrac{dC_S}{dt}\right)}{\partial C_X} & \dfrac{\partial\left(\dfrac{dC_S}{dt}\right)}{\partial C_S} \end{vmatrix} \tag{5.158}$$

$\therefore \quad a_{11} = \mu - D$

$$a_{12} = C_X \frac{d\mu}{dC_S}$$

$$a_{21} = -\frac{\mu}{Y_{X/S}}$$

$$a_{22} = -D - \frac{C_X}{Y_{X/S}}\frac{d\mu}{dC_S}$$

All are at steady states.

$$\therefore \quad J = \begin{vmatrix} \mu - D & C_X \dfrac{d\mu}{dC_S} \\ -\dfrac{\mu}{Y_{X/S}} & -D - \dfrac{C_X}{Y_{X/S}}\dfrac{d\mu}{dC_S} \end{vmatrix} = \begin{vmatrix} 0 & C_X \dfrac{d\mu}{dC_S} \\ -\dfrac{\mu}{Y_{X/S}} & -D - \dfrac{C_X}{Y_{X/S}}\dfrac{d\mu}{dC_S} \end{vmatrix} \text{ since } \mu = D, \quad a_{11} = 0 \qquad (5.159)$$

For the nontrivial steady state, stability is guaranteed if the following conditions are satisfied

Trace $J < 0$ and $\det J > 0$,

where trace is the sum of the diagonal elements.

So,

$$-D - \frac{C_X}{Y_{X/S}}\frac{d\mu}{dC_S} < 0$$

and

$$C_X \frac{\mu}{Y_{X/S}}\frac{d\mu}{dC_S} > 0$$

These are the conditions of stability with substrate inhibition.

Now, let us see how C_S and C_X vary with time. The equation which describes the substrate variation with time is given by Equation (5.160).

$$\frac{dC_S}{dt} = D(C_{S_0} - C_S) - \mu_{max} \frac{\left(1 - \dfrac{C_S}{C_S^*}\right)^n C_S}{K_S\left(1 - \dfrac{C_S}{C_S^*}\right)^m + C_S} \frac{C_X}{Y_{X/S}} \qquad (5.160)$$

In order to simplify this equation we have taken a species for which K_S is very small for which $\dfrac{C_S}{C_S^*} < 1$.

5.13 PHASE PLANE ANALYSIS

This technique is used to determine the stability of the multivariable system with non linear equations (Schügerl, 1995; Daugulis *et al.*, 2000). The phase plane is setup by drawing straight lines representing the two Eigen vectors which describes the conditions of stability (Bequette, 1998). The system either converges towards those lines or diverges away from them. Phase plane is represented by lines, not by direction of field dashes. For example, a phase plane for a two-state variable system consists of plot of one state variable against another state variable. Each curve in this plot is based on different initial conditions (Fig. 5.33).

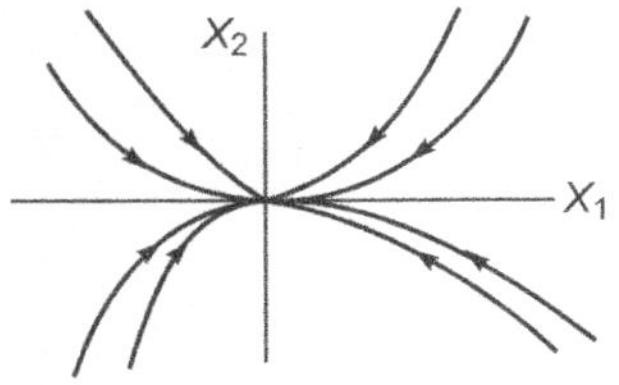

Figure 5.33 Phase plane plot for X_1 vs. X_2.

5.13.1 Generalized Phase-Plane Behavior

Linear two-state system is classified on the signs of Eigen vectors (λ_1 and λ_2). Table 5.1 summarizes the behavior.

For non-linear systems, the phase-plane behavior is like the linear systems as the model is linearized about the equilibrium point. An ideal biological reactions following Monod's kinetics, the phase-plane behavior of cell concentrations (C_x) and reactant concentration can be represented by Figure 5.40.

The signs of Eigen value will indicate how the system's phase plane behaves.

- If Eigen values have opposite signs, the intersection of the Eigen vectors is a saddle point (unstable equilibrium point).
- If the signs of Eigen values are both positive, the Eigen vectors represent unstable situations and the system diverges away. The intersection is an unstable node.
- If the signs of both Eigen values are negative, the system is a stable node and shows convergence of the lines.

5.13.2 Development of Phase Plane Diagram for a Bioreactor

The development of phase plane analysis can be done as follows.

(1) Use Eigen values and Eigen vectors of the Jacobian matrix to characterize the phase plane behavior.

(2) Predict the phase plane behavior close to an equilibrium point, based on the linearized model at that equilibrium point.

(3) Predict qualitatively the phase plane behavior of the non linear system, when there are multiple equilibrium points.

Two equations, which defines the system, are

$$\frac{dC_X}{dt} = (\mu(C_S) - D)C_X \tag{5.161}$$

$$\frac{dC_S}{dt} = -D(C_S - C_{S_0}) - \frac{\mu}{Y_{X/S}} C_X \tag{5.162}$$

where $D = \dfrac{F}{V}$, dilution rate

Table 5.1 *Summary of generalized phase-plane behavior*

Type	*Signs of Eigen values*	*Imaginary or Real*	*Phase-plane concept*
Stable node	Both λ_1 and λ_2 are negative ($\lambda_1 < 0$, $\lambda_2 < 0$)	Real	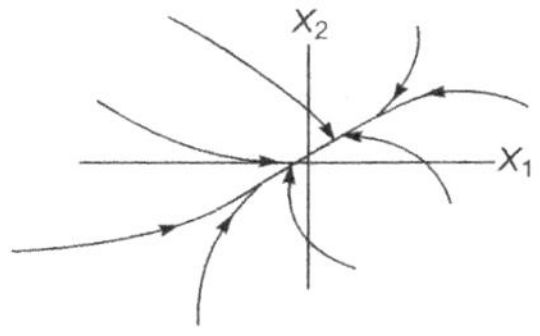 **Figure 5.34**
Unstable node	Both λ_1 and λ_2 are positive ($\lambda_1 > 0$, $\lambda_2 > 0$)	Real	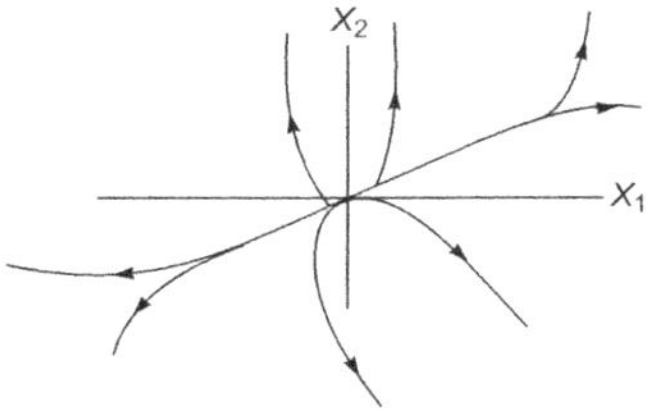 **Figure 5.35**
Saddle point	One positive and another negative ($\lambda_1 < 0$, $\lambda_2 > 0$)	Real	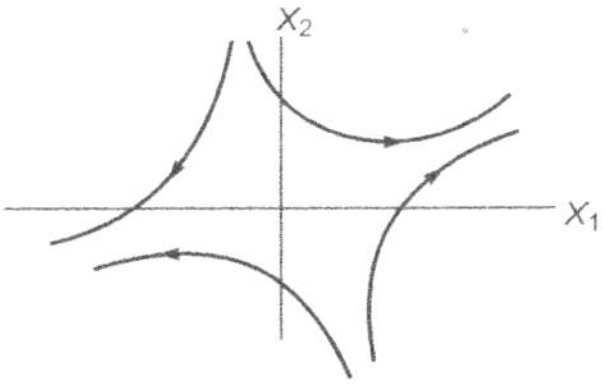 **Figure 5.36**
Stable focus	Real part of λ_1 is less than zero (Re (λ_1) < 0)	Complex conjugates	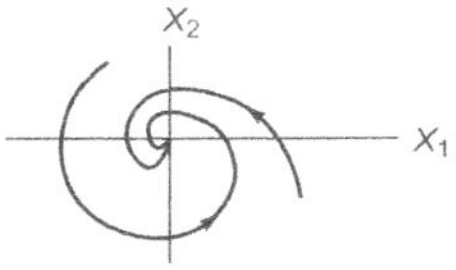 **Figure 5.37**
Unstable focus	Real part of λ_1 is greater than zero (Re λ_1) > 0)	Complex conjugates	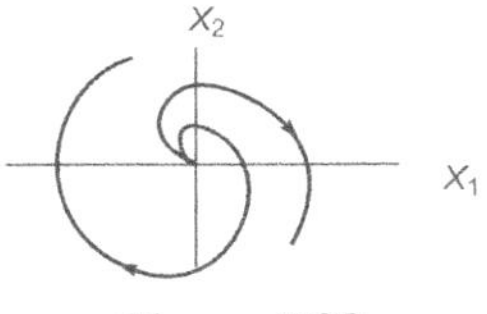 **Figure 5.38**
Center	Real part of Eigen value is zero	Complex conjugates	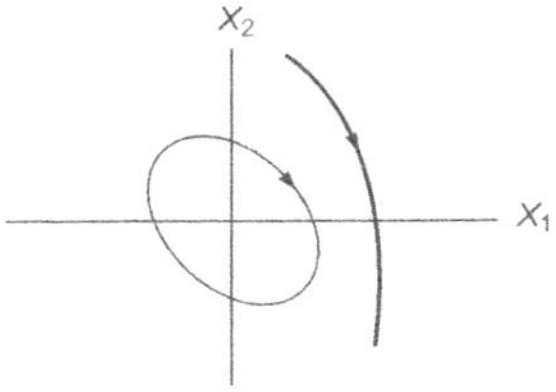 **Figure 5.39**

Following Monod's kinetics

$$\mu = \frac{\mu_{\max} C_S}{K_S + C_S} \qquad (5.163)$$

Now Equations (5.161) and (5.162) become

$$\frac{dC_X}{dt} = \left[\frac{\mu_{\max} C_S}{K_S + C_S} - D\right] C_X \qquad (5.164)$$

$$\frac{dC_S}{dt} = -D(C_S - C_{S_0}) - \frac{\frac{\mu_{\max} C_S}{K_S + C_S}}{Y_{X/S}} C_X \qquad (5.165)$$

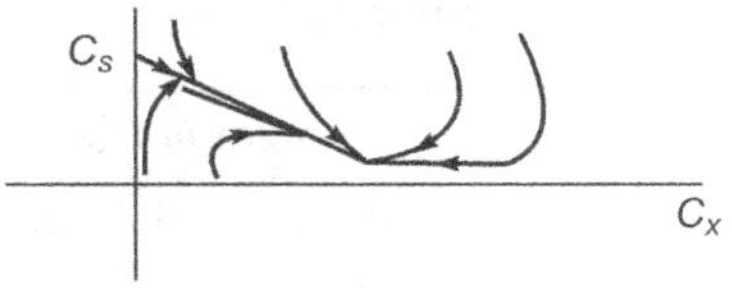

Figure 5.40 Phase-plane for cell mass concentration and reactant concentration. Cell obeys Monod's equation.

The variation of C_S and C_X is discussed qualitatively. Experimental data and similar cases in cell growth kinetics emphasize the need of reproduction–constant to be almost linear for small positive values of 'S', but it requires an upper limit for μ so that,

$$\mu(S) \rightarrow \mu_{\max}$$
$$\lim C_S \rightarrow \infty$$

The constant K_S in Equation (5.164) corresponding to the concentration at which $\mu = \frac{1}{2}\mu_{\max}$. For low '$C_S$', $\mu(C_S) \approx \frac{\mu_{\max} C_S}{K_S}$.

Thus if $C_S \ll K_S$, μ is almost linear. The maximum μ, i.e., $\mu_{\max}$, is never reached. No matter how great C_S will be, $\mu < \mu_{\max}$.

The maximum reproduction rate is achieved when the nutrient is unlimited.

With this $\mu(C_S)$, the model becomes

$$\frac{dC_X}{dt} = \left[\frac{\mu_{\max} C_S}{K_S + C_S} C_X - DC_X\right] \qquad (5.166)$$

$$\frac{dC_S}{dt} = -\alpha \frac{\mu_{\max} C_S}{K_S + C_S} C_X - DC_S + DC_{S_0} \qquad (5.167)$$

Considering a more general CFSTBR,

$$\frac{dC_X}{dt} = (\mu(C_S, \gamma) - D)C_X + DC_{X_0} \qquad (5.168)$$

$$\frac{dC_S}{dt} = -\alpha\mu(C_S, \gamma)C_X - D(C_S - C_{S_0}) \qquad (5.169)$$

where γ corresponds for other possible variables (C_X, products, pH, etc.) that may influence the reproduction rate, C_{X_0} (stands for organism's density inserted via in-flow) or is the inlet biomass concentration.

A number of important and necessary parameters, viz., $\mu_{\max}$, K_S, D, α and C_{S_0} exist in the above equations. To reduce the number of parameters, let us consider the method of non-dimensional analysis of the equations.

To simplify, C_S, C_X, t are replaced with $C_S \cdot \hat{C}_S, C_X \cdot \hat{C}_X, t \cdot \hat{t}$ where $\hat{C}_S, \hat{C}_X, \hat{t}$ correspond to the unit-dimension. The variables in Equations (5.166) and (5.167) will be replaced with these variables.

$$\frac{dC_X}{dt}\cdot\frac{\hat{C}_X}{\hat{t}} = \left[\frac{\mu_{max} C_S \cdot C_S}{K_S + C_S \cdot C_S} - D\right] C_X \cdot C_X \tag{5.170}$$

$$\frac{dC_S}{dt}\cdot\frac{\hat{C}_S}{\hat{t}} = -\alpha \frac{\mu_{max} C_S \cdot C_S}{K_S + C_S \cdot C_S} C_X \cdot C_X - DC_S \cdot C_S + DC_{S_0} \tag{5.171}$$

To get a dimensionless form, Equation (5.170) and Equation (5.171) are multiplied with $\frac{\hat{t}}{\hat{C}_X}$ and $\frac{\hat{t}}{\hat{C}_S}$, respectively.

Then fix $\hat{t} = \frac{1}{D}$,

$$\hat{C}_S = K_S$$

$$\hat{C}_X = \frac{\hat{C}_S}{\alpha \mu_{max} \hat{t}} = \frac{K_S D}{\alpha \mu_{max}}$$

and replace

$$\mu_{max} t = \frac{\mu_{max}}{D} = \theta_1$$

$$\frac{C_{S_0}}{\hat{C}_S} = \frac{C_{S_0}}{K_S} = \frac{\hat{t}\, DC_{S_0}}{\hat{C}_S} = \theta_2$$

Then

$$\frac{dC_X}{dt} = \theta_1 \frac{C_S}{1 + C_S} C_X - C_X \tag{5.172}$$

$$\frac{dC_S}{dt} = -\frac{C_S}{1 + C_S} C_X - C_S + \theta_2 \tag{5.173}$$

Above equations show that the chemostat depends on two parameters θ_1 and θ_2. For the chemostat, it is possible to find equilibrium solutions, null-clines, and to investigate interesting values of the parameters θ_1 and θ_2. Now linearize Equations (5.172) and (5.173) around equilibrium points and find an invariant line.

Equilibrium solution of the chemostat

To find the equilibrium solution, Equations (5.172) and (5.173) are solved by equating them to zero.

$$\frac{dC_X}{dt} = \theta_1 \frac{C_S}{1 + C_S} C_X - C_X = 0 \tag{5.172a}$$

$$\frac{dC_S}{dt} = -\frac{C_S}{1 + C_S} C_X - C_S + \theta_2 = 0 \tag{5.173a}$$

One trivial solution is $C_X = 0$. Inserting this into Equation (5.173a) gives $C_S = \theta_2$. Thus one equilibrium solution is

$$(C'_{X_0}, C'_{S_0}) = (0, \theta_2)$$

This is our first hint indicating that θ_2 has a meaning of dimensionless stock-nutrient concentration. The other (non-trivial) solution is

$$C_X = \theta_1 \frac{C_S}{1+C_S} C_X \quad \text{for} \quad C_X \neq 0$$

$$C_S = \frac{1}{\theta_1 - 1} \tag{5.173b}$$

From Equations 5.173(a) and (b),

$C_X = \theta_1 \left(\theta_2 - \dfrac{1}{\theta_1 - 1} \right)$ which indicates the limit for population to exist as $\theta_2 > \dfrac{1}{\theta_1 - 1}$

So, two equilibrium points are:

$$(C'_{X_0}, C'_{S_0}) = (0, \theta_2)$$

$$(C'_{X_1}, C'_{S_1}) = \left[\theta_1 \left(\theta_2 - \frac{1}{\theta_1 - 1} \right), \frac{1}{\theta_1 - 1} \right] \tag{5.174}$$

Null clines are the lines where $\dfrac{dC_X}{dt} = 0$ and $\dfrac{dC_S}{dt} = 0$, used for investigating the dynamic systems. They are easy to draw in a plot. In this analysis, the null-clines are given by

$$\dot{C}_X = 0 \quad \Rightarrow \quad C_X = 0 \quad \text{or} \quad C_S = \frac{1}{\theta_1 - 1}$$

$$\dot{C}_S = 0 \quad \Rightarrow \quad C_X = \frac{(\theta_2 - C_S)(1 + C_S)}{C_S}$$

By drawing null clines (some vectors) in the first quadrant, one can get good idea of how a state $(C_X(t), C_S(t))$ moves around in the (C_X, C_S) plane over time.

There exists two equilibrium solutions in the positive quadrant of the X-S plane. To study the stability of the system at these points, one needs to study the response of the chemostat for small disturbances around the neighbourhood of these points.

In order to analyze the chemostat at equilibrium, the stability conditions are tr $(\underline{\underline{\mathbf{A}}}) < 0$ and det $(\underline{\underline{\mathbf{A}}}) > 0$. For the Eigen values of the linearization matrix $(\underline{\underline{\mathbf{A}}})$ to be real, - 4det $(\underline{\underline{\mathbf{A}}})$ + tr $(\underline{\underline{\mathbf{A}}})^2 > 0$. Such situation has been previously in this chapter (Section 5.11).

Invariant line

All the solutions asymptotically approach the invariant line. Also, the invariant line has the direction of one of the Eigen vectors. The invariant line for this example is $C_X = -\theta_1 C_S + \theta_1\theta_2$. From Equations (5.172) and (5.173), multiply (5.173) by θ_1, then add to (5.172), which results Equation (5.175).

$$\frac{d}{dt}(C_X + \theta_1 C_S)(t) = \theta_1\theta_2 - (C_X + \theta_1 C_S)(t) \tag{5.175}$$

This is an ordinary differential equation with function $(C_X + \theta_1 C_S)(t)$.

All solutions wandering in the positive quadrant of (X, S) plane asymptotically approach the invariant line.

Setting $C_X = 0$ will give $C_S = \theta_2$, the trivial equilibrium point. $C_S = 0$ gives $C_X = \theta_1\theta_2$. We can verify that this line passes through the non-trivial equilibrium point as well.

5.14 THE BIFURCATION ANALYSIS

Bifurcation generally tells the possibility of existence of two or more than two steady state conditions (Kuznetsov, 2004). Generally, this tool is used with dynamic simulation, whereas dynamic simulation is useful to predict the average behavior of a system in a particular range of system variable (Hale and Kocak, 1991). For example, in a CFSTBR, dilution rate is one of the major system variables which can be changed to see the various patterns in output variables, i.e., cell concentration or the product concentration with time. Firstly, we will explain what bifurcation is and then we will go through certain terms which is useful for bifurcation analysis in bioreactor.

5.14.1 Drawbacks of Dynamic Analysis

1. It appears inefficient and potentially inconclusive, especially when the model possesses slow dynamic models.
2. It happens necessarily incomplete since only a limited number of simulation tests can be performed and important dynamic behaviors may not be observed.
3. It does not easily reveal the model characteristics that lead to certain dynamic behaviors.

Therefore, the best way to simulate bioreactors is with bifurcation analysis vis-à-vis dynamic simulation.

5.14.2 What is Bifurcation?

Bifurcation means the possibility of existence of different steady state with slight change in the value of system variables. It is the mathematical study of changes in the qualitative or topological structure of the integral curves of a vector field or equivalently the solutions of a differential equation. Most commonly applied to mathematical study of dynamic systems, a bifurcation occurs when a small smooth change made to the parameter values (the bifurcation parameters) of a system causes a sudden 'qualitative' or topological change in its long term dynamic behavior (Bequette, 1998). Bifurcation occurs in both continuous systems (described by ODEs, DDEs, or PDEs) and in discrete systems(described by maps).

5.14.3 Different Terms Used In Bifurcation Analysis Pertaining to Bioreactor Analysis

1. Number of Equilibriums

Let X be the output variable of interest

$$\frac{dX}{dt} = X^2 + aX + b \tag{5.176}$$

So, equilibrium exists at $\dfrac{dX}{dt} = 0$

i.e., at

$$X^2 + aX + b = 0 \tag{5.177}$$

The roots are

$$X_1 = \frac{-a - \sqrt{a^2 - 4b}}{2}$$

$$X_2 = \frac{-a + \sqrt{a^2 - 4b}}{2}$$

Hence, there are two equilibrium points at X_1 and X_2.

2. Stability of the Equilibrium

As we have seen in the above equilibrium point, the sign scheme of $\frac{dX}{dt}$ is the following

$$\frac{dX}{dt} > 0 \text{ for } X < X_1 \text{ and } X > X_2$$

$$\frac{dX}{dt} < 0 \text{ for } X_1 < X < X_2$$

If X is slightly disturbed from its position from $X = X_1$, again it will return to same position. So, X_1 is a stable point.

On the other hand, at $X = X_2$, if it is slightly disturbed it will never return to its position. It is an unstable equilibrium point.

3. Chaos

Some nonlinear dynamic systems exhibit dynamic behavior which is highly sensitive to initial condition as a result of this sensitivity the dynamic system manifests itself as an exponential growth of perturbations in the initial conditions. The behavior of chaotic system appears to be random (Strogatz, 1994).

4. Multiple Modes of Behavior

It means the variation of nature of plot of output variable by slight change in parameter for the same non-linear system.

5. Limit Cycle

In the area of dynamic systems, a limit cycle on a plane or a two dimensional manifold is a closed trajectory in phase space having the property that at least one other trajectory spirals into it either as time approaches infinity or as time approaches minus infinity. Such behavior is exhibited in some non linear systems.

6. Local Bifurcation

A local bifurcation occurs when a parameter change causes the stability of an equilibrium (or fixed point) to change. In continuous systems, this corresponds to the real part of an Eigen value of an equilibrium point passing through zero. In discrete systems (those described by maps rather than ODEs), this corresponds to a fixed point having a flout multiplier with modulus equal to one. In both cases, the equilibrium is non-hyperbolic at the bifurcation point. The topological changes in the phase portrait of the system can be confined to arbitrarily small neighbourhoods of the bifurcating fixed points by moving the bifurcation parameter close to the bifurcation point (hence 'local')

5.14.4 Types of Local Bifurcation

Mainly three different types of bifurcations, viz., pitch-fork, saddle-node, and transcritical, are mentioned here. The general dynamic equations is

$$\dot{S} = f(S, b)$$

where, "S" is the state variable and "b" is the bifurcation parameter. At equilibrium, i.e., at steady-state

$$f(S, b) = 0$$

At the bifurcation point both the function and its first derivative are zero. Figure 5.41 is the example of pitch-fork bifurcation. Figure 5.42 is the saddle point bifurcation.

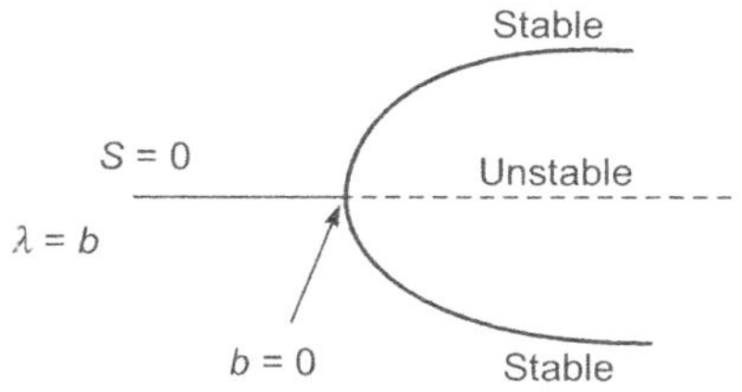

Figure 5.41 Example of Pitch-fork bifurcation.

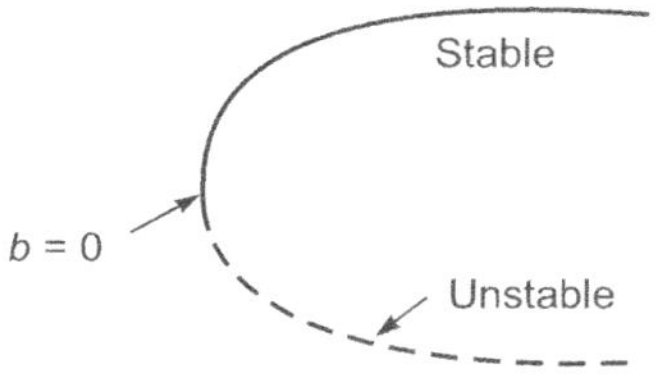

Figure 5.42 Saddle point bifurcation.

Transcritical bifurcation is illustrated in Figure 5.43.

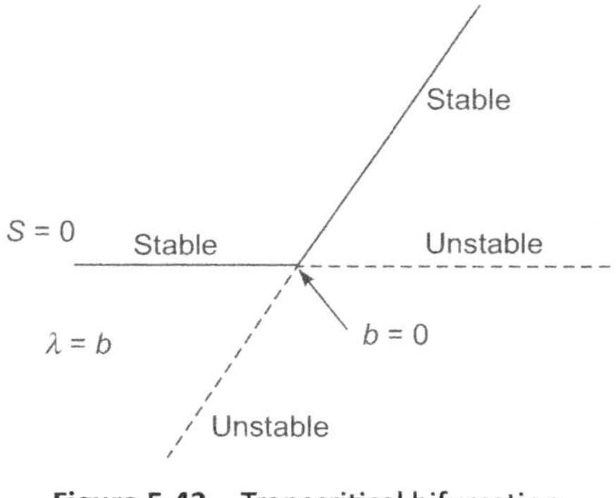

Figure 5.43 Transcritical bifurcation

The saddle point and Hopf bifurcations are explained with the examples.

1. Saddle-node Bifurcation

A saddle-node bifurcation or tangential bifurcation is a local bifurcation in which two fixed points (or equilibrium) of a dynamic system collide and annihilate each other. The term 'saddle-node bifurcation' is most often used with reference to the continuous dynamic systems. In discrete dynamic systems, the same bifurcation is often instead called a fold bifurcation. Another name is blue skies bifurcation in reference to sudden creation of two fixed points.

Consider the following one dimensional dynamic system depending on one parameter,

$$\frac{dX}{dt} = X^2 + p \tag{5.178}$$

At $p = 0$, this system has a non hyperbolic equilibrium at $X = 0$. The behavior of the system for all the other values of p not equal to zero is also clear. For $p < 0$, there are two equilibria in the system

$$X = \sqrt{p} \tag{5.179}$$

or

$$X = -\sqrt{p} \tag{5.180}$$

The equilibrium at $X = -\sqrt{p}$ is stable, while the other one is unstable. When p crosses zero from negative to positive values, the two equilibria collide with each other forming an equilibrium at $p = 0$ and then it disappears. This is called fold bifurcation or saddle-node bifurcation.

2. Hopf Bifurcation

A Hopf or Andronov-Hopf bifurcation is a local bifurcation (Kuznetsov, 2004). This occurs when a limit cycle form as a parameter varied. In this, a fixed point of a dynamic system loses stability, as a pair of complex conjugate Eigen values of the linearization around it crosses the imaginary axis of complex plane. Under reasonably generic assumptions about the dynamic system, one can expect to see a small amplitude limit cycle branching from the fixed point. The limit cycle is orbitally stable if a certain quantity called the first Lyapunov coefficient is negative, and the bifurcation is supercritical. Otherwise, it is unstable and the bifurcation is sub-critical.

Consider a system,

$$\frac{dX_1}{dt} = aX_1 + bX_2 + F_1(X, k) \tag{5.181}$$

$$\frac{dX_2}{dt} = cX_1 + dX_2 + F_2(X, k) \tag{5.182}$$

where F is a smooth vector function whose components F_1 have Taylor expansion in X starting with a least quadratic form.

Now we will form a matrix $\underline{\underline{\mathbf{A}}}$ with 2×2 dimensions. Along first row, we will have 'a' and 'b' and along second row will have 'c' and 'd'. Let its Eigen values are the roots of Equation (5.183).

$$p^2 - mp + G = 0 \tag{5.183}$$

where p is the Eigen value,

$$m = a + d \tag{5.184}$$

$$G = ad - bc \tag{5.185}$$

$$p_1 = 0.5\left[m + \sqrt{m^2 - 4G}\right] \tag{5.186}$$

$$p_2 = 0.5\left[m - \sqrt{m^2 - 4G}\right] \tag{5.187}$$

The Hopf bifurcation condition means that real part of p_1 and p_2 should be zero. So the condition for Hopf bifurcation is $m = 0$ and $G > 0$.

EXERCISES

5.1 For laminar flow system given below draw $E(t)$ and $F(t)$ vs. t plot for pulse and step input (Fig. 5.44).

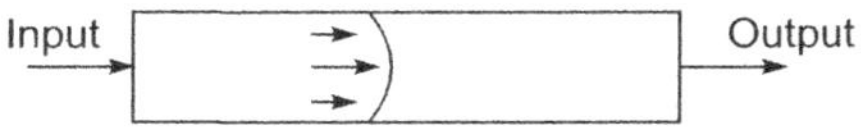

Figure 5.44 Laminar flow system in a pipe.

5.2 Baker's yeast was growing in two different reactors having following data.
Reaction volume (V) = 3.4 dm^3
Kinetic data for the organism:

$$K_s = 0.134 \text{ kg/m}^3$$
$$\mu_m = 0.352 \text{ h}^{-1}$$

Following non-ideal parameters were obtained in two reactors.

Parameter	Reactor 1	Reactor 2
α	0.110	0.197
β	0.040	0.060

Interpret the significance of these values in two reactors. Comment on the most non-ideal cases. (Data obtained from Gowthaman *et al.*, 1991)

5.3 The conversion of lactose to glucose and galactose by β-galactosidace was examined by a model system. This enzyme follows Michaelis-Menten kinetics with competitive product (galactose) inhibition. The kinetic constants for the hydrolysis are taken from the literature (Santos *et al.*, 1998; Abu-Reesh, 2000).

$$K_m = \exp\left(28.54 - \frac{10110}{T}\right) (\text{mol/l})$$

$$K_I = \exp\left(24.58 - \frac{9001}{T}\right) (\text{mol/l})$$

The enzyme is quite stable for temperatures up to 40°C. Typical lactose content of milk (C_{S_0}) is about 0.1462 mol/l. Two case studies were reported in the tubular reactor, i.e, Case 1: studied at 40°C and Case 2: at 7°C (Abu-Reesh and Abu-Sharkh, 2003). Calculate in each case K_m, K_I, $\frac{K_m}{C_{S_0}}$ (or K) and $K'\left(\text{or } \frac{K_m}{K_I}\right)$, conversion of reactants and yield.

5.4 Establish the dynamic response of continuous culture in ideally mixed stirred reactor with substrate limitation and inhibition using Howell model described below.

$$\mu = \frac{\mu_m C_S}{K_s + C_S + \dfrac{C_S{}^2}{K_I}}$$

5.5 How Bodenstain number influences RTD of a whole cell reactor?

5.6 How do you expect the influence of variation in *X-S* phase plane diagram in presence of wall growth?
What are the conditions of testing stability of non-linear system?
If you change the initial substrate condition, what will be the fate of the steady states? State with reference to separatrices.

5.7 Construct a plot of C_X and C_S vs. *D*. The growth rate is defined as

$$r_X = \frac{\mu_m C_S C_X}{K_s + C_S + C_i \dfrac{K_s}{K_I}}$$

where $C_{S_0} = 10$ kg/m^3, $C_{X_0} = 0$, $K_S = 1$ kg/m^3, $K_I = 0.01$ kg/m^3, $C_i = 0.05$ kg/m^3, $\mu_m = 0.5$ h^{-1}, $Y_{X/S} = 0.1$.

5.8 An organism is grown in a CFSTBR at a dilution rate of 0.23 h^{-1}. The organism follows substrate inhibition kinetics of the type

$$\mu = \frac{\mu_m (1 - C_I / C_I')^n C_s}{K_s (1 - C_I / C_I')^m + C_s}$$

The kinetic parameters are as follows:

$\mu_m = 0.1$ h^{-1}, $K_S = 0.001$ kg/m^3, $C_{S_0} = 15$ kg/m^3, $Y_{X/S} = 0.6$.

Assume other suitable data for this calculation. How many steady states are possible?

5.9 An organism is grown in a CFSTBR at a dilution rate of 0.20 h^{-1}. The organism follows substrate inhibition kinetics of Luoang model. The kinetic parameters are as follows: $\mu_m = 0.08$ h^{-1}, *Ks* = 0.005 kg/m^3, $C_{S_0} = 15$ kg/m^3, $Y_{X/S} = 0.4$. Assume other suitable data for this calculation. How many steady states are possible?

REFERENCES

Abu-Reesh IM (1997) "Predicting the performance of immobilized enzyme reactors using reversible Michaelis-Menten kinetics", *Bioprocess Engineering*, **17**(3), 131-137.

Abu-Reesh IM (2000) "Optimal design of CSTRs in series performing enzymatic lactose hydrolysis", *Bioprocess Engineering*, **23**(6), 709-713.

Abu-Reesh IM and Abu-Sharkh BF (2003) "Comparison of axial dispersion and tanks in series models for simulating the performances of enzyme reactors", *Industrial Engineering Chemistry Research*, **42**, 5495-5505.

Aris R (1959) "Notes on the Diffusion-Type Model for Longitudinal Mixing in Flow," *Chemical Engineering Science*, **9**, 266.

Bailey JE (1973) "Periodic operation of chemical reactors: a review", *Chemical Engineering Communication*, **4**, 111–124.

Bailey JE and Ollis DF (1986) *Biochemical Engineering Fundamentals* (2nd edn.) McGraw-Hill, New York.

Bequette BW (Ed.) (1998) *Process Dynamics-Modeling, Analysis, and Simulation*, Prentice-Hall, Inc., New Jersey.

Bourne JR, Crivelli E and Rys P (1977) Chemical Selectivities Disguised by Mass Diffusion V: Mixing-Disguised Azo Coupling Reactions, 6th Communication on the Selectivity of Chemical Processes, *Helvetica Chimica Acta.* **60**, 2944–2957.

Bourne JR (1984) Micromixing revisited, Proceedings of 8th International Symposium on reaction Engineering, Symposium Series, *Industrial Chemistry and Engineering* (London), **87**, 797-814.

Bryant J (1977) "The characterization of mixing in fermenter", *Advances in Biochemical Engineering*, **5**, 101-123.

Cholette A and Cloutier L (1959), "Mixing efficiency determinations for continuous flow systems". *Canadian Journal of Chemical Engineering*, **37**, 105-112.

Chowdhury R and Dutta Gupta A (2005) "Mathematical modelling of the transient behavior of a chemostat undergoing bio-desulfurization of model organosulfur compounds and diesel using pure and isolated strains", *International Journal of Chemical Reactor Engineering*, **3**, Article No: A60.

Daugulis AJ, McLellan PJ and Li J (2000) "Experimental investigation and modeling of oscillatory behavior in the continuous culture of *Zymomonas mobilis*", *Biotechnology and Bioengineering*, **56**, 99-105.

Danckwerts PV (1953) "Continuous flow systems distributions of residence time", *Chemical Engineering Science*, **2**(1), 1-18.

Doran PM (Ed) *Bioprocess Engineering Principles*, Academic Press, London, 1995.

Dunlop EH (1990) Bioreactor Intensification for the Production of Fuels and Chemicals-micromixing and Macromixing Effects, *In*: Energy Conversion Engineering Conference, 1990. IECEC-90. *Proceedings of the 25th Intersociety*, 12-17 Aug, Volume 3, pp. 500-504.

Elgeti K (1996) "A new equation for correlating a pipe flow reactor with a cascade of mixed reactors", *Chemical Engineering Science*, **51**, 5077-5080.

Fogler HS (Ed) (1992) *Elements of Chemical Reaction Engineering*, Prentice-Hall of India Pvt. Ltd, New Delhi.

Fogler HS and Gürman MN (Eds) (2008) *Elements of chemical reaction engineering*, University of Michigan.

Froment GF and Bischoff KB (Eds), (1990) *Chemical reactor analysis and design*, John Wiley & Sons.

Gowthaman MK, Rakshit SK, Krishnaiah K and Baradarajan A (1991) "Studies in RTD and continuous culture (SCP) in cylindrical and tapered reactors", *Bioprocess Engineering*, **7**, 41-46.

Hale J and Kocak H (1991) (Eds) *Dynamics and Bifurcations*, Springer-Verlag, New York.

Han K and Levenspiel O (1987) "Extended Monod kinetics for sub-strate, product, and cell inhibition". *Biotechnology and Bioengineering*, **32**, 430-437.

Howell JA, Chi CT and Pawlowsky U (1972), Effect of wall growth on scale-up problems and dynamic operating characteristics of the biological reactor, *Biotechnology and Bioengineering*, **14**, 253-265.

IMSL software, Visual Numerics, Inc., Huston, TX (1987) (support @imsl.com).

Kossen NWF and Oosterhuis NMG (1985) "Modelling and scaling–up of bioreactors". In: Rehm H-J, Reed G (Eds), *Biotechnology*, vol 2, pp. 571-605, VCH, Weinheim.

Krishnaiah K (1998) "Advanced bioreactor theory: Fundamentals of bioreactor design-I", Chapter 10, Lecture note in: Short Term Course on "Microbial Engineering" (May 23-June 04, 1988), Baradarajan A, Panda T (Eds), pp. 179-192.

Kuznetsov YA (2004) *Elements of Applied Bifurcation Theory*, Springer-Verlag, New York.

Levenspiel O (Ed) (1972), *Chemical Reaction Engineering*, 2nd Edn., Wiley, New York.

Liou C-T and Chien Y-S (1995) "The effect of micromixing on steady state multiplicity for autocatalytic reactions in a non-ideal mixing of CFST", *Chemical Engineering Science*, **50**(22), 3637-3644.

Lortie R (1994) "Evaluation of the performance of immobilized enzyme reactors with Michaelis-Menten kinetics". *Journal of Chemical Technology*, Biotechnology, **60**, 189-193.

Nagata S (1975) *Mixing: Principles and Applications*. Kodanska, Ltd., Tokyo, Wiley, New York.

Nauman EB (1987) *Chemical reactor design*, John Wiley & Sons, New York.

Pickett AM (1982) In *Microbial Population Dynamics*, Chapter 4 (Bazin M, Ed) CRC Press, Boca Raton, Florida.

Santes A., Ladero M. and Gdrcia-Ochoa F (1998) "Kinetic modelling of lactose hydrolysis by β-galactosidase from *Kluveromyces fragilis*", *Enzyme and Microbial Technology*, **22**, 558-567.

Schügerl K (Ed) (1985) *Bioreaction Engineering*, Vol 1, John Wiley & Sons, New York, p 172.

Strogatz SH (Ed) (1994) *Nonlinear dynamics and chaos*, Addison-Wesley, Reading, MA.

Wen CY and Fan LT (Eds) (1975) *Models for Flow Systems and Chemical Reactors*, Marcel Dekker, New York.

Chapter 6

Bioreactor Modeling

OBJECTIVES

- To acquire the concept of quantitative relationship between input and output of bioreactors at various conditions to formulate its procedure
- To give the concept of the design, modification, and control of bioreactors

6.1 MODEL—WHAT IS IT?

The word 'model' (derived from the Latin word "*modus*") means to measure some quantity in a sizeable representation of a planned or existing "object". Modeling is the mathematical representations of a system. More precisely, it is the representation of the phenomena in a set of mathematical equations. A comprehensive list of different tasks of modeling is given in Figure 6.1.

Model Hypothesis (Input: Prior information, new model assumptions, System of equations)

⇩

Experimental Design (Hypothesis verification, information optimization, numerical optimization, controller design)

⇩

Performing the Experiments (Process control systems)

⇩

Evaluation of Experiments (Improvement of process, Simulation of parameter estimation, model validation, tests of significance)

Figure 6.1 Different tasks of models.

A model of a process predicts the behavior of a process. In this regard, a set of equations that comprise the model is the best approximation of the true process. Engineers usually seek a compromise between time/cost or effort required to verify the model. If the model is represented by complex equations, more number of parameters are involved, which require proper evaluation. This may call for more experiments to verify the model.

6.1.1 Use of Models

Following are the important uses of models.

- Optimization of process operating conditions
- Better understanding of the process
- Proposing control strategy for a new process or bioreactor operation
- For bioreactor operation, control of heat is one of the important factors. This is a combination of heat generated by metabolic reaction as well as the contribution from mechanical sources (like agitation, etc.). The control of heat in the bioreactor can be done by controlling coolant temperature.
- Selection of control settings and proposing the control law
- Analysis of the safety aspect of the process
- Training the plant operating personnel
- Analysis of the marketability of the product

6.1.2 Classification of Models

Generally, models are developed from the first principles of physics and chemistry. They are classified into following categories (Thilakavathi *et al.*, 2007).

(a) Theoretical Models

They are developed from the principles of basic sciences. For example, the Henri-Michaelis-Menten kinetic model is derived from the first principles.

(b) Empirical Models

They are obtained from either mathematical or statistical analysis. These models might explain the experimental results, but they cannot be derived from the first principles (Sinclair and Kristiansen, 1987). For example, Monod's model is used to explain the growth kinetics of cells.

(c) Semi–empirical Models

This is a compromise between type (a) and type (b) models. One or two parameters can be evaluated from plant data. The modification of Han and Levenspiel model for the growth of *Trichoderma harzianum* growing on chitin to produce chitinase is an interesting example (Felse, 1999). Other examples are mixed vessels in series, plug flow with dispersion, etc. These models are proposed, but the validity of the models on a production scale is questionable. The basis of these models is RTD.

These models can be applied under dynamic (unsteady state) and steady state approaches.

Let us take an example of temperature control in a continuous stirred tank bioreactor model (Fig. 6.2).

For steady state approach, rate of accumulation term = 0.

Therefore,

$$wc(T_i - T) + Q = 0 \tag{6.1}$$

Unsteady state (Dynamic) approach

$$\begin{aligned}\text{Rate of energy accumulation} &= wc(T_i - T_{\text{ref}}) - wc\,(T - T_{\text{ref}}) + Q \\ &= wc(T_i - T) + Q \end{aligned} \tag{6.2}$$

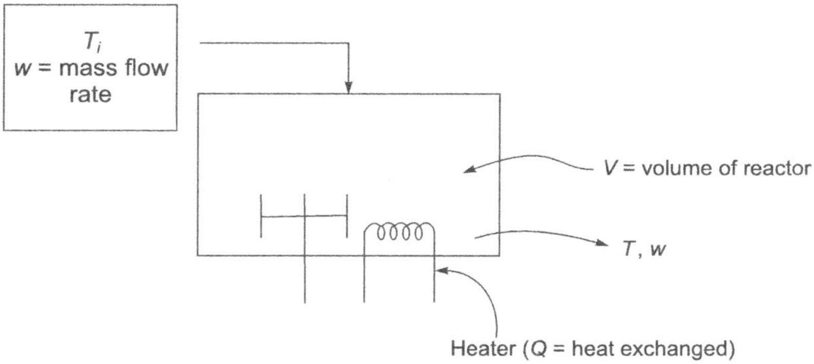

Figure 6.2 Schematic of a continuous strried tank bioreactor model temperature control.

where c is the specific heat capacity (energy/(unit mass)(unit temperature difference)). Q is the rate of heat transfer (energy/time).

Assuming density and volume to be constant,

$$\text{Rate of energy accumulation} = V\rho c \frac{dT}{dt} \tag{6.3}$$

Equating the above two expressions for the rate of energy accumulation,

$$V\rho c \frac{dT}{dt} = wc\,(T_i - T) + Q \tag{6.4}$$

No assumption, except the constant specific heat, has been made for the variables on the right hand side of the equation.

Relation between steady state and unsteady state approaches

One can derive the expression of steady state approach by considering accumulation term to be zero.

6.2 DEFINITION OF LUMPED AND DISTRIBUTED PARAMETER MODELS

Lumped Parameter Models

It assumes variation with time but not with space. For example, in unsteady state CFSTBR, concentration varies with time, but not with space. The model is represented by ordinary differential equations (ODEs).

Distributed Parameter Models

The assumed variation is with respect to both time and space. For example, in unsteady state packed bed reactor (PBR), concentration varies with time and distance. It has more than one independent variable. The model gives rise to partial differential equations (PDEs).

6.3 INTRODUCTION TO A FEW TERMINOLOGIES AND THEOREMS

(a) What is a System?

A combination of different pieces of equipment integrated to perform a specific function is known as system. For the present analysis, bioreactor and heat exchanger are the systems.

In micro concept, a single unit operation is a system having inputs, states, and outputs.

(b) What is Simulation?

Modeled equations have many variables. Changing some variables in the equations while keeping others constant (those variables called parameters) and finding their dependence on the phenomena is called simulation.

Simulation helps to understand the system behavior and must make sense to be useful for analysis.

(c) Techniques

Linear System Analysis

The mathematical tools that are used to study the problems of linear dynamic systems are known as system analysis techniques.

1. ***Laplace transform***

It is applied to analyze single, linear and the nth order ODEs. Annexure to this chapter summarizes a few Laplace transform of standard useful functions.

2. ***State-space techniques***

This is used to analyze the behavior of multiple first order linear differential equations. If the differential equations are non-linear, they can be linearized at a desired steady state point.

6.4 MODELING PRINCIPLES

In general, following principles are used (Bequette, 1998):

- Dynamic models consist of differential equations (ODEs or PDEs), often combined with one or more algebraic equations. For process control problems, dynamic models are obtained from the application of unsteady state conservation relations.
- Material/energy balances
- Force/momentum balances
- Algebraic equations in the process model arise from transport and thermodynamic relations.

Real model should include the important dynamic effects not that much complicated than needed and keeps the minimal number of equations and parameters. The model equation must provide unique relationship among all input and output variables. Degree of freedom is one of such approaches.

Assuming DF stands for degrees of freedom

NV stands for total number of variables (unspecified inputs and outputs)

NE stands for number of independent equations (both differential and algebraic),

$$DF = NV - NE$$

Cases:

(a) If $DF = 0$, $NV = NE$, then it is exactly determined process.

(b) If $DF > 0$, $NV > NE$, then it is under specified conditions.

(c) If $DF < 0$, $NV < NE$, it is over-specified conditions.

6.5 STEPS IN MODELING

1. Draw schematic diagram of process and label all process variables.
2. List all assumptions to be used in developing the model.
3. Determine whether independent variable other than time is required.
4. Write approximate dynamic balance (overall mass, component, energy, etc.).
5. Introduce other relations which may be from the thermodynamics.
6. Identify system parameters.
7. Identify model variables.
8. Calculate degrees of freedom.
9. Specify N_f, total number of input variables.
10. Solve the necessary equations and do the parameter sensitivity analysis.

To understand the procedure, let us take some examples.

Example 6.1 How will you control temperature in a stirred heating device required for a bioreactor?

Solution One may recall Figure 6.2 for this purpose.

Assumption: Hold-up is not constant.

Result: Inlet and outlet flows and temperatures differ.

Consideration: Unsteady state mass and energy balances are necessary to model the change in mass and temperature of the above problem.

Mass balance equation:

Rate of mass in – rate of mass out = rate of mass accumulated

$$w_i - w = \frac{d(V\rho)}{dt} \tag{6.5}$$

At constant density,

$$\rho\frac{dV}{dt} = w_i - w \tag{6.6}$$

The energy balance equation is

$$\begin{pmatrix}\text{Rate of energy in by}\\ \text{flow or convection}\end{pmatrix} - \begin{pmatrix}\text{Rate of energy out by}\\ \text{flow or convection}\end{pmatrix} + \begin{pmatrix}\text{Net rate of heat addition to}\\ \text{system from surroundings}\end{pmatrix}$$

$$= \text{Rate of energy accumulated}$$

$$w_i c(T_i - T_{\text{ref}}) - wc(T - T_{\text{ref}}) + Q = c\frac{d}{dt}[V\rho(T - T_{\text{ref}})] \tag{6.7}$$

where, T_i, T_{ref}, and T are the inlet, reference and outlet temperatures, respectively.

Equations (6.5) and (6.7) are the dynamic model systems. One needs to specify the conditions for volume and temperature, input variables (w_i, w, Q and T) as a function of time and relationships of an algebraic function (ρ, c).

From Equation (6.7)

$$c\frac{d}{dt}[V\rho(T - T_{\text{ref}})] = \rho c\frac{d}{dt}[V(T - T_{\text{ref}})] = \rho c\left[(T - T_{\text{ref}})\frac{dV}{dt} + V\frac{d(T - T_{\text{ref}})}{dt}\right]$$

$$= \rho c(T - T_{\text{ref}})\frac{dV}{dt} + \rho cV\frac{d(T - T_{\text{ref}})}{dt}$$

Substitute $\rho\frac{dV}{dt} = w_i - w$ **from Equation (6.6), Equation (6.8) is obtained:**

$$c\frac{d}{dt}[V\rho(T - T_{\text{ref}})] = c(T - T_{\text{ref}})(w_i - w) + \rho cV\frac{dT}{dt} \tag{6.8}$$

Since $$\frac{d(T_{\text{ref}})}{dt} = 0$$

Equating (6.7) and (6.8)

$$c(T - T_{\text{ref}})(w_i - w) + \rho cV\frac{dT}{dt} = w_i c(T_i - T_{\text{ref}}) - wc(T - T_{\text{ref}}) + Q$$

$$w_i c(T - T_i) - Q = -\rho cV\frac{dT}{dt}$$

$$\rho cV\frac{dT}{dt} = w_i c(T_i - T) + Q$$

Dividing by ρcV

$$\frac{dT}{dt} = \frac{w_i}{\rho V}(T_i - T) + \frac{Q}{\rho cV} \tag{6.9}$$

Example 6.2 Non-isothermal CFSTBR may be considered for sterilization system.
The system description is given below.

- The reaction is $A \xrightarrow{k_1} B$.
- This is n[th] order with respect to reactant A.
- α is the heat of reaction of A reacted (kcal/kmol).
- Heat generated is to be removed by cooling water in circulation.

Solution

1. ***Schematic representation*** of the system is given in Figure 6.3.
2. ***Assumptions***
   (a) Both the reactor contents and cooling jacket are perfectly mixed.
   (b) Heat loss is negligible to the surroundings.
   (c) Density can be assumed to be constant.
   (d) Jacket wall has negligible thermal inertia (assuming thin wall in the jacket).
3. ***Appropriate dynamic balances***
   (a) Total continuity equation in the bioreactor is described by Equation (6.10).

$$\frac{dV}{dt} = F_i - F \tag{6.10}$$

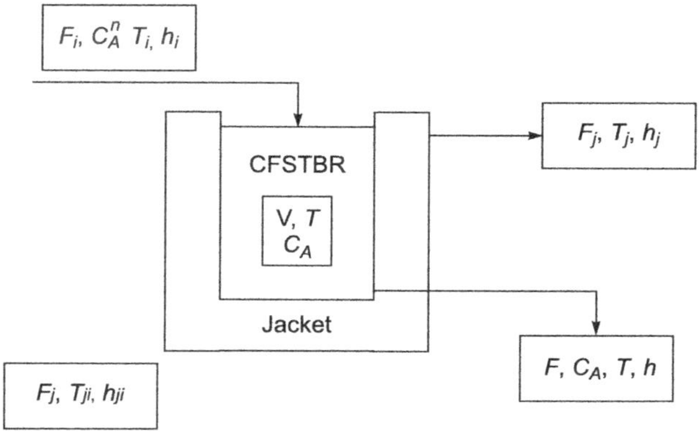

Figure 6.3 Non-isothermal CFSTBR.

(b) Continuity equation for component A

$$\frac{d}{dt}(VC_A) = F_iC_{Ai} - FC_A - VkC_A^n \tag{6.11}$$

where k is the reaction rate constant. kC_A^n = rate of reaction /unit volume.

(c) Energy balance equation for the reactor is described below.

$$\rho\,\frac{d}{dt}(VC_A) = \rho(F_ih_i - Fh) - \alpha VkC_A^n - Q \tag{6.12}$$

where h is the enthalpy Q is the heat removed from the system.

Balances for jacket

(i) Energy balance

Enthalpy accumulation,

$$\rho_jV_j\frac{dh_j}{dt} = F_j\rho_j(h_{ji} - h_j) + Q \tag{6.13}$$

(ii) Heat transfer from reactor to the cooling water

$$Q = UA_H(T - T_j) \tag{6.14}$$

where A_H is the heat transfer area.

4. Other balances

(i) Expression for rate constant

$$k = \beta\exp\left(-\frac{\Delta E}{RT}\right) \tag{6.15}$$

(ii) Hydraulic relationship

$$F = k_v(V - V_{\min}) \tag{6.16}$$

where k_v is valve coefficient and $(V - V_{\min})$ drop in volume across the valve.

This is not required if a pump is used for feeding.

Assuming enthalpy

$$h = c_p(T - T_{\text{ref}}) \text{ and } h_j = c_j(T_j - T_{\text{ref}})$$

5. Simplification of model equations

$$\frac{dV}{dt} = F_i - F \tag{6.17}$$

$$\frac{d}{dt}(VC_A) = F_iC_{Ai} - FC_A - V\beta e^{-E/RT}(C_A)^n \tag{6.18}$$

$$\rho c\frac{d(VT)}{dt} = \rho c(F_iT_i - FT) - \alpha\beta Ve^{-E/RT}(C_A)^n - UA_H(T - T_j) \tag{6.19}$$

$$\rho_j V_j c_j \frac{d(T_j)}{dt} = F_j \rho_j c_j (T_{ji} - T_j) - UA_H (T - T_j) \tag{6.20}$$

$$F = k_v(V - V_{\min}) \tag{6.21}$$

6. Analysis of degrees of freedom

Number of equations : 5

Number of variables : 10

Input Variables : $F_i, C_{Ai}, T_i, T_{ji}, F_j$

Output variables : V, C_A, T, T_j, F

Degrees of freedom = Number of variables – Number of equations

= 10 – 5

= 5

Hence, five variables have to be specified for the analysis.

6.6 FUNDAMENTAL LAWS USED IN PROCESS MODELING

- Continuity Equation – law of conservation of mass
- Conservation of energy – First law of thermodynamics
- Diffusion of mass – Transport law
- Heat transfer – Transport law
- Newton's law of viscosity – Transport law
- Equation of motion
- Equation of state (density, heat capacity)
- Rate of reaction
- Thermodynamic equilibrium

In bioreactor, dynamic analysis is more important for the design consideration (Fish *et al.*, 1989; Volesky and Votruba 1992). So the basic concept of first order and second order kinetics are described here.

6.7 FIRST–ORDER SYSTEMS

The schematic diagram is given in Figure 6.4.

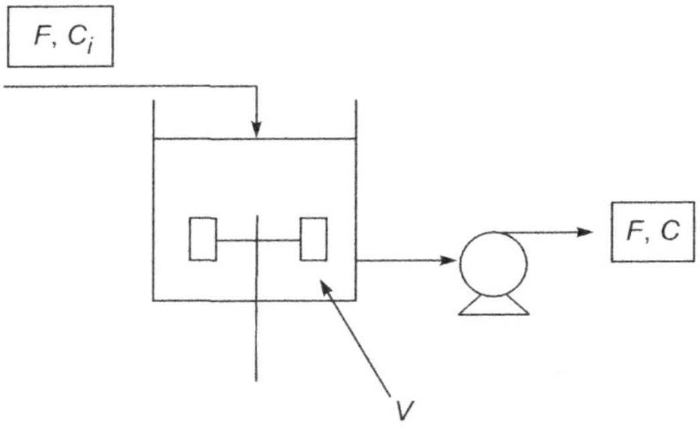

Figure 6.4 A typical CFSTBR.

The overall balance equations are given below.

$$\frac{dV}{dt} = F - F = 0 \qquad (6.\ 22)$$

Component balance is

$$\frac{d}{dt}(VC) = FC_i - FC \qquad (6.23)$$

At steady state, $C_i = C_i'$ and $C = C_S$. (6.24)

At steady state

$$\frac{d(C - C_S)}{dt} = \frac{F}{V}(C_i - C_i') - \frac{F}{V}(C - C_S) \qquad (6.25)$$

$$\frac{V}{F}\frac{d(C - C_s)}{dt} + (C - C_S) = (C_i - C_i') \qquad (6.26)$$

$$\tau\frac{dY}{dt} + Y = kU \qquad (6.27)$$

where,

$$\frac{V}{F} = \tau,$$

$$C - C_S = Y,$$

$$U = C_i - C_i'$$

Taking Laplace transform

$$\tau[sY(s) - Y(0)] + Y(s) = kU(s) \qquad (6.28)$$

$$\therefore \quad Y(s) = \frac{k}{(\tau s + 1)}U(s)$$

$$G(s) = \frac{Y(s)}{U(s)}$$, which is described below.

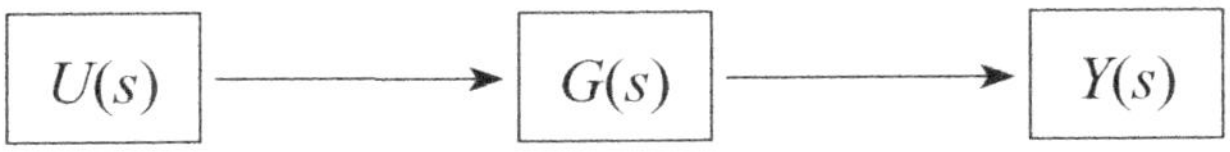

If the input change is a step change, $U(s) = \dfrac{A}{s}$

Then
$$Y(s) = \frac{A}{s}\left[\frac{k}{(\tau s+1)}\right] \tag{6.29}$$

Taking the inverse of Laplace transform
$$Y(t) = kA\,(1 - e^{-t/\tau}) \tag{6.30}$$

6.8 SECOND-ORDER SYSTEMS

This comes from a concept of two reactors in series. In this case, the overall transfer function of the system will be the product of each transfer function. The scheme is given in Figure 6.5.

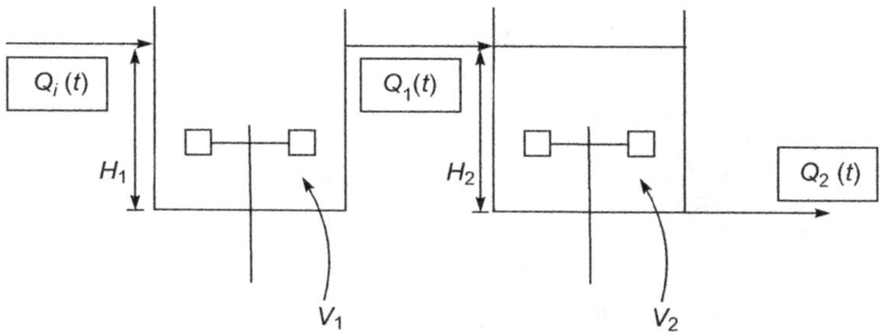

Figure 6.5 Schematic diagram of a second order system.

where $Q_i(t)$, $Q_1(t)$, $Q_2(t)$ are flow rates.
V_1, V_2 are volume of the reactors.
k, R_2 are gain.
H_1 and H_2 are height of the fluid in the first and second reactor, respectively.

Transfer function for a second-order system is given in Equation (6.31).

$$G(S) = \frac{Y(s)}{U(s)} = \frac{k}{(\tau s+1)} \tag{6.31}$$

Here, for reactor (1),

$$\frac{Y(s)}{U(s)} = \frac{Q_1(s)}{Q_i(s)} = \frac{k}{(\tau_1 s+1)}$$

Assuming gain, $k = 1$,

$$\frac{Q_1(s)}{Q_i(s)} = \frac{1}{(\tau_1 s+1)} \tag{6.32}$$

For reactor (2), Q_2 is dependent on H_2.

$$\frac{H_2(s)}{Q_1(s)} = \frac{R_2}{(\tau_2 s+1)}$$

$$\therefore \quad \frac{H_2(s)}{Q_i(s)} = \frac{R_2}{(\tau_1 s+1)(\tau_2 s+1)} = \frac{R_2}{(\tau_1\tau_2 s^2 + (\tau_1+\tau_2)s+1)} \tag{6.33}$$

Responses for a second-order system is described here.

Consider a linear second order ODE with constant parameters.

$$a_2 \frac{d^2y}{dt^2} + a_1 \frac{dy}{dt} + a_0 y = bu(t) \tag{6.34}$$

This equation can be written as

$$\tau^2 \frac{d^2y}{dt^2} + 2\xi\tau \frac{dy}{dt} + y = ku(t) \tag{6.35}$$

where,

$$a_0 \neq 0,$$

$$\tau^2 = a_2/a_0, 2\xi\tau = a_1/a_0, k = b/a_0 \tag{6.36}$$

where, ξ is the damping factor.

Taking Laplace transform for the above Equation (6.35)

$\tau^2 [s^2 Y(s)] + 2\xi\tau[sY(s)] + Y(s) = kU(s)$

Rearranging,

$$Y(s)[\tau^2 s^2 + 2\xi\tau s + 1] = kU(s)$$

$$\frac{Y(s)}{U(s)} = \frac{k}{\left[\tau^2 s^2 + 2\xi\tau s + 1\right]} = G(s) \tag{6.37}$$

The roots of the characteristic equation

$$[\tau^2 s^2 + 2\xi\tau s + 1] = 0 \text{ are} \tag{6.38}$$

$$s = \frac{-2\xi\tau \pm \sqrt{(2\xi\tau)^2 - 4\tau^2(1)}}{2\tau^2}$$

$$= \frac{-2\xi\tau \pm \sqrt{4\tau^2(\xi^2 - 1)}}{2\tau^2}$$

$$= \frac{-\xi \pm \sqrt{(\xi^2 - 1)}}{\tau}$$

$$\therefore \quad s_1 = \frac{-\xi}{\tau} + \frac{\sqrt{(\xi^2 - 1)}}{\tau} \text{ and}$$

$$s_2 = \frac{-\xi}{\tau} - \frac{\sqrt{(\xi^2 - 1)}}{\tau}$$

For step response, values of the roots depend on ξ, which are discussed by Coughanour (1991) with proper Equations (6.39)–(6.43).

When $\xi = 1$, the roots of the equation are real and equal.

$$Y(t) = 1 - \left(1 + \frac{t}{\tau}\right)e^{-t/\tau} \tag{6.39}$$

The response of the system is critically damped.

If $\xi < 1$, the roots of the equation are complex, indicating underdamped or oscillatory response.

$$Y(t) = 1 - \frac{1}{\sqrt{1-\xi^2}} e^{-\xi t/\tau} \sin\left(\sqrt{1-\xi^2}\,\frac{t}{\tau} + \tan^{-1}\frac{\sqrt{1-\xi^2}}{\xi}\right) \tag{6.40}$$

For $\xi > 1$, the roots of the equation are real, resulting in overdamped or non-oscillatory response.

$$Y(t) = 1 - e^{-t/\tau}\left(\cosh\sqrt{\xi^2-1}\,\frac{t}{\tau} + \frac{\xi}{\sqrt{\xi^2-1}}\sinh\sqrt{\xi^2-1}\,\frac{t}{\tau}\right) \tag{6.41}$$

For sinusoidal response

$$X(t) = A\sin(\omega t)$$

$$Y(t) = C_1\cos(\omega t) + C_2\sin(\omega t) + e^{-\xi t/\tau}\left(C_3\cos\sqrt{1-\xi^2}\,\frac{t}{\tau} + C_4\sin\sqrt{1-\xi^2}\,\frac{t}{\tau}\right) \tag{6.42}$$

$$\lim_{t\to\infty} Y(t) = C_1\cos(\omega t) + C_2\sin(\omega t) \tag{6.43}$$

In Chapters 4 and 5, we have discussed about batch and continuous flow bioreactors. Some of the mathematical treatments are directly used for bioreactor modeling. Here, a few specific examples are discussed in detail.

6.9 COMPLEXITY OF THE MODEL

Simple model is better to understand and is easily applied in modeling. This may not accommodate enough information of the system. For this reason, models are sometimes complicated which results in cumbersome solution. Such type of diffcult model may not find real applications. For examples, most of the models related to fungal morphological parameters are complicated models.

A suitable model should have a balance between simplicity and reliability.

To achieve the goals of a desired mathematical model, computer applications are necessary. A preferred applications of computer in this aspect is described here in the flow diagram.

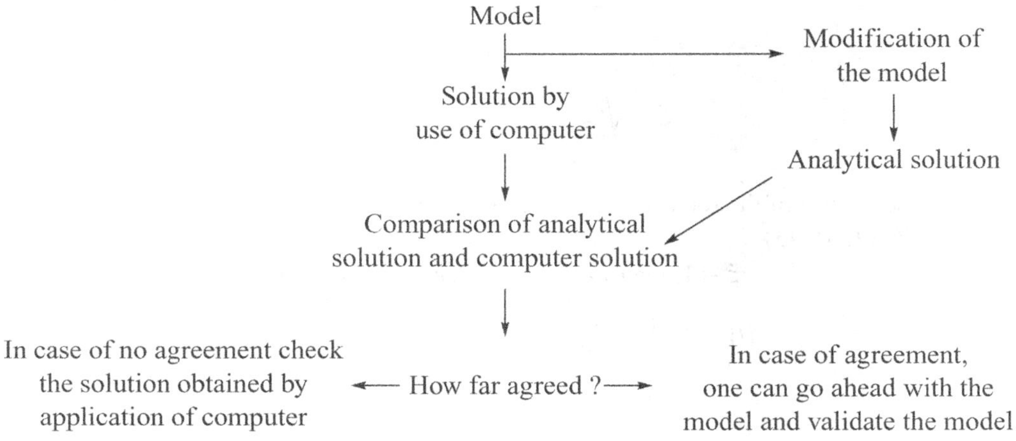

6.10 PARAMETER SENSITIVITY

A proper comparison between a mathematical model and experimental results can be made if the proper values are known for the different parameters in the model. Of course, the determination of parameters in the model is very much time consuming as there may have interaction of different parameters in the model.

Analysis to obtain an insight into the influence of parameters is called parameter sensitivity analysis. This guides us to avoid unnecessary efforts in obtaining accurate values of less important parameters. Simple way to do this analysis one may practice the following.

Step 1: To get first estimate of the values of the parameters in the model.

Step 2: Substitute these parameter values into the model and solve.

Step 3: Vary the values of the parameters and record the influence on the most important output variable of the model.

Kossen and Oosterhuis (1985) defined parameter sensitivity (PS) in the following way.

$$PS = \frac{\left(\dfrac{\text{Variation of the important variable due to variation of the parameter}}{\text{Value of this important variable at the original value of the parameter}}\right)}{\left(\dfrac{\text{Variation of the parameter}}{\text{original value of the parameter}}\right)}$$

Example 6.3 Batch growth with oxygen limitation

Considerations are listed below.

1. The concentration of oxygen in the liquid phase is zero in oxygen-limited growth.
2. Oxygen is transferred from the gas phase to the liquid phase by inter-phase mass transfer.

$$\text{The oxygen transfer rate} = k_L a C_L^* \tag{6.44}$$

where C_L^* is solubility of oxygen in the medium, and
$k_L a$ is the volumetric oxygen transfer coefficient.

The equation for growth is

$$\frac{dC_X}{dt} = \frac{1}{Y_{X/O}} k_L a C_L^* \tag{6.45}$$

Upon integration

$$C_X = C_{X_0} + \frac{1}{Y_{X/O}} k_L a C_L^* t \tag{6.46}$$

This equation suggests that the growth is linear. In real system, exponential growth occurs as the growth is limited by reactants.

Example 6.4 CFSTBR with oxygen limited growth

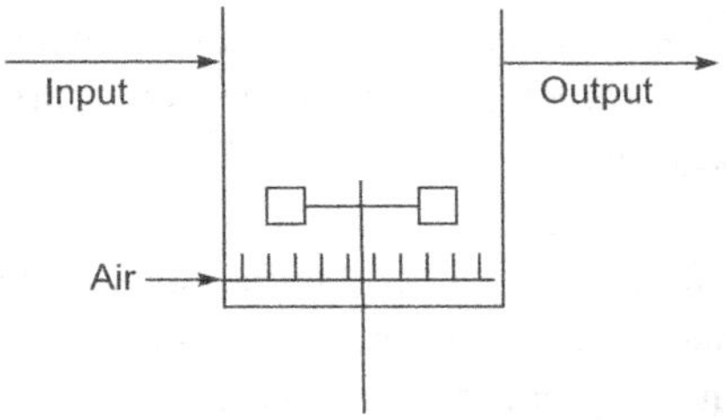

Figure 6.6 CFSTBR with oxygen limited growth.

Cell balance equation is

$$r_X - DC_X = 0 \tag{6.47}$$

Reactant balance equation is

$$-r_S - D(C_{S_0} - C_S) = 0 \tag{6.48}$$

Oxygen balance equation is

$$-r_0 - D(C_0 - C) + M_{i_0} = 0 \tag{6.49}$$

where,

$D = (F/V)$, and is the dilution rate,
M_{i_o} is the number of moles of oxygen transfered from gas phase to liquid phase,
C is the concentration of oxygen in the reaction fluid,
C_o is the concentration of oxygen in the medium inlet,
C_S is the concentration of reactant in the reaction fluid, and
C_{So} is the concentration of reactant in the inlet.

If oxygen is growth limiting

$$\mu = \frac{\mu_m C}{K_0 + C} \tag{6.50}$$

As $\mu = D$,

$$C = \frac{K_0 D}{\mu_m - D} \tag{6.51}$$

From oxygen balance, we get

$$M_{i_0} = k_L a(C^* - C) \tag{6.52}$$

where C^* = Concentration of oxygen in the bulk ($0 \le C^* \le C_L$). C_L is the saturation level of oxygen in reaction phase.

∴ From Equations (6.47), (6.49), and (6.52), Equation (6.53) is obtained

$$C_X = \frac{Y_{X/O}[D(C_0 - C) + k_L a(C^* - C)]}{D} \tag{6.53}$$

Example 6.5 Model for plug flow reactor

The model equation for ideal plug flow will be the same as the batch reactor equation except the real time is changed to residence time.

Example 6.6 Modeling of fed batch reactor

Fed-batch culture has two different concepts suggested by Yoshida *et al.* (1973) and Suga *et al.* (1980). In chemical engineering practice, this is referred to as semi-batch reactors. They are also called "extended cultures" and "repeated fed-batch cultures". We have discussed about fed batch reactor in Chapters 3 and 4.

For the modeling purpose, we consider three terminologies.

- Simple fed batch
- Repeated fed batch
- Dialysis culture

We need to develop model equations in this aspect to predict

- Cell concentration,
- Product concentration, and
- Residual reactant concentration as a function of time (Wang and Stephanopoulos, 1984; Roels, 1983).

Model equations are listed below.

Volume balance equation

$$\frac{dV}{dt} = F \tag{6.54}$$

Reactant, cell, and product balances are given as

$$\frac{d(VC_s)}{dt} = FC_{S_0} - r_S V \tag{6.55}$$

$$\frac{d(VC_X)}{dt} = FC_{X_0} + r_X V \tag{6.56}$$

$$\frac{d(VC_P)}{dt} = FC_{P_0} + r_P V \tag{6.57}$$

Under quasi-steady state conditions,

$$\frac{dC_X}{dt} = 0 \tag{6.58}$$

In a fed-batch reactor, the rate of consumption of reactants = rate of reactant addition.

Under quasi-steady state condition,

$$D = (F/V) \approx \mu \tag{6.59}$$

The dilution rate, D and, therefore, μ decreases with time in a fed batch culture.

To compare with other basic reactors let us consider the following figures.

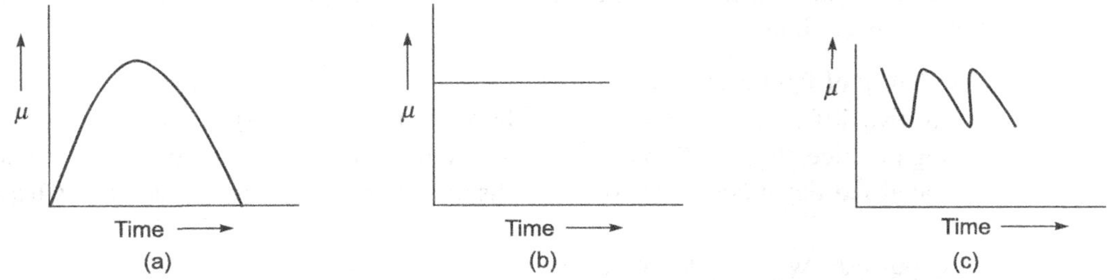

Figure 6.7 μ vs. time plot for various reactors. (a) Batch bioreactor, (b) CFSTBR (c) Fed-batch bioreactor.

Cell Concentration

$$C_{Xt} = C_{X_0 t} + FY_{X/S}C_{S_0 t} \tag{6.60}$$

where $C_X = \dfrac{C_{Xt}}{V}$. Subscript 't' denotes time function.

Product Concentration

Two expressions are suggested here.

(a) $Y_{P/S}$ is constant

Therefore, $$C_p = Y_{P/S}C_{S_0} \tag{6.61}$$

and hence, $$FC_p = FY_{P/S}C_{S_0} \tag{6.62}$$

Total product is $$C_{pt} = C_p V \tag{6.63}$$

(b) Specific product formation rate, q_p is constant

$$\frac{dC_{pt}}{dt} = q_p C_{X_t} \tag{6.64}$$

We know $$C_{Xt} = (V_i + Ft)C_{Xm}$$

Substituting C_{Xt} into Equation (6.64), we get

$$C_{P_t} = C_{Pi_t} + q_p\left(V_i t + F\frac{t^2}{2}\right)C_{Xm} \tag{6.65}$$

Therefore, the product per unit volume of culture from Equations (6.61) and (6.63)

$$C_P = \frac{C_{P_t}}{V} = \frac{C_{P_0}}{V} + q_p C_{Xm}\left(\frac{V_i}{V}t + \frac{F}{V}\frac{t^2}{2}\right) \tag{6.66}$$

where, V_i is the initial volume

C_{Xm} is the final cell concentration.

EXERCISES

6.1 The system is described with gas phase enzymatic reaction

$$2A \underset{k_2}{\overset{k_1}{\rightleftharpoons}} B$$

The CSFTBR is filled with gases. The forward and reverse reactions are of different orders. The mole fraction of A is x. Both pressure and x vary with time. The product stream flows into another reactor through a restriction valve. The second reactor is maintained at pressure of P_2 (absolute). The feed stream has a density of ρ_o and a mole fraction of y_i of reactant A. Assume both the gases are perfectly mixed and the system is isothermal. Molecular weights of A and B are M_A and M_B, respectively. Calculate degrees of freedom in the system.

6.2 Consider the statement given in worked out Example 6.2, but flow in the jacket is plug flow. Write the model equations and estimate the degrees of freedom.

6.3 Model the following CFSTBR with recycle (Fig. 6.8):

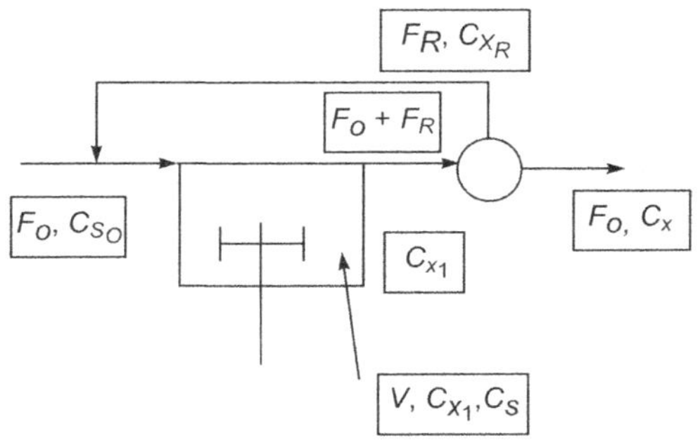

Figure 6.8 CFSTBR with recycle

where subscript 'o' for initial/input conditions and R for recycle condition.

6.4 Model the bioreactor given in Figure 6.9.

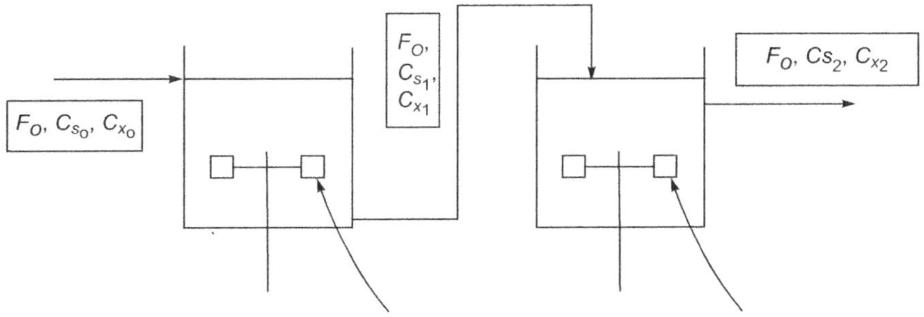

Figure 6.9 Two stage single input CFSTBR in series

6.5 In modeling of a bioprocess, the following equations are to be used

$$\frac{dX}{dt} = \mu(S)X$$

$$\frac{dP}{dt} = q_p(S)X$$

$$-\frac{dS}{dt} = \frac{\mu X}{Y_{X/S}} + mX + \frac{q_P X}{Y_{P/S}}$$

where

X is the cell concentration
S is the substrate concentration
P is the product concentration
μ is the specific growth rate
q_p is the specific production rate
$Y_{X/S}$, $Y_{P/S}$ are the yield coefficients

Discuss on the difficulties you expect in the estimation of the yield coefficients.

6.6 No mathematical model in bio-system is unique one. Explain

6.7 Comment on the following:
(a) Structured and unstructured models.
(b) Asymmetric dynamic response of cells to changes in substrate concentration.
(c) Relaxation – time criterion in modeling.

REFERENCES

Bequette BW (Ed.) (1998) *Process Dynamics: Modeling, Analysis, and Simulation,* Prentice-Hall Int., Inc., New Jersey.

Coughanour DR (Ed.) (1991) *Process System Analysis and Control*, 2nd edn., McGraw-Hill International edn.

Felse PA (1999) "Process development for extracellular chitinase production by *Trichoderma harzianum*", Ph.D. Thesis, IIT Madras, India.

Fish NM, Fox RI and Thornhill NF (Eds.) (1989) Computer applications in fermentation technology: Modeling and control of biological processes, *Society of Chemical Industry and Elsevier Applied Science*, London.

Kossen NWF and Oosterhuis, NMG (1985) "Modeling and scaling-up of bioreactors" in, *Biotechnology*, Vol 2, (Brauer H, Editor), VCH, Weinheim. pp 571-605,

Roels JA (Ed.) (1983) *Energetics and Kinetics in Biotechnology*, Elsevier Biomedical Press, Amsterdam.

Sinclair GG and Kristiansen B (1987) *Fermentation Kinetics and Modelling*, Bu'Lock JD (Ed.), Open University Press.

Suga K, Waki T, Kumano M, Chimange P, Shin SB and Ichikawa K (1980) Production of cellulose in fed-batch system, pp 371-392, *Proceedings of Bioconversion and Biochemical engineering,* Symposium 2, vol II, Ghose TK (ed) IITDelhi, New Delhi.

Thilakavathi M, Basak T and Panda T (2007) "Modeling of enzyme production kinetics", *Applied Microbiology and Biotechnology*, **73**, 991-1007.

Volesky B and Votruba J (Eds.) *Modeling and Optimization of Fermentation Processes*, Elsevier, Amsterdam (1992).

Wang NS and Stephanopoulos GN (1984) "Computer applications to fermentation processes". CRC critical reviews in *Biotechnology*, **2**, 1-103.

Yoshida F, Yamane T and Nakamoto K (1973) "Fed-batch hydrocarbon fermentation with colloidal emulsion feed", *Biotechnology and Bioengineering*, **15**, 257-270.

APPENDIX 6

LAPLACE TRANSFORM

Let us consider a time domain function $f(t)$. Laplace transform of $f(t) = L(f(t))$.

This transforms time domain to's' domain.

$$L(f(t)) \equiv F(s) = \int_0^\infty f(t) \exp(-st)\, dt$$

In other words, inverse of Laplace transform is

$$L^{-1}(F(s)) = f(t)$$

Application

Laplace transform is used for linear operation, to solve linear partial differential equations, exponential function, step function, time delay, integral and impulse functions. Some standard Laplace transforms are summarized in Table 6.1 to help solve the problems.

Table 6.1 *Some standard Laplace transforms*

f(t)	*F(s)*
αt	$\frac{a}{s^2}$
t^{n-1}	$\frac{(n-1)!}{s^n}$
$\exp(-mt)$	$\frac{1}{s+m}$
$1 - \exp\left(-\frac{t}{\tau}\right)$	$\frac{1}{s(\tau s+1)}$
$\sin \omega t$	$\frac{s}{s^2+\omega^2}$
$\cos \omega t$	$\frac{s}{s^2+\omega^2}$
$\exp(-mt) \sin \omega t$	$\frac{\omega}{(s+m)^2+\omega^2}$

CHAPTER 7

Transport Processes in Bioreactors

OBJECTIVE

- Performance of bioreactors is interlinked with the role of mass, energy, and momentum transfers during reactions. They are discussed with emphasis on mass and energy transfer aspects.

7.1 INTRODUCTION

Three aspects of transports are important both in chemical reactors and in bioreactors. We consider specially the bioreactors in this chapter for mass and energy transfer and their influence in bioreactor design.

7.1.1 Mass Transfer

In general, for all cellular reactions, reactants are transported to cell for the synthesis of product(s), and, in turn, product(s) are transported to the outside of the cells (Doran, 1995). This constitutes several processes. Time required for these steps are not negligible than those for cellular reactions. For this reason, mass transfer must be included for the analysis of bioreactors. We can divide the mass transfer phenomena in cellular processes into two categories, viz.,

- Gas-liquid mass transfer, and
- Molecular diffusion of medium components into the cell

As mass transfer is considered as the multistage process (Fig. 7.1), any of the steps might limit the mass transfer and hence, the production is influenced by this phenomenon.

Steps of mass transfer in biological systems are given below.

Step 1: Diffusion dominant step for reactant from liquid/gas/solid to the interface.

Step 2: Combined diffusion and convection phenomena in surrounding stagnant zone

Step 3: Phenomena of convection and turbulence in the bulk liquid phase to the surrounding of cell

Step 4: Diffusion dominant phenomenon across the stagnant layer to the cell surface

Step 5: Passive diffusion or enzyme mediated active transport across the cell membrane.

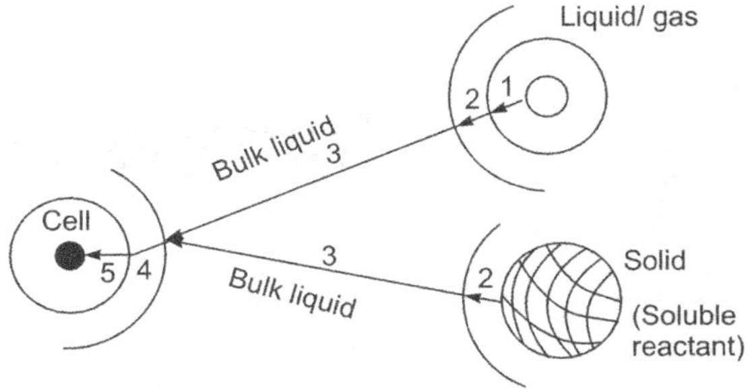

Figure 7.1 Steps of mass transfer from oxygen and reactants to cells.

(a) Gas–Liquid Mass Transfer

Gas–Liquid mass transfer is modeled by two-film theory (Whitman, 1923).

$$J_{A,g} = k_g(p_A - p_{Ai}) \tag{7.1}$$

$$J_{A,l} = k_L(C_{A,i} - C_A)$$

where $J_{A,g}$ and $J_{A,l}$ are the fluxes of the compound A through gas and liquid film, respectively. k_g is the gas phase mass transfer coefficient and k_L is the liquid phase mass transfer coefficient. The term in the parenthesis is the driving force. In dilute aqueous solution, the concentration on each side of gas-liquid interface is related to each other by Equation (7.2).

$$p_{A,i} = H_A C_{A,i} \tag{7.2}$$

which is called Henry's law. H_A is Henry's constant (atm. l/mole). It is difficult to measure the interfacial concentration directly. So the overall flux correlated to overall mass transfer and the driving force is considered in the liquid phase.

$$J_A = K_L(C_A^* - C_A) \tag{7.3}$$

where C_A^* is the saturation concentration in the bulk liquid corresponding to bulk gas phase. Therefore,

$$C_A^* = \frac{1}{H_A} p_A$$

$$\frac{1}{K_L} = \frac{1}{H_A k_g} + \frac{1}{k_L} \tag{7.4}$$

$$J_{A,g} = J_{A,l} = J_A$$

In biological systems, we mostly deal with oxygen and CO_2. Typically, $k_g >> k_L$. Oxygen and CO_2 have high H_A values. Therefore, gas-phase resistance in Equation (7.4) is negligible compared to liquid phase resistance. Hence, overall K_L is controlled by k_L.

To calculate the mass transfer rate per unit of bioreactor volume is expressed by $J_A a$, where

a = (gas-liquid interfacial area/unit liquid volume), (m^{-1})

$$\therefore \quad J_A a = K_L a(C_A^* - C_A) \tag{7.5}$$

$K_L a$ is the volumetric mass transfer coefficient.

Similarly we get from Fick's law of diffusion

$$J_A = -D_L \frac{dC}{dX}$$

$$= D_L \frac{(C_L^* - C_L)}{\delta} \tag{7.6}$$

J_A is mass flux, (kg mol/m^{2}s)
D_L is diffusivity in liquid phase (m^2/s)
δ is film thickness, (m).

It means that $k_L = \frac{D_L}{\delta}$. As per Danckwerts (1970),

$$k_L \propto D_L^{0.5} \tag{7.7}$$

$$\therefore \quad a = \frac{A}{V_G + V_L}$$

where a is specific interfacial area referred to dispersion (liquid) volume, (m^{-1}).

A is area (m^2) and V is volume (m^3).

The relative gas hold-up is defined by Equation (7.8).

$$\varepsilon_G = \frac{V_G}{V_T} = \frac{V_G}{V_G + V_L} \tag{7.8}$$

The relative liquid hold-up is given by, $\varepsilon_L = 1 - \varepsilon_G = \frac{V_L}{V_G + V_L}$

In addition, the following equation for 'a' is correlated with ε_G.

$$\therefore \quad a = \frac{6\varepsilon_G}{d_s} \tag{7.9}$$

where d_s is the Sauter mean diameter of the bubbles in (m) $= \frac{\sum_{1=1}^{n} n_i d_i^3}{\sum_{i=1}^{n} n_i d_i^2}$ (7.10)

n_i is the number of bubbles generated with diameter, d_i.

For bioreactor analysis and design, to understand the role of $k_L a$, we need to consider bubble coalescence. This influences mass transfer in the reactors, viz.,

- Coalescing liquid : poor mass transfer
- Non-coalescing liquid: highest mass transfer.

Let us consider a few typical reactor configurations, viz.,

- Bubble columns
- Airlift, and
- Stirred reactors.

Table 7.1 gives a comparison of such behavior of mass transfer in various reactors.

Table 7.1 *Comparison of mass transfer in bioreactors*

Bubble column	***Airlift***	***Stirred reactors***
Coalescing liquid: For non-viscous fluids, bubbles take the equilibrium size. *Non-coalescing liquid:* For small bubble size, k_La will be higher than the larger bubble size	Riser resembles the bubble column, but gas hold up is less than the predicted equation used in bubble column. This is true for k_La.	Flow phenomenon is difficult to predict, which is influenced by aeration and agitation
Correlation For coalescing bubbles_ • $k_La = 0.32\left(V_g \frac{p_0}{p}\right)^{0.7}$ • $\varepsilon = 0.6\left(V_g \frac{p_0}{p}\right)^{0.7}$ For non-coalescing bubbles, these values are higher.		*Correlation* For coalescing bubbles, $k_La = \left(\frac{P_s}{V_L}\right)^{0.4}\left(V_g \frac{p_0}{p}\right)^{0.5}$ $\varepsilon = 0.013\left(\frac{P_s}{V_L}\right)^{0.33}\left(\frac{V_g p_0}{p}\right)^{0.67}$ For non-coalescing bubbles, $k_La = 0.002\left(\frac{P_s}{V_L}\right)^{0.7}\left(\frac{V_g p_0}{p}\right)^{0.2}$

where V_g is the gas volume, p_0 is the reference pressure (1 bar), p is the pressure, P_S is the power input by the stirrer, V_L is the liquid volume, ε is the gas hold-up, and k_La is the volumetric mass transfer coefficient.

Correlations are based on Noorman (2001).

(b) Determination of Volumetric Mass Transfer Coefficient

A few possible measurements are given here with reference to the biological systems.

- Physical measurement technique
- Steady state bio-reaction technique
- Dynamic bio-reaction technique (or dynamic gassing out technique)

Physical measurement technique is based on continuous measurement of oxygen in the fluid. K_La is measured from the relation of

$$\frac{dC_L}{dt} = K_La(C^* - C) \tag{7.11}$$

Steady state bio-reaction technique is based on pseudo steady state system as growing microorganisms will continue to consume oxygen in the fermentation broth (Aiba *et al.*, 1984). Above equation can still be used to calculate K_La.

Dynamic gassing out technique is described in detail by Bandyopadhyay *et al.* (1967). The schematic diagram (Fig. 7.2) describes the method in detail.

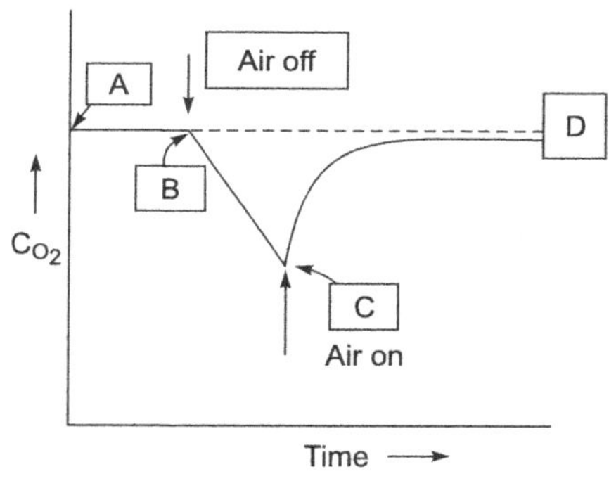

Figure 7.2 Schematic representation of dynamic gassing out technique.

As reaction progresses from A to B, suddenly for a brief time interval from B to C air supply is turned off. The oxygen concentration in liquid will steeply decrease to a critical value (C). Then air supply is turned on (at C). Concentration of oxygen will gradually increase, but it will try to attain the original value exponentially in infinite time.

Following measurements are made in this regard.

Section	*Value*
BC	$\frac{dC}{dt}$
CD	$\frac{dC_L}{dt}$ as a function of time

$$K_L a = \frac{\left(\frac{dC_L}{dt} - \frac{dC}{dt}\right)}{(C^* - C_L)}$$

(c) Empirical Correlations of $K_L a$

A few correlations are given earlier. $k_L a$ of bioreactors depends on a number of factors listed below.

- Vessel geometry and configuration
- Impeller characteristics (Mersmann *et al.*, 1975)
- Baffles
- Operating conditions : stirrer speed, gas flow rate, energy input
- Physico-chemical properties: density, viscosity, surface tension, etc.

Some important correlations of $k_L a$ are summarized in Table 7.2.

Table 7.2 *Important correlations of k_La*

Sl.No.	*Relations*	*References*
1	$k_La = C\left(\frac{P}{V_L}\right)^a\left(\frac{q_G}{A_R}\right)^b$ *or* $k_La = C\left(\frac{P}{V_L}\right)^a(u_G)^b$	Robinson and Wilke (1973) Moo-Young and Blanch (1981)
where P = power input (kW), V_L = volume, q_G = volumetric gas flow rate (m^3/s), A_R = reactor cross sectional area, u_G = linear gas velocity (= q_G/A_R), (m^3/s), a and b are adjustable constants.		
2	$k_La\left(\frac{v}{g^2}\right)^{0.33} = d\left(\frac{P}{V_L\rho(vg^4)^{0.33}}\right)^a\left(\frac{u_G}{gv^{0.33}}\right)^b$	Henzler and Kauling (1985)
where ρ = density, v = kinematic viscosity (m^2/s), a, b and d are constants.		
3	$k_La = C'\sqrt{D_L}\,\frac{\rho^{3/5}\left(\frac{P}{V_L}\right)^{13/20}}{\left(\frac{\mu}{\rho}\right)\sigma^{3/5}}\left(\frac{u_G}{u_t}\right)^{0.5}\left(\frac{\mu}{\mu_w}\right)^{-0.25}$	Kawase and Moo-Young (1985)
where μ = dynamic viscosity (Poise), u_t = terminal rise velocity of bubbles in Newtonian media, which is approximately 0.265 m/s. C' is a constant of value 0.675. D_L = diffusivity in liquid phase (m^2/s).		
4	$k_La\left(\frac{v}{g^2}\right)^{0.33} = C''\left(\frac{P/V_L}{p(vg^4)^{0.33}}\right)^a\left(\frac{V_G}{V_L}\left(\frac{v}{g^2}\right)^{0.33}\right)^b$	Schülter (1991)
where C'', a, and b depend on stirrer configurations. p = pressure. C'' is a constant of value 7.94×10^{-4} for a = 0.62 and b = 0.23 for Rushton turbines. $C'' = 5.89 \times 10^{-4}$ with a = 0.62 and b = 0.19 for intermig impellers.		
5	$Sh = 0.6Sc^{0.5}Bo^{0.62}Ga^{0.31}\varepsilon_G^{1.1}$	Akita and Yoshida (1973)
where $Sh = \frac{k_Ll}{D_L}$, l is characteristics length; $Bo = \frac{gD_R^2}{\sigma_L}$, $Sc = \frac{v}{D_L}$ and, $Ga = \frac{gD_L^3}{V_L^2}$; ε_G = gas hold-up and D_R = reactor diameter		

7.1.2 Mass Transfer Phenomena in Bioreactors

Three different mass transfer phenomena are encountered in bioreactors. They are mentioned below.

- **Liquid–solid mass transfer**
  (1) transport on to the surface of cells and
  (2) transport of reactant on the membrane of immobilized cells/enzymes

In this case,

volumetric rate of mass transfer $= k_L a\left(C_{A_{\text{bulk}}} - C_{A_{\text{int erface}}}\right)$ (7.12)

- **Liquid–liquid mass transfer**

Doran (1995) has suggested two overall mass transfer coefficients.

$$\frac{1}{K_{L_1}a} = \left(\frac{1}{k_{L_1}a} + \frac{m}{k_{L_2}a}\right) \tag{7.13}$$

and

$$\frac{1}{K_{L_2}a} = \left(\frac{1}{mk_{L_1}a} + \frac{1}{k_{L_2}a}\right)$$

where

$K_{L_1}a$ is the overall mass transfer coefficients based on bulk concentration in liquid 1.
$K_{L_2}a$ is the overall mass transfer coefficients based on bulk concencentration of liquid 2 and
m is the partition coefficient or distribution coefficient

- **Gas-liquid mass transfer:**

Oxygen, a sparingly soluble component in aqueous media, is considered as the limiting component in bio-reactions. In all aerobic reactions, distribution of oxygen in reactor and its availability to cell are the real challenges in bioreactor design.

A number of correlations exist for k_La using coalescing and non-coalescing systems in bioreactors. Some of those relations are given in the beginning of this chapter. Two important expressions are worth mentioning here.

(1) Oxygen transfer rate = OTR = $K_La\left(C_{O_{2eq}} - C_{O_2}\right)$ (7.14)

where OTR is in mol/(l)(h)

$C_{O_{2eq}}$ is equilibrium concentration of oxygen (8 mM at 25°C), i.e., C_L^*

C_{O_2} is concentration of oxygen at any time (moles).

This expression is exactly equivalent to

$$N_A = K_La\,(C^* - C_L) \tag{7.15}$$

(2) Oxygen uptake rate (OUR):

$$\text{OUR} = q_{O_2}\cdot C_X = \frac{k\mu}{Y_{X/O}}C_X \tag{7.16}$$

q_{O_2} is specific respiration rate (mM O_2 (g cells)$^{-1}$ h^{-1})

$$k = \text{mM } O_2/\text{g } O_2$$

$Y_{X/O}$ is yield coefficient of cells on oxygen (g cells/g O_2) as defined in Chapter 4.

$$\therefore \quad N_A = \frac{k\mu}{Y_{X/O}}C_X \tag{7.17}$$

As oxygen is the rate limiting component in aerobic reaction

$$\text{OUR} = \text{OTR}$$

$$\therefore \quad K_La\left(C_{O_{2eq}} - C_{O_2}\right) = \frac{k\mu}{Y_{X/O}}C_X \tag{7.18}$$

Cooney *et al.* (1969) suggested a relation for OUR and energy production.

$$Q = 0.12 \text{ OUR} \tag{7.19}$$

Q in k cal/(l)(h) and OUR= mM $O_2/(l)(h)$

Equation (7.18) can also be used to calculate C_{O_2} in liquid. By rearranging and assuming Monod's type of relation, Equation (7.18) becomes

$$\therefore \quad K_L a\left(C_{O_{2eq}} - C_{O_2}\right) = k\mu_{max} \frac{1}{Y_{X/O}} C_X \frac{C_{O_2}}{K_{O_2} + C_{O_2}} \tag{7.20}$$

With the assumption $C_{O_{2eq}} >> C_{O_2}$

$$C_{O_2} = C_{O_{2eq}} \frac{Y_{X/O} K_{O_2} K_L a / k\mu_{max} C_X}{1 - Y_{X/O} C_{O_{2eq}} K_L a / k\mu_{max} C_X} \tag{7.21}$$

7.2 HEAT TRANSFER

Microorganisms utilize energy for their growth, biochemical reactions, maintenance of cells and cellular components, and transport related work.

The oxidation of reactants supplies energy for biochemical reactions in cells (except for plants, algae and cyano bacteria). Energy is in the form of ATP in a cell. Some portion of this stored energy is released from the cell as heat.

If one considers the simple reaction of glucose to CO_2 and H_2O, heat generation is – 2870 kJ/mol of glucose. For biochemical conversion of glucose to CO_2 and H_2O,38 moles of ATP is generated. 1 mole of ATP produces 50 kJ under actual reaction conditions. The standard free enthalpy of reaction is – 31 kJ/mol.

$$\text{ADP} + P_i \rightarrow \text{ATP}$$

Part of free reaction enthalpy for ATP formation is

$$= (38 \text{ mol ATP/mol glucose})(50\text{kJ/mol ATP})$$
$$= 1900 \text{ kJ/mol glucose}$$

The other part of reaction enthalpy is $(-2870 + 1900) \dfrac{\text{kJ}}{\text{mol glucose}} = -970$ kJ/mol glucose.

This is liberated as heat by the oxidation process of glucose. Approximately 33% heat is removed from the reaction. This is a tedious method to evaluate the amount of energy to be removed for each individual component in the reaction. Some empirical relations might help in calculation. For example,

$$Q_{\text{respiration}} = F \times \text{OUR} \tag{7.22}$$

F is a factor defined by a number of authors (Table 7.3).

Table 7.3 *F values*

F values(kJ/m mol of O_2)	***References***
0.55	Cooney *et al.* (1969)
0.44	Luong and Volesky (1980)

For bioreactor design, one needs to consider heat balances which includes following additional components:

- Heat generation by biochemical reaction
- Energy input by aeration
- Energy input by agitation
- Heat loss by conduction, radiation, evaporation, etc.

Calculation of heat to be removed

Heat removal rate (kJ/h) is expressed by the following equation.

$$q = UA\Delta T \tag{7.23}$$

where,

A is heat transfer area

ΔT is temperature difference between medium and cooling liquid

U is heat transfer coefficient (kJ/(m^2)(h)(K)

$$\frac{1}{U} = \frac{1}{hc_W} + \frac{d_W}{\lambda_W} + \frac{1}{h_{MW}} \tag{7.24}$$

$\frac{1}{h_{CW}}$ accounts for heat transfer in cooling media for flat surface wall

$\frac{d_W}{\lambda_W}$ is heat conduction through the wall

$\frac{1}{h_{MW}}$ accounts for heat transfer in culture medium or wall

This is as per Peclet's law for plane surface whereas same law for tubes is given by Equation (7.25)

$$\frac{1}{Ud_W} = \frac{1}{h_o d_o} + \frac{\ln\left(\frac{d_o}{d_i}\right)}{2\lambda_W} + \frac{1}{h_i d_i} \tag{7.25}$$

d_i represents inner diameter of the tube

d_O represents outer diameter of tube

h_o, h_i is the heat transfer coefficient (kJ/(m^2)(h)(K))

d_W is the wall thickness

λ_W is the thermal conductivity of wall material (kJ/(m)(h)(K))

In practice, one can use the relations for Nussett number.

$$\text{Nu} = f(\text{Re, Fr, Pr}) \tag{7.26}$$

It is wise to refer to Perry's hand book in this regard.

where $\text{Nu} = \frac{hd}{\lambda}$

$\text{Re} = \frac{du\rho}{\mu}$

$$\mathrm{Pr} = \frac{C_P \mu}{\lambda}$$

$$\mathrm{Fr} = \frac{u^2}{gd}$$

where μ is the dynamic viscosity,

For stirred tank reactors, Kurpiers *et al.*, (1985) has compiled the correlations. For bubble column and air lift reactors, Deckwer (1980) has suggested the following relation to calculate wall/broth heat transfer coefficient

$$\mathrm{St} = \frac{h}{\rho C_P u_G} = 0.1\ (\mathrm{Re\ Fr\ Pr}^2)^{-0.25} \tag{7.27}$$

where, St is the Stanton number

u_G is the linear gas velocity $\left(\frac{q_G}{A_R}\right)$, m/s

q_G is the volumetric flow rate, (m^3/s)

A_R is the reactor cross sectional area (m^2)

ρ is the density (kg/m^3).

However, heat transfer coefficient is scale dependent. For large reactors, additional knowledge is required as the heat transfer area increases with D_R^2 (D_R = reactor diameter) whereas the heat generation is related to D_R^3.

For biological process:

$$q_{\text{generated}} = q\pi D_R^2 \frac{H}{4} \tag{7.28}$$

$$\begin{aligned} q_{\text{removal}} &= UA\Delta T \\ &= U\Delta T\pi H D_R \end{aligned} \tag{7.29}$$

For large reactors, it is necessary to install additional internal heat exchanger or an external heat exchanger to extract energy from the recycled reaction system (fermentation broth). The options will influence the mixing characteristics and mass transfer phenomenon. Again this simplified treatment assumes no gradient of heat inside the cell.

7.3 OTHER PARAMETERS INFLUENCING TRANSFER OPERATIONS

Power input and mixing time also influence the transfer operations in bioreactors.

7.3.1 Power Input

(a) Reactor with Mechanical Agitation

The power input in a bioreactor directly influences operating costs. Some process variables of stirred tank reactors, which influence transfer operations in the bioreactor, are given below.

(a) Impeller tip velocity is equal to ND_S.

(b) Pumping capacity of impeller is directly proportional to (tip velocity) and (impeller area).
(c) Power input (=P) is directly proportional to (kinetic energy) and (pumping capacity).
(d) Power number or Newton number (Ne) depends on the type of stirrer used in the bioreactor. Power number is also a function of Re for an impeller used.

Indirectly Ne is related to gas flow number $\left(=\dfrac{q_G}{ND_s^3}\right)$ for a stirrer. This relation is found directly related with the $P_{aerated}/P_{no\ aeration}$ and gas flow number for a stirrer.

For the selection of an agitated bioreactor, power characteristics of the various stirrer types under aerated and non-aerated conditions must be known or evaluated from the plot of Ne vs. Re. Then P is known from the relation of $P = \text{Ne}\rho N^3 D_S^5$. The value of P needs to be evaluated under aerated conditions from the plot of $P_{aerated}/P$ vs. gas flow number. In this regard, one needs additionally to calculate P/V. From the plot of P/V and V_R (= working volume of the reactor), one can decide how much power is really necessary for the operation of reactor. This gives an idea about the feasibility of the reactor for operation. Then we can also calculate specific energy dissipation rate (ε_d).

$$\varepsilon_d = \frac{\text{Ne}N^3 D_s^5}{V_R} \tag{7.30}$$

(b) Reactor without Mechanical Agitation

In bubble column and air lift reactors, the power input contains two components, viz.,

- kinetic energy and
- hydrostatic head (isothermal expansion energy). This is expressed by Equation (7.31).

$$\therefore \qquad P_G = \rho_G q_G \left(\frac{u_G^2}{2} + \frac{RT}{M} \ln \frac{p_1}{p_2} \right) \tag{7.31}$$

where, P_G = power input by the gas flow
M = molecular weight
ρ_G = gas density
q_G = volumetric gas throughput
p_1, p_2 = pressure
u_G = linear gas velocity in nozzle through which gas is distributed in the reactor.

7.3.2 Mixing Time

The distribution and contact of reactants with cells (or equivalent biocatalysts) and reduction of non-ideality in reactors can be quantified by mixing time. This influences the transfer operations in the bioreactors.

Following correlations are important in this aspect.

1. Dimensionless mixing time: $N\theta$
   where θ is mixing time. Again

$$N\theta = f(\text{Re}) \tag{7.32}$$

(a) For stirred bioreactor

In turbulent regime, Mersmann *et al.* (1975) proposed the relation

$$\frac{P\theta^3}{\rho D_s^5} = 300 \tag{7.33}$$

(b) For bioreactors without mechanical agitation

Airlift Reactor

Generally, liquid circulation time (t_c) is used to calculate the mixing behavior in such reactors. Kawase and Moo-Young (1985) proposed the following correlation for airlift reactor.

$$t_c = 8.3\frac{H}{u_G}\left(1+\frac{A_d}{A_r}\right)^{-1.3} \text{Fr}^{0.33} \tag{7.34}$$

where, $\text{Fr} = \dfrac{u_o^2}{D_R g}$ with

$$u_0 = 0.787(gDu_G)^{0.33} \tag{7.35}$$

A_d = Cross sectional area of the down comer.

A_r = Cross sectional area of the riser.

Mixing time (99% homogeneity) is related to the circulation time (Bello *et al.*, 1984).

Bubble Column Reactor

Mixing behavior is characterized by liquid dispersion coefficient (E_L). Many correlations are available in the literature. Deckwer (1992) proposed the following relations for E_L.

$$E_L = \frac{u_G D}{2.83Fr^{1/3}} \tag{7.36}$$

where $$\text{Fr} = \frac{u_G^2}{D_R g}$$

Example 7.1

Citric Acid is produced in a 5 m^3 bioreactor from molasses using the fungus *Aspergillus niger*. The oxygen uptake by the fungus is 1.1 kg/m^3 h. The agitator used has heat dissipation at the rate of 0.5 kW/m^3 of reactor volume. The heat is released due to reaction at a rate of 460 kJ/gmol of O_2 consumed. The water at 10°C is available for maintaining the reactor temperature at 30°C. Calculator $\Delta H_{r\times n}$ (i.e., ΔH for reaction) assuming most of the carbohydrate in molasses is in the form of glucose. Also, calculate the exit temperature of cooling water if the flow rate of water is 600 m^3/h.

Solution

Assumption: Process cooling is at steady state

Metabolic heat produced is given by

$$\begin{aligned}\Delta H_{r\times n} &= -460 \text{ kJ/gmol} \times 1.1 \text{ kg/m}^3\text{ h} \times 0.015 \\ &= 39.04 \text{ kJ/s} \\ &= 39.04 \text{ kW}\end{aligned}$$

The rate of heat load due to the agitator is

$$q_{\text{agi}} = 0.5\ \text{kW/m}^3 \times 5\ \text{m}^3 = 2.5\ \text{kW}$$

The total heat load is

$$\begin{aligned} q &= \Delta H_{r\times n} + q_{\text{agi}} \\ &= 39.04 + 2.5 = 41.54\ \text{kW} \end{aligned}$$

From the relation

$$\begin{aligned} q &= mCp\Delta T \\ \Delta T &= T_{c,\,\text{out}} - T_{c,\,\text{in}} \\ T_{c,\,\text{out}} &= T_{c,\,\text{in}} + q/mCp \\ &= 10 + 41.54/(4.18 \times 600/1000) \\ &= 26.56°\text{C} \end{aligned}$$

Table 7.4 *Dimensionless numbers used in this chapter*

Dimensionless number	***Definition***
Bo	Bond number $= \dfrac{gD_R^2}{\sigma_L}$
Fr	Froude number $= \dfrac{u_G^2}{gD_R}$
Ga	Galileo number $= \dfrac{gD_L^3}{V_L^2}$
Ne	Power or Newton number $= \dfrac{P}{\rho N^3 D_s^5}$
Nu	Nusselt number $= \dfrac{hd}{\lambda}$
Pr	Prandtl number $= \dfrac{C_P\mu}{\lambda}$
Re	Reynolds number $= \dfrac{du\rho}{\mu}$
Sc	Schmidt number $= \dfrac{v}{D_L}$
Sh	Sherwood number $= \dfrac{k_L l}{D_L}$
Sh’	Sherwood number $= \dfrac{Sh}{\varepsilon_L}$
St	Stanton number $= \dfrac{h}{\rho C_P u_G}$
Wi	Weissenberg number $= \dfrac{\sigma}{\tau}$

where,

ρ is density (kg/m^3)
μ is viscosity (Poise)
v is kinematic viscosity (m^2/s)
$u_G = \dfrac{q_G}{A_R}$ is linear gas velocity (m/s)
q_G is volumetric gas flow rate (m^3/s)
D_R is diameter of reactor (m)
D_S is stirrer diameter (m)
D_L is diffusivity in liquid phase (m^2/s)
A_R is reactor cross sectional area (m^2)
N is stirrer speed (revolution/min)
P is power input (kW)
V_L is liquid volume (m^3)
h is heat transfer coefficient (kJ/ (m^2)(h)(K))
τ is shear stress
σ is first normal stress differences
λ is heat conductivity (kJ/ (m^2)(h)(K)).

EXERCISES

7.1 Compare the OTR in a bubble column and stirred tank reactor having tank diameter of 4m, impeller diameter of 1.75 m and power input per unit volume = 2 kW/m^3. Other conditions are: coalescing liquids, reaction volume 105 m^3 and

$$(C_{equillibrium} - C_L) = 0.49 \text{ mol/ m}^3.$$

7.2 The dynamic method is used to measure k_La. The equilibrium concentration of oxygen in the broth is 7.9×10^{-3} kg/m^3 and this is 80% of air saturation. Following data are relevant in this regard.

Time(s)	**C_{AL} (% air solution) at 30°C**
10	43.5
20	60
30	67.5
100	75

Interpret the values of k_La during the fermentation.

7.3 The specific rate of oxygen uptake is 15 mmol/(g)(h). Using the data of Problem 7.2, calculate maximum possible cell concentration (kg/m^3) in the reaction?

7.4 Calculate k_La for a multi-turbine fermenter for the Nth stage. Following data are available for this purpose.

Parameters	$a = 42.167$
	$Pg(1) = 37.50$ kW/L
	$Q(1) = 8 \times 10^4$ standard L/min
	D_T = tank diameter = 15 ft.
	$P(1) = 16.77$ lb/in^2
Liquid volume in cell (l)	= 28522 L
Agitator speed	= 200 rpm

REFERENCES

Aiba S, Koizumi J, Shi RJ and Mukhopadhyay SN (1984) "The effect of temperature on k_La in thermophilic cultivation of *Bacillus stearothermophilus*", *Biotechnology and Bioengineering*, **26**, 1136–1138.

Akita K and Yohida F (1973) "Gas holdup and volumetric mass transfer coefficient in bubble columns", *Industrial Engineering Chemistry Process Design and Development*, **12**, 76-80.

Bello RA, Robinson CW and Moo-Young M (1984) "Liquid circulation and mixing characteristics of air-lift contactors". *Canadian Journal of Chemical Engineering*, **62**, 573-577.

Bandyopadhyay B, Humphrey AE and Taguchi H (1967), Dynamic measurement of the volumetric oxygen transfer coefficient in fermentation systems", *Biotechnology and Bioengineering*, **9**, 533-544.

Cooney CL, Wang DI and Mateles RI (1969), "Measurement of heat evolution and correlation with oxygen consumption during microbial growth", *Biotechnology and Bioengineering*, **11**(3), 269-281.

Danckwerts PV (1970) *Gas-Liquid Reactions*, McGraw-Hill, New York.

Deckwer W-D (1992) *Bubble Column Reactors*, John Wiley & Sons, New York.

Deckwer W-D (1980) "On the mechanism of heat transfer in bubble column reactors", *Chemical Engineering Science*, **35**, 1341-1346.

Doran PM (1995) *Bio Process Engineering Principles*, Academic Press, London.

Henzler H-J and Kauling J (1985) "Scale-up of mass transfer in highly viscous liquids". *Proc. 5th Europ. Conf. Mixing*. Wurzburg/Germany, Paper 30, pp. 303-312, Cranfield, BHRA.

Kawase A and Moo-Young M (1985) "Volumetric mass transfer coefficients in aerated stirred tank reactors with Newtonian and non-Newtonian media", *Chemical Engineering Research and Development*, **66**, 284-288.

Kurpiers P, Steiff A and Weinspach P-M (1985), "Heat transfer in stirred multiphase reactors", *German Chemical Engineering*, **8**, 48.

Luong JHT and Volesky B (1980). Determination of the heat of some aerobic fermentations, *Canadian Journal of Chemical Engineering*, **58**, 497.

Mersmann A, Einenkel WD, and Käppel M (1975) "Design and scale-Up of agitators", *Chem. Ing. Tech.* **47**, 953-964.

Moo-Young M and Blanch HW (1981) "Design of biochemical reactors-mass transfer criteria for simple and complex systems", *Advances in Biochemical Engineering*, **19**, 1-69.

Noorman HJ (2001) Mass transfer. In: C. Ratledge and B. Kristiansen (Eds.) *Basic Biotechnology*, Cambridge University Press, Cambridge, p. 178.

Robinson CW and Wilke CR (1973) "Oxygen absorption in stirred tanks: A correlation for ionic strength effects", *Biotechnology and Bioengineering*, **15**, 755-782.

Schlüeter V, Ph.D. Thesis, Technische Universitat Clausthal-Zellerfeld, 1991.

Thilakavathi M, Basak T, Panda T (2007) "Modeling of enzyme production kinetics", *Applied Microbiology and Biotechnology*, **73**, 991-1007.

Van't Riet C (1979) "Review of measuring methods and results in non-viscous gas-liquid mass transfer in stirred tanks". *Industrial Engineering Chemistry Process Design and Development*, **18**, 357-375.

Whitman WG (1923) "The two-film theory of absorption". *Chem. Met Eng.* **29**, 147.

Chapter 8

Controls in Bioreactors

OBJECTIVE

- Monitoring and control of biological reactions aim at improving the performance of bioprocesses and, in turn, ensure reliable bioreactor operation. This chapter deals with classical as well as advanced control systems for bioreactors. To achieve successful design, control action targeted through modeling is a better approach.

8.1 INTRODUCTION

In the beginning, some general features of bioreactors are highlighted with reference to control applications. Two main characteristics, which are important to know before designing a control system for bioreactors, are:

- Multivariable system, and
- Non-linear dynamics.

8.1.1 Multivariable Systems

Control of a bioreactor involves many variables. The technologies of measuring and controlling devices applied to a standard bioreactor are well known from classical process engineering. Examples of variables are

- motor power consumption
- motor current
- impeller tip speed
- liquid volume in reactor
- temperature of reaction fluid
- temperature in a jacket for energy transfer
- pressure in reactor head space.
- gas flow rate.

These variables are supplemented by suitable sensible sensors (probes), viz.

- pH value(as C_{H+})
- DOT (Dissolved oxygen tension as p_{O_2})

- Dissolved CO_2 level(as p_{CO_2})
- Redox potential

8.1.2 Nonlinear Dynamics

Hysteresis is a sure test for nonlinearity. For example, a step increase in reactor feed rate in case of CFSTBR will result in a transient state that will be different from the corresponding equivalent step decrease in feed rate. Even for the same initial conditions multiple steady states are observed and sometimes the behavior is like limit cycles and oscillatory transients. These two characteristics (multivariable systems and non-linear dynamics) are related to the complexities of the living cells in a bioreactor.

Important variables, which are desired for monitoring and control, can only be measured with substantial time lag or cannot be measured at all. For this reason, mathematical models and/or state estimation techniques are necessary in this field. We have discussed about the development of mathematical models in the previous Chapter 6. We would like to apply them in this chapter to understand its use in control design of bioreactor.

8.2 CONTROL TASKS IN A BIOREACTOR SYSTEM

The tasks can be divided in to four components.

Class of control	*What to control*	*Manner of control*
Environment control	Aeration, agitation, thermostat, etc.	Open loop control
Cell side control	Temperature in reaction fluid, pH of the broth, p_{O_2}, p_{CO_2}, oxygen and CO_2 mole fractions in gas phase measured in exit gas, pressure in reactor head space, reactant and product concentration in the liquid phase	Closed loop control, viz., temperature control
Control of cell activity	Cell concentration, specific rates of reaction with reference to cell and a large number of parameters.	Adaptive control systems
Control of cell supply and medium management	Feed rate, harvest rate, etc.	(a) Programmable logic control systems associated with operation state control (b) PID control systems

8.3 INSTRUMENTATION TO CONTROL A BIOREACTOR

A part of the instrumentation is given in Figure 2.2. They are listed below.

- Probes

- Control elements for continuous control: control valves, static converters, pumps, etc.
- Control elements for discontinuous control: valves, pumps, switches, etc.
- Alarm signals: steam emergency off, interlocking signals, etc.
- Control cabinet with standard control loops and remote control, control panel with state selection, indicator lamps, and programmable logic control driving the discontinuous final control elements.
- Gas analyzing system for O_2 and CO_2
- Facilities for control and data acquisition by super computers.

Detailed presentation of schematic diagrams is given by Erickson and Stephanopoulos (1985).

8.4 CONTROLLED VARIABLES AND MEASUREMENT DEVICES

Table 8.1 summarizes a few controllable parameters and their available measuring devices.

Table 8.1 *Parameters and measuring devices*

Parameter	*Measuring device*
Pressure	Diaphragm gauge
Temperature	Resistance thermometer
Agitator shaft power	Watt meter or strain gauge
Foam	Rubber-sheathed electrode
Flow rate of gas	Rotameter
Liquid level	Load cells
Turbidity	Photo meter
C_{H^+}	pH probe with pH meter
Viscosity	Rotational viscometer
Dissolved oxygen	DO probe with DO meter

Cell mass concentration, in a continuous system, can be a controllable parameter if proper algorithm is available. This is true for the measurement and control of volumetric mass transfer coefficient (k_La). Concentration of products and intermediates (metabolites) are difficult to control.

The classical control (conventional control) and advanced control techniques are introduced to the readers.

8.5 PROCEDURE FOR DESIGN OF EFFICIENT CONTROL SYSTEMS

For an efficient control system, three major steps can be followed.

- Calculation of number of steady states
- Characterization of nature of steady states and their stability
- Design of appropriate control loops

8.5.1 Linear Stability Analysis

The linearization of the system around a steady state and analyzing the Jacobian matrix for its Eigen values are the major criterion in this analysis. From this one can say, if Eigen value is negative, that the system is stable (Smith, 1995).

8.5.2 Bifurcation Analysis

Bifurcation analysis, described in Chapter 5, indicates how the system behavior changes with changing parameters (Kuznetsov, 2004).

These two techniques are discussed for the design of suitable controllers.

8.6 CONVENTIONAL CONTROL TECHNIQUES

Many bioreactors use the same type of control strategies as those used by common industrial chemical reactors (Bailey and Ollis, 1986). They include PI and PID control algorithms for SISO system approximation of bioreactor. The typical manipulated variables are sterile air flow, nutrient feed, pressure, and agitation rate while the controlled variables are pH, temperature, DO, and substrate concentration. We mention here PID, PI, and PD control laws.

The PID control law

$$p(t) = p_s + K_c\left[e(t) + \int_0^t \frac{(e(t))dt}{\tau_I} + \tau_D \frac{de(t)}{dt}\right]$$

The PI control law

$$p(t) = p_s + K_c\left[e(t) + \int_0^t \frac{(e(t))dt}{\tau_I}\right]$$

This expression is devoid of derivative mode.

The PD control law

$$p(t) = p_s + K_c\left[e(t) + \tau_D \frac{de(t)}{dt}\right]$$

This expression is devoid of integral mode.

where, $p(t)$ is the output of the controller

p_s is the nominal controller output corresponding to operation at undisturbed and design condition

$e(t)$ is the controlled variable's error = (set point value at time t – measured value at time t)

$\int (e(t))dt$ is the integral of $e(t)$ or integral of error

$\frac{de(t)}{dt}$ is the derivative of error

τ_I is integral time

τ_D is derivative time

K_c is controller gain (proportional gain).

Many bioreactors use the conventional techniques of control. However, these techniques suffer from severe disadvantages. For example, they are not very effective on processes with slow dynamics. They also cannot work in cases where there are frequent disturbances or when the multivariable interactions cannot be neglected. For simplicity, these techniques do not 'understand' the process that they are meant to control. The conventional techniques involve at the most 2 or 3 tuning parameters that are typically tuned offline in the industry. When the process goes away, the tuning parameters have to be returned manually. The tuning process itself is entirely empirical using Ziegler-Nichols or the Cohen-Coon techniques (O'Dwyer, 2000). In order to get better performance, the controller should be model based (i.e., based on a 'near to accurate model of the process'). This leads us to the use of advanced control techniques.

Terminologies with short description are given in the Appendix to this chapter.

8.6.1 Examples of Measurement and Control by Conventional Techniques

(a) The pH Measurement and Control

For a pH control, PID controller is used to minimize the deviation between the actual pH measurement (pH_m) and the corresponding set point (pH_s) (Coughanowr and Koppel, 1965). The controller output is converted into two pulse-modulated manipulation signals. Depending on the signs of the deviation, either acid ($pH_m > pH_s$) or alkali ($pH_m < pH_s$) is pumped into the reactor for a shorter duration (Fig. 8.1). An addition cycle is always accompanied by a mixing cycle during this period without the addition of any acid or alkali. Small fluctuation of process signal around the set point does not result in an unexpected value. The input line of controller has a dead band of deviation.

The regulator output is connected to the two demand-operated tubular pumps, each one for acid and alkali.

(b) The p_{O_2} Measurement and Control

The measuring system is based on amperometric principle and consists of a steam sterilizable p_{O_2} electrode and the amplifier (Wang and Stephanopoulos, 1974).

$$p_{O_2} \text{ is defined by } p_{O_2}(t) = \frac{p_{O_2L}|_t}{p_{O_2G}|_c} \tag{8.1}$$

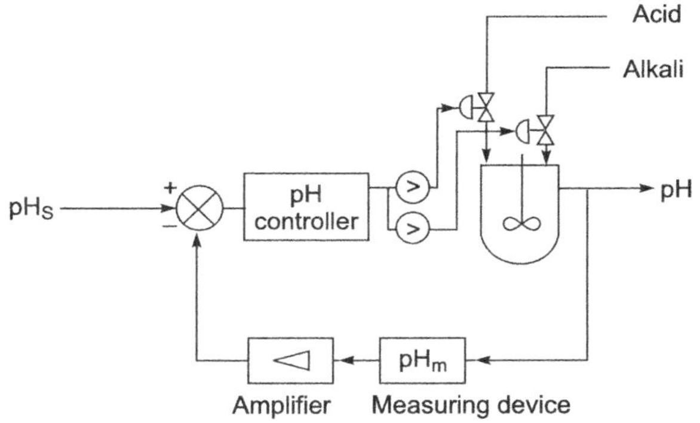

Figure 8.1 The pH measurement and control arrangement.

where $p_{O_2L}|_t$ stands for absolute dissolved oxygen partial pressure and
$p_{O_2G}|_c$ denotes certain calibration in the partial pressure of oxygen in the gas phase
$p_{O_2G}|_c = p_G|_c * x_{O_2G}|_c$
$x_{O_2G}|_c$ denotes oxygen mole fraction in gas phase
p_{O_2} is set to 100% saturation. If one selects a calibration for O_2 mole fraction,

$$x_{O_2G}|_c = x_{O_2}|_{air} = 0.2$$

The p_{O_2} reading is in air saturated values. The calibration in pure O_2 aeration,

$$x_{O_2G}|_c = 1$$

which leads to O_2 saturation defined measurements.

As we have discussed about the derived measurement of O_2, i.e., K_La in the batch reaction.

$$\frac{dC_{O_{2L}}|_t}{dt} = \text{OTR}(t) - \text{OUR}(t) \tag{8.2}$$

$$\text{OTR}(t) = K_La|_t \left(C^*_{O_2L}|_t - C_{O_2L}|_t\right) \tag{8.3}$$

$$C^*_{O_2L}|_t = \frac{p_{O_2G}|_t}{H_{O_2}|_t} = \frac{p_G|_t}{H_{O_2}|_t} x_{O_2G}|_t \tag{8.4}$$

$$C_{O_2L}|_t = \frac{p_{O_2L}|_t}{H_{O_2}|_t} = \frac{p_G|_c \times x_{O_2G}|_c}{H_{O_2}|_t} p_{O_2}|_t \tag{8.5}$$

H_{O_2} is the Henry's constant.

Therefore, Equation (8.3) is expressed in

$$\text{OTR}(t) = K_La|_t \frac{p_G|_t}{H_{O_2}|_t}\left(x_{O_2G}|_t - \frac{p_G|_c}{p_G|_t} x_{O_2G}|_c \; p_{O_2}|_t\right) \approx \text{OUR}(t) \tag{8.6}$$

This expression may be true for some cases.

The O_2 uptake is limited by the transfer step and cell growth changes from an exponential one to a linear one.

Possibilities of p_{O_2} control in bioreactors

Three possibilities are indicated below (Wernan and Wilke, 1973).

(1) p_{O_2} control with manipulation of agitation and/or aeration

The p_{O_2} signal measured is sent to the controller and is compared to the set point of p_{O_2}. The controller output is cascaded onto the set point inputs of the agitator speed controller and/ or aeration rate controller.

(2) p_{O_2} control with gas phase p_{O_2} manipulation

Controlling oxygen transfer at constant p_{O_2} is done by changing the saturation concentration, $C^*_{O_2L}$. Increase in oxygen partial pressure at this point produces various regulation possibilities.

(3) p_{O_2} control with reactant feed

This control method is carried out in a reactant fed batch procedure (Suga *et al.*, 1980). The OUR(t)and URR (uptake rate of reactant) have some kind of growth equilibrium. This is cell internal control with the suitable introduction of yield factor (Y).

$$Y_{X/O}|_t \cdot \text{OUR}(t) \cong Y_{X/S}|_t \cdot URR(t) \tag{8.7}$$

On the other hand, reactant feed rate

$$RFR(t) = \frac{F_R(t)C_{S_R}}{V_L(t)} \tag{8.8}$$

is in equilibrium with URR (t) and OUR(t) with oxygen transfer. Hence,

$$\frac{Y_{X/S}|_t}{Y_{X/O}|_t} \cdot \frac{F_R(t)C_{S_R}}{V_L(t)} \cong K_L a|_t \left(C^*_{O_{2L}}|_t - \frac{p_{O_{2G}}|_c}{H_{O_2}|_t} p_{O_t}(t) \right) \tag{8.9}$$

This suggests possible control mechanism through p_{O_2} and reactant. A certain increase in p_{O_2} is adjusted by a feed rate increase in order to enhance O_2 demand by cells. On the other hand, a decrease in p_{O_2} is accompanied by a reduction in feed rate. Similar principle is used in fed-batch reactor control operation.

(c) Exhaust Gas Balance – An Example of On-line Process Activity Monitoring

Exhaust gas balance method is described by Lee (1992).

Total specific gas mole flow rate into the reactor (Q_m in mol/ l h) is defined by

$$Q_{mi}|_t = \frac{F_{Gi}|_t p_{Gi}|_t}{V_L(t) \cdot R.T_{Gi}(t)} \tag{8.10}$$

where

- i stands for entry point
- F_{Gi} is the actual gas volume flow rate
- p_{Gi} is the inlet gas pressure
- T_{Gi} is the inlet temperature of gas.

Gas flow rate is used to calculate the normalized gas conditions or measured with a thermal mass flow meter.

Therefore,

$$Q_{mi}|_t = \frac{F_{Gin}|_t p_{Gn}}{V_L(t) \cdot R.T_{Gn}} \tag{8.11}$$

where

- 'n' represents normalized values
- $T_{Gn} = 273.15$ K
- $p_{Gn} = 1.01325 \times 10^5$ N/m^2
- $R = 8.314$ Nm/(mol)(K)
- V_L is measured in l.
- F_{Gin} is measured in l/h

Standard mole volume (V_{mn}) = 22.4 l/mol = $\dfrac{RT_{Gn}}{p_{Gn}}$

The liquid related specific mass flow rate Q_{jk} (g/l h) of a component j (j stands for nitrogen or oxygen or carbon dioxide) at the point k (k = entry or exit) is given by

$$Q_{jk} = Q_{mk}(t) \cdot M_j x_{jgk}(t) \tag{8.12}$$

where

M_J is the molecular weight of component j (g/mol)

x_{jgk} is the gas phase mole fraction of component j at point k (dimensionless).

Combining inlet components

$$x_{\text{inlet } g(i)} = 1 - x_{O_2 g(t)} - x_{CO_2 g(t)}$$

This is true for entry and exit points of gas.

Therefore, $Q_{g_{\text{inert } i(t)}} = Q_{g_{\text{inert } O(t)}}$

For exhaust gas balances one can get from the above relation

$$\frac{F_{go}(t) \times p_{go}(t)}{V_L(t) \cdot R.T_{go}(t)} = Q_{mi}|_t \times \frac{1 - x_{O_{2gi}}(t) - x_{CO_{2gi}}(t)}{1 - x_{O_2 go}(t) - x_{CO_{2go}}(t)}$$

The oxygen supply rate is

$$Q_{O_2}(t) = Q_{O_2 i}(t) - Q_{O_2 O}(t) \tag{8.13}$$

So the final expression is

$$Q_{O_2}(t) = Q_{mi} M_{O_2} \frac{x_{O_{2gi}}(t)\left[1 - x_{CO_{2go}}(t)\right] - \left[x_{O_{2go}}(t)\left(1 - x_{CO_{2gi}}(t)\right)\right]}{1 - x_{O_{2go}}(t) - x_{CO_2 go}(t)}$$

where

$x_{O_{2gi}}$ is the oxygen mole fraction in air inlet = 0.209

$x_{CO_{2gi}}$ is the CO_2 mole fraction in air inlet = 0.00033. They are known for air.

The related exit gas mole fractions $x_{O_{2go}}$ and $x_{CO_{2go}}$ are measured online with the exit gas analyzer. CO_2 evolution rate is given below.

$$Q_{CO_2}(t) = Q_{CO_2 O}(t) - Q_{CO_2 i}(t)$$

Similarly

$$Q_{CO_2}(t) = Q_{mi}(t) M_{CO_2} \frac{x_{CO_{2go}}(t)\left(1 - x_{O_{2gi}}(t)\right) - x_{CO_{2gi}}(t)\left(1 - x_{O_{2go}}(t)\right)}{1 - x_{O_{2go}}(t) - x_{CO_{2go}}(t)} \tag{8.14}$$

Therefore, respiratory quotient,

$$RQ(t) = \frac{Q_{CO_2}(t) M_{O_2}}{Q_{O_2}(t) M_{CO_2}} \tag{8.15}$$

This is used for online process activities.

8.7 ADVANCED CONTROL TECHNIQUES

Some of the advanced control techniques are given below:

- adaptive control
- cascade control

- fuzzy logic control
- feed forward control
- internal model control
- model predictive control
- neural networks

8.7.1 Adaptive Control and Online Estimation

In adaptive control system, the controller parameters are adjusted automatically in a manner to compensate for the variation in characteristics of the process. Adaptive control systems find extensive application in control of bioreactors since they are suitable for non-linear non-stationary systems (Basković, 1995). Typically, bioreactors are also non- stationary since their characteristics and dynamics are susceptible to change with time.

Adaptive controllers can be classified into two categories

(a) programmed or scheduled adaptive control (Fig. 8.2).
(b) self-adaptive control (Fig. 8.3).

(a) Programmed or Scheduled Adaptive Control

This technique is useful when the process is well known and when adequate mathematical model is available (Shi *et al.*, 2004). It essentially involves a parameter adjustment mechanism, which is based on the measurement of auxiliary variables in the process. The auxiliary variable chosen for this process must be correlated to the change in process dynamics.

(b) Self Adaptive Control

In case of bioreactors most often self- adaptive control is used (Bastin and Dochain, 1990). In such a scheme, online estimation of the objective is done and the adaptation mechanism will change the process parameters in such a way that an objective function is optimized.

Some of the common objective functions are one quarter decay ratio, integral of the square error and gain and phase margins. The objective functions can also be based on cost reduction and product yield and so on.

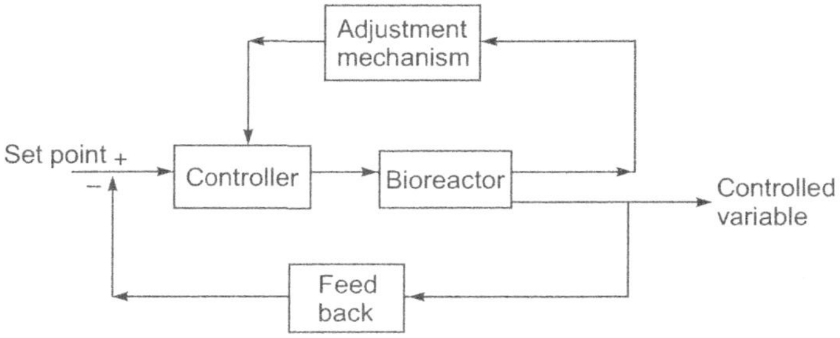

Figure 8.2 Programmed adaptive control.

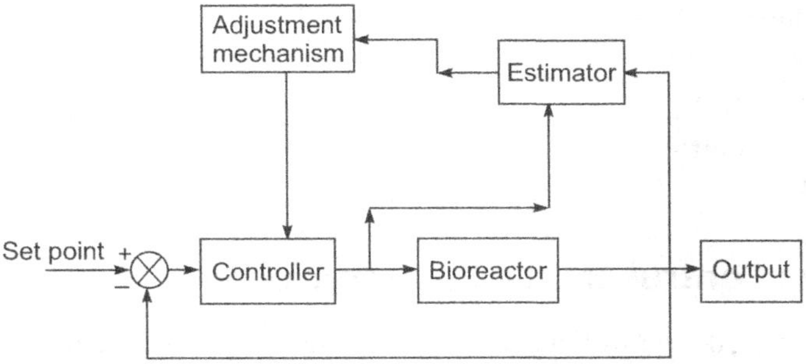

Figure 8.3 Self-tuning regulator.

The online estimation of the parameters is done based on the recursive least squares (RLS) algorithm.

An insight into the applicability of adaptive control has been shown by Rhyneir *et al.* (1992) and Mailleret *et al.* (2008). The use of adaptive control has been applied to anaerobic digestion (Rhyneir *et al.* 2004). In this particular exercise, a model with two equations of state are used which has the following forms (Equations 8.16 and 8.17).

$$Y_1 = b_1 u(t-1-d_1) + c_1 \quad (8.16)$$

$$Y_2 = b_2 u(t-1-d_2) + c_2 \quad (8.17)$$

where Y_1 and Y_2 are the methane production rate and the organic acids production rate, respectively.
u is the feed flow rate
d_1 and d_2 are the delay indices.

The performance index or the optimizing criteria is given by

PI = methane production rate – constant × (total organic acids concentration). This criterion was maximized using method of steepest descent.

The performance index is essentially a function of input feed rate and the outputs. The optimized values of these input feed rate and the output are fed to the process and online estimation of parameters was done using RLS.

In another approach (Mailleret *et al.*, 2008), a non linear adaptive algorithm was used to control a simple class of bioreactors to describe the growth of single species of microorganisms.

$$ks \xrightarrow{r} X + hP \text{, represents the reaction.} \quad (8.18)$$

where the parameters k and h are the yield coefficients. r is reaction rate.

$$\frac{dC_X}{dt} = r - \alpha D C_X$$

$$\frac{dC_s}{dt} = D(C_{S_0} - C_S) - kr$$

$$\frac{dC_P}{dt} = hr - \beta D C_P$$

where α = constant related to free cells
β = concentration of production – fixation.

The above relations are state equations,

where C_X is biomass concentration
C_S is substrate concentration
C_P is product concentration.

Assuming two hypotheses for the reaction rate

(1) reaction rate is non-negative and is a function of C_X and C_S
(2) a quantity Y_1 is measured as reaction rate.

The nonlinear adaptive algorithan is explained by Mailleret *et al.* (2004).

8.7.2 Model Predictive Control (MPC)

Model predictive control is used for handling constraints especially for multivariable systems (Dougherty and Cooper, 2003).

In model predictive control, at each time step, k, an objective function that uses the model predicted values for a range of P time steps, is formulated. This function is then optimized for M control moves. Although M control moves are optimized, only the first move is implemented. Thus the first controller response (u_k) is implemented for which the output is measured as y_k. The objective function is then optimized for the next M control moves using the model predicted values from $(k + 1)^{th}$ time step. The P prediction steps are called the prediction horizon. M control moves are called the control horizon. M and P together are the parameters of the *MPC*.

As stated earlier, the "Model predictive control" uses an objective function that has to be optimized in the process. There are several different choices for the objective function, but the most commonly used one is the least squares or the quadratic objective function.

It is of the form of Equation (8.19)

$$\phi = (r_{k+1} - y_{k+1})^2 + (r_{k+2} - y_{k+2})^2 + (r_{k+3} - y_{k+3})^2 + \nabla u^2{}_k + \nabla u^2{}_{k+1} \quad (8.19)$$

where ϕ is the cost function
y_k is the output
u_k is the first controller response

The objective function given in (8.19) has been written for a prediction horizon of 3 and a control horizon 2, which can be extended further.

8.7.3 Artificial Neural Networks (ANN)

An artificial neural network scheme uses a set of interconnected neurons that work based on a computed model (Thibault et al, 1990). The neurons are some variables of the system which depend on each other as well as on the process parameters through some complex relationship. In many ways the ANN system has been evolved from the biological neural network in the sense that the 'nodes' or the processing elements are connected together to form a complex network of nodes. The operation does not involve a clear delineation of subtasks rather the operation is parallel and functions are performed collectively. The modern approaches to ANN are based on signal processing and statistical techniques.

A typical neural network model can be a class of functions in the form of

$$F : X \rightarrow Y \tag{8.20}$$

where X and Y represent the nodes of the system.

The word 'network' in the term 'artificial neural network' arises because the function $f(x)$ is defined as a composition of other functions $g_i(x)$. This can be conveniently represented as a network structure, with arrows depicting the dependencies between variables. A widely used type of composition is the nonlinear weighted sum, where K is some predefined function like the hyperbolic tangent.

$$f(x) = K\left(\sum_i w_i g_i(x)\right) \tag{8.21}$$

There are two views to the neural network problems, viz., 1) functional view and 2) the probabilistic view.

In the functional view, the nodes X should be decomposed into a set of vector functions which are further transformed into other vector functions.

In case of the probabilistic view, the random variable X is supposed to depend on another random variable say $h(x)$. Then $h(x)$ depends on another random variable and so on.

Figure 8.4 shows the ANN dependency diagram as illustrated by the functional model. It can be seen that the elements of the different layers do not depend on each other, but they depend only on the elements in the preceding and succeeding layers.

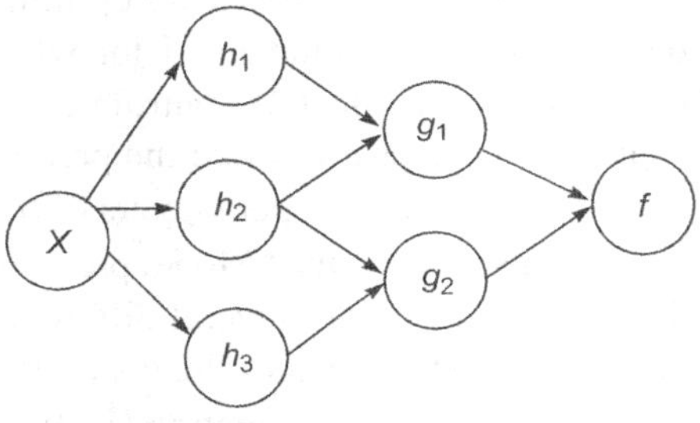

Figure 8.4 Feed forward network of ANN.

The primary advantage of neural network model is 'to learn'. That is the model is tutored to choose the optimal model from a set of models by using some optimization criterion using 'learning algorithms'. Most of the training algorithms are essentially optimization problems.

There are different kinds of neural networks. The simplest is the feed forward network which is explained earlier. It can be either a single layer network or a multi layer network. ADALINE (Adaptive Linear Neuron or later Adaptive Linear Element) is another network algorithm which uses a linear relationship between different nodes and then minimizes the mean square error (Valluvan and Natarajan, 2008). It is essentially a simple linear regression logic that is used for formulating the objective function.

There are other types of networks called the recurrent network, the self-organizing network, the cascading, and the modular neural network.

A neuro fuzzy network uses a fuzzy logic based inference system and is finding increased applications in bioreactor control (Chidambaram, 2008).

A neural network model has been developed based on the back propagation algorithm for a continuous bioreactor (Leigh, 1987). The back propagation term is most useful for feed forward networks. Using this algorithm, the network has been trained and an ANN model developed which fits the reactor response both in the open loop and in closed loop conditions. An effective PI controller has been developed for this model.

8.7.4 Internal Model Control

Internal model control (IMC) is based on the inverse of model transfer function which may not be realized at all due to the presence of positive zero or delay in process model (Fig. 8.5). Hence, the design of internal model control involves the factorization of the process model into invertible and non-invertible elements (Wu and Chang, 2007)). The inverse of the invertible part is then multiplied with an appropriate filter and is then implemented as the controller transfer function.

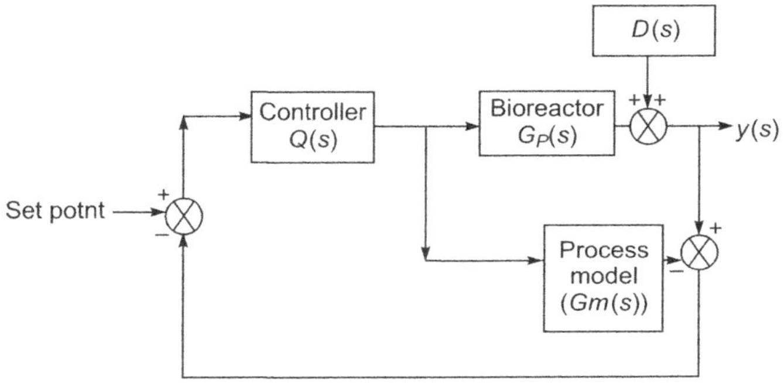

Figure 8.5 IMC.

The difference of model and plane output is used as feed back signal. This feed back structure can be reformulated to give equivalent standard PID structure. Hence, this design procedure can be used for tuning PID controller.

It is important to note that parameters of the PID controller depend on the filter tuning parameters and, therefore, the tuning parameters have to be carefully chosen in order to achieve desirable controller response. The IMC based tuning procedure is faster and more reliable than the other conventional tuning methods. The operating conditions also play a major role in selecting the parameters of PID controller, because there may be stable and unstable operating points which yield completely different responses.

(a) The IMC Based PID Procedure

In the previous section, a brief introduction is given to the IMC based PID tuning structure. The standard IMC structure described above can be reduced as follows (Fig. 8.6).

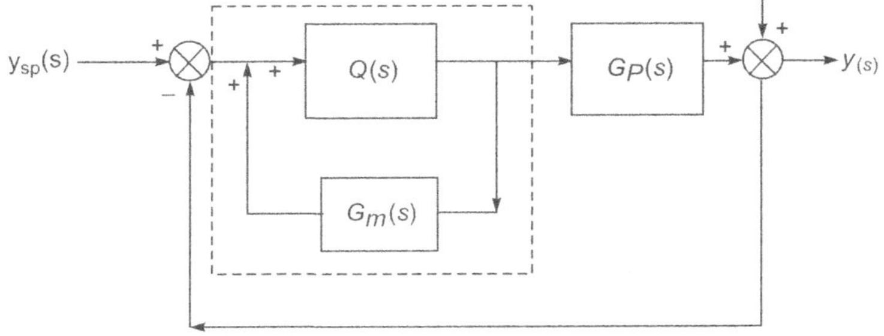

Figure 8.6 Modified IMC based PID tuning structure.

In Figure 8.6, $Q(s)$ is the ideal open loop controller and $G_P(s)$ is the transfer function representing the process and $Gm(s)$ is the transfer function representing the process model. The dashed box in Figure 8.6 represents the reduced controller that can be obtained by using Equation (8.22).

$$Q_c(s) = \frac{Q(s)}{(1 - Q(s)G_m(s))} \tag{8.22}$$

This leads to the following reduced feed back control loop (Bailey and Ollis, 1986) (Fig. 8.7).

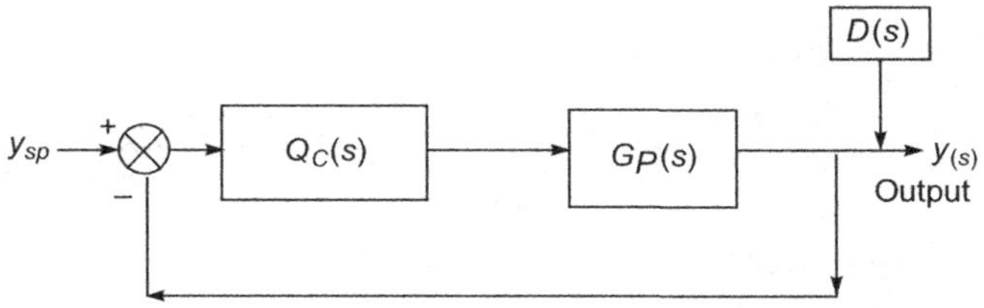

Figure 8.7 Feedback control loop.

In the above control loop, the controller transfer function ($Q_c(s)$) can be reformulated to the standard PID structure. Memory-based IMC tuning of PID controllers for nonlinear systems is described by Takao *et al.* (2006).

(b) Sequence of Steps for IMC Based PID Design

(1) Find the realized ideal open loop controller $Q(s)$ (i.e., by factorization, inversion and addition of a filter to the model transfer function).

(2) Find the equivalent feed back controller using the formula in Equation (8.22).

(3) The $Q_C(s)$ is obtained as a ratio of two polynomials. This can be expressed in terms of PID transfer functions and is given by the Equation (8.23) (Coughanowr, 1991).

$$G_{PID} = K_c \frac{(\tau_i \tau_d s^2 + \tau_i s + 1)}{\tau_i s} \tag{8.23}$$

Usually the Qc(s) can be split into two factors, one of which is a PID transfer function and the other one is usually a filter.

(4) Closed loop simulations of the resulting control loop are carried out and the value of the filter parameter is obtained as a trade off between performance and robustness.

8.7.5 Feedback Control

When we speak about process control, we normally mean feed back control. In this type of control, measurements from the process are used to find out a suitable control signal (Wang *et al.*, 1999). The online measured values are fed back to the process input. The scheme of feed back control is given in Figure 8.8.

A feed back controller responds to disturbances and, therefore, it reduces deviations from the set point. As illustrated in Figure 8.8, the process may be suitably improved by feed back control, because the controller can respond to an increased signal by suitable action.

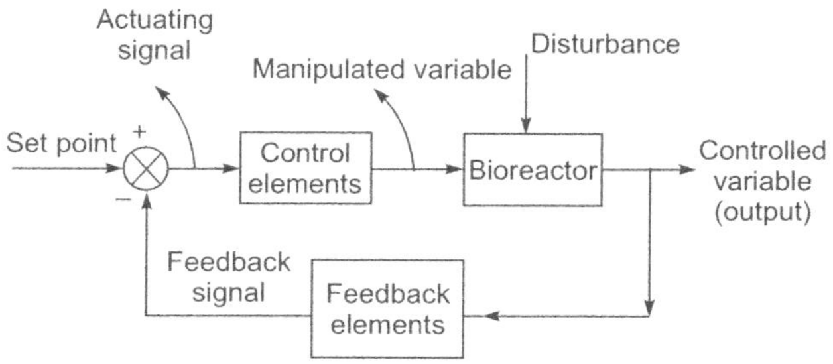

Figure 8.8 Feedback control scheme.

8.7.6 Feed Forward Control

Feed forward control can be applied without measuring the variable of interest (Jyothi and Chidambaram, 2001). One can develop the control relationship between the extent of disturbances and the correction factor. The feed forward control along with feed back scheme is shown in Figure 8.9.

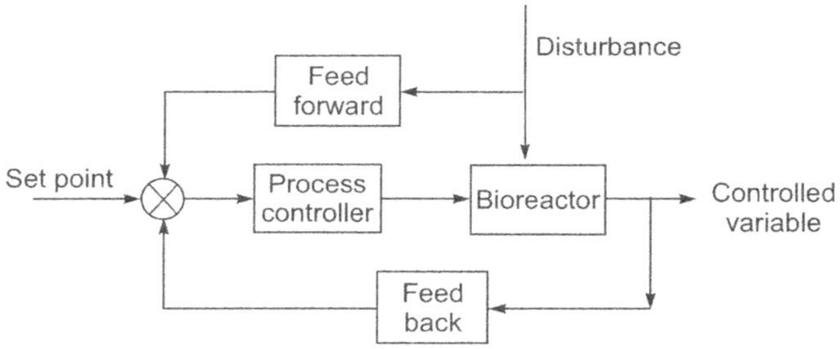

Figure 8.9 Feed forward control scheme.

It is important to realize that feed forward control is a model based control. Sometimes, it is better to combine feed forward control with feed back control in order to compensate for small deviations in the model and unpredicted disturbances.

8.7.7 Cascade or Supervisory Control

The term cascade control indicates a special set up of different control loops (Chidambaram, 2000). An outer control dictates the set point of an inner control. This set up is used to make a controller more suitable for start up periods. The use of feed forward control reduces the need for cascade control. However, cascade control can be implemented simply by means of some standard analog devices, while feed forward control usually involves a digital computer and a model of the system. Cascade control is very efficient if the process can be split up into fast and slow components. The cascade control is schematically represented by Figure 8.10.

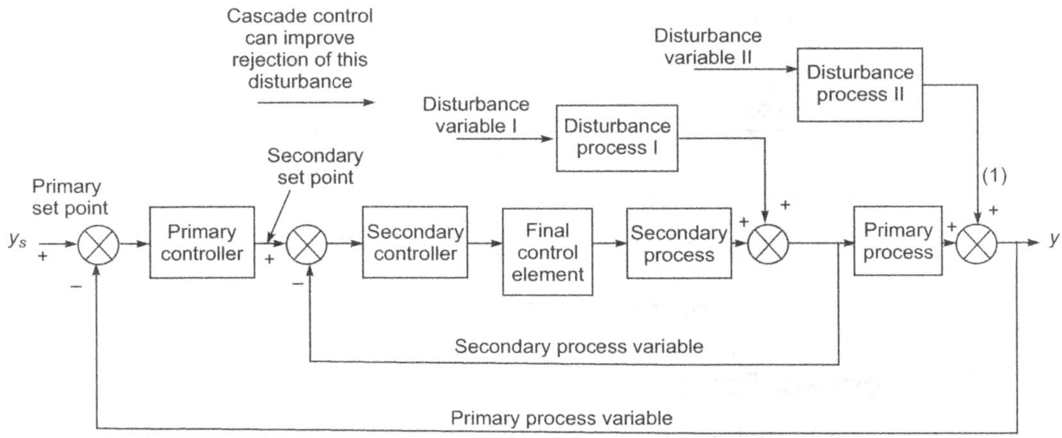

Figure 8.10 Cascade control scheme.

'(1)' indicates primary process variable.

8.8 CONSISTENCY CHECKS ON MEASUREMENTS

Even if process control is based either on direct measurements or on state estimators, in both cases the measurements should be reliable. For some processes, it may, therefore, be better to perform some measurements more than once, in order to improve the reliability on them (e.g., by using more than one sensor or analyzing duplicates). This kind of redundant information can be used to check on the measurements, based on the law of conservation for elements (or energy). Still redundancy is the key.

Let us assume the following conversion takes place in the bioreactor (only the overall conversion is stated).

Glucose + NH_3 + O_2 → biomass + water + ethanol + carbon dioxide

We can write the stoichiometry as follows:

$$r_s\, CH_2O + r_N\, NH_3 + r_O\, O_2 \rightarrow r_X\, CH_{1.8}O_{0.5}N_{0.2} + r_H\, H_2O + r_P\, CH_3O_{0.5} + r_C\, CO_2 \tag{8.24}$$

where r is the conversion rates.

r_s, r_N and r_O are negative since the compounds involved are being consumed during the processes.

Note that we have described the biomass containing the elements C, H, O, and N in the proportions of 1 : 1.8 : 0.5 : 0.2.

Since carbon can neither be created nor be destroyed during the conversion, we can write the carbon balance represented by Equation (8.25).

$$-r_s = r_X + r_P + r_C \tag{8.25}$$

Similar equations can be written for other elements. Thus, for hydrogen the balance is

$$-2r_s - 3r_N = 1.8r_X + 2r_H + 3r_P \tag{8.26}$$

and for oxygen, it is

$$-r_s - 2r_O = 0.5r_X + r_H + 0.5r_P + 2r_C \tag{8.27}$$

For nitrogen:

$$-r_N = 0.2r_X \quad \text{or} \quad 0.2\, r_X + r_N = 0 \tag{8.28}$$

In this system, we have 6 conversion rates (i.e., r_S, r_N, r_O, r_X, r_P, and r_C). We also have four equations from the elemental balances which relate the rates to each other. Thus, in principle, if we measure two of the rates, the remaining rate values can be calculated using these balance equations.

However, if more than two rates are measured, the possibility arises to check the measurement set, because the given balances may not be violated for such analysis. Then the system is said to be over determined in that situation.

Let us take an example. We will use the bioconversion system described above in which glucose is converted to biomass, ethanol and carbon dioxide.

In our system four rates have been determined.

These are, for example,

- The rate of reactant consumption, $r_S = -61 \times 10^{-2}$ C-mol l^{-1} h^{-1};
- The rate of production of biomass, $r_X = 40.6 \times 10^{-2}$ C-mol l^{-1} h^{-1};
- The rate of carbon dioxide production, $r_C = 20.4 \times 10^{-2}$ C-mol l^{-1} h^{-1};
- The rate of oxygen consumption, $r_O = -20.5 \times 10^{-2}$ C-mol l^{-1} h^{-1};

Our task is to use the oxygen consumption rate to describe if there are errors in our measurements.

We will begin by explaining two of the values. The reactant consumption rate (r_S) is given in units of moles of substrate carbon per liter per hour. Likewise, the rate of biomass production is given by moles of carbon per liter per hour. The important thing to bear in mind is that we must use the same units for each term, otherwise we may get into difficulty.

Essentially the question is to use r_S, r_X and r_C to calculate r_O. r_O is also measured in the process. If they are different, we can say that there are some errors in the stoichiometry used to describe the conversion or there is an error in one, or more, of the measured r-values.

Let us do the necessary calculations from Equations (8.25) to (8.28).

First we can calculate $r_N = -0.2\, r_X = -8.12 \times 10^{-2}$ C-mol l^{-1} h^{-1}

We can calculate r_P from $-r_S = r_X + r_P + r_S$

From which we get $r_P = 10^2$ (C-mol) l^{-1} h^{-1}

We can now calculate r_H using the equation,

$$-2r_S - 3r_N = 0.5\, r_X$$

Thus $r_H = 36.64 \times 10^{-2}$ C-mol l^{-1} h^{-1}

Finally, we can calculate r_O using Equation (8.27).

So $r_O = -18.37 \times 10^{-2}$ C-mol l^{-1} h^{-1}.

This value is different from the actual measured value of r_O (-20.5×10^{-2} C-mol l^{-1} h^{-1}).

Hence, there is some error in at least one of the measured rates (i.e., error in r_O is -2.13×10^{-2} C-mol l^{-1} h^{-1}).

One way is to measure the rates, for example, r_N. This would enable us to use a variety of combinations of values to calculate the remaining rates. In this way, we would be able to determine which measured r-values are consistent with each other. Also, we would be able to determine, which measured r-value is erroneous.

Example 8.1

IMC BASED PID DESIGN OF BIOREACTOR WITH MONOD'S KINETICS

Problem definition

The problem deals with the design of a IMC based PID controller for a bioreactor which follows Monod's kinetics. The parameters are given below.

$$\mu_{\max} = 0.53 \text{ h}^{-1}$$
$$K_s = 0.12 \text{ kg/m}^3$$
$$Y_{X/S} = 0.4$$
$$D_S = 0.3 \text{ h}^{-1}$$
$$C_{S_O} = 4.0 \text{ kg/m}^3$$

Linearization of the model

The Monod's kinetic models are discussed in the section, 'dynamics of bioreactor', the parameters are given above. The modeling equations are given in Equations (8.29) and (8.30).

$$\frac{dC_X}{dt} = (\mu - D)C_X = f_1 \tag{8.29}$$

$$\frac{dC_s}{dt} = D(C_{S0} - C_S) - \mu\frac{C_X}{Y_{X/S}} = f_2 \tag{8.30}$$

The above equations are non linear in nature and hence they have to be linearized in this case. The linearization procedure involves the Taylor function expansion of the equations and truncating all the higher order terms.

In concise form, the linearized system can be written as

$$\mathbf{X}' = \underline{\underline{\mathbf{A}}}\mathbf{XS} + \underline{\underline{\mathbf{BD}}}$$

The term **XS** is the vector of state variables $[X\ S]^T$

XS$'$ indicates the derivatives of the state variables $[X'S']^T$.

$\underline{\underline{\mathbf{A}}}$ is the Jacobian matrix of functions f_1 and f_2.

$\overline{D}$ is the dilution rate.

For the above case, the $\underline{\underline{\mathbf{A}}}$ matrix is given by

$$\underline{\underline{\mathbf{A}}} = \begin{pmatrix} 0 & \dfrac{\mu_{\max} K_S C'_X}{(\mu_{\max} + C'_S)^2} \\ \dfrac{-\mu_{\max} C'_S}{K_S + C'_S} & D_s - \dfrac{\mu_{\max} K_S C'_X}{Y_{X/s}(\mu_{\max} + C'_S)^2} \end{pmatrix}$$

$$\underline{\underline{\mathbf{B}}} = \begin{bmatrix} -C'_X \\ C_{S_0} - C'_S \end{bmatrix}$$

The superscript ' ' ' in the above equations represents the corresponding values at the steady state. The linearized set of equations is in terms of deviation of variables, i.e., deviation from steady state and can be expressed as state space equations of the form

$$\mathbf{XS}' = \underline{\underline{\mathbf{A}}}\mathbf{XS} + \underline{\underline{\mathbf{B}}}\mathbf{U}$$

$$\mathbf{Y} = \underline{\underline{\mathbf{C}}}\mathbf{XS} + \underline{\underline{\mathbf{D}}}\mathbf{U}$$

Y is the output variable.

U is the forcing function.

$\underline{\underline{\mathbf{A}}}$ and $\underline{\underline{\mathbf{B}}}$ are matrices already obtained while the $\underline{\underline{\mathbf{C}}}$ matrix is given by [1 0] and $\underline{\underline{\mathbf{D}}}$ is a null a matrix.

Steady states of the model

The steady states of the model are obtained by setting Equations (8.29) and (8.30) to zero. Substituting the values of the parameters stated in the problem definition and keeping the value of the steady state dilution rate as 0.3 hr $^{-1}$, we get two steady states for the Monod's kinetics.

First steady state is given by,

$$\mu_S = D_S = 0.3$$

$$C_S' = \frac{D_S K_s}{(\mu_m - D_S)} = 0.156522 \text{ kg/m}^3$$

$$C_X' = Y_{X/S}(C_{S_0} - C_S') = 0.0973 \text{ kg/m}^3$$

Second steady state is as follows,

$$C_X' = 0 \text{ kg/m}^3$$

$$C_S' = 4 \text{ kg/m}^3$$

The second steady state signifies a condition called wash out in the bioreactor. None of the reactant is being used up for the production of cells and they exit the reactor in the unreacted form. This typical condition occurs at high dilution rate as it will be seen through bifurcation analysis.

Bifurcation analysis and stability

Bifurcation analysis has been performed on the model in order to study the dynamics of the bioreactor at various values of the dilution rate (D) (Wang *et al.*, 1999).

A computer programme can be written to calculate the steady states and assess the stability of each steady state. The assessment of the stability is done by checking the real part of the Eigen values. Negative real part of Eigen values indicate stability. The typical bifurcation curve is shown in Figure 8.11.

Figure 8.11 shows the variation of substrate concentration with dilution rate. The plot with 'circles' indicates the unstable region while the curved line indicates the stable region. It is to be noted here that wash out steady state becomes a stable steady state for a particular dilution rate. However, reactors are hardly operated at this condition and no yield of cells is obtained. Thus high dilution rates should be avoided in the bioreactor operation using free cells. This justifies our choice of D_s as 0.3 hr $^{-1}$.

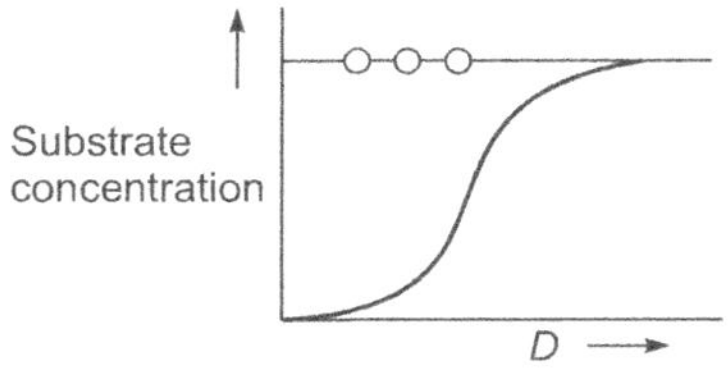

Figure 8.11 Plot of reactant concentration vs. dilution rate.

Transfer function form of the linearized model

The state space equations are given here.

$$XS' = \underline{\underline{A}}XS + \underline{\underline{B}}U$$

$$Y = \underline{\underline{C}}XS + \underline{\underline{D}}U$$

$$\underline{\underline{A}} = \begin{vmatrix} 0 & 0.1278 \\ -0.75 & -3.4968 \end{vmatrix}$$

$$\underline{\underline{B}} = \begin{vmatrix} -1.53739 \\ 3.8434 \end{vmatrix}$$

$$\underline{\underline{C}} = \begin{bmatrix} 1 \\ 0 \end{bmatrix}$$

$$\underline{\underline{D}} = [0]$$

The $\underline{\underline{A}}$ and $\underline{\underline{B}}$ matrices have been derived from the formula given earlier at steady state (8.29). The transfer function for this system can be obtained using MTALAB's ss2tf function.

The syntax of the ss2tf command is

$$T_f = \text{ss2tf (A, B, C, D)}$$

$$\frac{Y(s)}{D(s)} = \frac{(-1.537s + 0.4641)}{s^2 + 3.497s + 0.9588}$$

The above transfer function is transformed into

$$\frac{Y(s)}{D(s)} = \frac{[-1.537(s + 0.3)]}{(s + 3.1977)(s + 0.2988)}$$

It is observed that a zero of the numerator and the pole of the denominator are approximately equal and can be cancelled. This is called pole zero cancellation.

Thus the final transfer function becomes

$$\frac{Y(s)}{D(s)} = \frac{-1.537}{(s + 3.1977)} = Gp(s) \tag{8.31}$$

The resulting system is first order stable system with negative gain.

Design of IMC based controller

The ideal open loop controller $Q(s)$ is given by

$$Q(s) = \frac{Gm(s)^{-1}}{(\lambda s + 1)} \tag{8.32}$$

We assume that the model used in this design is nearly accurate, i.e., there is no model plant mismatch. Thus

$$Q(s) = \frac{Gp(s)^{-1}}{(\lambda s + 1)} = \frac{-(s + 3.1977)}{1.537(\lambda s + 1)} \tag{8.33}$$

where λ is the filter parameter that has been added to the $Q(s)$ to make it physically realizable quantity.

The IMC based PID controller has been derived in Equation (8.22) and is given by Equation (8.34).

$$Q_c(s) = \frac{Q(s)}{[1 - Q(s)Gm(s)]} \tag{8.34}$$

Substituting the values of $Q(s)$ and $Gm(s)$ from Equations (8.32) and (8.31),
We get the controller transfer function as

$$Q_c(s) = \frac{-(s+3.1977)}{\lambda s^* - 1.537} \tag{8.35}$$

Now the controller transfer function is to be expressed in a PID form cascaded with or without a filter. From Equation (8.35), we get Equation (8.36).

$$Q_c(s) = \left(\frac{-1}{1.537\lambda}\right)\left(\frac{-2.080}{\lambda s}\right) \tag{8.36}$$

The above equation is for the *PI* controller without any filter element. To clarify further, Equation (8.36) is compared with the usual *PI* controller which is given by Equation (8.37).

$$Q_{PI}(s) = K_c\left(1 + \frac{1}{\tau_1 s}\right) \tag{8.37}$$

Thus the value of gain and integrator time of the *PI* controller are obtained as functions of the filter parameter λ as follows

$$K_c = \frac{-1}{1.537\lambda}$$

$$\tau_1 = 0.3127$$

In order to find the tuning parameter λ, simulations have to be carried out and it is a tedious procedure to analyze the robustness and performance. This yields good results in the case where the model matches plant data. In the current example, there is no such mismatch and the filter time constant can be chosen to be approximately one-third of the dominant time constant.

The advantage of this method of using PID controller is that the number of tuning parameters is significantly reduced and the controller tuning is not an empirical-hit and trial tuning method, but it is based on the model of the process.

The method also suffers from some inadequacies. It is suitably applied to linear systems and stationary processes. Many bioreactors cannot be treated in this simple manner and have complicated kinetics and dynamic behavior. In such cases, the virtues of MPC and adaptive control are put to effect to obtain better performance.

8.9 ADAPTIVE ONLINE OPTIMIZING CONTROL OF BIOREACTOR SYSTEM

Many fermentation industries demand the control of fermentation processes at their optimal states accurately to reduce their production cost, to increase the yield, and to maintain the quality of the metabolic products.

To make it real, one must take into account the lack of accurate mathematical models which describe the cell growth and the metabolite production, the transient nonlinear nature of the system and also the lack of reliable on-line sensors which detects variables like concentrations of the cells and the metabolites.

To cope with the above problems, people have proposed many solutions (Shi *et al.*, 2004). If we use computer, it provides fast and efficient data acquisition and is able to monitor and control fermentation conditions. The objective of on-line optimizing control of bioreactors is to keep tracking the optimal state in enzymatic deactivation and environmental changes. Here, we will discuss efficient on-line optimization control strategies for bioreactor systems.

8.9.1 Online Optimization Control for Bioreactor

The dynamics of bioreactor system is nonlinear and time dependent (Bastin and Dochain, 1990). It shows slow behavior. In online optimizer, the optimal steady state values are calculated online using steady state model and information from process and compensator. This information is fed as set point to the control loop. For the system to follow optimal steady state path in the presence of disturbances and measurement noise, a control system is needed which tracks the process to optimal set point value. Due to the unavailability of online measurement of many of the variables, it has to be estimated using dynamic model of the process using the information of measured output and controlled input.

Figure 8.12 shows the block diagram of online optimizing control. The online optimizer has the following component.

1. Fermentor—This is the process to be controlled.
2. Dynamic model identifier—It is also called observer, which is used for the estimation of unmeasured states.
3. Compensator—It is the controller which implements the optimal input value to the fermentor.

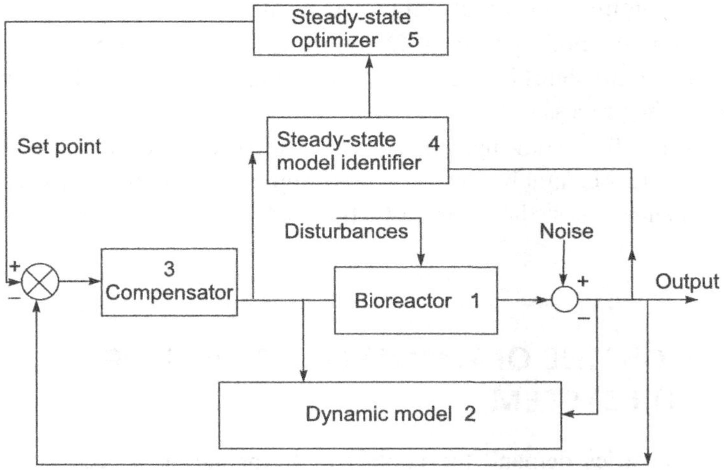

Figure 8.12 Block diagram for on-line optimization control.

4. Steady state model identifier—Steady state model of the process is used to calculate various steady state values.
5. Steady state optimizer—It calculates the optimal steady state points and feed it as the set point to the control loop.

For a continuous bioreactor system, the solution for on-line optimization is given by Shi *et al.*,(1989).

Example 8.2

Cell recycle system for lactic acid fermentation.

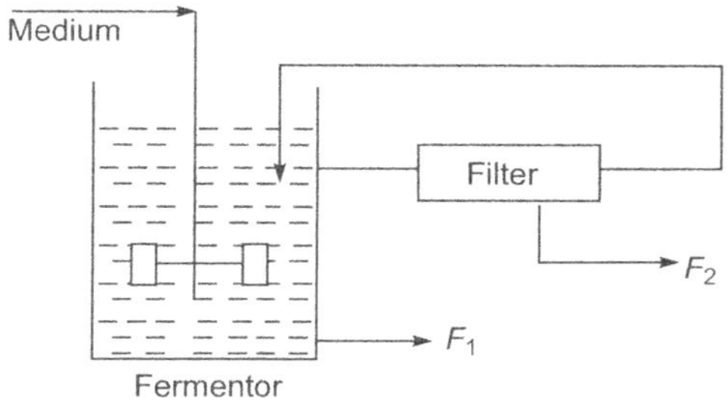

Figure 8.13 Block diagram for cell recycle system for lactic acid fermentation.

To attain the maximum production rate of lactic acid during fermentation the development of efficient on-line measurement system with powerful computer is essential.

The mathematical equations for Figure 8.13 are (8.38a)–(8.38c).

$$\frac{dC_X}{dt} = \mu(C_S, C_P)C_X - D_1C_X \tag{8.38 a}$$

$$\frac{dC_s}{dt} = (D_1 + D_2)(C_{s_0} - C_s) - \frac{\mu(C_s, C_p)}{Y_s}C_X \tag{8.38 b}$$

$$\frac{dC_P}{dt} = -(D_1 + D_2)C_P + (\alpha\mu(C_S, C_P) + \beta)C_X \tag{8.38 c}$$

where

C_X is the cell mass concentration.
C_S is the reactant concentration.
C_{S_o} is the initial reactant (substrate) concentration.
C_P is the product concentration.
D_1 and D_2 are dilution rates $\frac{F_1}{V}$ and $\frac{F_2}{V}$, respectively.
F_1 is the flow rate of output of the fermentor.
F_2 is the flow rate of output from the filter.
V is the volume of the fermentor.

All parameters in the above example should be found and then we should do on-line optimization to maximize the lactic acid productivity. In this regard, the performance index for the bioreactor system must be defined accurately. Shi *et al.* (2004) expresses this parameter as J defined by Equation (8.39).

$$J = J(y, x, p) \tag{8.39}$$

where y stands for output,
x stands for input, and
p stands for parameter vectors.

J must be optimized by changing the inputs (x) and relationship of the non-linear process $y = g(x, p)$.

Since those are not known accurately, proper algorithm must be developed for this analysis. To understand this problem, Shi *et al.*, (1989) has suggested the solution.

EXERCISES

8.1 Define feed forward control mechanism in biological system.

8.2 Do you believe PID can be successfully applied to biochemical system? If it is not possible, give the reasons for the same.

8.3 Suppose the specific growth rate is changed by (a) 10%, (b) 20% (c) 30% at 60 h, 100 h and 120 h, respectively. Where do you obtain fastest adaptation to a new optimal steady state? Assume the growth model is expressed by

$$\frac{dC_X}{dt} = \phi_1(C_X, C_S, D) = \mu(C_S)\, C_X - DC_X$$

$$\frac{dC_S}{dt} = \phi_2(C_X, C_S, D)$$

$$= -\frac{\mu(C_S)}{Y} C_X + D(C_{S_o} - C_S)$$

$$\mu(C_S) = \frac{\mu_m C_S}{K_S + C_S}$$

The model parameters are μ_m, K_S and Y. Suggest your solution by nonlinear programming technique.

REFERENCES

Bailey JE and Ollis DF (Eds.) (1986) *Biochemical Engineering Fundamentals*, 2nd edn., McGraw-Hill Inc., Chemical Engineering Series, New York, p. 658.

Bastin G and Dochain D (Eds.) (1990) *On-line Estimation and Adaptive Control of Bioreactors*, Elsevier Science Publications, Amsterdam.

Bosković JD (1995) "Stable adaptive control of a class of first-order nonlinearly parametrized plants", *IEEE Transactions on Automatic Control*, **40**, 347-350.

Chidambaram M (Ed.) (2008) *Computer Control of Processes*, Narosa Book Distributors Pvt. Ltd., India.

Coughanowr DR and Koppel LB (Eds.) (1965) *Process Systems Analysis and Control*, McGraw-Hill, New York.

Dougherty D and Cooper D (2003) "A practical multiple model adaptive strategy for multivariable model predictive control", *Control Engineering Practice*, **11**, 649-664.

Erickson LE and Stephanopoulos G (1985) "Biological Reactors", (Chapter 13) in *Chemical Reaction and Reactor Engineering*, Carberry JJ and Varma A (eds.), Marcel Dekker Inc., New York.

Jyothi SN and Chidambaram M (2001) "Non linear feed forward control of bioreactors with input multiplicities", *Bioprocess and Biosystems Engineering*, **24**, 123-129.

Kuznetsov YA (2004) *Elements of Applied Bifurcation Theory*, Springer-Verlag, New York.

Lee JM (Ed.) (1992) *Biochemical Engineering*, Prentice-Hall, Inc., Englewood Cliffs, New Jersey.

Leigh JR (Ed.) (1987) "Modelling and control of fermentation processes", *IEE Control Engineering* Series 31, Peter Peregrinus Ltd, London, United Kingdom.

Mailleret L (2004) "Nonlinear adaptive control for bioreactor with unknown kinetics", *Automatica*, **40**, 1379-1385.

Mailleret L, Gouzé J-L and Bernard O (2008) "Global stabilization of a class of particularly known non-negative systems", *Automatica*, **44**, 2128-2134.

O'Dwer A (Ed) (2000) *Handbook of PI and PID Controller Tuning Rules*, Taylor & Francis.

Shi Z, Shimizu K, Watanabe N and Kodayashi T (2004) "Adaptive on-line optimizing control of bioreactor systems", *Biotechnology and Bioengineering*, **33**, 999-1009.

Smith HL (1995) "Monotone dynamical systems, an introduction to the theory of competitive and cooperative systems". In: *Mathematical Surveys and Monographs*, American Mathematical Society.

Thibault J, Berusegem VV, and Cheruy A (1990) "On-line prediction of fermentation variables using neural networks", *Applied Microbiology and Biotechnology*, **37**, 756-761.

Valluvan KR and Natarajan AM (2008) "Implementation of ADALINE algorithm on a FPGA for computation of total harmonic distortion of load current", *ICGST-ACSE Journal*, **8**, 35-41.

Wang NS and Stephanopulos G (1974) "Computer applications to fermentation processes", *CRC Critical Reviews in Biotechnology*, vol. 2, p1, CRC Presss Inc.

Wang H-H, Krstic M and Bastin G (1999) "Optimizing bioreactors by extremum seeking", *International Journal of Adaptive Control and Signal Processing*, **13**, 651-669.

Wernan WG and Wilke CR (1973) "New method for evaluation of dissolved oxygen response for k_La determination", *Biotechnology and Bioengineering*, **15**, 571.

Wu W and Chang H-Y (2007) "Output regulation of self-oscillating biosystems: Model-based proportional-integral/proportional-integral-derivative (PI/PID) control approaches", *Industrial Engineering Chemistry & Research*, **46**, 4282-4288.

APPENDIX 8

COMMON TERMINOLOGIES FOR CONTROL OF BIOREACTORS

Set Point

Set point is the target value that an automatic control system, for example, a PID controller will aim to reach. For example, a particular temperature can be a set point in a boiler control system.

Offset

Offset is a sustained deviation of the process variable from the set point. In the temperature control of a bioreactor, if the system is holding the process fluid at 30.5°C consistently, even though the set point is 30°C, then an offset of 0.5°C exists.

Closed Control Loop

A closed loop exists where a process variable is measured, compared to a set point and action is taken to correct any deviation from the set point.

Open Loop Control

An open control loop exists where the process variable is not compared and action taken is not in response to feed back on the condition of the process variable, but instead without regard to process variables. For example, a water valve may be opened to add cooling water to a process to prevent the process fluid from getting too hot, based on a preset time interval, regardless of the actual temperature of the process fluid.

Transducer

A transducer is a device that translates a mechanical signal to an electrical signal. For example, inside a capacitance pressure device, a transducer converts changes in pressure into a proportional change in capacitance.

Over Shoot

When a signal from a measured output deviates from the set point, this deviation will result in a modification of the input signal by the controller. The effect of this, however, may not be apparent in the system output before some time has elapsed (due to system dynamics). Therefore, the deviation may still grow for some time, causing the controller to modify the process input even more. As the result becomes apparent it may become clear that the total modification on the input is very strong. A deviation to the other side may be the result. This common phenomenon is known as overshoot.

Delay Time

A period in which the system does not respond to a change in the input signal, is delay time.

RQ Control

Frequently the state of a culture is measured by the respiratory quotient (RQ) of the culture. A respiratory quotient is determined by the ratio of CO_2 generated to oxygen consumed. The RQ can be derived on-line by calculating it from CO_2 and O_2 measurements in the exhaust gases. Since RQ can be determined directly on-line, RQ control can be operated.

State Estimators

State estimators (or observers of filters as they are sometimes described) use a model to calculate variables based on the online measurement of one (or more) or the other variables.

SISO

In biotechnological processes, there are many inputs, viz., oxygen supply rates, stirrer speed, flow rate of cooling water, temperature of cooling water, etc. In many instances, however, a single input and signal output are considered. Usually, the most logical inputs are chosen. For example temperature may be controlled by the flow rate of cooling water. Such systems in which only and one out put is considered is often referred as SISO (single input single output) systems.

MIMO

If the different control loops cannot be studied separately or if the number of inputs do not equal to number of outputs, then we have Multiple Inputs Multiple Output (MIMO)

Drift

If the output signal slowly changes independent of the measured property, this is defined as drift.

On Line Measurement

In order to be able to perform automatic process control, direct process information in the form of measurements is essential. On line measurements are performed automatically and the results are directly available for the control. Online measurements are usually regular and frequent (or even continuous) with time.

Off Line Measurement

Off line measurements require human interface, viz., for taking samples, performing a (partly) manual analysis and/ or feeding the result into the controller. Usually they are much less frequent and are irregular.

Chapter 9

Case Studies

OBJECTIVE

- To highlight a few important bioreactor designs.

9.1 INTRODUCTION

Other chapters in this book describe the general theory of bioreactor analysis and design. It is not possible to highlight some specific information in those chapters. This chapter describes some of those specific information while discussing a few important reactors, viz.,

- Packed bed bioreactors
- Airlift bioreactors
- Hollow fiber bioreactors
- Plant cell bioreactors
- Animal cell bioreactors
- Bioreactor for solid state fermentation

9.2 DESIGN OF PACKED BED BIOREACTOR

Packed bed bioreactors are applied to various fields of biotechnological processes, viz., immobilized enzyme, immobilized cells, waste management/treatment, etc. We restrict our discussion to immobilized enzyme and cells systems.

9.2.1 Design of a Packed Bed Reactor for a Bio-Film Growth on Support System

Consider a differential section of a packed column (Fig. 9.1). Mass balance on the rate-limiting substrate over a differential element is described by Equation (9.1).

$$-FdC_{So} = N_s aAdZ \tag{9.1}$$

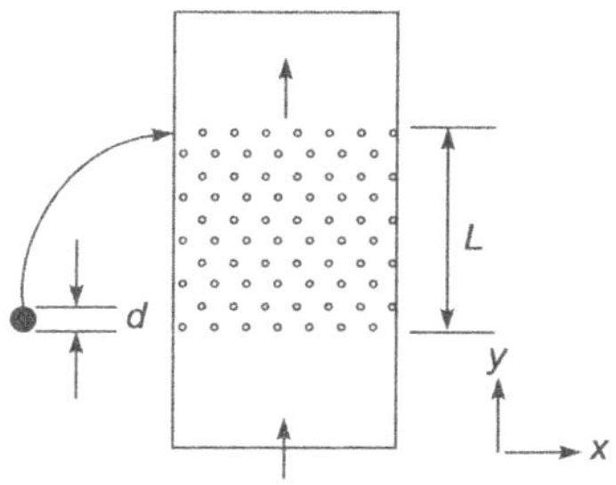

Figure 9.1 Packed bed reactor.

where

C_{S0} is the bulk liquid phase substrate concentration (kg reactant/m^3).
F is the liquid nutrient flow rate (m^3/h).
N_s is the flux of substrate into the biofilm (kg/m^{2}h)
a is the biofilm or support particle surface area per unit reactor volume (m^2/m^3).
dZ is the differential height of an element of the column (m).
A is the cross sectional area of the reactor (m^2)

Using the kinetic equation of an immobilized biocatalyst,

$$-F\frac{dC_{S_0}}{dZ} = \eta\frac{R_m C_{S_0}}{K_s + C_{S_0}} LaA \tag{9.2}$$

where η is the effectiveness factor. Upon integration it yields,

$$K_S \ln\frac{C_{S_{0i}}}{C_{S_0}} + (C_{S_{0i}} - C_{S_0}) = \frac{\eta R_m LaA}{F} h \tag{9.3}$$

where C_{S0i} is the inlet bulk substrate concentration (kg/m^3).
R_m is the maximum rate of reaction (kg/m^{3}h).
L is the bio-film thickness or the characteristic length of support material = $(V/A)_p$ (m)
'h' is the height of the packed bed bioreactor (m).
C_{S0} is unknown in the above equation. We can evaluate it from Equation (9.3).

9.2.2 Specific Design

Monod's model is used to describe the biomass growth kinetics. The kinetic model does not consider substrate inhibition effects. Three phases are considered in the system, viz., liquid, bio-film, and solid. Mass balance equations are derived based on the following assumptions.

- The flow pattern of the bulk liquid is plug flow thus avoiding back mixing.
- The bed is composed of spherical packings of uniform radius.

- The aggregates are non-porous, homogeneous, and isotropic. The adsorption process and the biomass formation into it may be assumed negligible.
- There is no accumulation at the liquid-bio-film interface.
- Bio-film is formed only on the external surface of the packing material. This may be considered as a surface phenomenon.
- The system is free from any kind of inhibition.

Consider a small section of packed tower of length 'dy' (cf. Fig. 9.1). Number of packing material in this section can be written in the form of Equation (9.4).

N = (total solid volume/volume occupied by one particle) × (volume of small section/total volume of tower).

Thus,

$$N = \frac{3(1-\varepsilon)\,Ady}{4\pi R_p^3} \tag{9.4}$$

where

ε is the porosity of the packing.
A is the cross-sectional area of the reactor.
R_P is the particle radius.

Also

$$\frac{1}{K_b} = \frac{1}{k_b} + \frac{H}{k_L} \tag{9.5}$$

where K_b and k_b are overall and liquid phase mass transfer coefficient of substrate in the bio-film, respectively. This is a similar expression given in Chapter 7. H is Henry's constant defined by Equation (9.12).

(a) Mass Balance for the Substrate in Liquid Phase Neglecting the Diffusive Transport in Liquid Phase

$$U_L A C_{S_L}|_y - U_L A C_{S_L}|_{y+dy} - K_L(C_{S_L} - C_{Sb})\, NA_b = Ady\varepsilon\left(\frac{\partial C_{S_L}}{\partial t}\right) \tag{9.6}$$

$$\left(\frac{U_L A C_{S_L}\big|_y - U_L A C_{S_L}\big|_{y+dy}}{Ady\varepsilon}\right) - \left(\frac{K_L(C_{S_L} - C_{S_b})NA_b}{Ady\varepsilon}\right) = \left(\frac{\partial C_{S_L}}{\partial t}\right)$$

$$-U_L\frac{\partial C_{S_L}}{\partial y} - \varepsilon\frac{\partial C_{S_L}}{\partial t} = \frac{K_L\left(C_{S_L} - C_{S_b}\right)A_b}{Ady}\left(\frac{3(1-\varepsilon)Ady}{4\pi R_p^3}\right) \tag{9.7}$$

$$\frac{1}{K_b} = \frac{1}{k_b} + \frac{H}{k_L}$$

where

U_L is superficial liquid velocity (length / time)
V_b is The bead volume (length^3)
C_{S_L} is reactant concentration in liquid (mass / volume)

C_{S_b} is reactant concentration in bio-film phase (mass / volume)
K_L is liquid phase mass transfer coefficient (length / time)
y is position along the length of the reactor (length)
ε is porosity of the bed (dimensionless)
A is cross-sectional area of the tower (length^2)
A_b is surface area of the bio-film (length^2)
Initial condition: at $t = 0$, $C_{S_L} = 0$ for all y
Boundary condition: at $y = 0$, $C_{S_L} = C_{S_{L0}}$

The above equation can be solved for $-U_L \dfrac{\partial C_{S_L}}{\partial y}$ under steady state.

(b) Mass Balance for the Substrate in Bio-film Phase Considering Lumped Parameter Model in Bio-film Phase

$$[K_b(C_{S_b} - C_{S_{b0}}) \times N \times A_b] - rV_b \times N = V_b \times N \times \left(\frac{\partial C_{S_b}}{\partial t}\right) \tag{9.8}$$

$$[K_b(C_{S_b} - C_{S_{b0}}) \times A_b] - rV_b = V_b \times \left(\frac{\partial C_{S_b}}{\partial t}\right)$$

$$\left(\frac{\partial C_{S_b}}{\partial t}\right) = \frac{[K_b(C_{S_b} - C_{S_{b0}}) \times A_b]}{V_b} - r$$

$$r = \frac{\mu C_{S_b} C_{X_b}}{Y_{x/s}(C_{S_b} + K_b)} \tag{9.9}$$

$$\frac{\partial C_{S_b}}{\partial t} = \frac{3k_b k_L (C_{S_L} - C_{S_b})}{(R_p + T_b)(k_b H + k_L)} - \frac{\mu C_{S_b} C_{X_b}}{Y_{x/s}(C_{S_b} + K_b)} \tag{9.10}$$

where T_b is thickness of bio-film on the particle of radius (R_P). $A_b = 4\pi (R_P + T_b)^2$ and $V_b = \frac{4}{3}\pi (R_P + T_b)^3$

Under steady state conditions, the above equation can be converted into

$$\frac{3k_b k_L (C_{S_L} - C_{S_b})}{(R_p + T_b)(K_b H + k_L)} - \frac{\mu C_{S_b} C_{X_b}}{Y_{x/s}(C_{S_b} + K_b)} = 0 \tag{9.11}$$

With initial condition: at $t = 0$, $C_{S_b} = 0$ for all y,
Henry's constant is described by similar Equation (9.12) mentioned by Chisti (1989)

$$\text{Henry's constant, } H = \left(\frac{C_{S_{b0}}}{C_{S_b}}\right) \tag{9.12}$$

The two partial differential equations obtained from Equation (9.7) at steady state and Equation (9.11) can be solved by numerical techniques.

9.2.3 Design of Packed Bed Bioreactor Packed with Immobilized Whole Cell Catalysts

Rational reactor design requires knowledge of reaction kinetics and flow pattern in immobilized catalytic bed reactor as well as diffusion of substrate through the immobilized support. Hence, analyzing diffusion affects kinetics of substrate uptake and flow pattern in the vessel from the strategy of modeling process.

(a) Effectiveness Factor

Effectiveness factor (η) is the ratio of rate of substrate uptake in the presence of diffusion and the rate of uptake of substrate under same conditions without mass transfer effects (having no diffusion resistance).

The mathematical expression of effectiveness factor is described by considering the following assumptions.

- The concentration of substrate at the surface of the immobilized particle can be taken as equal to the bulk substrate concentration.
- No transport of reactant occurs by means other than radial diffusion. This implies diffusion occurs symmetrically from all directions. The concentration profile is flat at the center.
- Uniform distribution of active reaction center is through out the immobilized particle.
- Fick's law is valid.
- The diffusion rate is equal to the reaction rate.

We consider spherical catalyst particles.

Consider the flow of substrate into a spherical particle at steady state.

Rate of diffusion of substrate into the catalyst at $r = R$ = rate of substrate consumption by cell in the catalyst.

$$= (4\pi R^2)\, D_S \left(\frac{dC_s}{dr}\right)_{r=R}$$

Rate of substrate consumption in the absence of diffusion will occur at $C_s = C_{sb}$

$$= \left(\frac{4}{3}\pi R^3\right) r_S\,(C_{S_b})$$

So, effectiveness factor (η) defined by Melick *et al.* (1987) is described by Equation (9.13).

$$\eta = \frac{(4\pi R^2) D_S \left(\dfrac{dC_s}{dr}\right)_{r=R}}{\left(\dfrac{4}{3}\pi R^3\right) r_S (C_{S_b})} \tag{9.13}$$

where C_s is concentration of reactant (mass/unit volume)
D_S is diffusivity of reactant in matrix
r_s is rate of reactant uptake by the cells at pH of interest.
C_{S_b} is concentration of reactant in bulk conditions.

Expressing the effectiveness factor in terms of the following dimensionless variables

$$S = \frac{C_S}{C_{S_b}} \tag{9.14}$$

$$\bar{x} = \frac{r}{L} \tag{9.15}$$

$$r'_S(S) = \frac{r_S(C_S)}{r_S(C_{S_b})} \tag{9.16}$$

L = characteristic length of the bead

$$= \text{volume/ area} = \frac{\frac{4}{3}\pi R^3}{4\pi R^2}$$

$$= \frac{R}{3} \text{ for a sphere.}$$

Substituting these variables in Equation (9.13) we get,

$$\eta = \frac{(4\pi R^2) D_S \frac{C_{S_b}}{L}\left(\frac{dS}{d\bar{x}}\right)_{\bar{x}=3}}{\left(\frac{4}{3}\pi R^3\right) r_s(C_{S_b})}$$

$$= \frac{D_S C_{S_b}}{\left(\frac{R}{3}\right)^2 r_S(C_{S_b})}\left(\frac{dS}{d\bar{x}}\right)_{\bar{x}=3}$$

$$\eta = \frac{1}{\Phi^2}\left(\frac{dS}{d\bar{x}}\right)_{\bar{x}=3} \tag{9.17}$$

where Φ is the Thiele modulus, defined by $\Phi^2 = \dfrac{\left(\frac{R}{3}\right)^2 r_S(C_{S_b})}{D_S C_{S_b}}$ (9.18)

Φ is completely specified by bead surface conditions. So, the evaluation of effectiveness factor now requires the determination of dimensionless substrate flux, $\left(\frac{dS}{d\bar{x}}\right)_{\bar{x}=3}$. Continuity equation for bead is written for this determination.

Rate of input – Rate of output – Rate of consumption = Rate of accumulation

At steady state, rate of accumulation = 0

Therefore,

Input rate – Output rate = Rate of consumption

$$\left((\pi r^2) D_S\left(\frac{dC_S}{dx}\right)_{x=x+dx}\right) - \left((\pi r^2) D_S\left(\frac{dC_S}{dx}\right)_{x=x}\right) = (2\pi r\Delta x) r_s(C_s) \tag{9.19}$$

Above balance is written for flat plate geometry instead of spherical geometry. This modification greatly simplifies the continuity equation and it does not appreciably change the numerical value of the effectiveness factor.

We know that,

$$\lim \Delta x \to 0 \frac{\left(\frac{dC_S}{dx}\right)_{x=x+dx} - \left(\frac{dC_S}{dx}\right)_{x=x}}{\Delta x} = \frac{d^2C_S}{dx^2} \tag{9.20}$$

So, the continuity equation in dimensionless form becomes

$$\frac{d^2S}{d\bar{x}^2} - \Phi^2 r_S(S) = 0 \tag{9.21}$$

By applying variable transformation suggested by Bischoff (1965), gives Equation (9.22).

$$\frac{dS}{d\bar{x}} = \Phi\left[2\int_{S_i}^{S} r_S'(S)dS\right]^{1/2} \tag{9.22}$$

where S_i is the dimensionless substrate concentration at the center of the bead.

Substituting Equation (9.22) in Equation (9.17), η is expressed by Equation (9.23).

$$\eta = \frac{2}{\Phi}\left[\int_{S_i}^{1} r_S'(S)dS\right]^{1/2} \tag{9.23}$$

Thiele modulus is at substrate conditions. One can measure it. The remaining obstacle to find η is the evaluation of S_i, the dimensionless substrate concentration at the center of the bed.

Integrating continuity equation twice and solving for Φ gives Equation (9.24). The solution is reported by Melick *et al.* (1987).

$$\Phi = \frac{1}{3\sqrt{2}}\int_{S_i}^{1} \frac{dS}{\left[\int_{S_i}^{S} r'(S)dS\right]^{\frac{1}{2}}} = \frac{R}{3}\left[\frac{r_S(C_{S_b})}{D_S C_{S_b}}\right]^{\frac{1}{2}} \tag{9.24}$$

We have to assume S_i to perform the integration in Equation (9.24) and check for convergence by computing ϕ at the substrate conditions. To evaluate the outside integral, we need to evaluate the inside integral first. In order to evaluate this integral, the form of rate equation chosen, consists of the classical Monod's equation, modified for product inhibition as proposed by Levenspiel (1980). That is

$$-r_S = \frac{\mu C_X C_S \rho_S}{Y_{X/S}(K_S + C_S)}\left(1 - \frac{C_P}{C_p^*}\right)^n \tag{9.25}$$

C_S = reactant concentration

C_X = cell concentration, g cell/g matrix or support material

ρ_S = matrix density, g matrix/ml.

K_S, n and μ = parameters

$Y_{X/S}$ = cell mass yield coefficient.

C_P = product concentration varies with the bead and needs to be expressed in terms of C_S.

C_P^* = product concentration at which cell metabolism ceases.

n = empirical constant

From Equation (9.22), after substitution of r_s following integration, one can calculate the design length of the reactor.

(b) Incorporation of Flow Pattern

Due to the lack of turbulent flow in the column it is not reasonable to assume perfect "plug flow". To account deviations from plug flow, the generalized dispersion model is expressed by Equation (9.26). This is analogous to the equation described by Levenspiel (1980).

$$\left(\frac{D_S}{UL}\right)\frac{d^2C_S}{dz^2}-\frac{dC_S}{dz}-\tau\varepsilon(1-\varepsilon_c)(-r_S)=0 \tag{9.26}$$

z = dimensionless length.

ε_c = column void fraction.

$\left(\frac{D_S}{UL}\right)$ = dispersion number.

τ = residence time.

Equation (9.26) can be solved by Runge-Kutta method.

(c) Determination of Diffusivity

To evaluate diffusivities of reactant and product, the unsteady state diffusion equation is given in Equation (9.27). This equation is described by Crank (1975).

$$\frac{C_S}{C_{S_i}}=\frac{\alpha}{\alpha+1}+\sum_{n=1}^{\infty}\left[\frac{6(1+\alpha)\exp(-D_S/q_n tR^2)}{9+9\alpha+q_n^2\alpha^2}\right] \tag{9.27}$$

α is volume of beads/volume of solution

C_S is concentration of compound of interest at time τ.

C_{S_i} is initial concentration

q_n are the nonzero positive roots of $\tan q_n=\dfrac{3q_n}{3+q_n}$ (9.28)

Example 9.1

The bioconversion of a reactant to product is carried out in a packed bed, immobilized cell bioreactor containing cells entrapped in Ca-alginate beads. The rate limiting reactant is glucose and its concentration in the feed bulk liquid is $C_{S_{oi}}$ = 5 g/l. The reactant flow rate is F = 2 l/min. The particle size of Ca-alginate beads is D_P = 0.5 cm. The rate constants for this conversion is r_m = 100 mg reactant/ (cm^3) (h), for the following rate expression, $-r_S = \dfrac{r_m S}{K_S + S}$

The surface area of the alginate beads of the bed per unit reactor volume is 'a' = 25 cm^2/cm^3 and the cross-sectional area of the bead is, A = 100 cm^2. Assuming Monod's kinetics, determine the required bed height for 80% conversion of reactant to product at the exit stream. Other relevant data are K_S = 10 mg S/cm^3, and $D_S = 10^{-6}$ cm^2/s, $\dfrac{V_p}{A_p} = \dfrac{0.5\ \text{cm}^3}{6\ \text{cm}^2}$, and effectiveness factor = 0.3.

Solution

$$-F\frac{dC_{S_0}}{dZ} = \eta\frac{R_m C_{S_0}}{K_S + C_{S_0}} LaA$$

Upon integration,

$$K_S \ln\frac{C_{S_{0i}}}{C_{S_0}} + (C_{S_{0i}} - C_{S_0}) = \frac{\eta R_m LaA}{F} H$$

where

$C_{S_{oi}}$ is the inlet bulk substrate concentration

L is the bio-film thickness or the characteristic length of support material = $(V/A)_p$

H is the height of the packed bed bioreactor

On substituting the values,

$$10\left(\frac{\text{mg s}}{\text{cm}^3}\right)\ln\frac{5}{1} + (5-1)\left(\frac{g}{l}\right) = H\frac{0.3\times\left(100\,\dfrac{\text{mg s}}{\text{cm}^3\,\text{h}}\right)\times\left(\dfrac{0.5\ \text{cm}^3}{6\ \text{cm}^2}\right)\times\left(25\,\dfrac{\text{cm}^2}{\text{cm}^3}\right)}{2\left(\dfrac{1}{\text{min}}\right)}100\ \text{cm}^2$$

Hence, the height of the packed bed required is 3.858 meters.

9.3 AIRLIFT BIOREACTORS

Air-lift bioreactor is a gas-induced circulation of liquid system. The reactor consists of two interconnecting zones—riser and down-comer. In the riser, gas is sparged. The down-comer receives gas from the riser (Fig. 2.29).

9.3.1 Classification of Airlift Reactors

Important classes of airlift reactors are

1. No gas injection in the down-comer, and
2. Entrained gas in the down-comer

Detailed analysis of airlift bioreactor is discussed by Chisti (1989). They are also classified as internal loop and external loop reactors.

9.3.2 Main Design Criterion

Gas-induced fluid circulation is the main design aspect.

What can we vary?

- Width of riser and down-comer
- Configuration of gas-liquid separator
- Location of the sparger
- Liquid level in the gas-liquid separator

9.3.3 Type of Analysis

One may consider airlift reactor

- as a single unit without the effect of recirculation
- as a single unit with the effect of recirculation
- considering the effects of the riser, down-comer, and the gas-liquid separator independently.

9.3.4 What are the Parameters to Measure?

(a) Superficial Liquid Velocity

Chisti and Moo-Young (1993) suggested the following relationship.

For riser:

$$v_{lr} = \left[\frac{2gh_D(\in_r - \in_d)}{\dfrac{F_T}{(1-\in_r)^2} + F_B\left(\dfrac{A_r}{A_d}\right)^2 \dfrac{1}{(1-\in_d)^2}} \right]^{0.5} \tag{9.29}$$

where

r for riser
d for down-comer
T for top of the reactor
B for bottom of the reactor
l for liquid in subscripts
g for gas in subscripts
g acceleration due to gravity
v is superficial velocity
h_D is gas-liquid dispersion height in the reactor
F is friction loss coefficients in top and bottom zones
$\in$ is fractional gas hold-up
A is cross-sectional area

(b) For Internal Loop, F_T <<< F_B

F_B is expressed by the following empirical relation (Chisti, 1989; Chisti *et al.*, 1988).

$$F_B = 11.402\left(\frac{A_d}{A_r}\right) \tag{9.30}$$

(c) For External Loop System, $F_T = F_B$

h_D is given by Chisti (1989)

$$h_D = \frac{h_L}{1 - \in_1} \tag{9.31}$$

where

h_L is ungassed liquid height

$\in_1$ is overall fractional gas hold-up

$$v_{lr} A_r = v_{ld} A_d = v_{lh} DW \tag{9.32}$$

where

v_{lr} is superficial liquid velocity in the head region

D is depth of liquid in the head region

W is width of head region of the reactor

$$v_{lh} = \frac{v_{lr} A_r}{DW} \tag{9.33}$$

Bubble rise velocity (v_b) is

$$\frac{1}{v_b} \leq \frac{\pi L D_c}{4 V_{lrm} A_r} \tag{9.34}$$

where

L is length of connecting pipe

D_c is diameter of connecting pipe

v_{lrm} is maximum anticipated liquid circulation velocity

(d) Gas Hold-up

$$\in_1 = \frac{\in_r A_r + \in_d A_d}{A_r + A_d} \tag{9.35}$$

The riser gas hold-up ($\in_r$) is given by Chisti and Moo-Young (1993).

$$\in_r = \frac{v_{gr}}{0.24 + 1.35(v_{gr} + v_{lr})^{0.93}} \tag{9.36}$$

and

$$\in_d = k \in_r \tag{9.37}$$

(e) Overall Mass Transfer Coefficient

Bello *et al.*, (1985) suggested following correlations for mass transfer coefficient

$$(k_L a_D)_{\text{Total}} = \frac{(k_L a_l)_{\text{Total}} (V_L)_{\text{Total}}}{(V_D)_{\text{Total}}} \tag{9.38}$$

where

$(k_L a_l)_{Total}$ = volumetric mass transfer coefficient based on total contactor liquid volume

$$= -\alpha_1 \ln\left(\alpha_2 - \frac{\beta}{\alpha_3}\right)$$

where

α_i' s and β **are parameters defined by Bello** *et al.*, (1985).

$(V_L)_{Total}$ = total liquid volume in contactor

$(V_D)_{Total}$ = total dispersed volume in contactor

Znad *et al.* (2004) have used different unsteady balance equations for the components in an airlift bioreactor, viz.,

- Bottom section
- Riser
- Top section
- Down-comer

for microorganism, reactant, product, and oxygen. They have described kinetic model and **hydrodynamic and mass transfer correlations. Hydrodynamic and mass transfer correlations are also** the same as described earlier in this section. The typical unsteady state mass balance equations are written for different parts of the airlift reactor. The reactor is divided into '*n*' number of stages. Four sections of airlift reactor are represented by bottom (stage number, $i = 1$), riser (stage number, $i = 2 \ldots (M_r - 1)$), top (stage number, $i = M_r$) and down corner (stage number, $j = M_r + 1 \ldots n$). Final expressions in different sections for cell, reactant, and product and dissolved oxygen levels are described in the following sections (Znad *et al.*, 2004).

1. Bottom Section

For cell, reactant and product: Znad *et al.* (2004) have suggested Equation (9.39).

$$\frac{dC_{b1}}{dt} = \frac{Q}{V_b(1-\in_{gr})}C_n + \frac{\beta' Q}{V_b(1-\in_{gr})}C_2 - \frac{(1+\beta')Q}{V_b(1-\in_{er})}C_1 + R_{C,b1} \quad (9.39)$$

where

b = bottom section

C = concentration of X, S, P

r = riser section

d = downcomer section

$$\beta' = \text{backflow parameter} = \frac{\text{back flow rate}}{\text{net forward liquid flow rate}}$$

Q = gas flow rate

$\in$ = fractional gas hold-up

V_b = volume of bottom section

V_r = volume of riser section

V_T = volume of top section

M_r = Number of stages in the riser
n = number of stages in the reactor
R = rate

For dissolved oxygen,

$$\frac{dC_{O_{2b,1}}}{dt} = \frac{Q}{V_b(1-\epsilon_{gr})}C_{O_{2,n,1}} + \frac{\beta' Q}{V_b(1-\epsilon_{gr})}C_{O_{2,2,1}} - \frac{(1+\beta')Q}{V_b(1-\epsilon_{gr})}C_{O_{2,1,1}} + K_L a(C^*_{O_{2b,1}} - C_{O_{2b,1}}) + R_{O_2 b,1} \tag{9.40}$$

where $C^*_{O_{2b}}$ is the equilibrium concentration of oxygen.

2. Top Section

For X, S, and P

$$\frac{dC_{M_r}}{dt} = \frac{(1+\beta')Q}{V_T(1-\epsilon_{gr})}(C_{M_{r-1}} - C_{M_r}) + R_{C,M_r} \tag{9.41}$$

For dissolved oxygen

$$\frac{dC_{O_2,M_r}}{dt} = \frac{(1+\beta')Q}{V_T(1-\epsilon_{gr})}(C_{O_2,M_r-1} - C_{O_2,M_r}) + K_L a\left(\frac{C_{P,M_r}}{H}Y_{M_r} - C_{O_2,M_r}\right) + R_{O_2,M_r} \tag{9.42}$$

3. Riser Section

For X, S, and P

$$\frac{dC_i}{dt} = \frac{\beta' Q}{\left(\frac{V_r}{M_r-2}\right)(1-\epsilon_{gr})}(C_{i+1} - C_i) + \frac{(1+\beta')Q}{\left(\frac{V_r}{M_r-2}\right)(1-\epsilon_{gr})}(C_{i-1} - C_i) + R_{C,i} \tag{9.43}$$

For dissolved oxygen

$$\frac{dC_{O_{2,i}}}{dt} = \frac{\beta' Q}{\left(\frac{V_r}{M_r-2}\right)(1-\epsilon_{gr})}(C_{O_{2,j+1}} - C_{O_{2,i}}) + \frac{(1+\beta')Q}{\left(\frac{V_r}{M_r-2}\right)(1-\epsilon_{gr})}(C_{O_{2,j-1}} - C_{O_{2,i}}) + K_L a\left(\frac{C_{P_i}}{H}Y_i - C_{O_{2,i}}\right) + R_{O_{2,i}} \tag{9.44}$$

4. Down-comer Section

For X, S, and P

$$\frac{dC_i}{dt} = \frac{Q}{(V_d/(n-M_r))(1-\epsilon_{gd})}(C_{i-1} - C_i) + R_{C,i} \tag{9.45}$$

For dissolved oxygen

$$\frac{dC_{O_{2,i}}}{dt} = \frac{Q}{(V_d/(n-M_r))(1-\epsilon_{gd})}(C_{O_2,j-1} - C_{O_{2,i}}) + K_L a\left(\frac{C_{P_i}}{H}Y_i - C_{O_{2,i}}\right) + R_{O_2,i} \tag{9.46}$$

It is preferred to calculate total pressure in different sections of the reactor. They are described by Ho *et al.* (1977). Total pressure in the bottom section

$$P_b = P_{atm} + \alpha\left[h_t(1-\epsilon_{gt}) + (h_r + h_b)\times\left(1 - \frac{1}{2(M_r-1)}\right)(1-\epsilon_{gr})\right] \tag{9.47}$$

Total pressure in the riser

$$P_r = P_{\text{atm}} + \alpha\left[h_t(1-\epsilon_{gt}) + (h_r + h_b)\times\left(1 - \frac{2i-1}{2(M_r-2)}\right)(1-\epsilon_{gr})\right] \tag{9.48}$$

Total pressure in the down-comer

$$P_d = P_{\text{atm}} + \alpha\left[h_t(1-\epsilon_{gt}) + (h_r + h_b)\times\left(\frac{2i-1}{2(n-M_r)}\right)(1-\epsilon_{gd})\right] \tag{9.49}$$

Total pressure in the top section

$$P_{Mr} = P_{\text{atm}} + 0.5\alpha\, h_{Mr}(1-\epsilon_{gr}) \tag{9.50}$$

where

h_r is height of riser

h_b is height of bottom section

P_{atm} is atmospheric pressure

α is conversion factor = 0.000968 atm/cm H_2O

For details, one can refer Znad *et al.*, (2004) and Siegel *et al.*, (1986).

9.4 HOLLOW FIBER BIOREACTOR (HFBR)

This reactor is used for free microbial cells, free enzyme, immobilized system, and mammalian cells.

A schematic representation of a conventional HFBR is given below (Fig. 9.2).

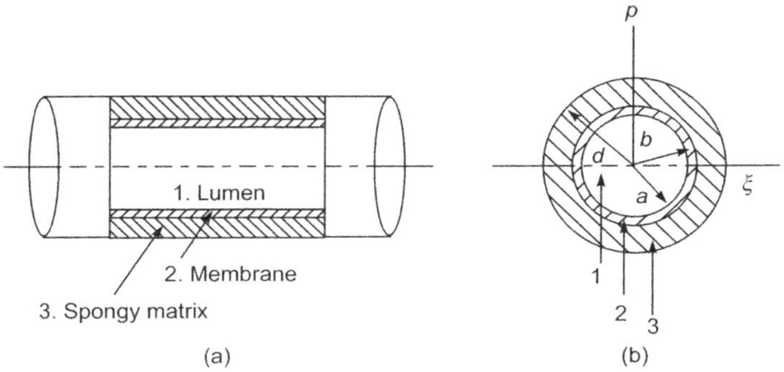

Figure 9.2 Schematic diagram of the hollow fiber bioreactor with laminar velocity profile in the lumen region: (a) axial cross-section, and (b) radial cross-section.

The reactor is divided into the following three regions.

- Region 1 is the lumen (inner tube),
- Region 2 is the permeable membrane (wall of the inner tube), and
- Region 3 is the spongy matrix (annular region)

An active biocatalyst in the form of either enzymes or live cells is immobilized inside the spongy matrix. The substrate solution is fed through the inner tube and it diffuses through the permeable membrane to react with the active enzymes or live cells supported on the spongy matrix. The product diffuses back to the lumen and flows downstream.

Following assumptions may be considered for the design purpose. Isothermal and steady-state conditions are followed.

Assumptions for region 1:

- laminar flow with parabolic and fully developed velocity profile,
- axial dispersion is neglected.
- Fick's diffusion is in the radial direction.
- homogeneous reaction is absent in the bulk.
- all physical properties are constant.

Assumptions for region 2:

- membrane is not reactive.
- axial diffusion and convective fluxes are negligible.
- mass transfer coefficients and partition coefficients are constants.

Assumption for region 3:

- axial diffusion and convective fluxes are neglected.
- kinetic model is described as a function of the reactant concentrations.

The mass balance differential equations for the 'i'th species are the following (Cabrera *et al.*, 2001).

For region 1:

Diffusive flux = Convective flux

$$D_{1,i}\frac{1}{r}\frac{\partial\left[r\dfrac{\partial C_{Si,1}(r,z)}{\partial r}\right]}{\partial r}-2v_z\frac{\partial(C_{Si,1}(r,z))}{\partial z}=0,\ \text{for } 0<r<a,\ 0<z<L$$

where

suffices 1 for region 1
2 for Region 2
3 for Region 3

C_{Si} is the concentration of compound 'i' (mass/vol.)

γ is radial coordinate (length unit)

z is axial coordinate (length unit)

a is inner radius of member are (length unit)

L is hollow fiber length (length unit)

V_Z is the velocity of reaction at any point in the hollow fiber (length / time).

D_i is diffusion coefficient of reactant (length^2/time)

$$D_{1,i}\frac{1}{r}\frac{\partial\left[r\dfrac{\partial C_{Si,1}(r,z)}{\partial r}\right]}{\partial r}-2v_{\max}\left[1-\frac{r^2}{a^2}\right]\frac{\partial(C_{Si,1}(r,z))}{\partial z}=0,\ \text{for } 0<r<a,\ 0<z<L \quad (9.51)$$

where $v_{\max}$ is maximum rate (length / time).

For region 2:
Diffusive flux = 0

$$D_{2,i}\frac{1}{r}\frac{\partial\left[r\frac{\partial C_{Si,2}(r,z)}{\partial r}\right]}{\partial r}=0,\text{ for } a<r<b,\ 0<z<L \tag{9.52}$$

where b is outer radius of the membrane (length unit).

For region 3:
Diffusive flux = Rate of reaction

$$D_{3,i}\frac{1}{r}\frac{\partial\left[r\frac{\partial C_{Si,3}(r,z)}{\partial r}\right]}{\partial r}-Yv_3\mathbf{C}(r,z)=0,\text{ for } b<r<d,\ 0<z<L \tag{9.53}$$

where Y is stoichiometric coefficient
v_3 is the reaction rate (time^{-1})
d is outer radius of spongy matrix wall (length)
$\mathbf{C}$ is concentration profiles in spongy matrix region (mass / unit volume).

Boundary conditions are suggested by Cabrera *et al.*, (2001).

$$D_{1,i}\left[\frac{\partial C_{Si,1}(0,z)}{\partial r}\right]=0 \tag{9.54}$$

$$D_{1,i}\left[\frac{\partial C_{Si,1}(a,z)}{\partial r}\right]=D_{2,i}\left[\frac{\partial C_{Si,2}(a,z)}{\partial r}\right] \tag{9.55}$$

with

$$P_a^i=\frac{C_{Si,2}(a,z)}{C_{Si,1}(a,z)},$$

where P_i^a = lumen to membrane partition coefficient (dimensionless).

$$D_{2,i}\left[\frac{\partial C_{Si,2}(b,z)}{\partial r}\right]=D_{3,i}\left[\frac{\partial C_{Si,3}(b,z)}{\partial r}\right] \tag{9.56}$$

$$P_i^b=\frac{C_{Si,3}(b,z)}{C_{Si,2}(b,z)}$$

where P_i^b = membrane to spongy matrix partition coefficient (dimensionless).

$$D_{3,i}\left[\frac{\partial C_{Si,3}(d,z)}{\partial r}\right]=0,\text{ if } C_{si,3}(d,z)\geq 0 \tag{9.57a}$$

$$D_{3,i}\left[\frac{\partial C_{Si,3}(r_c,z)}{\partial r}\right]=0,\text{ if } C_{Si,3}(r_c,z)=0,\text{ for } b\leq r_c<d \tag{9.57b}$$

Initial conditions

$$C_{Si,1}(r, 0) = C_{Si0,1}(r),\ i \text{ stands for reactant} \tag{9.58a}$$

$$C_{Si,1}\ (r, 0) = 0,\ \ i \text{ is not reactant} \tag{9.58b}$$

where r_c is the radius at the critical boundary.

$C_{S_{i0}}$ is the initial concentration of the substrate which reacts in the reactor.

Non-dimensionalizing,

$$\frac{C_{Si}}{C_{Si0}} = A_i \text{ is the dimensionless concentration} \tag{9.59}$$

$$\rho = \frac{r}{a},\ \text{dimensionless radial coordinate for lumen}$$

$$\rho = \frac{r}{b},\ \text{dimensionless radial coordinate for annular region} \tag{9.60}$$

$$L' = \frac{z}{L},\ \text{dimensionless transformed axial coordinate} \tag{9.61}$$

$$Ge = \frac{a}{L},\ \text{geometrical parameter} \tag{9.62}$$

$$P_{ei} = \frac{v_{\max} a}{D_{i1}},\ \textbf{Peclet member} \tag{9.63}$$

$$K_L = -\frac{D_{i3}}{D_{i1}} \frac{K_i^a}{\ln(a/b)},\ \text{membrane mass transfer coefficient (dimensionless)} \tag{9.64}$$

Non-dimensionalization is useful for further modeling purposes because the number of differential equations to be solved reduces from three to two in the domain $0 \le \rho \le \beta$ and $0 \le L' \le 1$, where $\beta = d/b$, a dimensionless parameter.

Equation (9.51) becomes

$$\frac{1}{\rho}\frac{\partial\left[\rho\dfrac{\partial A_{i,1}(\rho, L')}{\partial \rho}\right]}{\partial \rho} - 2GePe(1-\rho^2)\frac{\partial A_{i,1}(\rho, L')}{\partial L'} = 0, \text{ for } 0 < \rho < 1, 0 < L' < 1 \tag{9.64}$$

Equation (9.53) becomes

$$\frac{1}{\rho}\frac{\partial\left[\rho\dfrac{\partial A_{i,3}(\rho, L')}{\partial \rho}\right]}{\partial \rho} - Y\kappa_3 A_{i,1}\ (\rho, L') = 0, \text{ for } 1 < \rho < \beta, 0 < L' < 1 \tag{9.65}$$

where $\beta = \dfrac{b}{a} = \dfrac{d}{a}$ = dimensionless parameter

$$\kappa = \frac{b^2 v_3}{D_{3,i}} = \text{dimensionless reaction rate}$$

By non-dimensionalizing the boundary conditions,
Equation (9.54) becomes

$$\frac{\partial A_{i,1}(\rho, L')}{\partial \rho} = 0 \text{ at } \rho = 0 \tag{9.66}$$

Equations (9.56) becomes

$$\left.\frac{\partial A_{i,2}(\rho, L')}{\partial \rho}\right|_{\rho=1} - K_{Li}\,[A_{i,1}(1, L') - K_i A_{i,3}(1, L')] = 0 \tag{9.67}$$

Equation (9.57a) becomes

$$\left.\frac{\partial A_{i,3}(\rho, L')}{\partial \rho}\right|_{\rho=\beta} = 0, \text{ if } A_{i,3}(\beta, L') \geq 0 \tag{9.68}$$

Equation (9.57b) becomes

$$\left.\frac{\partial A_{i,3}(\rho, L')}{\partial \rho}\right|_{\rho=\rho_c} = 0, \text{ if } A_{i,3}(\rho_c, L') = 0, \text{ for } 1 \leq \rho_c < \beta \tag{9.69}$$

At initial conditions, Equation (9.58a) becomes

$$A_{i,1}(\rho, 0) = A_{i0,1}(\rho) \tag{9.70}$$

for i is reactant, and Equation (9.58b) becomes

$$A_{i,1}(\rho, 0) = 0 \tag{9.71}$$

for i is not a reactant.

Derivation of design equations in terms of integral equations is described by Cabrera *et al.* (2001). Appendix to this chapter gives useful information for the design of hollow fiber bioreactor.

9.5 PLANT CELL BIOREACTOR

The production of chemicals from plant cell culture offers a number of important advantages. They are, for example,

- Control and supply of products are independent of the availability of plant itself.
- Cultivation is under controlled and optimized conditions.
- Strain improvement with programs is analogous to those used for microbial systems.

However, the role of bioreactors in plant tissue culture is important to generate elite plants as well as chemicals.

9.5.1 Bioreactor Considerations

Plant cells are used in various forms in the reactors:

- Suspension culture
- Immobilized cells
- Organized tissues

Reactor design should consider the following problems associated with the need of above forms in reactors.

(a) For Suspension Culture

1. Shear Force

Plant cells are larger than microbial cells. When they are exposed to turbulent shear, the eddy size approaches the cell size. The exposed cells are treated under this motion. So the cells are damaged by shear forces. Lower levels of shear appear to affect cell surface receptors and nutrient transport. Hence, the stirred tanks designed for the bacterial culture are not good choices for plant cell culture, but modified stirred tank is suitable in this regard.

2. Mixing

Mixing plays an important role in shear damage. Mixing depends on a combination of sparging and mechanical agitation. Oversparging can be a problem for plant cell culture. Even though photoautotrophic growth is not important in most cultures, higher concentration of CO_2 can enhance productivity. For reactor design, optimization of gas consumption, sparging rate and the degree of mechanical agitation is an important challenge.

3. Growth Rates

Low growth rates of plant cells create problems of maintenance of aseptic conditions and reduced volumetric productivities.

4. Genetic Instability

Suspension and callus culture will produce a desired compound initially, but after a certain period, the capacity for synthesis by the same compound decreases considerably. This might be due to inadequate nutrition or genetic or epigenetic basis. So, one can think of two-phase culture. The first phase targets for optimal growth, whereas the second phase is for optimal product synthesis.

(b) For Immobilized Cells

This is used for continuous operation for extra-cellular products. High cell density and protection of cell from shear forces are the major advantages of immobilization. However, a few considerations may be adopted for reactor selection and design. They are:

- Large scale aseptic immobilization procedure
- Mass transfer limitation and cell metabolism
- Products must be targeted from non-growing cells
- Influence on cell physiology.

(c) For Organized Tissues

In many cases, neither suspension nor immobilized cell cultures will produce satisfactory amounts of a desired metabolite. However, organ cell cultures from the same plant may give good yields.

Advantages

- Biosynthetic capacity often returns upon organogenesis.
- Product spectrum for flavors and fragrances returns to that of parent plant.

- Product secretion is enhanced in many cases.
- Genetic stability, under growth conditions, is greatly improved over callus or suspension culture.
- Self immobilization provides more optimal mixing of cell types.

Disadvantages

- Growth rate may be lower than suspension cultures.
- Efficient, scalable reactors for organized tissues need to be developed.
- Control of micro-environmental conditions is often more difficult.

The primary disadvantages for organ cultures are their apparent slow growth rates and difficulties in designing effective bioreactors to handle organized tissues. Recent advances suggest that both limitations may be circumvented, and organ cultures that grow more rapidly than their corresponding suspension cultures have been found in several cases.

One advantage of using organ cultures over whole plants is the possibility of using precursor feeding and elicitors. For example, with species of onion and garlic, the use of precursors can greatly enhance the formation of flavor compounds. Different precursors give different levels of enhancement to particular components to the flavor spectrum. By using combinations of chemical precursors, it may be possible to custom-make flavors for specific applications.

Shoot cultures present additional problems. Light may be required for some shoot cultures, while roots can be grown easily in the dark. Shoot cultures may be mixotrophic, involving the exogenous supply of sugars as well as some photosynthesis. In some cases, control of the photoperiod (hours of exposure to light per day) is important. However, the most crucial role for light is in cellular regulation. Exposure of light of certain wavelengths is essential to induce synthesis of some enzymes. In at least some cases, these enzymes play a crucial role in secondary metabolism. The maintenance of uniform light intensity in a large reactor is a challenging and a partially unsolved problem.

One other perceived limitation on shoot cultures is that they could not be grown under totally submerged conditions. Recent experiments have shown that tobacco shoots will grow well in submerged culture as in a standard shake flask, provided that the liquid phase is well-mixed.

Although organ cultures have not yet been extensively exploited, they are the promising vehicles for the production of some important secondary metabolites.

9.5.2 Classes of Bioreactors for Plant Cell Growth

Like in most bioreactor applications, plant cell bioreactors are also divided into two broad categories depending on the mode of operation. This is discussed in Chapter 2.

- Batch reactors
- Continuous flow reactors

There are currently two bioreactor systems used for cell suspension culture.

- Stirred-tank reactors
- Air-lift reactors

For example, stirred-tank bioreactor has been used for large-scale ginseng production using *Panax ginseng*, while air-lift bioreactors have been used for alkaloid production with *Catharanthus roseus* or

Thallctrum rugosum. The two systems have also been used with success for *Nicotiana tabacum* plant cell culture, but it is generally admitted that that the air-lift bioreactor has the advantage of having a low shear, low energy requirement and a simple design.

The continuous flow type is largely a patented invention made by Maillard *et al.*, (2003). This design offers an advantage of continuous operation up to 3–4 weeks based on continuous flow through a fermentation chamber.

(a) Batch Reactors

All batch reactors, generally, deal with three phase reactor design. Three phase reactors are more difficult to design, because the mass transfer of components among the three phases must be controlled. Inadequate tools for complex fluid mechanics in such systems coupled with complex nature of cells as reactive solids make the prediction of system performance from first principles very difficult.

(b) Stirred Tank Reactors

This is the most traditional mode of plant cell reactor used in early stages. The main virtues of such systems are extremely flexible and can provide high $k_L a$ values for gas transfer.

Gas under pressure is supplied to a sparger. The size of gas bubbles and their dispersion throughout the tank are critical to reactor performance. Gas dispersion is also a function of the impeller. The impeller must provide sufficiently large agitation to disperse the bubbles throughout the tank, to increase their residence time within the liquid and to shear larger bubbles into smaller ones. Too much shearing can be detrimental, owing to shear-sensitivity exhibited to varying degrees by some cells and the stratification of reactor contents with multiple impeller systems. Although a wide variety of impeller designs have been proposed, the predominant choices are disc- and turbine-type impellers.

The vessel is, generally, made of stainless steel. Type 316 is used on all wetted parts. Type 304 is used on covers and jackets. However, in more recent versions of plant cell reactor design, glass vessels with stainless steel cover plates are used. Most stirred tank reactors for plant cell synthesis are low volume vessels, 50 – 500 l.

(c) Loop Reactors

The most common mode of operation used is an air-lift type. Here, the motion of the gas carries fluid and cells up a draft tube. At the top, gas disengages from the liquid, and the degassed liquid (which is denser than the gassed liquid) descends in the annular outside the tube. At the bottom of the reactor, the descending liquid again encounters a gas stream and is carried back to a draft tube. Airlift systems can, generally, handle somewhat more viscous fluids than bubble columns. Coalescence is not a problem. With large volume reactor designs, the use of non-mechanically agitated designs is preferred, as high oxygen transfer rates and better cooling can be attained. With an airlift reactor design, the interchange of material between fluid elements is small. So the transient time to circulate biomass from the bottom of the draft tube to the top and back again is important. Three basic systems for circulating biomass are known in an airlift bioreactor.

Jet Loop Bioreactor (JLR)

This is one of the unique designs of a plant cell bioreactor used on a large scale. It consists of double column cylindrical vessel in which a nozzle or a frit, together with pump hydrodynamic effect disperses air bubbles. Different types of JLR configuration used for plant cells are:

- An external- loop air-lift reactor, whose risers and down-comers are separate tubes connected by short horizontal sections at the top and bottom.
- An internal-loop air-lift reactor, in which the riser and the down-comer are separated by an internal baffle or draft tube, and
- A bubble column reactor, which has the advantage of being easily sterilized using an autoclave, and for which maintenance is minimal.

One of the problems of such a bubble column system is that little is known about fluid flow within it, and mixing is non-uniform. For example, Kato *et al.*, (1976) have observed a reduced growth rate of *N. tabacum* cells cultivated in a bubble column bioreactor because of insufficient mixing in the reactor.

9.5.3 Design of Bioreactor

(a) Batch Reactor

Let

C_X be concentration of cells,
C_S be concentration of key reactant, and
C_P be concentration of product.

Mass balance:

The general form of the mass balance for any closed or open system is given by

Rate of accumulation = Rate of input – Rate of output + Net rate of generation

(A) For Cells:

$$V\frac{dC_X}{dt} = (r_g - r_d)V \tag{9.72}$$

where

r_g is growth rate of cells
r_d is death rate.

(B) For Reactant:

$$V\frac{dC_S}{dt} = [Y_{S/X}(-r_g) - mC_X]V \tag{9.73}$$

(C) For Product:

$$V\frac{dC_P}{dt} = Y_{P/X}(r_g)V \tag{9.74}$$

Consider Monod's kinetics

$$r_g = \frac{\mu_{\max} C_S}{K_S + C_S} C_X \tag{9.75}$$

$$r_d = k_d C_X \tag{9.76}$$

From Equation (9.72),

$$\frac{dC_X}{dt} = \frac{\mu_{max} C_S}{K_S + C_S} C_X - k_d C_X \quad (9.77)$$

From Equation (9.73),

$$\frac{dC_S}{dt} = -Y_{S/X}\left(\frac{\mu_{max} C_S}{K_S + C_S}\right) C_X - mC_X \quad (9.78)$$

From Equation (9.74),

$$\frac{dC_P}{dt} = Y_{P/X}\left(\frac{\mu_{max} C_S}{K_S + C_S}\right) C_X \quad (9.79)$$

For a batch reactor, the most critical parameter is the batch reaction time, t_r.

By assuming an ideal batch reactor, we get

$$t_r = \left(\frac{1}{\mu_{max} Y_{X/S}}\right)\left(\left(\frac{K_S}{C_{S_0}}\right)\ln\left(\frac{C_{S_0}}{C_S}\right) - \left(\frac{K_S + C_{S_0}}{C_{S_0}}\right)\ln(C_{S_0} - C_S)\right) \quad (9.80)$$

9.6 DESIGN OF BIOREACTORS FOR SOLID STATE FERMENTATION (SSF)

SSF processes involve reactions in absence of free water. So the treatments are entirely different from the SLF processes. To carry out these reactions, varieties of configurations are available in the literature. Mitchell *et al.* (2006) described four classes of bioreactors:

- Unaerated, unmixed bioreactors, *viz.*, tray type
- Forcefully aerated bioreactors without mixing, viz. packed bed bioreactors
- Rotating drum and stirred drum bioreactors
- Continuously – mixed/intermittently mixed forcefully-aerated bioreactors

(a) Unaerated, Unmixed Bioreactors

Operating variables are

- Temperature
- Humidity
- Flow rate of air
- Velocity of circulated air
- Cooling medium temperature.

In these reactors, critical depth of the tray (D_c) is more important. D_c is expressed by Raghavendrarac *et al.* (1993) by the following Equation (9.81).

$$D_c = \sqrt{\frac{2DCY_{X/O}}{R_{XM}}} \quad (9.81$$

where

$$Y_{X/O} = \frac{g \text{ dry biomass}}{g \text{ oxygen}}$$

D = effective diffusivity of oxygen in the bed (cm^2/h)

C = concentration of O_2 in the surrounding atmosphere (g/cm^3)

$$R_{XM} = \text{maximum growth rate} \left[\frac{g \text{ dry biomass}}{(\text{cm}^3 \text{ of bed})\,(\text{h})}\right]$$

Mitchell *et al.* (2006) expressed the temperature difference between the bottom of the solid bed and tray surface without any other heat loss (θ) by the following equation:

$$\theta = \frac{R_Q Z^2}{2k} \tag{9.82}$$

R_Q = volumetric heat production rate (W/m^3)

Z = total bed height (m)

k = thermal conductivity of the bed (W/(m)(°C)).

(b) Forcefully Aerated Bioreactors without Mixing

Operating variables are

- Aeration rate
- Temperature of inlet air
- Temperature of the surroundings

(c) Rotating Drum and Stirred Drum Bioreactors

Operating variables are

- Rotation speed or stirring speed
- Particle strength
- Aeration rate and humidity of incoming air

For the design, Ishikawa *et al.* (1982) have mentioned about the critical rotational speed (N_c) of the particles held against the inside of the drum wall [Equation (9.83)].

$$N_c = \frac{42.3}{\sqrt{D_t}} \tag{9.83}$$

where D_t is the diameter of the drum.

(d) Continuously Mixed, Forcefully Aerated Bioreactors

Operating variables are

- Inlet air flow rate
- Temperature
- Humidity
- Rate of mixing
- Addition of water, etc.

(e) Intermittently Mixed Forcefully-aerated Bioreactors

Operating variables are

- Air flow rate
- Temperature
- Mixing rate
- Extent of mixing

Mitchell *et al.*, (2006) have described bioreactor design of solid state fermentation in details.

9.7 MAMMALIAN CELL BIOREACTOR DESIGN

Variety of bioreactors is employed in the mammalian growth. HFBR mentioned earlier in this chapter is also applied for the specific production using mammalian cells. One of the important aspects of the mammalian cell culture for its optimal performance is scheduling of the reaction process.

9.7.1 Fermentor Balancing for Semi-continuous Multi-tank Mammalian Cell Culture Process

The configuration considered for the case study consists of a solera (series of vessels used for ageing liquids) tank that feeds cells to either of two equal volume production vessels or to the drain for decontamination and disposal in the biological kill system. Here, we are considering two production schedules for this configuration.

Schedule 1:

- duration of solera cycle (S) = two days
- duration of production cycle (P) = three days
- time for harvesting turn around of each production vessel = one day

Schedule 2:

- duration of solera cycle (S) = two days
- duration of production cycle (P) = four days
- time for harvesting turn around of each production vessel = two days

Different durations of production cycle affect the efficiency of utilization of solera batches. For the first pattern, the availability of fresh cell inocula from the solera vessels precisely matches the demand from purification vessels. No solera batch is disposed directly to the kill system. No production vessel is idle. This production pattern results in a balanced utilization of fermentors.

In the other production patterns, every third solera batch must be disposed of to the kill system directly, because no production vessel is available for seeding. When production vessel becomes available, there will be no solera batch at the correct phase of its culture cycle to provide cells. Here, production vessel will remain unproductive for two days between the batches.

Each production pattern is characterized by repeat duration (R), a certain time after which the sequence of operation is repeated. During repeat period, the number of harvest (h) is equal to the number of production vessels (Dey *et al.*, 1997).

$$R = (h + d) \times S$$

d is non-productive disposals of solera vessel to the kill system in a repeat period R.

S is the duration of solera cycle.

For any pattern, we have $R - P \geq T$

P is the duration of production phase.

T is the minimum time required to clean and sterilize.

In general, for operational consideration $T = 1$

So, $(h + d) \times S - P \geq 1$

h and d are integers.

The product accumulation with time may be considered as a linear process or a non linear process with time.

For linear antibody production with a single day production vessel turn round capability ($T = 1$), the fermenter scheduling prefers longest P, subjected to the condition that $R\text{-}P = T$ and $P <$ (maximum duration of a production batch). This type of fermenter scheduling results in balanced utilization of fermentor vessels.

The optimal production schedules are specific to $T = 1$. $T = 1$ is reasonable and commonly achievable condition. Longer turn round times is selected on the basis of the optimal production pattern.

For the nonlinear antibody production, the balanced fermenter configuration, when harvesting is carried out during the period of peak antibody accumulation rate, is analyzed against operability consideration. The unbalanced line operation is not acceptable. Dey *et al.* (1997) have analyzed this behavior in detail.

EXERCISES

9.1 An industrial waste water stream is fed to a stirred tank reactor continuously and the cells are recycled back to the reactor from the bottom of the sedimentation tank placed after the reactor. The following data are given for the system:

$$F = 100 \text{ l/h}, C_{So} = 5000 \text{ mg/l}, \mu_m = 0.25 \text{ h}^{-1}$$
$$K_S = 200 \text{ mg/l}, \alpha = \text{recycle ratio} = 0.6, e = \text{cell concentration factor} = 2$$
$$Y_{x/s} = 0.4$$

The effluent concentration is desired to be 100 mg/l. Determine the required reactor volume.

9.2 Develop an empirical equation for computing bed height of a carrier- bound-enzyme column taking into consideration all the parameters that matter much in the case of a conversion system which should operate under equilibrium conditions.

9.3 Critically examine the basis for consideration of two regions in transport of air in bubble column reactor.

9.4 What are the advantages of multi-turbine bioreactor over bubble column reactor?

9.5 It is desired to test the activity of an enzyme in a reactor of the following configuration in Figure 9.3.

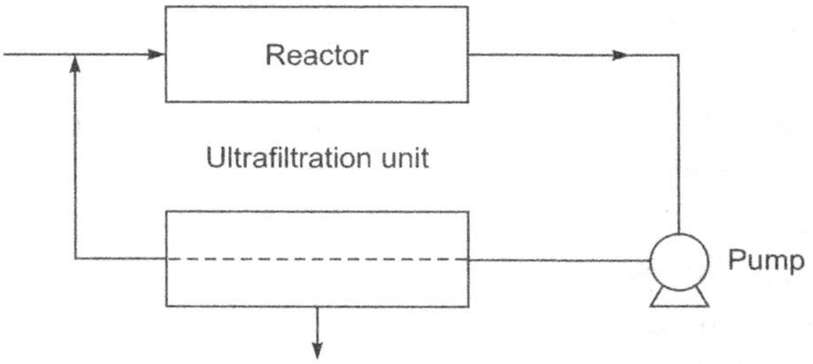

Figure 9.3 Ultrafiltration reactor.

Write a transient mass-balance for the reactor. Input conditions are flow rate = v and reactant concentration C_{So}.

9.6 Derive an expression of dilution rate for a fed-batch culture.

9.7 Critically justify the following statements:

(a) Critical flow rate of O_2 through an orifice to the bubble column reactor is indirectly related to the growth rate of disturbance in the system.

(b) Production of α-amylase under immobilized condition by *Bacillus* sp. suffers substrate independent temperature inactivation in CFSTBR. Volume of the reactor in this cannot be specifically defined.

(c) Effects of the degree of coalescence tendency of the medium and sparger design (pore or hole size) on the bubble size in a bubble column reactor.

9.8 Pressure drop through a 5 ft. high column of glucose isomerase immobilized in glass beads was found to be 8 psi. A superficial mass velocity of 0.5 lbft^2 sec^{-1} was used for the conversion of 30% of a 60% (w/v) glucose feed to fructose at 60°C. Particle size was 700 μm and shape factor and void fraction were 0.6 and 0.3, respectively. Verify that whether the pressure drop is within the limit or not?

9.9 Derive a conclusion for the following statement: Competition between *E.coli* and *Azotobacter vinelandii* in continuous culture for multiple substrates is assumed. The competitive elimination principle of Gauss applies in terms of D_{crit} and μ_m when two organisms competing for one substrate. If two substrates are present, microbial growth depends on both. What are the parameters for their coexistence?

9.10 Citric acid is produced by the fermentation route. For this purpose, 200 m^3 of fermenter is used for the production of 18,000 kg per batch. The fermenter is on-stream 95% (including turnaround) of the total available time. Following this production strategy, the net earnings after taxes (50%) during first year is ₹ 80,00,000 whereas net earnings after taxes in a period of 2-15 years is ₹ 96,00,000. Also, the original fixed capital investment is ₹ 328,00,000. It is also observed that the organization has used 25% of net sales as working capital, i.e., ₹ 119,00,000. What could be the return on investment? State that with this return on investment the organization will not face any lock-out or closure.

9.11 In PFR, having dispersed plug flow, prove that $\dfrac{C_{exist}}{C_{inlet}} = f(Bs\gamma)$

where $\gamma = \dfrac{V_R}{V}$ and Bs = Bodenstein number.

REFERENCES

Abramowitz M and Stegun IA (Eds.) (1965) *Handbook of Mathematical Functions*, Dover, New York.

Baksalary JK and Tabis Z (1987) "Conditions for the robustness of block designs against the unavailability of data". *J Statist Plann Inf*, **16**, 49-54.

Bello RA, Robinson CW and Moo-Young M (1985) Gas hold up and overall volumetric oxygentransfer coefficient in air lift contactors, *Biotechnology and Bioengineering*, **27**, 369-381.

Bischoff KB (1965), "Effectiveness factors for general reaction rate forms", *AIChE J*, **11**, 352.

Cabrera MI, Luna JA, Grau RJ (2001) "Solving design equations for a hollow fiber bioreactor with arbitrary kinetics", *Chemical Engineering Journal*, **84**, 445-461.

Chisti Y (Ed) (1989) *Air lift bioreactors*, Elsevier Applied Science, New York.

Chisti MY, Halard B and Moo-Young M (1988) "Liquid circulation in air lift reactors", *Chemical Engineering Science*, **43**, 451-457.

Chisti Y, (Ed) Moo-Young M (1993) "Improve the performance of airlift reactors", *Chemical Engineering Progress*, 38-45, June.

Crank J (Ed.) (1975), *The Mathematics of Diffusion*, (2nd Edn., Oxford) Clarendon Press.

Dey S, Marshall C, Slater NKH (1997) "Fermenter balancing for semicontinuous multi-tank mammalian cell culture processes", *Chemical Engineering Journal*, **65**, 123-132.

Ho CS, Erickson LE and Fan LT (1977) "Modeling and simulation of oxygen transfer in airlift fermentors", *Biotechnology and Bioengineering*, **14**, 1503–1522.

Ishikawa Y, Kase M, Sasaki M, Satoh K, and Sasaki S (1982) recent progress in the sintering technology-high reproducibility and improvement of fuel consumption. *Iron making conference proceedings* 41, 80-89.

Kato K, Shiozawa Y, Yamada A, Nishida K and Noguchi M (1977) A jar fermenter culture of *N. tabucum* L cell suspensions, *Agricultural and Biological Chemistry*, **36**, 899–904.

Levenspiel, O (1980), The Monod Equation: A Revisit and a Generalization to Product Inhibition Situations. *Biotechnology and Bioengineering*, **22**, 1671.

Maillard P, Wallis G and Langlais A (2003) Continuous plant cell bioreactor and method of continuous plant cell culture, European Patent, Number EP 03291790.8, Jan. 19, 2005.

Melick MR, Karim MN, Linden JC, Dale BE and Mihaltz P (1987) Mathematical modeling of ethanol production by immobilized *Zymomonas mobilis* in a packed bed fermenter, *Biotechnology and Bioengineering*, **29**, 370-382.

Mitchell DA, Krieger N and Berovic M (Eds) (2006) *Solid state fermentation bioreactors- fundamentals of design and operations*, Springer-Verlag, Berlin, Heidelberg.

Raghavendrarao KSMS, Gowthaman MK, Ghildyal NP and Karanth NG (1993) *Bioprocess Engineering*, **8**, 255-262.

Siegel MH, Merchuk JC and Schuegul K(1986) "Air lift reactor analysis: interrelationships between riser, downcomer, and gas-liquid separator behavior including gas recirculation effects", *AIChE Journal*, **32**, 1585-1596,

Znad H, Bales V and Kawase Y (2004) Modeling and scale-up of air lift bioreactor, *Computers and Chemical Engineering*, **28**, 2765-2777.

APPENDIX 9

Useful Information for the Design of HFBR

Baksalary and Tabis (1987) and Abramowitz and Stegun (1965) have given details for the functions used in HFBR design.

1. Indicator Function $\Theta(\rho c - \rho)$

$\Theta(\rho_c - \rho)$ is an indicator function which is defined as

$$\Theta(\rho_c - \rho) = \begin{cases} 1 \text{ if } \rho_c \geq \rho \\ 0 \text{ if } \rho_c < \rho \end{cases}$$

$\Theta(\rho_c - \rho)$ being a Heaviside step function whose derivative is given by

$$\frac{d}{d\rho}\Theta(\rho_c - \rho) = -\delta(\rho_c - \rho)$$

where $\delta(\rho_c - \rho)$ is the generalized delta function.

2. Green's Function

Green's function is the following Fourier expansion

$$G^1(\rho, \bar{L}/\bar{\rho}, \bar{L}') = \sum_0^{\infty} A_n^{-2} \exp\left[-\frac{\lambda_n^2}{2\,Ge\,Pe}(L' - \bar{L}')\right]\phi_n(\rho)\phi_n(\bar{\rho})\Theta(L' - \bar{L}')$$

where A_n is normalization constant for the nth Eigen function (dimensionless) and $\phi_n(\rho)$ is called orthogonal Eigen functions,

$$\phi_n(\rho) = \exp\left[-\frac{\lambda_n \rho^2}{2}\right] M\left(\frac{2-\lambda_n}{4}, 1, \lambda_n \rho^2\right)$$

λ_n is called orthogonal Eigen values,

$$\lambda_n\left[\left(1 - \frac{\lambda_n}{2}\right) M\left(\frac{6-\lambda_n}{4}, 2, \lambda_n\right) - M\left(\frac{2-\lambda_n}{4}, 1, \lambda_n\right)\right] = 0$$

Cabrera *et al.*, (2001) have defined M as confluent hypergeometric function. With A_n^2 normalization coefficients defined by

$$\int_0^1 d\bar{\rho}\,\bar{\rho}(1 - \bar{\rho}^2)\phi_n(\bar{\rho})\phi_m(\bar{\rho}) = A^2{}_n \delta_{om}$$

where M is Kummer's function and δ_{nm} is the Kronecker delta.

Chapter 10

Application of Computational Fluid Dynamics in Bioreactor Analysis and Design

OBJECTIVE

- To explore the role of computational fluid dynamics (CFD) in bioreactor analysis and design.

10.1 INTRODUCTION

Computational fluid dynamics (CFD) is the analysis of reaction systems in bioreactors using computer-based simulation codes. Here, the numerical techniques are employed to solve the Navier-Stokes equations for given flow geometry and thereby implementing the models for various aspects, viz., turbulence, heat transfer, mass transfer, etc. CFD appears to be the cost-effective way to understand the complex fluid-dynamic situation in multiphase flows (Mathiesen *et al.*, 2000). Computational fluid dynamics is an alternative tool for supporting process design and optimization. The models developed using CFD can be used to obtain a greater degree of details of flow required in design. Contact pattern in bioreactor designs is an important factor which influences flow of reactants and cells in the reaction medium. Combined phenomenon of flow of components in the reactor and the role of various transfer operations are required to be evaluated simultaneously for the analysis and design of bioreactors. For this reason, CFD application in biological reactions is gaining momentum. Biological reactions involve complex interrelationship of transport processes and reaction kinetics. If one considers spatial variations in transport and reactions in biological systems, the problem becomes quite a challenging job due to dynamic situation. Similar situation happens in a bioreactor. To get some concept of such dynamic situation, two typical models may be considered, viz., reactor flow model and CFD including turbulence models. In this chapter, we try to highlight the application of CFD with the help of the state equations for mass, momentum, kinetic energy and energy dissipation which are associated with bioreactor analysis and design.

There are a lot of discrepancies in the phenomenon of modeling and simulation. Probably they are dimensionality of simulation (asymmetric or three-dimensional, turbulence modeling, modeling approaches and the accuracy of numerical predictions which depends on the grid-size).

10.1.1 Modeling Approaches

Let us consider an example of stirred bioreactor. Modeling approaches are experimental based, sliding-mesh techniques, and snap-shot technique.

Experimental Based

A real bioreactor experiment is necessary to specify the boundary conditions for the impeller or mixing elements (Aubin *et al.*, 2004). This method gives clear picture about mixing elements without much intervention of computational resources (Joshi and Ranade, 1990).

Advantages

(1) Reliable description of outflow region of the impeller, i.e., circumferential and time-averaged radial-tangential jet.
(2) Reduced computational expenses for stationary simulation.

Disadvantage

It is difficult to quantify other details in fluid, viz., movement of cells, insoluble reactants, local accumulations of cells and vortex for motion, etc.

Sliding-mesh Technique

This technique is used to avoid the disadvantage in the previous method. This requires successful computational resources for unsteady flow within and outside the impeller swept region (Lane and Koh, 1997).

Snapshot Technique or Multiple Reference Frame Method

This technique is used to simulate flow details between impeller blades. It is also called multiple reference frame method for which experimental data are not required to specify the boundary conditions.

Advantage

Full time-dependent transport equations are not necessary to solve for understanding the problem (Bujalski *et al.*, 2002).

10.1.2 Dimensionality of Simulation

In axisymmetric simulation of liquid flow in stirred reactor using experimental data as the boundary condition, the drag force is used for modeling. Three dimensional simulations can show details in the vessel. Of course, selection of the turbulence model is important. The models are the following.

- Standard $k - \in$ model–This over-estimates energy loss.
- Optimized Chen-Kim $k - \in$ model–This fits experimental data well (Schügerl and Bellgardt, 2000).
- RNG $k - \in$ model–It shows higher value than the measured data (Yakhot and Smith, 1992).
- $k_p - k_T - \in$ or algebraic stress model

We discuss their application to single phase and two-phase flows. For two-phase flow modeling, two approaches are Eulerian and Lagrangian approaches (Schügerl and Bellgardt, 2000).

10.1.3 Difference between Lagrangian and Eulerian Approaches

(1) In Lagrangian approach, the continuous phase is treated as a continuum whereas dispersed gas bubbles are considered as particles.
(2) In the Eulerian approach, dispersed phase is also considered as a continuum resulting in so-called two fluid models.

10.2 FLUID DYNAMIC MODELING

Various approaches to model dispersed multiphase flows have been developed depending on the required resolution. The two-fluid models are divided into the Euler-Euler and Euler-Lagrange models where the difference is based on how the dispersed phase is considered in the system. For example, when it is essential to resolve small-scale fluid dynamics around individual bubbles, it is necessary to use volume of fluid (VOF) approach. With VOF, it is possible to resolve small-scale vortices behind bubbles, bubble-bubble interactions (coalescence/breakup), and mass and heat transfer between bubbles and surrounding liquid. However, application of VOF is usually restricted to simulations of a few bubbles due to the huge computational requirements. If it is reasonable to model the small-scale flow around individual bubbles using lumped parameters such as drag coefficient or mass transfer coefficient, it is necessary to simulate trajectories of individual bubbles for which a better approach is a Euler-Lagrangian analyses. This approach allows one to simulate bubble-scale phenomena accurately, but it becomes computationally too demanding if millions of bubbles are to be simulated over a period of time. In such cases, Eulerian-Eulerian approach is suggested for analysis (Ranade and Utikar, 1999).

10.2.1 Euler–Lagrange Approach

The Euler-Lagrange model considers the liquid phase to be continuous while the gas phase is discrete. The discrete gas phase is accounted for the calculation of the hydrodynamics, assuming quasi-homogenous gas-liquid phase within which flow of the bubbles are closely followed. Individual bubble is modeled in the multiphase flow. So the direct use of bubble-bubble, bubble-liquid interactions, mass transfer with and without biological reactions, bubble coalescence and redispersion is allowed by this model without any errors into the gas phase calculations (Sokolichin and Eigenberger, 1994).

10.2.2 Eulerian–Eulerian Approach

In Eularian-Eularian model, liquid and gas phases are interacting continua, assuming that each element of the finite volume of the space domain contains a respective fraction of the continuous and dispensed phases (Sokolichin and Eigenberger, 1994). Balance equations for gas and liquid are solved simultaneously. Following are the requirement for use of the model.

- Storage space and computational power depend on finite volume elements but not on number of bubbles.
- To avoid error in the gas phase calculation, the equation for the gas phase may be computed on a finer grid than the liquid phase.

Recent advances in computational fluid dynamics encourage vigorous application of CFD to model flows by Eulerian-Eulerian approach in bubble column and airlift reactors. Initial results of such CFD approaches are available in the literature. Apart from Ranade (1995), Jakobsen (2001) has reviewed some of the recent modeling concepts in this field.

10.2.3 Model Equations and Averaging Methods

Generally, a two-fluid approach is used to derive governing continuity and momentum transport equations for multiphase flows. Invariably, some kind of averaging method needs to be employed to derive these equations. Several different averaging methods are suggested by Roco (1993). The starting point for the derivation of governing equations is careful definition of the control volume. In Eulerian-Eulerian approach, control volume is assumed to be large enough to define local phase volume functions. Various averaging procedures (volume averaging, ensemble averaging, time averaging, etc.), their advantages and disadvantages are explained by Jakobsen (2001) and other authors (Banerjee and Chan, 1980; Roco 1993).

10.2.4 Hydrodynamic Parameters

The gas-hold up and liquid circulation velocity are important parameters which provide the link between engineering design parameters, viz., reactor geometry, gas flow rate, etc., and biological parameters.

(a) Gas-hold Up

This is an index of mean residence time of the gas phase. Gas-liquid mass transfer efficiency and liquid circulation velocity are affected by the gas hold up.

(b) Liquid Circulation Velocity

In airlift reactor, liquid circulation velocity, being a major design characteristics, determines the residence time of the various zones and controls the performances of the reactor in terms of mass transfer, heat transfer, mixing, and turbulence.

10.2.5 Hydrodynamic Model

The example of annulus-sparged internal-loop airlift reactor is considered to understand the approach in hydrodynamic model formation. The Euler-Euler model is applicable in this case with the following assumptions:

- The reactor is assumed to be operated at isothermal conditions. If the reactor is operated under non-isothermal conditions, the energy equation should also be considered in the calculations, for which the initial temperatures and temperatures at the boundaries of the two phases have to be specified for the analysis. If information about this is not available in the experimental results, relevant literature (Wongsuchoto and Pavasant, 2004) may be considered for this purpose.
- Due to low gas fractions present in the airlift reactor, the liquid phase is assumed to be unaffected by the presence of the bubbles and, therefore, liquid density is assumed to be constant.

- The gas phase is considered as an incompressible ideal gas, the density of which is given by the ideal gas equation $\rho_g = \frac{P_{op}}{RT}$. This assumption has been widely used which is also applicable to CFD simulations of two phase reactors. If test simulations are carried out for the reactor, neglecting this assumption, will result in severe convergence difficulties.
- Each bubble has a constant bubble diameter. If bubbles are of different diameters, the governing equations have to be written separately for each class of bubbles with identical diameters.
- Mass transfer among the phases is negligible, i.e., all bubbles are generated with constant mass at the sparger and retain their mass as long as they are in the two phase domain.
- The airlift reactor is assumed to be operated at low gas-holdups which indicates relatively large distances between the bubbles. Thus it has negligible interactions such as bubble coalescence and re-dispersion.

(a) Governing Equations of the Model

Turbulence is incorporated in the models developed for the simulation of flows in the two phase reactors, since turbulence has been found to be inherent in these flows. Simulation of two phase flows, considering the flow of both phases as laminar, has found to produce large errors in the simulation results (Pfleger *et al.*, 1999). A test simulation is necessary assuming laminar conditions in both the phases, where convergence occurs in the first few time steps followed by divergence. In the model, turbulence has been considered in both phases.

Turbulent flows are characterized by fluctuating velocity fields. These fluctuations cause other transport quantities such as momentum, energy and species concentration to fluctuate as well. The fluctuations in the quantities are taken into account by expressing instantaneous values of the quantities to be the sum of a mean part (denoted by $< >$) and a time varying fluctuating part (denoted by prime). Therefore, the instantaneous velocity ($\vec{v}_i$) and the instantaneous values of the scalar quantities (Φ) are expressed in Equations (10.1) and (10.2), respectively.

$$\vec{v}_i = <\vec{v}_i> + \vec{v}_i' \qquad (10.1)$$

$$\Phi = <\Phi> + \Phi' \qquad (10.2)$$

With these fluctuations, the instantaneous conservation equations can be written for each phase and simulations carried out to describe the hydrodynamics of the two phase flow. Since these fluctuations can be of small scale and high frequency, they are computationally extensive to simulate directly. Therefore, the complete time-dependent solution of the exact governing equations for high Reynolds number turbulent flows, including the fluctuations in all the quantities, would be highly complex.

For practical engineering situations, where the temporal or spatial fluctuations of the quantities are much smaller than their average value, the predictions of the average values are found to be sufficient (Banerjee and Chan, 1980). The average values of the quantities can be obtained by solving the averaged governing equations obtained by averaging out the local conservation equations in time, volume or over an ensemble, or combinations of these (Banerjee and Chan, 1980; Roco, 1993; Jakobsen, 2001).

In this model, for example, local instantaneous equations of the single-phase are first multiplied using a single ensemble averaging operator known as the phase indicator function (Roco, 1993). Then the equations are subjected to ensemble averaging. Since terms containing averages of the products of the dependent variables are formed, the equations cannot be solved directly. Therefore, to obtain a solvable set of equations, the averages of products have to be related to expressions containing products of averages only, for which the variables are weighted with a respective weighted average (Roco, 1993; Jacobsen, 2001). It should be noted that the closure laws found in the literature are not valid for any model formulation, but only for the model approach in which these are derived. The velocities, $\vec{v}_i$, in the continuity and momentum equations given below, represent the mass weighted (or Favre′) average values by Equation (10.3).

(b) Equation of Continuity

The continuity equation describes the mass flux into and out of a control volume and its integral change of mass. The continuity equations governing the turbulent multiphase flow (Roco, 1993) is given by Equation (10.3).

$$\frac{\partial}{\partial t}(\alpha_i \rho_i) + \nabla \cdot (\alpha_i \rho_i \vec{v}_i) = \sum_{j=l,g} \dot{m}_{ji},\ i = l, g \tag{10.3}$$

where $\dot{m}_{jj} = 0$, $\dot{m}_{ji} = -\dot{m}_{ij}$, α_i is volume fraction (dimensionless), ρ is the density (kg/m^3), $\vec{v}_i$ is the velocity vector (m/s), and $\dot{m}_{ji}$ is mass transfer rate per unit volume from phase j to i (kg/m^3).

On the left hand side of above Equation (10.3), the first term describes the integral change of mass over time, while second term describes the convective flux crossing the boundaries of the volume. The term on the right hand side of Equation (10.3) describes the mass transfer from phase 'j' to 'i'.

If only a two-step averaging process of the governing equations (i.e., substituting the quantities in terms of their mean and fluctuating components and then ensemble averaging) are employed, this would lead to the appearance of source terms on the right hand side of the continuity equation, containing fluctuations in the volume fractions. The expression of the ensemble averaged terms by their respective weighted averages in addition to the two step averaging process (where the phase indicator function is also used), lead to no such source terms in the continuity equation (Jakobsen, 2001). In the simulations, mass transfer occurring between the phases are assumed negligible ($\dot{m}_{ji} = 0$). Hence, the continuity equations become the Equation (10.4).

$$\frac{\partial}{\partial t}(\alpha_i \rho_i) + \nabla \cdot (\alpha_i \rho_i \vec{v}_i) = 0,\ i = l,\ g \tag{10.4}$$

where 'l' and 'g' are liquid and gas phases, respectively.

Detailed derivation of Equations (10.4) and (10.5) is given in the appendix of this chapter.

(c) Equation of Momentum

The momentum conservation equations governing multiphase turbulent flow (Roco, 1993) are given by Equation (10.5).

$$\frac{\partial}{\partial t}(\alpha_i \rho_i \vec{v}_i) + \nabla \cdot (\alpha_i \rho_i \vec{v}_i \vec{v}_i) = -\alpha_i \nabla P + \nabla \cdot \bar{\bar{\tau}}_i + \alpha_i \rho_i \vec{g} + \sum_{j=l,g} \dot{m}_{ji} \vec{v}_{ji} + \vec{F}_{l,i},\ i = l,\ g \tag{10.5}$$

where P is pressure (N/m^2), $\vec{F}$ is force (N/m^2), $\vec{g}$ is acceleration due to gravity and $\bar{\bar{\tau}}$ is shear stress (N/m^2).

The terms on the right hand side of the above equation describe all the forces acting on the phase 'i' fluid element in the control volume. These are the overall pressure gradient, the viscous stresses, the gravitational force and the interphase momentum forces (accounting for the momentum exchange terms between the phases) combined in the term $\vec{F}_{l,k}$. The pressure defined is to be shared by the two phases. It is defined to be equal in both phases. Since the mass transfer occurring between the phases has been assumed to be negligible, Equation (10.6) becomes the equation of momentum.

$$\frac{\partial}{\partial t}(\alpha_i \rho_i \vec{v}_i) + \nabla \cdot (\alpha_i \rho_i \vec{v}_i \vec{v}_i) = -\alpha_i \nabla P + \alpha_i \nabla \cdot \overline{\overline{\tau}}_i + \alpha_i \rho_i \vec{g} + \vec{F}_{l,i},\ i = l, g \tag{10.6}$$

The averaging of the governing equations leads to the formation of terms containing the fluctuating components of the velocity. These terms (i.e., terms containing $-\rho_i \overline{v'_{i,x} v'_{i,x}}$, $-\rho_i \overline{v'_{i,x} v'_{i,y}}$ and $-\rho_i \overline{v'_{i,y} v'_{i,y}}$) constitute the terms of the Reynolds stress tensor, $\overline{\overline{\tau}}''_i$. The $(k, l)^{th}$ component of $\overline{\overline{\tau}}''_i$ is given in terms of the fluctuating components of the velocity as per Equation (10.7).

$$\left(\overline{\overline{\tau}}''_i\right)_{k,l} = -\rho_i \overline{v'_{i,k} v'_{i,l}} \tag{10.7}$$

The Reynolds stress tensor is then related to the mean velocity gradients according to the Boussinesq hypothesis (Jayanta *et al.*, 1999) as

$$\overline{\overline{\tau}}'' = \mu_{i,\text{tur}}\left(\nabla \vec{v}_i + \nabla \vec{v}_i^T - \frac{2}{3}\nabla \cdot \vec{v}_i \overline{\overline{I}}\right) \tag{10.8}$$

An effective viscosity $\mu_{i,\text{eff}}$ is then defined to take into account the laminar and turbulent contributions of the stress tensor (Ranade, 2002) as follows:

$$\mu_{i,\text{eff}} = \mu_{i,\text{lam}} + \mu_{i,\text{tur}}$$

where μ is dynamic viscosity (kg/m·s). (10.9)

The shear stress term in Equation (10.6), is given by

$$\overline{\overline{\tau}}_i = \mu_{i,\text{eff}}\left(\nabla \vec{v}_i + \nabla \vec{v}_i^T - \frac{2}{3}\nabla \cdot \vec{v}_i \overline{\overline{I}}\right) \tag{10.10}$$

The volume fraction of each phase in the governing equations has to satisfy Equation (10.11).

$$\sum_{i=l,g} \alpha_i = 1 \tag{10.11}$$

(d) Interphase Force Terms

The interfacial force term describes the interaction forces between the continuous and the dispersed phase. In a motionless liquid, when the bubble is not in motion, the only forces acting on the bubble are the pressure and gravity forces. When the bubble is in motion, there is a relative motion between the bubble and the liquid, and hence the liquid flow around individual bubbles leads to local variations in pressure and the shear stress. The resulting interaction forces due to these variations are approximated by empirical correlations, since these cannot be considered in detail within the frame of a two fluid model. The various interfacial forces mostly considered in literature, for simulating multiphase flows (Clift *et al.*, 1978; Sokolichin and Eigenberger, 1994; Sokolichin *et al.*, 1997; Crowe *et al.*, 1998; Krishna *et al.*, 2000; Mudde and van den Akker, 2001) are the drag, virtual mass, lift, the turbulent pressure forces, and the turbulent interphase momentum transfer.

Drag Force

When a bubble moves at a uniform velocity in a stagnant liquid, it accelerates part of the liquid around it. This, in turn, slows down the bubble. This force exerted on the bubble in a uniform flow field is called as the drag force (Sokolichin and Eigenberger, 1994). The drag force which is the dominant contributing interfacial force is described by Equation (10.12).

$$\vec{F}_{D,g} = \frac{3}{4}\alpha_g \mu_l C_D d_B^2 (\vec{v}_l - \vec{v}_g) \tag{10.12}$$

$$\vec{F}_{D,l} = -\vec{F}_{D,g}$$

where d_B is bubble diameter.

The drag coefficient C_D is given by the model of Schiller and Naumann (1933) referred by Spidla *et al.*, (2005).

$$C_D = \begin{cases} \frac{24}{\text{Re}}(1 + 0.15\,\text{Re}^{0.687}) \text{ for } (\text{Re} \le 1000) \\ 0.44 \text{ for } (\text{Re} > 1000) \end{cases}$$

The drag force depends primarily on the bubble diameter.

Virtual Mass or Added Mass Force

The drag force takes into account the interaction forces between the bubbles and the liquid in a uniform flow field under non-accelerating conditions. If the bubbles are accelerated relative to the liquid, part of the surrounding liquid has to be accelerated as well. The contribution of additional force is called as the "added mass or virtual mass force" (Sokolichin and Eigenberger, 1999). The virtual mass or added mass force term is given by

$$\vec{F}_{VM,g} = -C_{VM}\alpha_g \rho_l \frac{D}{Dt}(\vec{v}_g - \vec{v}_l) \tag{10.13}$$

$$\vec{F}_{VM,l} = -\vec{F}_{VM,g}$$

The added mass coefficient, C_{VM}, corresponds to the volume fraction of liquid which is accelerated with the bubble. The added mass effect can be neglected if it is assumed that the slip velocity between both phases is constant. This assumption of constant slip velocity is not true for regions where the liquid flow changes the direction as in vortices and at the ends of a loop reactor.

Small bubbles (for example, diameter between 1 mm and 6 mm) are either spherical or ellipsoidal depending on the physical properties of the liquid while large bubbles (for example diameter between 20 and 80 mm) are in the spherical cap regime. The large bubbles undergo frequent bubble coalescence and break up. The small bubbles have a closed wake and the large bubbles donot have a closed wake. So, for the small bubbles, the consideration of the added mass contributions is necessary (Krishna *et al.*, 2000). For these reasons, the virtual mass may be required in all the simulations.

Lift Forces

The lift forces, that give rise to lift on single bubbles in liquids, can roughly be divided into three types: Magnus lift force, Saffman lift force, and the turbulent wake force. If a particle with a rigid surface moves in a non-uniform flow field, the flow field may induce a particle rotation (around its own axis) perpendicular to the main flow direction causing a lift force to act on the bubble. This is caused due to an unsymmetrical pressure field that is created by the interaction between the flow field and the fluid motion. This lift force is called the Magnus lift force.

The non-uniform flow field around the bubble produces a non-uniformity in the shear field acting on the bubble, causing a lift force to act on it. This lift force is called as the Saffman lift force. Since bubbles tend to deform under various flow conditions producing wakes in the process, a slanted wake behind the distorted bubble in a shear field causes a transversal lift force to act on the bubble. This lift force is called the turbulent wake force. The transverse lift forces have been considered in this model. An average model for these forces is given by

$$\vec{F}_{L,g} = C_L \alpha_g \rho_l \left(\vec{v}_g - \vec{v}_l\right) \times \left(\nabla \times \vec{v}_l\right) \tag{10.14}$$

$$\vec{F}_{L,l} = -\vec{F}_{L,g}$$

For large bubbles, a negative lift coefficient has to be used, while a positive lift coefficient has to be used for small bubbles (Krishna *et al.*, 2000).

Turbulent Pressure

The effects of the turbulent pressure accounts for the correlation between the instantaneous distribution of the particles and the undisturbed fluid pressure fluctuations. This effect can be taken into account by adding an additional term of ($-\rho_l <v_l^n v^n_g> \Delta\alpha_g$) in the interfacial momentum term, where the term $<v_l^n v_g^n>$ is the fluid-bubble velocity covariance of the dispersed phase (Oey *et al.*, 2001).

Interphase Turbulent Momentum Transfer

The turbulent fluctuations in the volume fraction of the phases, arising due to turbulence, result in an additional term in the momentum equations called the interphase turbulent momentum transfer term. This effect is taken into account by an additional term $-K_{ij}\vec{v_{dr}}$ to the interfacial term. The exchange coefficient, K_{ij}, is given by Equation (10.15).

$$K_{ij} = \frac{18\mu_j \alpha_i C_D \mathrm{Re}}{24 d_B^2} \tag{10.15}$$

10.2.6 Turbulence Modeling

In order to close the set of equations in the Eulerian-Eulerian multiphase model, it is necessary to specify the additional variable, turbulent viscosity, $\mu_{i,\,\mathrm{tur}}$. For this, a turbulence model is to be specified from which the turbulent velocity can be determined. For the CFD simulations, turbulence has been considered in the liquid phase as well as in the gas phase. A modified $k - \in$ model has been used to describe the turbulence in the continuous phase, whereas the turbulence of the dispersed phase has been described by extended version of Tchen's theory (Ranade, 2002).

(a) Modified k-ε Turbulence Model

Though the $k - \varepsilon$ turbulence model can model only isotropic turbulence (turbulence viscosity is isotropic, i.e., it is same in all directions), it is by far the most widely accepted and used turbulence model. In the standard $k - \in$ turbulence model, two additional transport equations, one for the turbulent kinetic energy (k) and the other for the rate of dissipation of turbulent kinetic energy ($\in$) are introduced into the calculations. The turbulent kinetic energy, 'k_i'(m^2/s^2) is defined by Equation (10.16).

$$k_i = \frac{1}{2}\left(\overline{v_{i,x}'^2} + \overline{v_{i,y}'^2}\right) \tag{10.16}$$

One can use a modified $k - \in$ model to describe the turbulence in the liquid phase, while for the gas phase, turbulence is described using an extended version of Tchen's theory if dispersion of discrete particles caused by homogeneous turbulence. In this theory, the turbulence kinetic energy and turbulence dissipation rate of the dispersed phase are obtained using algebraic equations and are functions of the turbulent kinetic energy and turbulent dissipation rate of the continuous phase (Oey *et al.*, 2001).

The modified $k - \varepsilon$ model is the standard $k - \varepsilon$ model supplemented with extra terms to include interphase turbulence momentum transfer, i.e., terms containing the correlation between the instantaneous distribution of the dispersed phases and the turbulent fluid motion. Usually the standard $k - \varepsilon$ model is used to describe the turbulence in two phase flows (Ranade and Tayalia, 2001). The use of the modified $k - \in$ model along with Tchen's theory instead of the standard $k - \varepsilon$ model, removes the need of an additional equation for ε to be solved along with other governing equations, thus saving computational time. In literature, the use of the modified $k - \varepsilon$ turbulence model along with the extended version of Tchen's theory has been found to produce good results (Jayanta *et al.*, 1999).

(b) Turbulence in Liquid Phase

The conservation equations for turbulent kinetic energy 'k' and turbulent dissipation 'ε' of the liquid phase (Mudde and van Den Akker, 2001) are given by the following equation.

$$\frac{\partial}{\partial t}(\alpha_l \rho_l k_l) + \nabla \cdot (\alpha_l \rho_l \vec{v}_l k_l) = \nabla \cdot \left(\alpha_l \left(\mu_{l,\text{lam}} + \frac{\mu_{l,\text{tur}}}{\sigma_{k,l}}\right)\nabla k_l\right) + \alpha_l G_{k,l} - \alpha_l \rho_l \varepsilon_l + \alpha_l \rho_l \Pi_{k,l} \tag{10.17}$$

$$\frac{\partial}{\partial t}(\alpha_l \rho_l \varepsilon_l) + \nabla \cdot (\alpha_l \rho_l \vec{v}_l \varepsilon_l) = \nabla \cdot \left(\alpha_l \left(\mu_{l,\text{lam}} + \frac{\mu_{l,\text{tur}}}{\sigma_{\varepsilon,l}}\right)\nabla \varepsilon_l\right) + \alpha_l \frac{\varepsilon_l}{k_l}(C_{1\varepsilon} G_{k,l} - C_{2\varepsilon}\rho_l \varepsilon_l) + \alpha_l \rho_l \Pi_{\varepsilon,l} \tag{10.18}$$

where $\in$ is turbulent energy dissipation (m^2/s^2) and $\Pi_{k,l}$ is momentum inter phase transfer.

The generation of turbulence for the continuous phase due to the mean velocity gradients, $G_{k,l}$, is modeled according to the Boussinesq hypothesis (Roco, 1993). The terms $\Pi_{k,l}$ and $\Pi_{\varepsilon,l}$ represent the influence of the interphase between the gas phase and the liquid phase on the liquid phase turbulence. The term $\Pi_{k,l}$ is given by Equation (10.19).

$$\Pi_{k,l} = \sum_{j=l,g} \frac{K_{g,l}}{\alpha_l \rho_l}(< v_l'' v_g'' > - 2\kappa_l + \vec{v}_{gl} \cdot \vec{v}_{dr}) \tag{10.19}$$

where the suffix 'dr' is drift.

The fluid bubble velocity covariance $< v_l''\ v_g'' >$ of the dispersed phase is modeled using the extended version of Tchen's theory (described under the section below). $\vec{v}_{gl}$ is the relative velocity between the gas phase and the liquid phase. The drift velocity, $\vec{v}_{dr}$, arising due to the turbulent fluctuations in the volume fractions of the phases (i.e., due to bubble dispersion caused by the turbulent fluid motion), is given by Equation (10.20).

$$\vec{v}_{dr} = -\left(\frac{D_g}{\sigma_{gl}\alpha_g}\nabla\alpha_g - \frac{D_l}{\sigma_{gl}\alpha_l}\nabla\alpha_l\right) \tag{10.20}$$

where the diffusivities of the continuous phase and the dispersed phase, D_l and D_g, are assumed to be equal to $\frac{1}{3}k_{gl}\tau_{t,gl}$. The term $\Pi_{\varepsilon,l}$ in the turbulence equations of the liquid phase is given by

$$\Pi_{\varepsilon,l} = C_{3\varepsilon}\frac{\varepsilon_l}{k_l}\Pi_{k,l} \tag{10.21}$$

where $C_{3\varepsilon} = 1.2$.

The turbulent viscosity of the liquid phase is given by

$$\mu_{l,\,\mathrm{tur}} = \rho_l C_\mu \frac{k_l^2}{\varepsilon_l} \tag{10.22}$$

The constants used in the modified $k - \varepsilon$ turbulence model (Ranade, 2002) are: $C_\mu - 0.09$, $C_{\varepsilon 1} - 1.4$, $C_{\varepsilon 2} - 192$, $\sigma_k - 1.0$, $\sigma_\varepsilon - \frac{\kappa^2}{(C_{\varepsilon 2} - C_{\varepsilon 1})\sqrt{C_\mu}}$, and $\kappa - 0.4187$.

where C_μ, $C_{\in 1}$, $C_{\in 2}$, k, $\sigma_\in$, σ_k are dimensionless constants in $k - \in$ model.

(c) Turbulence in Gas Phase

Assuming steady and homogeneous fluid turbulence, Tchen's theory of turbulence (Mudde and van Den Akker, 2001) has been extended to predict turbulence in the dispersed phase. For this, three time scales characterizing the interaction between the fluctuating motions of the bubbles are defined. First time scale is the characteristic time of the energetic turbulent eddies, $\tau_{t,l}$, which is given by

$$\tau_{t,l} = \frac{3}{2}C_\mu \frac{k_l}{\varepsilon_l} \tag{10.23}$$

where $\tau_{t,l}$ is characteristics time in Tchen's theory.

The second time scale is the bubble relaxation time, $\tau_{F,gl}$, which is the characteristic time of bubble entrainment by the fluid motion connected with the inertial effects acting on the dispersed phase.

$$\tau_{F,gl} = \alpha_g \rho_l K_{gl}^{-1}\left(\frac{\rho_g}{\rho_l + C_V}\right) \tag{10.24}$$

where $\tau_{F,gl}$ is characteristics time in Tchen's theory.

The third time scale is the eddy-bubble interaction time, $\tau_{t,gl}$, given by

$$\tau_{t,gl} = \frac{\tau_{t,l}}{\sqrt{(1 + C_\beta)\xi^2}} \tag{10.25}$$

where ξ is given by

$$\xi = \frac{|\vec{v}_{gl}|\tau_{t,l}}{L_{t,l}} \tag{10.26}$$

The parameter C_β relates to the Lagrangian over Eulerian characteristic length-scale ratio, and this is described by Equation (10.27).

$$C_\beta = 1.8 - 1.5 \cos^2 \theta \tag{10.27}$$

where θ is the angles in radian.

The ratio between the characteristic time of the energetic turbulent eddies and the characteristic bubble entrainment by the carrier fluid motion is given by η_{gl} (Equation 10.28).

$$\eta_{gl} = \frac{\tau_{t,gl}}{\tau_{F,gl}} \tag{10.28}$$

The turbulent kinetic energy of the dispersed phase k_g and the fluid is equal to the bubble velocity covariance of the dispersed phase $<v_l'' v_g''>$, which are obtained as a function of the carrier-phase kinetic energy using the following equations.

$$k_g = k_l \left(\frac{b + \eta_{gl}}{1 + \eta_{gl}} \right) \tag{10.29}$$

where

$$<v_l'' v_g''> = 2k_l \left(\frac{b + \eta_{gl}}{1 + \eta_{gl}} \right) \tag{10.30}$$

where

$$b = (1 + C_{VM}) \left(\frac{\rho_l + C_{VM}}{\rho_g} \right)^{-1} \tag{10.31}$$

This results in a full set of continuity equations for mass, momentum, and turbulence energy plus closure terms which are used for numerical solution.

10.3 SIMULATION

The commercial computational fluid dynamics software package (for example, Fluent 6.2.18 from Fluent Inc.) can be used for modeling the hydrodynamics of an annulus-sparged internal loop airlift reactor. The experiments carried out by Wongsuchoto and Pavasant (2004), in an annulus-sparged internal loop airlift reactor, is considered as an example for CFD simulation.

10.3.1 Computational Domain

Experimental Statement

The experiments of Wongsuchoto and Pavasant (2004) were carried out in an annulus-sparged internal loop airlift reactor of height 1.2 m and diameter 0.137 m. The draft tube inserted into the reactor had a height of 1 m, inner diameter of 0.034 m, outer diameter of 0.04 m and a clearance of 5 cm from the reactor base. The unaerated water level was controlled at 3 cm above the top of the draft tube. The air sparger used has perforated rings made of a 0.8 cm diameter tubing with 14 orifices of 1 mm diameter. The sparger is located at the base of the annulus section.

The CFD simulation can be set up in 2D Cartesian co-ordinates. A rectangular computational domain of breadth (length in the horizontal direction) equal to the diameter of the reactor can be chosen for this purpose. If the height of the computational domain is taken to be equal to the reactor height and the top of the computational domain is defined as an outlet (i.e., allowing the gas-liquid interface to be within the computational domain), then severe convergence difficulties occur. So, an alternative approach to model the outlet of the reactor needs to be undertaken. An alternative approach found in the literature assumes a reasonable solution domain height and then to take the top surface of the solution domain to be coinciding with the free surface of the dispersion (Mudde *et al.* 2001; Ranade, 2002). The free surface is also to be assumed as flat. In accordance with this approach, the height of the rectangular computational domain may be fixed at 1.1 m for all the simulations. The top section of the computational domain is then defined as a velocity inlet for both the phases. Ranade and Tayalia (2001) used a similar approach to simulate the flow in a shallow bubble column, but they assumed that the entire column to be filled with gas-liquid dispersion.

Modification of the Problem

The ring sparger used in the experimental setup is modeled by defining two sections at the bottom of the computational domain as sparger inlets. These sections, which have a thickness of 0.008 m, are placed at a distance of 0.111 m apart from each other. The draft tube is included in the computational domain by removing two rectangular faces from the face of the computational domain. The faces, which are removed has a height of 1 m, width 0.003 m, are separated by a distance of 0.034 m and are located at a distance of 0.05 m from the base of the domain. The new edges, that formed in the process, are then defined as walls. The right hand side most edge, the left hand side most edge and the bottom of the computational domain (excluding the sparger inlets) are also defined as walls.

10.3.2 Geometry Grid Generation

The correct choice of an appropriate grid is of crucial importance for a numerical solution procedure in CFD simulations. The numerical solution procedure implemented in the simulations is the finite volume scheme, which makes the discretization of the flow domain into sufficiently small grid cells as necessary. The mesh width (or grid cell size) must not be too large so that errors due to the grid selected, are significant. The choice of smaller grid size yields to large number of cells for which the discretized equations have to be solved and this also leads to the requirement of smaller time steps to correctly calculate the fast variations in the local vortices. These lead to immense computational demand with respect to the processor time and memory usage. A structured mesh is used to generate the grid. The grid generation for simulations can be carried out using a commercial software (for example, GAMBIT 2.2.30)

10.3.3 Initial Conditions

The various initial conditions, are specified by Ranade (2002), which may be used for the simulation. These are: pressure- 106710.788 Pa, X-component and Y-components of velocity of water as well as air are zero m/s, turbulent kinetic energy and turbulent dissipation rate for energy of water – 0.03 m^2/s^2, volume fraction of air-zero. The specification of the initial values for the turbulent kinetic energy and the turbulent dissipation rate for energy of water led to the turbulent viscosity to be initially about 2.7 times the laminar viscosity. So the simulation started with fully developed turbulent flow, which is required for the modified $k - \varepsilon$, used as the turbulence model, to be valid.

Test simulations with different initial values specified for the turbulence quantities could show no significant differences in the simulation results.

10.3.4 Boundary Conditions

The various boundaries that exist in the computational domain are the top surface, sparger inlets, and the walls.

(a) Top Surface

For multiphase calculations, FLUENT 6.2.16 or equivalent software allows only the use of the segregated solver. In Fluent, when the segregated solver is used and the flow exits the computational domain through a velocity inlet boundary condition, it is required to specify only the normal velocity component (component in the y-direction) since all other flow conditions, except the normal velocity component, are assumed to be equal to that in the upstream cell. Hence, at the top surface, only the normal components of the velocity of the air and water are specified (Joy and Panda, unpublished results).

Since the top surface of the computational domain coincides with the free surface of the dispersion, no amount of water can leave the computational domain. To ensure this, the normal component of water at the top surface has been specified as zero. To ensure that the air bubbles leave the computational domain at their terminal velocity, the normal component of air at the top surface is specified to be equal to the terminal velocity. The consideration of the top surface as a velocity inlet requires the specification of the velocity component normal to the top surface, with the direction of the normal pointing towards the computational domain (i.e., negative y-direction). Therefore, the normal component of air ($v_{g,\,top}$), at the top surface, is specified to be equal to the negative of the bubble terminal velocity ($U_{b,\infty}$) The bubble terminal velocity is calculated from Equation (10.32) (Ranade and Utikar, 1999).

$$U_{b,\infty} = \sqrt{\left(\frac{2\sigma}{\rho_l d_B} + \frac{g d_B}{2}\right)} \tag{10.32}$$

Such a specification for the top surface computational domain has been reported in literature as a reasonable assumption (Ranade and Tayalia, 2001). The volume fraction of air at the top surface ($\alpha_{g,\,top}$) is then calculated from Equation (10.33).

$$\alpha_{g,\,top} = \frac{v_{g,\,superficial}}{v_{g,\,top}} \tag{10.33}$$

Since no additional governing equations for k and ε are specified for the dispersed phase, only k and ε of water at this boundary need to be specified in this case. This has been done by specifying the turbulent intensity (defined as the ratio of the root-mean-square of the velocity fluctuations to the mean flow velocity) and the turbulent viscosity ratio. Since the turbulence conditions are not known, we have used the generally defined types of turbulence, such as the low, medium and high turbulence. The turbulence intensities and the turbulent viscosity ratios for these types of turbulence are described by Ranade (2002).

The turbulent kinetic energy is then calculated using Equation (10.34) and the turbulent dissipiation rate is calculated using Equation (10.35)

$$k = 1.5\,(I v_{avg})^2 \tag{10.34}$$

$$\varepsilon = \frac{C_\mu k^2}{\mu_{\text{lam}}} \left(\frac{\mu_{\text{tur}}}{\mu_{\text{lam}}} \right)^{-1} \qquad (10.35)$$

where the ratio $\frac{\mu_{\text{tur}}}{\mu_{\text{lam}}}$ is the turbulent viscosity ratio and v_{avg} constitutes the average velocity of flow.

(b) Sparger Inlets

At the sparger inlet, both the components of the water velocity are specified to be zero. Medium turbulence boundary conditions are specified for water at the sparger inlets. The velocity of air (velocity vector normal to the sparger inlet) is calculated using Equation (10.36).

$$v_{g,\text{ inlet}} = \frac{v_{g,\text{ sup}}}{\alpha_{g,\text{inlet}}} \frac{A_{\text{reactor}}}{A_{\text{sparger}}} \frac{P_{\text{op}}}{P_{\text{sparger}}} \qquad (10.36)$$

where A is the area (m^2).

The volume fraction of air at this boundary may be assured at 0.25. The term $\frac{P_{\text{op}}}{P_{\text{sparger}}}$ is called as the pressure correction term. Since the superficial gas velocity has been specified with respect to the top of the reactor, where the pressure is the atmospheric pressure and the pressure at the sparger P_{sparger} is higher than this value, it is necessary to include this correction term while calculating the velocity of air at the sparger. In the simulations, P_{sparger} is calculated using Equation (10.37).

$$P_{\text{sparger}} = P_{\text{op}} + h\rho_l g \qquad (10.37)$$

(c) Walls

The walls that are considered in the simulation are impermeable, which does not allow liquid or heat into its surface. The base of the reactor, sides of the reactor and the internal draft tube (except the sparger inlets and top outlet) are treated as walls. Standard wall functions are used as wall boundary conditions to make all these walls as impermeable.

(d) Material Properties and Operating Conditions

The operating pressure (P_{op}) is specified as atmospheric pressure (1.01325×10^5 Pa) and the reference pressure location is fixed at $x = 0.0$ m and $y = 1.1$ m (Wongsuchoto and Pavasant, 2004). Gravity forces are included in the calculations, where the x-component of the acceleration of gravity is taken to be 0 and the y-component is specified to be -9.81 m/s^2. The density of water is taken to be 98.2 kg/m^3 and its viscosity is specified as 0.001003 kg/(m)(s). The density of air is calculated using the ideal gas equation, while its viscosity is specified to be 1.7894×10^{-5} kg/(m)(s).

For solving the governing equations, the segregated solver is used, where the formulation of the linearized equations is done using the implicit formulation. The staggered grid formulation is used, which means that the scalar quantities are attached to the centers of the control volume, and the velocity components are calculated for the centers of the control cells. A second order accurate central-difference scheme may be employed for the discretization of the diffusion terms. A fully implicit backward difference scheme is used for the time integration. The pressure velocity coupling is obtained by using the PHASE COUPLED SIMPLE algorithm.

The QUICK scheme is used for the discretization of the convective terms in the continuity equation. The POWER LAW scheme may be employed for the discretization of the convective terms in the momentum and turbulence equations. Higher order schemes such as the SECOND ORDER and the QUICK scheme are also used, but this leads to large convergence difficulties. The use of the POWER LAW gives sufficiently accurate results. Detailed explanations of the algorithms and the discretization schemes can be found in literature (Versteeg and Malalasekara, 1993; Ranade, 2002).

The simulations can be carried out till a quasi-steady state is achieved, i.e., when the volume- and time-averaged quantities such as the liquid velocities in the riser and down-comer and the gas-holdups in the riser and down-comer sections (quantities of interest) attain a constant value. For each time step, the convergence criteria for the scaled residuals can be set to 1×10^{-3}. The under relaxation forces can be set to 0.5 for pressure and 0.3 for the velocities. The use of higher values under relaxation forces for velocities might give rise to convergence difficulties. A simulation run for a few hundred seconds in a suitable processor may continue for several hours.

10.3.5 Evaluation of Design Parameters

If transient experimental data are not available, simulations may be done by calculating time averaged values followed by volume-averaged values which is finally used for comparisons. For simulations, we try to give steps to calculate important design parameters. Certain condition in steps might vary depending on the problem for simulation. Following steps with conditions are specific for annular-sparged internal-loop airlift reactor.

Step 1: Selection of geometry grid
Test simulations are necessary for a specific superficial velocity of air with a time step which may be 0.015. Initially grid structure will be coarse. For example, with a superficial velocity of air 0.03 m/s and time step of 0.01s, 29 cells in the horizontal direction and 109 cells in the vertical directions (i.e. grid size: 29×109) can be considered for initial coarse grid (Joy and Panda, unpublished results).

Coarse grid, thus results, requires refinement to obtain optimal gird. In each step of refinement in both the horizontal and vertical directions, it is necessary to carry out test simulations. For example, the experimental results of velocity of liquid in the down comer (v_{ld}) and mass fractions of gas in the down comer (α_{gd}) may be useful for comparison. One can find no improvement in refinement either in horizontal or in vertical direction. If the refinement gives positive improvement in one direction, the refinement is done only in that direction.

Step 2: Selection of time step
The choice of time steps starts from the initial value for coarse grid selection. In this case, it is 0.01s. If the step value is higher than 0.01s, it might cause an increase of turbulent viscosity ratio. One needs to specify certain value of viscosity ratio, for example; 10^5 for simulation. If the time step selected does not change the results appreciably, this might cause massive increase in computation time. In each simulation, the simulated values of $v_{r,d}$ and $\alpha_{g,d}$ are compared with this corresponding experimental values.

Step 3: Turbulence boundary conditions

In absence of experimental data for turbulence boundary conditions, test simulations are generally carried out using three different turbulence conditions, viz., low, medium and high turbulence conditions at the top of the computational domain and at the sparger inlets. Tchen's theory might be useful to describe the turbulence in the dispersed phase. Turbulent boundary conditions for the continuous phase (fermentation medium) requires selection. For guidance, one can assume turbulence intensity of 0.05 and turbulent viscosity ratio of 10. Average values of gas-holdups, liquid velocities, and turbulent kinetic energies are calculated in each case of simulation.

Step 4: Studies on important design parameters

(a) Effect of interfacial forces: Various interfacial forces, viz., drag force, drag and virtual mass force, lift forces with different lift coefficients are considered individually for separate test simulations with the refined conditions of steps 1–3. One can calculate turbulent kinetic energy (k, m^2/s^2) in the riser and in the down comer for each test simulation. Test simulation can be carried out by considering turbulent momentum transfer, turbulent pressure, and lift forces terms individually.

(b) Flow regimes: The knowledge of flow regimes in airlift reactor in particular is essential for the design. The simulations are generally considered under transient conditions till a quasi-steady state is attained. Quasi-steady state is defined when time averaged values of the main flow variables reach a constant value. For this reason, time averaged values need to be continuously monitored during simulation. Variables of interest are gas-holds ups and liquid velocity in riser and in down comer sections of the reactor. To determine the flow regimes, simulations run at various superficial velocities. If at a particular superficial velocity no gas entrains in the down comer, the reactor is said to operate in flow regime I. If some gas entrains in to the down comer at a particular superficial gas velocity, the reactor is believed to operate in flow regime II.

EXERCISES

10.1 For a stirred tank bioreactor (Fig. 12.1), carry out complete CFD simulation of flow.

10.2 For a bioreactor with Rushton turbine impeller following properties associated with the bioreactor are necessary in the design context.

(a) Finite volume grid analysis
(b) Radial profiles of the time averaged tangential velocity component
(c) Vertical profiles of velocity vectors through the center of the tank.

Dimensions of the reactor is given in Figure 12.1

REFERENCES

Aubin J, Fletcher DF and Xuereb C (2004) "Modeling turbulent flow in stirred tanks with CFD: The influence of the modeling approach, turbulence model and numerical scheme", *Experimental Thermal and Fluid Science* **28**, 431-435.

Banerjee S and Chan AMC (1980) "Separated flow models-I, Analysis of the averaged and local instantaneous formulations". *International Journal of Multiphase Flow*, **6**, 1-24.

Bujalski W, Jaworski Z and Nienow AW (2002) "CFD Study of Homogenization with Dual Rushton Turbines—Comparison with Experimental Results: Part II: The Multiple Reference Frame", *Chemical Engineering Research and Design*, **80**, 97-104.

Clift R, Grace JR and Weber ME (Eds) (1978) *Bubbles, drops and particles*, Academic Press, London.

Crowe C, Sommerfield M and Tsuji Y (Eds) (1998) *Multiphase flows with drops and particles*. CRC Press LLC, Florida.

Jakobsen HA (2001) "Phase distribution phenomena in two-phase bubble column reactors". *Chemical Engineering Science*, **56**, 1049-1056.

Jayanta S, Vasquez S, Roy S and Dudukoic, MP (1999) "Numerical simulation of gas-liquid dynamics in cylindrical bubble column reactor". *Chemical Engineering Science*, **54**, 5071-5086.

Joshi JB and Ranade VV (1990) *Trans I Chem. Eng*, **68**, 19.

Krishna R, van Baten JM and Urseanu MI (2000) "Three-phase Eulerian simulations of bubble column reactors operating in the churn turbulent regime: a scale up strategy". *Chemical Engineering Science*, **55**, 3275-3286.

Lane GL and Koh PTL (1997) "CFD simulation of a Rushton turbine in a baffled tank", *International Conference on CFD in Mineral & Metal Processing and Power Generation*, CSIRO, pp. 377-38.

Mathiesen V, Solberg T and Hjertager BH (2000) "An experimental and computational study of multiphase flow behavior in a circulating fluidized bed", *International Journal of Multiphase Flow*, **26**, 387-41.

Mudde RF and van Den Akker HEA (2001) "2D and 3D simulations of an internal airlift loop reactor on the basis of a two-fluid model", *Chemical Engineering Science*, **56**, 6351-635.

Oey RS, Mudde RF, Portela LM and van den Akker HEA (2001) "Simulation of a slurry airlift using a two-fluid mode". *Chemical Engineering Science*, **56**, 673-68.

Pfleger D, Gomes S, Gilbert N and Wagner HG (1999) "Hydrodynamic simulations of laboratory scale bubble columns". *Chemical Engineering Science*, **54**, 5091-509.

Ranade VV (1995) "Computational fluid dynamics for chemical reactor engineering", *Reviews in Chemical Engineering*, **11**, 229-28.

Ranade VV (Ed) (2002) *Computational Flow Modeling for Chemical Reactor Engineering*, Academic Press, London.

Ranade VV and Tayalia Y (2001) "Modelling of fluid dynamics and mixing in shallow bubble column reactors: Influence of sparger design" *Chemical Engineering Science*, **56**, 1667-1675.

Ranade VV and Utikar RP (1999) "Dynamics of gas-liquid flow in bubble column reactors". *Chemical Engineering Science*, **54**, 5237-5244.

Roco MC (Ed) (1993) *Particulate two-phase flow*, Reed Publishing Inc., USA.

Schiller L, Neumann Z (1933) Z. Vev. Dtsclh. Ing., 318.

Schügerl K and Bellgaradt KH (Eds) (2000) *Bioreaction engineering: Modeling and control*, Springer-Verlag Berlin, Heidelberg.

Sokolichin A and Eigenberger G (1994) "Gas-liquid flow in bubble columns and loop reactors Part 1 Detailed modelling and numerical simulation". *Chemical Engineering Science*, **49**, 573-5746.

Sokolichin A, Eigenberger G, Lapin A and Luebert A (1997) "Dynamic numerical simulation of gas-liquid two phase flows Euler/Euler versus Euler/Lagrange. *Chemical Engineering Science*, **52**, 611-626.

Špidla M, Mŏstěk M, Sinevič V, Jahodo M and Machon V (2005) "Experimental assessment and CFD simulation of local solid concentric profiles in a pilot-scale stirred tank", *Chem. Pap.* 59(6a) 386-393.

Versteeg HK and Malalasekera W (Eds) (1993) *An introduction to computational fluid dynamics: The finite volume method*, Longman Group Ltd., England.

Wongsuchoto P and Pavasant P (2004) "Internal liquid circulation in annulus sparged internal loop airlift reactor. *Chemical Engineering Journal*, **100**, 1-9.

Yakhot V and Smith LM (1992) "The renormalization grouts, E-expansion and derivation of turbulence models" *Journal Scientific Computing*, **7**, 35-61.

APPENDIX 10

Detailed derivation of Equations (10.4) and (10.5)

The balance equation of motion is given below.

$$\frac{\partial \rho}{\partial t} + \nabla \cdot (\rho \vec{\upsilon}) = 0 \qquad \text{(Continuity equation)}$$

$$\frac{\partial \rho \vec{\upsilon}}{\partial t} + \nabla \cdot \rho \vec{\upsilon}\vec{\upsilon} = \nabla \cdot \tau + \rho \vec{g} - \nabla P + \vec{F} \qquad \text{(Momentum conservation equation)}$$

Averaging procedure adopted is summarized here.

Step 1: Instantaneous values = mean part + time varying fluctuating part

$$\vec{\upsilon}_i = \langle \vec{\upsilon}_i \rangle + \vec{\upsilon}_i'$$

$$\phi = \langle \phi \rangle + \phi'$$

Step 2: Multiplying exact equations for each phase with phase indicator function X_k,

Then employing "Ensemble averaging procedure",

$$\text{Ensemble average} = \frac{\Sigma \text{ values of variables}}{\text{Number of observations}}$$

Step 3: Treat velocities in Favre's approach and use phase averaged values for quantities other than velocity (Favre weighted average).

Continuity equation:

$$\frac{\partial \rho}{\partial t} + \nabla \cdot (\rho \vec{\upsilon}) = 0 \qquad \text{(Let the equation represent continuity of phase } k\text{)}$$

write

$$\vec{\upsilon} = \upsilon + \upsilon'$$

$$\frac{\partial \rho}{\partial t} + \nabla \cdot \rho(\upsilon + \upsilon') = 0$$

$$\frac{\partial \rho}{\partial t} + \nabla \cdot \rho\upsilon + \nabla \cdot \rho\upsilon' = 0$$

$$X_k \frac{\partial \rho}{\partial t} + X_k \nabla \cdot \rho\upsilon + X_k \nabla \cdot \rho\upsilon' = 0$$

$$\overline{X_k \frac{\partial \rho}{\partial t}} + \overline{X_k \nabla \cdot \rho\upsilon + X_k \nabla \cdot \rho\upsilon'} = 0$$

$$\overline{X_k \frac{\partial \rho}{\partial t}} + \overline{X_k \nabla \cdot \rho\upsilon} + \overline{X_k \nabla \cdot \rho\upsilon'} = 0$$

but

$$\frac{\overline{\partial \rho X_k}}{\partial t} = \overline{X_k \frac{\partial \rho}{\partial t}} + \overline{\rho \frac{\partial X_k}{\partial t}}$$

$$\partial\frac{\overline{\rho X_k}}{\partial t} = \overline{X_k\frac{\partial\rho}{\partial t}} + \overline{\rho\frac{\partial X_k}{\partial t}}$$

and
$$\nabla\cdot\overline{X_k\rho\upsilon} = \overline{X_k\nabla\cdot\rho\upsilon} + \overline{\rho\upsilon\nabla\cdot X_k}$$

$$\nabla\cdot\overline{X_k\rho\upsilon} = \overline{X_k\nabla\cdot\rho\upsilon} + \overline{\rho\upsilon\nabla\cdot X_k}$$

also
$$\nabla\cdot\overline{X_k\rho\upsilon'} = \overline{X_k\nabla\cdot\rho\upsilon'} + \overline{\rho\upsilon'\nabla\cdot X_k}$$

$$\therefore\quad \frac{\partial\overline{\rho X_k}}{\partial t} - \frac{\overline{\rho\partial X_k}}{\partial t} + \nabla\cdot\overline{X_k\rho\upsilon} - \overline{\rho\upsilon\nabla\cdot X_k} + \nabla\cdot\overline{X_k\rho\upsilon'} - \overline{\rho\upsilon'\nabla\cdot X_k} = 0$$

Mean of fluctuating quantity = 0

and $X_k = 0$ or 1

Therefore, the equation simplifies to $\dfrac{\partial\overline{\rho X_k}}{\partial t} + \nabla\cdot\overline{X_k\rho\upsilon} = 0$

Here, we introduce weighted average values.

Phase average weighted: $\rho^{\omega} = \dfrac{\overline{X_k\rho}}{\overline{X_k}} = \dfrac{\overline{X_k\rho}}{\alpha_k}$, where $\alpha_k = \overline{X_k}$ = volume fraction of phase k

Velocity is Favre (mass) weight average: $\upsilon^m = \dfrac{\overline{X_k\rho\upsilon}}{\overline{X_k\rho}} = \dfrac{\overline{X_k\rho\upsilon}}{\alpha_k\rho^{\omega}}$

Therefore, $\dfrac{\partial\overline{\rho X_k}}{\partial t} + \nabla\cdot\overline{X_k\rho\upsilon} = 0$ becomes

$$\frac{\partial\alpha_k\rho^{\omega}}{\partial t} + \nabla\cdot\alpha_k\rho^{\omega}\upsilon^m = 0$$

Momentum conservation equation is given below with $P = P + P'$ (consider for a phase k).

$$\frac{\partial(\rho\vec{\upsilon})}{\partial t} + \nabla\cdot(\rho\vec{\upsilon}\vec{\upsilon}) = -\nabla P + \nabla\cdot\overline{\overline{\tau}} + \rho\vec{g}$$

$$\frac{\partial(\rho\upsilon)}{\partial t} + \frac{\partial(\rho\upsilon')}{\partial t} + \nabla\cdot\rho(\upsilon+\upsilon')(\upsilon+\upsilon') = -\nabla P + \nabla\cdot\overline{\overline{\tau}} + \rho\vec{g} - \nabla P'$$

$$\frac{\partial(\rho\upsilon)}{\partial t} + \frac{\partial(\rho\upsilon')}{\partial t} + \nabla\cdot\rho\upsilon\upsilon + 2\nabla\cdot\rho\upsilon\upsilon' + \nabla\cdot\rho\upsilon'\upsilon' = -\nabla P + \nabla\cdot\overline{\overline{\tau}} + \rho\vec{g} - \nabla P'$$

$$X_k\frac{\partial(\rho\upsilon)}{\partial t} + X_k\frac{\partial(\rho\upsilon')}{\partial t} + X_k\nabla\cdot\rho\upsilon\upsilon + 2X_k\nabla\cdot\rho\upsilon\upsilon' + X_k\nabla\cdot\rho\upsilon'\upsilon'$$
$$= -X_k\nabla P + X_k\nabla\cdot\overline{\overline{\tau}} + X_k\rho\vec{g} - X_k\nabla P'$$

$$\overline{X_k\frac{\partial(\rho\upsilon)}{\partial t}} + \overline{X_k\frac{\partial(\rho\upsilon')}{\partial t}} + \overline{X_k\nabla\cdot\rho\upsilon\upsilon} + 2\overline{X_k\nabla\cdot\rho\upsilon\upsilon'} + \overline{X_k\nabla\cdot\rho\upsilon'\upsilon'}$$
$$= -\overline{X_k\nabla P} + \overline{X_k\nabla\cdot\overline{\overline{\tau}}} + \overline{X_k\rho\vec{g}} - \overline{X_k\nabla P'}$$

but

$$\overline{\frac{\partial (X_k \rho \upsilon)}{\partial t}} = \overline{X_k \frac{\partial (\rho \upsilon)}{\partial t}} + \overline{\rho \upsilon \frac{\partial X_k}{\partial t}}$$

$$\frac{\partial}{\partial t}(\overline{X_k \rho \upsilon}) = \overline{X_k \frac{\partial (\rho \upsilon)}{\partial t}} + \overline{\rho \upsilon \frac{\partial X_k}{\partial t}}$$

and

$$\overline{\nabla \cdot X_k \rho \upsilon \upsilon} = \overline{X_k \nabla \cdot \rho \upsilon \upsilon} + \overline{\rho \upsilon \upsilon \nabla \cdot X_k}$$

$$\nabla \cdot \overline{X_k \rho \upsilon \upsilon} = \overline{X_k \nabla \cdot \rho \upsilon \upsilon} + \overline{\rho \upsilon \upsilon \nabla \cdot X_k}$$

also

$$\nabla \cdot \overline{X_k \rho \upsilon \upsilon'} = \overline{X_k \nabla \cdot \rho \upsilon \upsilon'} + \overline{\rho \upsilon \upsilon' \nabla \cdot X_k}$$

$$\nabla \cdot \overline{X_k \rho \upsilon' \upsilon'} = \overline{X_k \nabla \cdot \rho \upsilon' \upsilon'} + \overline{\rho \upsilon' \upsilon' \nabla \cdot X_k}$$

On substitution

$$\frac{\partial}{\partial t}(\overline{X_k \rho \upsilon}) - \overline{\rho \upsilon \frac{\partial X_k}{\partial t}} + \frac{\partial}{\partial t}(\overline{X_k \rho \upsilon'}) - \overline{\rho \upsilon' \frac{\partial X_k}{\partial t}} + \nabla \cdot \overline{X_k \rho \upsilon \upsilon} - \overline{\rho \upsilon \upsilon \nabla \cdot X_k} + 2\nabla \cdot \overline{X_k \rho \upsilon \upsilon'}$$

$$-2\overline{\rho \upsilon \upsilon' \nabla \cdot X_k} + \nabla \cdot \overline{X_k \rho \upsilon' \upsilon'} - \overline{\rho \upsilon' \upsilon' \nabla \cdot X_k} = -\alpha_k \nabla P + \nabla \cdot \overline{X_k \overline{\overline{\tau}}} + \alpha_k \rho^{\omega} \vec{g}$$

Ensemble average of fluctuating quantity = 0

The equation simplifies to,

$$\frac{\partial}{\partial t}(\overline{X_k \rho \upsilon}) + \nabla \cdot \overline{X_k \rho \upsilon \upsilon} + \nabla \cdot \overline{X_k \rho \upsilon' \upsilon'} = -\alpha_k \nabla P + \nabla \cdot \overline{X_k \overline{\overline{\tau}}} + \alpha_k \rho^{\omega} \vec{g}$$

The term $\nabla \cdot \overline{X_k \rho \upsilon' \upsilon'}$ contains fluctuating parts of velocity.

$$\overline{\overline{\tau}}_{\text{total}} = \overline{\overline{\tau}}_{\text{lam}} + \overline{\overline{\tau}}''_{\text{turb}}$$

$$\overline{\overline{\tau}}_{ij} = \mu_{\text{lam}} \frac{\partial \upsilon_i}{\partial X_j} - \overline{\rho \upsilon'_{k,i} \upsilon'_{k,j}}$$

Reynolds stress term $(\tau'')_{k,l} = -\overline{\rho_i \upsilon'_{i,k} \upsilon'_{i,l}}$

Defining phase weighted averages for laminar (molecular) and turbulent stresses

$$\overline{\overline{\tau}}^{\omega} = \frac{\overline{X_k \overline{\overline{\tau}}}}{\overline{X_k}} = \frac{\overline{X_k \overline{\overline{\tau}}}}{\alpha_k}$$

$$(\overline{\overline{\tau}})^{\omega}_{\text{turb}} = -\frac{\overline{X_k \rho \upsilon' \upsilon'}}{\overline{X_k}} = -\frac{\overline{X_k \rho \upsilon' \upsilon'}}{\alpha_k}$$

So the equation becomes

$$\frac{\partial}{\partial t}\alpha_k \rho^{\omega} \upsilon^m + \nabla \cdot \alpha_k \rho^{\omega} \upsilon^m \upsilon^m = -\alpha_k \nabla P + \alpha_k \nabla \cdot \overline{\overline{\tau}}^{\omega} + \alpha_k \rho^{\omega} \vec{g} - \nabla \cdot \overline{X_k \rho \upsilon' \upsilon'}$$

$$\frac{\partial}{\partial t}\alpha_k\rho^{\omega}\upsilon^{m}+\nabla\cdot\alpha_k\rho^{\omega}\upsilon^{m}\upsilon^{m} = -\alpha_k\nabla P+\alpha_k\nabla\cdot\overline{\overline{\tau}}^{\omega}+\alpha_k\rho^{\omega}\vec{g}+\alpha_k\nabla\cdot\overline{\overline{(\tau)}}_{\text{turb}}^{\omega}$$

Therefore, $\frac{\partial}{\partial t}\alpha_k\rho^{\omega}\upsilon^{m}+\nabla\cdot\alpha_k\rho^{\omega}\upsilon^{m}\upsilon^{m} = -\alpha_k\nabla P+\alpha_k(\nabla\cdot\overline{\overline{\tau}}^{\omega}+\nabla\cdot\overline{\overline{(\tau)}}_{\text{turb}}^{\omega})+\alpha_k\rho^{\omega}\vec{g}$

$$\frac{\partial}{\partial t}\alpha_k\rho^{\omega}\upsilon^{m}+\nabla\cdot\alpha_k\rho^{\omega}\upsilon^{m}\upsilon^{m} = -\alpha_k\nabla P+\alpha_k\nabla\cdot\overline{\overline{\tau}}_{\text{total}}+\alpha_k\rho^{\omega}\vec{g}$$

∴ The final averaged equations for phase 'i'

$i = p$ for gas, q for liquid

Continuity equation is: $\frac{\partial(\alpha_i\rho_i)}{\partial t}+\nabla\cdot(\alpha_i\rho_i\vec{\upsilon_i}) = 0$

Momentum equation is: $\frac{\partial}{\partial t}(\alpha_i\rho_i\vec{\upsilon_i})+\nabla\cdot(\alpha_i\rho_i\vec{\upsilon_i}\vec{\upsilon_i}) = -\alpha_i\nabla P+\alpha_i\nabla\cdot\overline{\overline{\tau}}_i+\alpha_i\rho_i\vec{g}+\vec{F_{I,i}}$

$\vec{F_{I,i}}$ represents the interfacial force terms (e.g., drag, lift, virtual mass forces, etc.). $\vec{\upsilon_i}$ is Favre average value of velocity of phase 'i'. $\rho_i, \overline{\overline{\tau}}_i$ are phase weighted averages of density and shear stress of phase 'i'.

Drag force is:

$$\vec{F_{D,g}} = \frac{3}{4}\alpha_g\alpha_l\mu_l C_D d_B^2(\vec{\upsilon_l}-\vec{\upsilon_g})$$

Also, $\vec{F_{D,l}} = -\vec{F_{D,g}}$

Drag coefficient (C_D) is modeled by Schiller & Nauman (1933)

Virtual mass force is

$$\vec{F}_{\upsilon M,g} = -C_{\upsilon M}\alpha_g\rho_l\frac{D}{Dt}(\vec{\upsilon_g}-\vec{\upsilon_l})$$

$$\vec{F}_{\upsilon M,l} = -\vec{F}_{\upsilon M,g}$$

Lift force is:

$$\vec{F}_{L,g} = C_L\alpha_g\rho_l(\vec{\upsilon_g}-\vec{\upsilon_l})\times(\nabla\times\vec{\upsilon_l})$$

Average model:

$$\vec{F_{L,l}} = -\vec{F_{L,g}}$$

Turbulent pressure

Additional term: $-\rho_l k_{pq}\nabla\alpha_g$

wher k_{pq} = fluid bubble velocity covariance $= \langle v_l'' v_g'' \rangle$

Interphase turbulent momentum transfer is

$$= K_{ij}\vec{\upsilon_{dr}}$$

$$K_{ij} = \frac{18\mu_j\alpha_i C_D \,\text{Re}}{24 d_B^2}$$

Other terms encountered in modeling are:

Generation of kinetic energy: $G_k = -\rho\overline{\upsilon_i'\upsilon_j'}\frac{\partial \upsilon_j}{\partial X_i}$ (Boussinesq's hypothesis)

Exchange coefficient:

$$K_{pq} = K_{qp} = \frac{\alpha_q \alpha_p \alpha_p f}{\tau_p}$$

$$\tau_p = \frac{\rho_p d_p}{18\mu_l}$$

$$D_p = D_q, = D_t, p_q = \frac{1}{3} K_{pq} \tau_t, _{pq}$$

where

τ_p is point relaxation time
f is the drag function (from various model).
D_p and D_q are momentum diffusivities.

CHAPTER 11

Scale-up of Bioreactors

OBJECTIVE

- Scale-up and scale-down are indispensable tools for designers engaged in industrial processes together with various transport processes. These processes are scale-dependent implying different behavior of laboratory model or full-scale plants. This chapter describes sequential components in scale-up with proper numerical examples.

11.1 INTRODUCTION

Biological and chemical processes are developed in the laboratory, and can be carried out in small unit to yield small amount of the product. However, small scale production is not sufficient to meet the demands of the product. One needs to produce at a larger scale. The translation of laboratory information to a desired larger scale is called scale-up of the process. The objective of the scale-up in bioreactor design is to determine a criterion or a set of criteria which are important in smooth translation of process information. It is difficult to define additional steps to gather all the information as quickly as possible at minimum cost. The methodology of process development leading to scale-up is the main factor for the success of the operation. In general, experiments are classified into the laboratory, pilot plants, and demonstration units.

In laboratory type experiments, certain aspects of the process are investigated by handling relatively small amount of raw materials to reduce the material constraints to a minimum. A series of measurements are taken concerning all the mechanisms that are independent of size (viz., thermodynamics and chemical/biological kinetics). A number of physical properties such as densities, viscosities, specific heats, and phase equilibria which are involved in the model must be ascertained throughout the operating conditions of the process.

Pilot plant experiments vary over a wide range of variables, accounting industrial constraints (e.g., duration of operation, control parameters, equipment reliability, and impurities in the raw materials). Scale-up problems are investigated during pilot plant experiment. A pilot plant is an experimental rig, which displays the part of operation that corresponds to an industrial plant. It allows for simultaneous analysis of the physical, chemical and biochemical parameters. A pilot plant is indispensable for measuring the extent of the possible interactions among the various parameters. It can be small to minimize extraneous costs such as the total operation cost as well as other constraints.

Experiments at the level of a demonstration unit apply to the construction of a first industrial unit, but on a modest scale. This step can be very costly, but it proves to be indispensable.

The important phenomena in bioreactor design are:

- Thermodynamics
- Micro kinetics – kinetic rate process, reaction mechanisms (biochemical pathways)
- Hydrodynamics/Fluid Mechanics (flow rates and their hold-ups, RTDs)
- Transport phenomena (heat transfer, inter phase and internal/external mass transfer)

Of these, first two are independent of scale. Scale-up problems exist when there is a transport of heat, mass or momentum in a system.

11.2 ADDITIONAL SCALE-UP PROBLEMS IN BIOREACTORS

- Screening practice, viz., shake flask vs. stirred reactor
- Quality and quantity of inoculum
- Interaction of genotype plus environment

Montante *et al.* (2003) have described the scale-up criteria for solids distribution in slurry reactors.

11.3 CRITERIA OF SCALE-UP

The process characteristics constant during scale-up is classified into two categories.

1. Single constant criteria
2. Combination criteria

11.3.1 Single Constant Criteria

The single constant criteria of scale-up include the following (Broadkey and Hershey, 1988).

- Reactor geometry
- Power per unit volume of liquid
- Volumetric mass transfer coefficient
- Volumetric gas flow rate per unit volume of liquid (Q/V or volume per volume per minute)
- Superficial gas velocity (V_s)
- Impeller Reynold's number, Re, $\left(\mathrm{Re}_i = \dfrac{\rho N D_i^2}{\mu}\right)$
- Maximum shear
- Mixing time
- Momentum factor
- Gas phase oxygen partial pressure

Table 11.1 gives certain aspect of these criteria which can be considered to control during scale-up.

Table 11.1 *Problems to tackle for scale-up*

Criteria	*Problems to tackle*
Mixing time	Poor mixing in large scale reactor
Volumetric power input	Reduction of power and in proper gas dispersion
Reactor geometry	Validity of correlations for D_i/D_t for good gas dispersion where D_i = impeller diameter and D_t = reactor diameter
Maximum shear	For shear-sensitive organisms
Power per unit volume	Successful parameter for mixing in shear-sensitive processes, crucial for aerobic processes
vvm(volume of gas per volume of medium per minute) and V_s	Critical for aerobic processes as vvm influences mass transfer in reactors. High V_s causes overloading
Gas phase oxygen partial pressure	For precise translation of optimal results from laboratory scale to higher scale

11.3.2 Combination of Criteria

Combination of criteria may be divided into three types.

Type I:

- Geometric similarity, constant k_La, constant vvm
- Determination of tip speed from k_La relations

Type II:

- Geometric similarity, constant k_La, constant impeller tip speed
- Q calculated from k_La relation

Type III:

- Constant impeller tip speed (ND_i), constant Q/V and k_La
- D_i/D_t adjustment

One can generate other combinations from the individual criteria suggested.

If following conditions of similarity can be achieved in the translation of a smaller scale to a larger scale, scale-up can be complete (Zlokarnik, 2006),

- Geometrical similarity: For reactor geometry, for example, $\left(\dfrac{\text{reactor diameter}}{\text{reactor volume}}\right)^{\frac{1}{3}}$ and $\left(\dfrac{\text{reactor volume}}{\text{liquid volume in ractor}}\right)^{\frac{1}{3}}$
- Hydrodynamic similarity: Constant ungassed or gassed power input per liquid volume

- Mechanical similarity: Constant impeller tip speed, e.g., $\frac{N_2}{N_1} = \frac{D_{T_1}}{D_{T_2}}$
- Thermal similarity: Heat transfer rates.
- Chemical similarity, for example: Constant oxygen uptake rate (OUR).
- Biochemical similarity, for example: Number of generations through which the organism will progress during the production. Trilli (1986) suggested the following relations for successful scale-up. Number of generations = 1.44 ($\ln V_L + \ln X_f - \ln X_0$), where V_L is working volume of bioreactor (m^3), X_f stands for final cell mass or number and X_0 is the initial cell mass or number.

11.4 SIMILARITY CRITERIA

Some physical processes occurring in a single phase may be scaled up using the principle of physical modeling. This is based on the criteria of geometric and chemical similarity derived from differential equations, which describe the process, or from dimensional analysis of the process variable (Ju and Chase, 1992). The process of interest is reproduced on different scales, and the effect of physical features and linear dimensions are analyzed in physical modeling. Experimental data are reduced to relationships involving dimensionless groups composed of various combinations of physical quantities and linear dimensions. The relationships can be classified into dimensionless groups or similarity criteria (Lee, 1992).

Physical modeling involves searching for the same or nearly the same similarity criteria for the model and the real process. The full scale process is modeled on an increasing scale with the principal linear dimensions scaled up in proportion, based on the similarity principle. For relatively simple systems, the similarity criteria and physical modeling are acceptable because the number of criteria involved is limited for the study. For complex systems and processes involving a complex system of equations, a large set of similarity criteria is required, which are not simultaneously compatible and, as a result, cannot be realized for scale-up study.

The objectives are not realized when physical modeling are applied to complex processes. However, consideration of the appropriate differential equations at steady state for the conservation of mass, momentum, and thermal energy has resulted in various dimensionless groups. These groups must be equal for both the model and the prototype for complete similarity to exist in scale-up.

11.4.1 Scale-up Based on Constant Power Per Unit Volume

When the geometric similarity is maintained, dynamic similarity could be achieved by having the same gassed power per unit volume in the two scales (Broadkey and Hershey, 1988). Under these conditions, the impeller tip speed is related to the impeller diameter by the equality of power numbers, i.e.,

$$\left(\frac{P_o}{N^3 D_i^5}\right)_1 = \left(\frac{P_o}{N^3 D_i^5}\right)_2 \tag{11.1}$$

Subscripts '1' and '2' stand for Larger- and smaller-scales, respectively.

$$P_o = N_p N^3 D_i^5 \rho \tag{11.1a}$$

where P_o stands for ungassed power, N_p for power number, ρ for density, N for impeller tip speed, and D_i for diameter of the impeller.

Rearranging Equation (11.1) gives

$$\left(\frac{P_o}{D_i^3}\right)_1 = \left(\frac{P_o}{D_i^3}\right)_2 \left(\frac{N_1}{N_2}\right)^3 \left(\frac{D_{i1}}{D_{i2}}\right)^2 \quad (11.2)$$

For geometrically similar vessels and constant $\left(\frac{P_o}{D_i^3}\right)$

$$\left(\frac{N_1}{N_2}\right)^3 = \left(\frac{D_{i2}}{D_{i1}}\right)^2 \quad (11.3)$$

Therefore,

$$N_1 = N_2\left(\frac{D_{i2}}{D_{i1}}\right)^{2/3} \quad (11.4)$$

Equation (11.4) gives the impeller speed for any particular change in diameter of the impeller so as to maintain dynamic similarity.

11.4.2 Scale-up Based on K_La

The volumetric oxygen transfer coefficient can be used for scale-up of bioreactors. The following relation for the estimation of K_La is proposed in Equation (11.5) (Junker, 2004).

$$K_La = K_o\left(\frac{P_g}{V_L}\right)^a (V_S)^b \quad (11.5)$$

K_o is a proportionality constant. P_g/V_L is gassed power per unit volume. V_S is gas superficial velocity. a and b are exponents. Values of a and b are: for laboratory reactor $a = 0.95$, $b = 0.67$; for pilot scale a and b are 0.67, and for production scale a and b are 0.50. The value of K_o depends on specific process and the units of $\frac{P_g}{V_L}$, K_La, and V_s.

K_La may also be estimated by any standard techniques such as sulphite oxidation, dynamic gassing out, etc. (Chapter 7 for detailed discusion). The K_La between two different scales is maintained constant by varying agitation rate, aeration rate or both.

11.4.3 Constant Mixing Time

A number of biopolymer fermentations, on scale-up, in larger vessels exhibit non-uniform characteristics. While small fermentor is usually well mixed, large vessels may be poorly mixed leading to internal gradients in dissolved oxygen, substrate, cell, and product concentration.

The mixing time in turbulent regime for a stirred tank is given by Equation (11.6).

$$\frac{t_{m1}}{t_{m2}} = \left(\frac{N_2}{N_1}\right)^{2/3} \left(\frac{D_1}{D_2}\right)^{1/6} \quad (11.6)$$

Subscripts '1' and 2 are for larger- and smaller-reactors, respectively. Equation (11.6) gives the change in impeller speed (N) or diameter (D) that is required to maintain a constant mixing time (t_m). This means that scale-up can be achieved by maintaining estimated mixing time constant. Of course, there are several relations reported in the literature as well.

11.5 SCALE-UP METHODS

It is not possible to apply similarity theory in a complete form to the scale-up of bioreactors. This theory has guided the formulation of a series of rules-of-thumb. In fact, some of them are the result of regime analysis.

Scale-up methods are classified in the following ways.

(1) Fundamental Methods

Differential balances of heat, mass, and momentum require detailed knowledge of all transport effects, their interactions, and characteristic coefficients. Structured models constitute the fundamental methods for scale-up (Catapano *et al.*, 2008).

(2) Semi-fundamental Methods

Modeling is based on the basis of simplified balances (Lonsane *et al.*, 1997).

(3) Regime Analysis

Recognition of rate limiting steps is of interest (Pollard *et al.*, 2007).

(4) Dimensional Analysis

This also includes regime analysis (Lee, 1992).

(5) Rules-of-thumb

"Know-how" is the guidelines.

(6) Trial and Error Method

This depends on the experience to adjust the scale-up factors (Fuchs *et al.*, 1971).

(7) Combination of Methods

This suggests no one rule for scale-up (Sweere *et al.*, 1987). Application of any suitable contribution of methods mentioned above is a better choice.

Let us discuss them in detail.

(1) Fundamental Method

The method is used to solve momentum, mass, and heat transfer balances for the system in micro scale. This has some complications when used for scale-up. They are:

- The balances should contain terms for transport in all three directions and the boundary conditions are very complicated.
- The balances are coupled, i.e., the solution of momentum balance gives the flow components that has to be used in mass and heat balances.
- The momentum balance assumes homogeneous fluid which is not realistic.

- It is difficult to solve the micro momentum balance.
- For a well defined flow pattern this method is used in scale-up problems as it is easy to solve the microbalances.

(2) Semi-fundamental Method

Scale-up can be based on simplified flow models in order to avoid the use of too complex micro momentum balance equations. Most of the models are bulk flow and do not give information about flows near important regions like stirrer blades, cooling coils, and vessel walls. Generally, flow models are of three types: plug flow, plug flow with dispersion or well mixed systems. Combinations of these models are also used for scale-up. These bulk flow models in small equipment scale-up are based on extrapolation. So the concept of similarity is introduced and quite often it is used in scale-up. Four similarities for scale-up, which are important in chemical engineering, are geometrical, mechanical, thermal, and chemical similarities. Similar bulk flow profiles at laboratory and production scale imply geometric similarity of both scales.

(3) Regime Analysis

Regime analysis refers to the dominance of a particular mechanism in bioreacter design. For example, kinetic regime and transport regime are important in the performance of bioreactors. It can be dominated by one regime or the combination of the regimes which suggests proper characterization of the regimes, viz., rate determination and the dependence of regimes or scale.

There are various ways to do the regime analysis, viz., experimental methods, theoretical methods including numerical techniques. Experimental methods depend on change of velocity, change of concentration, change of temperature, etc. Theoretical methods are of analytical methods (time constant, dimensionless number) and numerical methods by parameter sensitivity analysis.

One of the theoretical methods, i.e., time constents, is described here.

Regime analysis is based on the use of so called characteristic times. Characteristic time is a measure of the rate of a mechanism or sub-processes and can be considered as the time needed by that mechanism to reach a certain percentage of its final value after a change. A low value of a characteristic time means a fast mechanism whereas a high value means a slow mechanism. The use of these characteristic times can also give an insight into the complexity of the process. When different time constants are of same order of magnitude, it leads to a mixed regime. In this case, scale-up of the process cause problems.

For estimation, Equation (11.7) is used for simplicity.

$$t_{char} = \text{capacity/flow} \tag{11.7}$$

where t_{char} is the characteristic time.

For example: for growth, $t_{char} = 1/\mu_{max}$ and for oxygen transfer $t_{char} = 1/k_L a$.

By knowing all the sub-processes and calculating the characteristic times, the rate limiting mechanism can be determined for a particular process. This should be done not only for the final production scale but also for the laboratory and the pilot plant scale to predict the possibility of the regime changing on scale-up. From the knowledge originating from the regime analysis, it can be concluded which mechanism needs further investigation on a small scale.

A drawback with regime analysis is the lack of information of production scale systems. In many cases, regime analysis is only possible using non accurate model equations which are not validated by large scale experimental information.

A first estimate of the value is normally sufficient enough to identify the rate limiting mechanisms and to predict whether there will be a change in the regime if the process is scaled up. The characteristic times have been separated for transport and conversion phenomena. The characteristic times for transport phenomena are dependent on reactor type while those for conversion phenomena are found to be independent on reactor type.

(4) Dimensional Analysis

Another approach to scale-up problem is dimensional analysis. This is widely used in the scale-up of chemical engineering problems, which can be very useful for scale-up of microbial processes also.

Dimensional analysis is a technique by which dimensionless groups of parameters, for example, the Fr and Re number can be obtained from the dimensional momentum, mass and heat balances, as well as their boundary and initial conditions in the dimensionless form. Dimensional analysis is, however, of such importance and open to misinterpretation that it is essential to review this approach in order to show how it may be employed in scale-up problems.

The technique of dimensional analysis is driven by the need for dimensional consistency and the constraints in the places on functional relationships between variables. Essentially this technique allows us to group a number of variables in a problem to form dimensionless groups. In general, dimensionless numbers are ratios between two fundamental properties.

For example: R_e = (inertial forces)/(viscous forces).Other useful dimensionless numbers are given in Table11.3. Conventions used in Table 11.3 are given below.

r is the rate of reaction, kg/m^3. s

R is the geometrical factor (for spherical particle $R = \frac{dp}{3}$), m

dp is particle diameter, m

d stands for inner pipe diameter, m

σ is the surface tension, N/m

v is the velocity of fluid, m/s

v_S is the speed of sound in the fluid medium, m/s

ω is angular speed, radian/s

A_p is projected area of solids, m^2

v is kinematic viscosity, m^2/s

α is thermal diffusivity, m^2/s

k_L is thermal conductivity of fluid medium, J/s m K

k_s is thermal conductivity of solid, J/s m K

L is characteristic length applicable for the transfer process, m

β is coefficient of volume expansion, 1/K

C_A is concentration of reactant, kg/m^3

t is time, s

λ is latent heat of condensation, J/gm

Q is volumetric flow rate of fluid, m^3/s

Q_P is mass velocity, kg/m^2.s

Table 11.3 *Common dimensionless groups and their significance*

Type of transport phenomena	*Dimensionless group*	*Symbol*	*Equation*	*Significance*	*Use*
Momentum	Reynolds number	Re	$\frac{dv\rho}{\mu}$	$\frac{\text{Inertial Forces}}{\text{Viscous forces}}$	Decides the nature of flow and is useful in momentum, heat and mass transfer
Momentum	Friction factor	f	$\frac{\Delta p}{\rho v^2}\frac{d}{2L} = Eu\frac{d}{2L}$	$\frac{\text{Pressure forces}}{\text{Inertial forces}}\left(\frac{d}{2L}\right)$	Measure of frictional loss in a fluid flow through a pipeline
Momentum	Euler number	Eu	$\frac{\Delta p}{\rho v^2}$	$\frac{\text{Pressure forces}}{\text{Inertial forces}}$	In fluid friction calcula-tions in pipelines
Momentum	Froude number	Fr	$\frac{v^2}{gL}$	$\frac{\text{Inertial forces}}{\text{Gravitational forces}}$	In mixing operations and particle mechanics in fluid
Momentum	Weber number	We	$\frac{d\rho v^2}{\sigma}$	$\frac{\text{Inertial forces}}{\text{Surface tension forces}}$	In flow calculations in-volving large surface forces like bubbling in liquids, jet flow, etc.
Momentum	Mach number	Ma	$\frac{v}{v_s}$	$\frac{\text{Inertial forces}}{\text{Elastic forces}}$	In compressible fluid flow studies
Momentum	Power number	N_P	$\frac{P}{\rho\omega^3 D^5}$	$\frac{\text{(Drag forces on mixer blades)}}{\text{Inertial forces}}$ or $\frac{\text{Total dissipated power}}{\text{Power due to inertia}}$	Power calculations in mixing operations
Momentum	Bond number	B_0	$\frac{\Delta\rho d^2 g}{\sigma}$	$\frac{\text{Gravitational forces}}{\text{Surface tension forces}}$	For large surface area of liquid calculations
Momentum	Drag coefficient	C_D	$\frac{F_D/A_P}{\rho v^2/2}$	$\frac{\text{Drag forces}}{\text{Inertial forces}}$	In particle mechanics where fluids flow
Heat	Prandtl number	Pr	$\frac{C_P\mu}{k} = \frac{\nu}{\alpha}$	$\frac{\text{Viscous forces}}{\text{Thermal forces}}$	In forced convection heat transfer calculations

Contd.

Table 11.3 *Contd.*

Type of transport phenomena	*Dimensionless group*	*Symbol*	*Equation*	*Significance*	*Use*
Heat	Grashof number	Gr	$\frac{L^3\rho^2\beta g\Delta T}{\mu^2}$	$\left(\frac{\text{Buoyant forces}}{\text{Viscous forces}}\right)$ Re	Applied in natural convection heat transfer
Heat	Fourier number	Fo	$\frac{kt}{C_P\rho L^2} = \frac{\alpha t}{L^2}$ and $\frac{D_v t}{L^2}$	$\left(\frac{\text{Length of heat transfer}}{\text{Length of object}}\right)$ at time 't'	Applied in transient (unsteady) state heat transfer calculations in conduction heat transfer and mass transfer
Heat	Condensation number	N_{co}	$\frac{h}{k_L}\left(\frac{v^2}{g}\right)^{1/3}, \frac{\rho C_P v L}{k}$	$\frac{\text{Number of molecules condensing on the surface}}{\text{Total number of molecules touching that surface}}$	Used in condensation calculations in heat transfer
Heat	Peclet number	Pe_{heat} Pe_{mass}	= (Re) (Pr) $\frac{vL}{D_v}$ = (Re) (*Sc*)	$\frac{\text{Heat transfer by convection}}{\text{Heat transfer by conduction}}$ $\frac{\text{Mass transfer by convection}}{\text{Mass transfer by diffusion}}$	In forced convection heat transfer calculations and mass transfer calculations
Heat	Stanton number	St	$\frac{h}{\rho C_p v} = \frac{\text{Nu}}{(\text{Re})(\text{Pr})}$	$\frac{\text{Nusselt number}}{\text{Peclet number}}$	In forced heat transfer calculations
Heat	Vapour condensation number	*vc*	$\frac{K_C}{V} = \frac{S_h}{R_e S_c}, \frac{\rho^2 L^3 g\lambda}{k_L \mu \Delta T}$	$\frac{\text{Sherwood number}}{\text{Peclet number}}$	In condensation heat transfer calculations
Heat	Graetz no.	Gz	$\frac{Q\rho C_p d}{k_L L} = \frac{\pi D}{4L}\text{Pe}$		In forced convection calculations in heat transfer.
Heat	Rayleigh no.	Ra	$\frac{L^3\rho^2\beta g\Delta T C_p}{\mu k_L}$ = Pr Gr	$\frac{\text{Buoyant forces}}{\text{Product of thermal and momentum diffussivities}}$	In boundary layer calculations in natural convection.

Contd.

Table 11.3 *Contd.*

Type of transport phenomena	*Dimensionless group*	*Symbol*	*Equation*	*Significance*	*Use*
Heat	Heat transfer factor (Colburn)	j_H	$\frac{h}{C_p Q_p}\left(\frac{C_p\mu}{k}\right)^{2/3} = St\, Pr^{2/3}$	**Dimensionless heat** transfer coefficient	Obtained from analogy between heat and momentum transfer and is used for heat transfer calculations
Heat	Biot number	Bi $_{mas}$ Bi$_{heat}$	$\frac{hL}{k_L}, \frac{K_C L}{D_v}$	External mass transfer/internal mass transfer Or External heat transfer/internal heat transfer	**For heat or mass transfer** calculations between different phases.
Heat	**Nusselt number**	Nu	$\frac{hd}{k_L}$	$\frac{\text{Bulk fluid heat transfer resistance}}{\text{Surface film heat transfer resistance}}$	**Used in heat transfer calcula**tion
Heat	Lewis number	Le	$\frac{k_L}{\rho C_p D_v} = \frac{\alpha}{D_v} = \frac{\text{Sc}}{\text{Pr}}$	$\frac{\text{Thermal diffusivity}}{\text{Mass diffusivity}}$	In humidification calculation which is a simultaneous heat and mass transfer operation.
Mass	Schmidt number	Sc	$\frac{v}{D_v}$ $\frac{\mu}{\rho D_v}$	$\frac{\text{Momentum diffusivity}}{\text{Mass diffusivity}}$ $\frac{\text{Hydrodynamic boudary layer}}{\text{Mass transfer boundary layer}}$	In mass transfer calculations.
Mass	Sherwood number	Sh	$\frac{K_c L}{D_v}$	$\frac{\text{Bulk mass transfer resistance}}{\text{Surface mass transfer resistance}}$ or $\frac{\text{Total mass transfer}}{\text{mass transfer by diffusion}}$	In mass transfer calculations.

Contd.

Table 11.3 *Contd.*

Type of transport phenomena	*Dimensionless group*	*Symbol*	*Equation*	*Significance*	*Use*
Mass	Mass transfer factor (Colburn)	jM	$\frac{K_c}{v}\left(\frac{\mu}{\rho D_v}\right)^{2/3} = St\, Sc^{2/3}$		Obtained from analogy studies and used in such calculations.
Mass	Dispersion number	Di	$\frac{D_Q}{uL}$	$\frac{\text{Rate of transport by diffusion}}{\text{Rate of transport by convection}}$	**To find out nature of flow in reactors. The value of Di** indicates deviation from plug **flow**
Chemical reaction	Damköhler number	Da	$\frac{kLC_A^{n-1}}{v}$	$\frac{\text{Chemical reaction rate}}{\text{Mass transfer by convection or diffusion rate}}$	In rate calculations in catalytic reactions.
Chemical reaction	Knudsen number	Kn	$\frac{\lambda_m}{L}$	$\frac{\text{Molecular mean free path length}}{\text{Representative physical scale length scale}}$	**For calculations involving** highly porous catalyst with reactions at low pressure.
Chemical reaction	Thiele Modulus	ϕ	$R\sqrt{r/D_V C_A}$	$\left(\frac{\text{Chemical reaction rate in perticle}}{\text{Diffusion in particle}}\right)^{0.5}$	

Contd.

K_G is gas film mass transfer coefficient, m/s
D_v is diffusivity (mass), m^2/s
K_c is mass transfer coefficient, m/s
λ_m is mean free path of molecules
D_e is dispersion coefficient, m^2/s
k is reaction rate constant

Summary of some relations encountered in various topics of Chemical Engineering involving the dimensionless groups are as follows. Some of the groups can be considered or neglected in biological systems.

For Momentum transport: The general relation for the variables involved during the flow of fluids in closed conduits and solid particles settling in a fluid can be written as

$$E_u = [\text{Re, Fr, We, Ma}, (2L/d), (dp/d)] \qquad (11.8)$$

In mixing or agitation operations, N_P is defined by Equation (11.9).

$$N_P = f(\text{Re, Fr}) \qquad (11.9)$$

For heat transfer: For heat transfer in fluids, which occurs by natural or forced convection, Equation (11.10) is the general form.

$$\text{Nu} = f\left[\text{Re}, \text{Pr}, \text{Gr}, \left(\frac{d}{L}\right), \left(\frac{\mu}{\mu_w}\right)\right] \qquad (11.10)$$

where μ_w is the viscosity of the fluid at wall temperature.

For mass transfer: For mass transfer in fluids, occurring in natural and forced convection, Equation (11.11) is the general form.

$$\text{Sh} = f\left([\text{Re}, Sc, \text{Gr},], \frac{L}{d}\right) \qquad (11.11)$$

The key point to remember about the application of dimensionless numbers in scale-up is that each of these numbers measures the ratio of two fundamental parameters. In principle, one can aim to keep these ratios constant during the scale-up of a process.

Once the dimensionless groups are obtained their use for the proper set up of scale-up or scale-down experiments, at least in principle, is rather simple. Equal values of these groups are used for both the model scale and the prototype scale systems. In practice, the situation is more complicated, because often it is not possible to keep all dimensionless groups constant during scale-up.

However, if a change in the regime takes place during scale-up, the formal dimensional analysis method is invalid. Another disadvantage of the dimensional analysis can be that the formal application of dimensional analysis leads to technically unrealistic situations like too large power consumption, high stirrer speeds, etc. For heterogeneous systems, during scale-up, usually one may not be interested in geometric similarity for dispersed particles present in the heterogeneous system which violates the concept of dimensional analysis. Geometric similarity and also the choice of the parameters involved in dimensional analysis is not always obvious and sometimes rather arbitrary.

Dimensional analysis offers a considerable potential for scale-up by reducing the number of variables and the number of experiments at laboratory scale, which are necessary to predict the system behavior at production scale, but one must be aware of its limitations.

Dimensional analysis uses the concept of grouping the quantities into dimensionless numbers. There are a number of methods available to do this analysis. They are:

- Gauss–Jordan reduction method (Junker, 2004)
- Rayleigh's method (Portillo *et al.*, 2008)
- Gukhman's method (Gukhman, 1987)
- Buckingham-Pi method (Buckingham, 1914).

Buckingham-Pi Method

This is based on a general rule called Buckingham-Pi theorem which enables one to predict the number of dimensionless groups (d), parameters (p) and number of basic units (b). As per the theorem,

$$d = p - b \qquad (11.12)$$

(5) Rules-of-thumb

In biochemical engineering, 'rules-of-thumb' combined with extrapolation to larger scale are commonly applied as a scale-up procedure. A small range of scale-up criteria have been used by an industry. Following are the common criteria which may be considered constant:

- power per unit volume
- volumetric mass transfer coefficient
- tip speed of the impeller
- dissolved oxygen content
- Reynolds number
- mixing time
- momentum factor
- shear rate, etc.

The exact procedure to be used for scale-up depends on the characterization of the particular system. For example, Reynolds number is used for scale-up of anaerobic system, power per unit volume in presence of gas in the medium for organic acid production, constant k_La for antibiotic fermentation, etc. For biopolymer fermentation, prominent scale-up method is by keeping P_g/V, k_La, impeller tip speed and mixing time constant.

11.6 GENERALIZED APPROACHES TO SCALE-UP IN COMBINATION OF METHODS

11.6.1 Hubbard Method (1987)

Step 1:

Laboratory scale generates data on the following information.

(i) air flow rate, impeller speed, yield, reactor geometry

(ii) determination of basic fermented medium properties

- oxygen diffusion coefficient, surface tension, viscosity, density

(iii) Based on these, following values can be calculated, viz.,
Reynolds number, impeller tip speed, aeration number, Q/V_L.

Step 2:
Volume of the higher scale reactor is selected based on yield and capacity of the plant.

Step 3:
Geometric similarity is used to calculate reactor dimensions.

Step 4:
Scale-up is based on constant K_La.

Technique (a): Calculate flow rate by maintaining Q/V_L or V_S constant.
Then calculate impeller speed (N) based on P_g/V_L vs K_La correlations
or

Technique (b): From constant impeller tip speed (ND_i), calculate N. Then calculate Q from P_g/V_L and K_La correlations

11.6.2 Method of Wang *et al* (1979)

Step 1:
Two key values (K_La and impeller tip speed) are maintained constant.

Step 2:
Then D_1/D_T is adjusted to a reasonable value for the complete design.

This method does not guarantee for the geometrical similarity, but it provides information on dispersion. This method is a flexible one.

11.6.3 Ettler's Method (1992)

Step 1:
This method evaluates key parameters for scale-up based on two variables, viz., impeller speed and impeller diameter.

Step 2:
In this case, related dimensionless groups are impeller Reynolds number, Froude number, Weber number, power per unit volume of liquid in the presence of gas in liquid, impeller tip speed, etc.

Step 3:
The parameters combine relevant dimensionless group analysis with appropriate empirical parameters.

11.6.4 Other Methods

Junker (2004) suggested another scale-up strategy based on oxygen uptake rate (OUR) and K_La. Following relations can be considered useful for scale-up.

$$\text{OUR} = A_1 K_La + B_1 \tag{11.13}$$

where $$\log K_L a = A_2 \log (\rho_g / V_L) + B_2 \tag{11.14}$$

N is impeller speed, and

D_i is diameter of impeller.

Assuming geometrical similarity,

$$\frac{D_{T_1}}{D_{T_2}} = \left(\frac{V_{T_1}}{V_{T_2}}\right)^{1/3}, \tag{11.15}$$

The scale-up based on constant ungassed $\left(\frac{P_o}{V_L}\right)$, power per unit volume and similarly gassed one $\left(\frac{P_g}{V_L}\right)$ simplifies to

$$\frac{N_2}{N_1} = \left(\frac{V_{T_1}}{V_{T_2}}\right)^{2/9} \quad \text{(Junker, 2004)} \tag{11.16}$$

(iv) Similar impeller Reynolds number

Re defined for impeller

$$\text{Re}_{\text{imp}} = \frac{\rho N D_i^2}{\eta} \tag{11.17}$$

where η = broth viscosity.

For same broth,

$$\frac{N_2}{N_1} = \left(\frac{D_{i1}}{D_{i2}}\right)^2 \tag{11.18}$$

11.7 EXAMPLES

(a) Scale-up for Microbial Systems

(i) Geometric similarity of bioreactor

This can be expressed in the form of Equation (11.19) (Ju and Chase, 1992)

$$\frac{\text{Diameter of the bioreactor vessel } (D_T)}{(\text{Liquid volume in the bioreactor vessel } (V_L))^{1/3}} = \text{Constant} \tag{11.19}$$

(ii) Constant impeller tip speed

$$\text{Impeller tip speed} = \pi N_i D_i \tag{11.20}$$

where N_i is impeller speed and

D_i is impeller diameter

(iii) Constant power input/unit volume

This can be gassed $\left(\frac{P_g}{V_L}\right)$ or ungassed $\left(\frac{P_o}{V_L}\right)$.

P_o is defined by

$$P_o = N_P N^3 D_i^5 \rho \qquad (11.21)$$

where N_P = Power number

ρ = broth density.

(b) Application of Gauss–Jordan Reduction Method

This method is described in detail by Pawlowski (1971). An example of reactor is considered with a few important parameters. Following steps are summarized for this purpose.

Step 1 Identification of parameters

Reactor: h (height of the reactor), d_i (internal diameter)

Liquid in the reactor: v (velocity), ρ (density), η (dynamic viscosity)

Bed in the reactor packing: d_p (particle diameter), $\Delta\rho$ (density gradient), τ (yield stress)

Gas inside: u_s (superficial velocity)

General parameters: g (acceleration due to gravity), t (time)

So we have 11 parameters for this reactor. As per procedure of Pawlowski we have to write dimensionless groups.

Step 2: Variables and their dimension in matrix

Dimension	*Variables*										
	Core matrix			*Residual matrix*							
	ρ	h	t	g	d_i	v_s	dp	$\Delta\rho$	τ	v	η
M	1	0	0	0	0	0	0	1	1	0	1
L	–3	1	0	1	1	1	1	–3	–1	1	–1
t	0	0	1	–2	0	–1	0	0	–2	–1	–1

Left part of the matrix is to be written in the form of a diagonal matrix

Step 3: Conversion of left matrix to a diagonal matrix

We need to add three times the first row to the second row.

Dimension	*Variables*										
	Core matrix			*Residual matrix*							
	ρ	h	t	g	d_i	u_s	dp	$\Delta\rho$	τ	v	η
M	1	0	0	0	0	0	0	1	1	0	1
L	0	1	0	1	1	1	1	0	2	1	2
t	0	0	1	–2	0	–1	0	0	–2	–1	–1

Step 4: Formation of dimensionless groups

Group 1 $\cong \dfrac{g}{h}t^2$. ρ does not appear here as in g-column, which corresponds to ρ_1 column in the left matrix, having zero. Other dimensionless groups are:

Group 2 $\cong \dfrac{d_i}{h}$; Groups 5 $\cong \dfrac{\Delta\rho}{\rho}$ Group 8 $\cong \dfrac{\eta t}{\rho h^2}$

Group 3 $\cong \dfrac{u_s t}{h}$; Group 6 $\cong \dfrac{\tau t^2}{\rho h^2}$

Group 4 $\cong \dfrac{dp}{h}$; Group 7 $\cong \dfrac{vt}{h}$

Total number of dimensionless groups are eight which is in agreement as per Buckingham-Pi method. If these groups are possible to regroup for calculation purposes, it is done accordingly. The dimensionless groups are evaluated in two different scales. If they are maintained same value (or values), the scale-up procedure is perfect.

(c) Solution of Problem Stated in (b) by Rayleigh's Method

Solution of the problem can be written in the following form

$$v = h^{\alpha}\, \rho^{\beta}\, di^{\gamma}\, \Delta\rho^{\delta}\, \eta^{\theta}\, d_p^{\phi}\, \tau^{\Psi}\, u_s^{\zeta}\, g^{\kappa}\, t^{\lambda}$$

By the use of dimensions, this relation is modified to

$$\left(\frac{L}{t}\right) = mL^{\alpha}\left(\frac{M}{L^3}\right)^{\beta}(L)^{\gamma}\left(\frac{M}{L^3}\right)^{\delta}\left(\frac{M}{Lt}\right)^{\theta}(L)^{\phi}\left(\frac{M}{Lt^2}\right)^{\Psi}\left(\frac{L}{t}\right)^{\zeta}\left(\frac{L}{t^2}\right)^{k}(t)^{\lambda}$$

We have three dimensions M, L and t. As there are eleven parameters, total number of dimensionless groups will be 8 from the analysis of Buckingham-Pi method.

For the three dimensions, the physical meaningful relation will be homogeneous. Therefore,

For M: $0 = \beta + \delta + \theta + \Psi$ ... (a)

For L: $1 = \alpha - 3\beta + \gamma - 3\delta - \theta + \phi - \Psi + \zeta + \kappa$ ... (b)

For t: $-1 = -\theta - 2\Psi - \zeta - 2\kappa + \lambda$ ... (c)

From the definition of Reynolds number, Re $= \dfrac{d_p v\rho}{\eta}$. It is better to express β, θ and ϕ in terms of others unknowns. From equations (a) to (c) following equations are obtained.

$$\beta = -(\delta + Q + \Psi) \quad \text{... (d}$$

$$\phi = 1 - (\alpha - 3\beta + \gamma - 3\delta - \theta - \Psi + \zeta + \kappa) \quad \text{... (e}$$

and

$$\theta = 1 - 2\Psi - \zeta - 2\kappa + \lambda \quad \text{... (f}$$

From equations (d) and (f),

$$\beta = -1 - \delta + \Psi + \zeta + 2\kappa - \lambda \quad \text{... (g}$$

From equations (d) to (f) we find ϕ.

$$\begin{aligned}\phi &= 1 - (\alpha + 3\delta + 3\theta + 3\Psi + \gamma - 3\delta - \theta - \Psi + \zeta + \kappa)\\ &= 1 - (\alpha + 2\theta + 2\Psi + \gamma + \zeta + \kappa)\\ &= 1 - (\alpha + 2 - 4\Psi - 2\zeta - 4\kappa + 2\lambda + 2\Psi + \gamma + \zeta + \kappa)\\ &= -1 - \alpha + 2\Psi + \zeta + 3\kappa - 2\lambda - \gamma\end{aligned}$$

Therefore,

$$v = m\left(\frac{\eta}{d_p\rho}\right)\left(\frac{h}{dp}\right)^{\alpha}\left(\frac{d_i}{d_p}\right)^{\gamma}\left(\frac{\Delta\rho}{\rho}\right)^{\delta}\left(\frac{\tau\rho d_p^2}{\eta^2}\right)^{\psi}\left(\frac{u_s\rho}{\eta}d_p\right)^{\zeta}\left(\frac{g\rho^2 d_p^3}{\eta^2}\right)^{\kappa}\left(\frac{t\eta}{\rho d_p^2}\right)^{\lambda}$$

Hence
$$\left(\frac{d_p v\rho}{\eta}\right) = m\left(\frac{h}{d_p}\right)^{\alpha}\left(\frac{d_i}{d_p}\right)^{\gamma}\left(\frac{\Delta\rho}{\rho}\right)^{\delta}\left(\frac{\tau\rho d_p^2}{\eta^2}\right)^{\psi}\left(\frac{u_s\rho d_p}{\eta}\right)^{\zeta}\times\left(\frac{g\rho^2 d_p^3}{\eta^2}\right)^{\kappa}\left(\frac{\eta t}{\rho d_p^2}\right)^{\lambda}$$

So eight dimensionless groups are found by this method. For explanation one can regroup the dimensionless groups. m is proportionality constant. We know only three equations, but number of unknowns are more. It is required to express three variables in terms of seven variables.

EXERCISES

11.1 Theoretical analysis and experimental results show that the power used for agitation in a stirred bioreactor depends on the dynamic viscosity and the density of the fermented broth, acceleration due to gravity, speed of rotation of the impeller, agitator diameter and other geometrical characteristics of the bioreactor. Find the general form of the dimensionless numbers which can be used for the scale-up of the bioreactor. For scale-up, follow the principle of Buckingham-Pi method.

11.2 Write the procedure for scale-up of bioreactor used for solid state fermentation.

11.3 Air is sparged into a tank reactor through a sparger at the rate of 0.06 m^3/s. The sparger has 50 holes of diameter 0.42 cm. Holes are uniformly distributed across the cross-seetional area of the tank reactor. Tank reactor is of 2 m high with a diameter of 1.5 m. Gas properties are: flow rate 0.06 m^3/s, viscosity 1.84×10^{-5} Ns/m^2, diffusivity 2.25×10^{-9} m^2/s. Liquid properties are: density 1000 kg/m^3, viscosity 0.001 Ns/m^2, surface tension 0.07 N/m. Calculate important parameters used for scale-up: bubble diameter generated at the orifice, superficial gas velocity, P/V, K_La, mixing time. Using these parameters, how will you proceed to scale-up the bioreactor based on geometrical and mechanical similarity criteria?

11.4 A large stirred tank reactor of liquid capacity of 20 m^3 is used for penicillim fermentation. A few important data are available for evaluating scale-up parameters. They are air flow rate 0.25 vvm, liquid density 1.1 g/cm^3, surface tension 52 dynes/cm, diffusivity of oxygen in the broth 2×10^{-5} cm^2/s. Tank height to diameter ratio is 2:1 and that of the ratio of impeller diameter to tank diameter is 0.5. Calculate K_La, ungassed power requirement, N_P and mixing time. How will you proceed to scale-up this reactor based on mixing time concept?

REFERENCES

Broadkey RS and Hershey HC (Eds) (1988) *Transport Phenomena: A unified approach*, McGraw-Hill.

Buckingham E (1914) "On physically similar systems; illustrations of the use of dimensional equations". *Physical Review*, 4, 345-376.

Catapano G, Czermak P, Eibl R, Eibl D and Pörtner R (Eds) (2008) "Bioreactor Design and Scale-Up" In *Cell and Tissue Reaction Engineering, Principles and Practice*, Springer, Berlin, Heidelberg, 173-259.

Doran PM (Ed) (1995) *Bioprocess Engineering Principles*, Academic Press, New York.

Ettler P (1990) "Scale-up and scale-down techniques for fermentations of polyene antibiotics", *Collect. Czech. Chem. Commun.*, **55**, 1730-1740.

Fuchs R, Ryu DDY and Humphrey AE (1971), "Effect of Surface Aeration on Scale-Up Procedures for Fermentation Processes," *Industrial Engineering and Chemical Process Design and Development*, **10**, 190–196.

Gukhman AA (1987) "Review of work on the development and application of methods of generalized analysis", *Journal of Engineering Physics and Thermophysics*, **53**, 1233-1239.

Hubbard DW (1987) "Scale-up strategies in bioreactors". In: Ho, C.S. and Oldslune, J.Y. (Eds): *Biotechnology Processes, Scale-up and mixing*, pp. 168-184, AIChE, New York.

Ju LK and Chase GG (1992) "Improved Scale-up strategies of bioreactors", *Bioprocess Engineering*, 8, 49-53.

Junker BH (2004) "Scale-up methodologies for *Escherichia coli* and yeast fermentation processes", *Journal of Bioscience and Bioengineering*, **97**, 347-364.

Lee JM (Ed) (1992) *Biochemical Engineering*, Prentice Hall, New Jersey (USA), 268-274.

Lonsane BK, Saucedo-Castaneday G, Raimbault M, Roussos S, Viniegra-Gonzalez G, Ghildyal NP, Ramakrishna M, and Krishnaiah MM (1992) "Scale-Up Strategies for Solid State Fermentation Systems", *Process Biochemistry*, **27**, 259-273.

Montante G, Pinelli D and Magelli F (2003) "Scale-up criteria for the solids distribution in slurry reactors stirred with multiple impellers", *Chemical Engineering Science*, **58**, 5363-5372.

Mooyman JG (1987) "Scaling oxygen mass transfer in agitated fermentors". *Biotechnology and Bioengineering*, **29**, 180-186.

Nielsen J and Villadsen J (Eds) (2003) *Bioreaction Engineering Principles*, Plenum Press, New York.

Pawlowski J (1971) Die Ähnlichkeitstheorie in der physikalisch-technischen. Forshchung, Springer Verlag, Berlin.

Pollard DJ, Kirschner TF, Hunt GR, Tong I-T, Stieber R and Salmon PM (2007), "Scale-up of a viscous fungal fermentation: Application of scale-up criteria with regime analysis and operating boundary conditions", *Biotechnology and Bioengineering*, **96**, 307-317.

Portillo PM, Ierapetritou M, Tomassone S, McDade C, Clancy D, Avontuur PPC, and Muzzio FJ (2008) "Quality by Design Methodology for Development and Scale-up of Batch Mixing Processes", *Journal of Pharmaceutical Innovation*, **3**, 258-270.

Sweere APJ, Luyben KCAM and Kossen NWF (1987) "Regime analysis and scale-down: Tools to investigate the performance of bioreactors", *Enzyme and Microbial Technology*, **9**, 386-398.

Trilli A (1986) Scale-up of fermentation, pp 277–307, In Demain AL, Solomon NA (Eds) "Manual of industrial microbiology and biotechnology," *American Society for Microbiology*, Washington DC.

Wang DIC, Cooney CL, Demain AL, Dunnill P, Humphrey AE and Lilly MD (Eds) (1979) *Fermentation and Enzyme Technology*, John Wiley & Sons Inc., New York.

Wolf K-H (1997) "Statistical planning and analysis of experiments, and scaling-up", In: *Methods in Biotechnology*, Schmauder, H-P (Ed.) Taylor and Francis, London, pp 233-252.

Zlokarnik M (Ed) (2006) *Scale-up in Chemical Engineering* (Extended edition), Wiley-VCH, Verlag GmbH & Co, Weinheim.

CHAPTER 12

Mechanical Aspects of Bioreactor Design

OBJECTIVE

- To understand the basic mechanical aspects of design of bioreactors.

12.1 INTRODUCTION

This chapter discusses materials of construction of various reactors and accessories with advantages and disadvantages, specific mechanical design aspect, plant practices in bioreactors and other equipments used in biochemical plants, e.g., motors, pumps, air compressor, air filtration devices, etc.

12.2 REQUIREMENTS FOR CONSTRUCTION OF A BIOREACTOR

For the construction of a bioreactor the following requirements are necessary.

System Sterility

For monoseptic conditions, easy sterilization, clean-in-place (CIP), and non-contamination of components must be assured by proper material selection, and construction and selection of seals. Additional components are heat exchanger for sterile operation and filter for maintaining sterility.

Mixing of Bioreactor Content

Agitator selection and agitator surface are important criteria. Additional components are optimal controller operation (viz., pH, temperature, pO_2, etc.) for thorough mixing condition, and the design of suitable exhaust gas coolers or fresh gas saturator to prevent evaporation loss.

Temperature Control

Selection of suitable heat exchanger and control circuits for effective temperature regulation are necessary.

pH Control

The pH value is preferred within a narrow band width for efficient bioreactor operations. Additional components are valves, sensors, and pumps.

User Friendly Design

Equipment design and layout must meet minimal work for the start up of a process, execution of the process, and in the product harvest.

12.3 GUIDELINES FOR BIOREACTOR DESIGN

(1) Connections between sterile and non-sterile parts of the system may be avoided in the bioreactor.
(2) Minimization of flange connection is necessary to avoid possibility of future contamination in the process.
(3) Welded parts need to be polished to avoid accumulation of any component.
(4) Dead spaces, crevices, and like factors are to be avoided in the bioreactor.
(5) Options are necessary for independent sterilization of various parts of the system.
(6) Any connection to the vessel is required to be steam-sealed. Easy cleaning for valves used in the system is preferable.
(7) Positive pressure is to be maintained in the bioreactor during the process.

12.3.1 Pereferred Materials for Bioreactor Design and Fabrication

For quality design and better operation of bioreactors, selection of suitable materials is necessary. Table 12.1 summarizes a few attempts in this area. There are scopes for the development of new materials.

12.3.2 Welding Techniques

For internal weld, welded surface must be ground smooth. Some bioreactor jackets are provided along with the body which is welded. Also for better heat transfer, hemispherical coils and coiled-coil structure immersed in the media are welded around the vessel.

A number of factors are necessary to consider for welding of high alloy austenitic steels. To avoid chrome carbide formation due to welding, stabilization is done with *Ti* or *Ta* or *Mo* or with low *C* steel (Ruge, 1974) for improving the quality of weld.

(1) Cr–Ni Steel Welding

This process is done depending on the wall thickness.

(2) Surface Treatment

Electroplating, pickling and passivation are adopted to reduce surface roughness (Klapp, 1980).

12.4 BIOREACTOR VESSELS

Variation in sizes and their configurations of bioreactors are described in Chapter 2. The combination of glass and stainless steel are common materials for small bioreactor vessel. Stainless steel materials are used for the construction of larger bioreactor vessels. Table 12.2 gives a few examples of materials used in bioreactors and its accessories (Lydersen *et al.*, 1994).

Table 12.1 *Materials used with advantages and disadvantages for the fabrication of bioreactor and its accessories (Solomon, 1969; Pandeya and Shah, 2006)*

Material	*Category*	*Advantages/Disadvantages*	*Use in bioreactor/accessories*
Copper	Non-ferrous metals	Advantage: Good resistance to steam Disadvantage: Acids and alkalis will cause corrosion	Tubes for carrying non-sterile gases for pneumatic control system and water for heat exchanger coil
Brass, bronze, gun metal	Non-ferrous metals	Advantage: Can be used in combination with Stainless steel (SS)	- Valves for duty with air, water and steam and machine parts not in contact with medium - Bearing, housing for stirrer shafts - Locking rings for rubber diaphragms - End-caps for air filters
Titanium	Non-ferrous metals	Advantages: Non-toxicity, mechanical strength, excellent resistance to acid	pH control systems Thin-walled tubing
Platinum	Non-ferrous metals	Advantage: Resistant to acids	Wires Pt-Ag as in dissolved O_2 probe Pt-Rh for hypodermic needles
Aluminum	Non-ferrous metals		1. Specialized welding
Mild steel	Ferrous metals	Disadvantage: Less corrosion resistance	Low carbon steel for fermentation vessel Piping and in jackets of SS vessels
Stainless steel	Ferrous metals	Advantage: Better acid resistance Disadvantages: Expensive machining, liable to weld decay, and cost	Major construction material for bioreactors
Plastics	Polythene, polypropylene, polyvinyl chloride, nylon, fluorinated plastics (PTFE, FEP, Kel F), oxy resins	Advantages: Non-toxic, and withstand corrosion Disadvantage: Heat-sensitive	Not for primary fabrication
Rubbers	Natural rubber, butyl-, neoprene-, nitrile-, silicone, fluorosilicones-, fluorinated- rubbers	Advantages: Non-toxic, withstand corrosion Disadvantage: Heat-sensitive	O-rings Sampling systems Lining of reactors

Table 12.2 *Materials used in bioreactors and its accessories*

Components	*Norm*	*Material No.*	*Alloy (% weight)*				
			Max. C	*Cr*	*Ni*	*Mo*	*Min. Ti*
Contact with the medium	AISI	316L	0.03	16	10	2	-
	SIS	2343	0.07	16.5	11.5	2.5	-
	DIN	1.4435	0.03	16.5	12.5	2.5	-
Other materials like DIN 1.4301, DIN 1.4306, DIN 1.4541, AISI 304 and AISI 304L are used for the components where there is no direct contact with the medium.							

Generally, Mo is not used in alloy steel for the construction of components not in direct contact with the medium.

12.4.1 Geometry of Reactor Vessel

Generally, vertical cylinder type bioreactor is used for SLF. Of course, there are some variations in the configuration (refer Chapter 2). We take vertical, cylinder type for our discussion. The ratio of height (H) to diameter (D) of the reactor is between 1 : 1 and 3 : 1. Typical dimension of a stirred vessel is represented in Figure 12.1. Dimensions are given below.

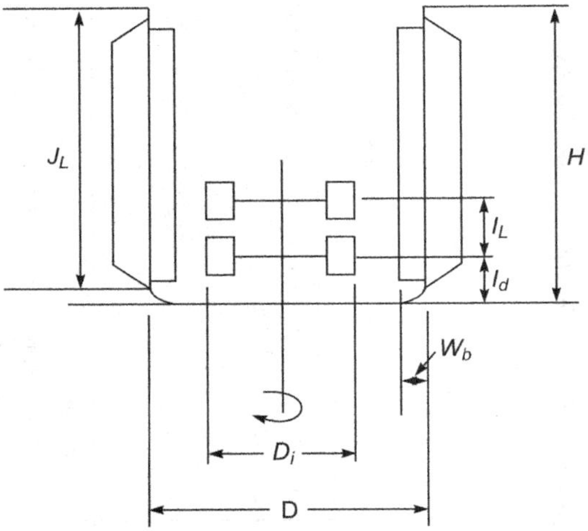

Figure 12.1 Typical stirred vessel with dimensions.

$$\frac{H}{D} = 1:1$$

$$\frac{D_i}{D} = 0.3 \text{ where } Di \text{ is the impeller diameter}$$

W_b = (0.08–0.1) D_i, where W_b is the width of the baffle

J_L = ¾ th of H, where J_L is the length of the jacket

$$\frac{I_L}{D_i} = 1\text{–}1.2, \text{ where } I_L \text{ is the vertical distance between the adjacent stirrer blades.}$$

For stainless steel (SS) vessel, the body is constructed from a cylinder formed from sheet metal by welding along one edge. The bottom consists of a dished end (Fig. 12.2) welded to the cylinder.

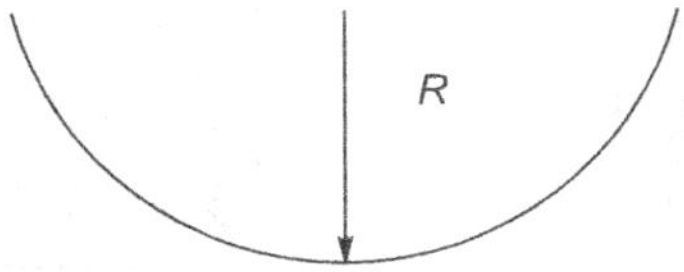

Figure 12.2 Dished-end of a reactor.

Other end of the cylinder is welded to a flange which allows the head plate to be bolted on. In some cases, a top dished-end plate is welded to the vessel directly instead of welding to a flange.

12.4.2 Components in Bioreactor Vessel

Following is the comprehensive list with the omission of a few minor important components for a special bioreactor vessel (Fig. 2.2). Components are:

- Bioreactor vessel
- Upper head plate
- Jacket
- Bearing case
- Mechanical seal
- Stirrer shaft
- Extra port
- Antifoam electrode port
- Handle
- Locking bolt
- Viewing window
- Lamp housing
- Water inlet/Steam outlet
- Water outlet/Steam inlet
- Air feed
- Base input
- Acid input
- Sampling device
- Impeller
- Mechanical foam breaker
- Baffles
- Sparger
- Inoculation port
- Pressure gauge
- pH electrode port
- Temperature sensor port
- Dissolved oxygen port
- Control flange for agitator shaft
- Drain valve

12.4.3 Size of the Vessel

Following considerations may be used in determining the size of the vessel used in a bioreactor.

- Projected process capacity
- Greater risk of loss due to contamination
- Fabrication capability on site or future shipping and installation.

In general, bioreactor vessels withstand sterilization temperature and pressure. In that context, if one considers the reactor vessel design, one can find that it may be categorized as a cylindrical design (Pandeya and Shah, 2006). For thin cylinder, usually t/D ratio is $\leq 1/10$. **In terms of pressure, it may be considered as 140 kgf/cm^2(** $\equiv$ 2000 psi). It is further assumed that the stresses are distributed uniformly over the wall thickness.

12.4.4 The Design Procedure of Vessel Wall of Bioreactor

The cylindrical vessel used for reactor fabrication may fail along the longitudinal section (seam) or in transverse section. σ_t is the intensity of induced tangential stress in the reactor vessel wall. Following steps may be followed in this regard.

Steps

(1) The design is on the basis of σ_{th}, which is computed by dividing σ_t or σ_c, whichever is less for the given material, by the factor of safety.

Therefore,
$$\sigma_{thw} = \text{working } \sigma_{th} = \frac{\sigma_t \text{ or } \sigma_c}{\text{Factor of safety (FS)}}$$

It is suggested that the factor of safety of the shell preferably is not less than 4 (IBR regulation 564).

(2) Thickness of the cylinder ($= t_d$)

$$t_d = \frac{pD}{2\sigma_{thw}\eta} + C_1$$

where p is working pressure,
η is efficiency of the longitudinal joint
C_1 is constant for corrosion allowance.

(3) Checking with specific standard or regulations for the design. Equations (12.1) and (12.2) are suggested by the Indian Boiler Regulations (1950).

Working pressure is
$$WP = \frac{(t-2)\,SC}{C_2 D} \tag{12.1}$$

where

WP is the working pressure, lb/in^2
S is ultimate tensile strength, tons/in^2

If this is not available, one can get from the standard table for the properties of the materials.
Thickness of the vessel plate is 32nd of an inch.
D is inner diameter of the cylinder, inches
C is a constant, which depends on the class of the vessel. A few values are given in Table 12.3 as the possible guidelines.
C_2 is a constant. This may be taken as 2.75 for the longitudinal seam made up of double butt wrap (Pandeya and Shah, 2006).

Table 12.3 *Values of C for different class of boiler*

Class of boiler	*Value of C*
I	32
II	27
III (Stress relieved)	23
III (Stress not relieved)	21

$$WP = \frac{(t-1.5)SC}{0.7D} \tag{12.2}$$

where

t is minimum thickness of vessel plate, mm

WP is working pressure, kgf/cm^2

S is minimum tensile strength of plate, kgf/mm^2

D is maximum internal diameter, mm

C is a constant whose value is given in Table 12.3.

(4) If $t_d > t$ obtained from step 3, t_d is alright and this 't' value will be whole number (not a fraction!). This is in conformity to the 'standard'.

(a) Accessories for the End Plates

For Dished-end Plate

The thickness of the dished-end plate according to specific regulations/standard design is calculated from the following Equation (12.3).

$$\text{Working pressure} = WP = \frac{(t_e - 8)30S'}{r} \tag{12.3}$$

where

WP is in lb/in^2

S' is minimal tensile strength of the plate, tons/in^2

r is inner radius of curvature of the end plates which should not exceed 'D'. 'D' is diameter of the shell and it should not be less than 4 times the shell wall thickness in any case.

The guideline is $4t < r < D$.

Thickness of end plate $= t_C$ in $\frac{1}{32}$ inch.

The schematic diagram of dished-end plate is given in Figure 12.3.

t is the thickness of the shell plate.

(b) For Flat-end Plates

Figure 12.4 shows the sketch for flat-end plate.

Thickness of the flat-end plate is expressed by Equation (12.4).

$$t_c = R_p\sqrt{\frac{P}{\sigma_{bw}}} \tag{12.4}$$

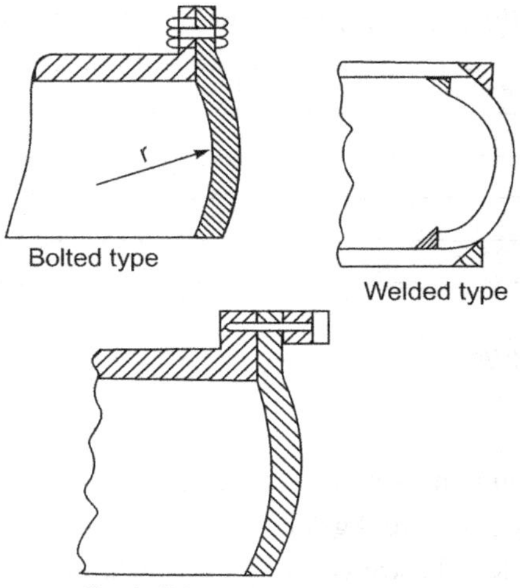

Figure 12.3 Schematic of dished-end cover for bioreactor.

where σ_{bw} is working bending stress (i.e., lower value between σ_t and σ_c from the standard table divided by factor of safety).

If elastic limit or yield point stress is given, the factor of safety is the half of the value as used to calculate ultimate tensile strength (σ_{ult}).

R_p is described by Equation (12.5).

$$R_p = \frac{D_i}{2} + t + 2\delta \tag{12.5}$$

R_p is the pitch circle radius. It may be considered as $4t < R_p <$ diameter of the shell.

δ is the diameter of the bolt and not less than that of M16 bolt.

t is the thickness of the shell plate.

12.4.5 Design of Flange

For bolted on flat-end plates or dished-end plates, flange may be designed in the following manner.

Step 1

The resultant load on the bolt is described by Equation (12.6) (Pandeya and Shah, 2006).

Resultant load on the bolts, $F = F_1 + KF_2$ (12.6)

where

K is gasket constant (range is between 0 and 1). For soft thick gasket, the value is between 0.75 and 1.

F_1 is initial tightening load on the bolt, and

F_2 is the load due to fluid pressure.

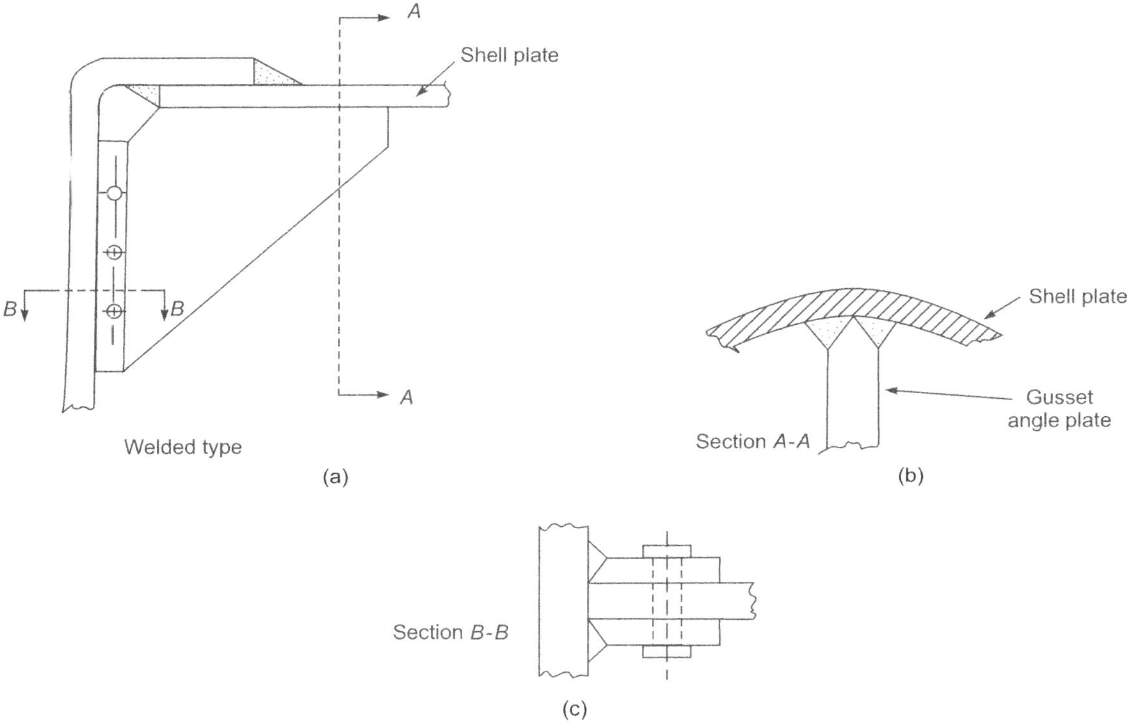

Figure 12.4 Sketch of welded type flat-end plate.

Step 2(a)

F_1 is load for pre-stressing = 284 δ in kgf (12.7)

δ is the bolt diameter of M16 bolt, for example.

Step 2(b)

$$F_2 \text{ is the load due to fluid pressure } = \frac{\pi}{4} D_e^2 p \quad (12.8)$$

where D_e is the effective diameter of the reactor cylinder

$$= d + 2t + 3\delta$$

P is the pressure of steam in the reactor cylinder

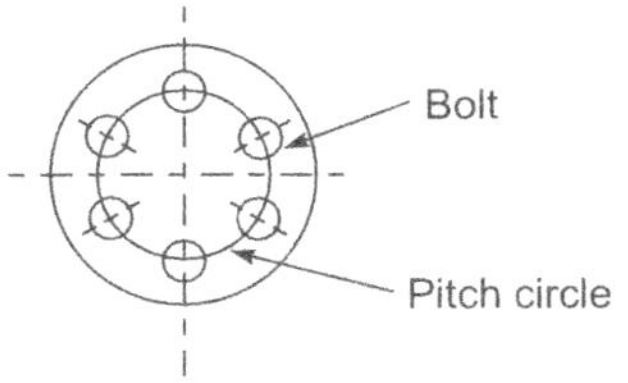

Figure 12.5 Pitch circle.

Step 3

D_P is pitch circle diameter (Fig. 12.5).

Step 4

Number of bolts,
$$n = \frac{\pi D_P}{B_P} \quad (12.9)$$

where B_P is spacing for bolt. The guideline is $3\delta < B_P < 5\delta$.

3δ is minimal for wrench operation. If spacing is more than 5δ, the joint may not be leak-proof.

If the calculated value of 'n' is obtained as a fraction, the next higher integer is considered which is at least a multiple of 4.

Step 5 Calculation of stressed area

Figure 12.6 shows the stressed area of the bolt.

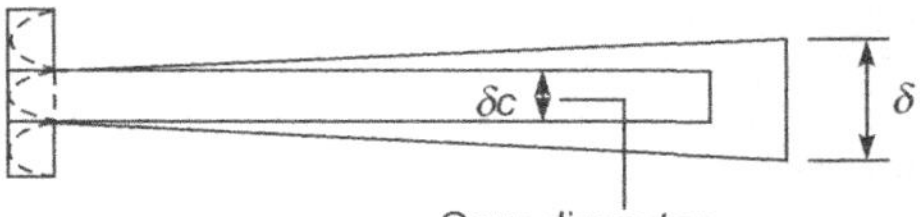

Figure 12.6 Core diameter of a bolt.

The stressed area, $$A_s = \frac{\pi}{4}\delta_c^2 = \frac{F}{n\delta_{tw}} \quad (12.10)$$

where

F is external load

δ_c is core diameter of the bolt

δ_{tw} is working tensile strength of the bolt material

n is number of bolts.

Step 6 Comparison with the standard table

The values are compared with the 'Metric Coarse Thread' chart and next higher value of A_S is considered From this value, the size of the bolt is *MXY*.

Step 7 Design of flange

Let t_1 be the thickness of the flange.

Load per bolt is expressed earlier as F/n.

$$t_1 = \sqrt{\frac{3F}{n\sigma_{bw}}} \quad (12.11$$

where σ_{bw} is working bending stress of the flange material.

12.4.6 Design of Shaft

Shaft carries impeller blades, pulleys, gears, etc. This is supported on bearings. Generally, it is rollin type distributed load. The calculation of its diameter involves two criteria, viz.,

(A) Torque is important.

(B) Bending moment is not neglected in the design calculation.

In each case, the design steps are considered separately.

(a) Design Steps for Shaft Considering Torque

Torque is important than the bending moment induced by the pulleys, gears, or agitator blades and be tension.

Step 1

Torque is calculated from metric horse power by Equation (12.12).

$$MHP = \frac{2\pi TN}{4500} \tag{12.12}$$

where T is torque in kgfm.

Step 2

(a) For solid shaft,

$$T \times 100 = \frac{\pi}{16} d^3 \tau_w \tag{12.13}$$

'd' is diameter of the shaft in cm. τ_w is working shear stress of the shaft material.

(b) For hollow shaft,

$$T \times 100 = \frac{\pi}{16}\left(\frac{d_1^4 - d_2^4}{d_1}\right)\tau_w \tag{12.14}$$

Generally, $d_1 = 1.5\ d_2$. d_1 and d_2 are outside and inside diameters, respectively.

Another value of 'd' is calculated from the angle of twist (= θ in radian) and G is modulus of rigidity.

$$\theta = \frac{2\tau_w l}{Gd} = \frac{32Tl}{\pi G d^4} \tag{12.15}$$

where

l is the length of shaft
T is torque on shaft
G is modulus of rigidity
d is shaft diameter.

Then d is calculated.

The greater value of 'd' is determined from the two calculations of 'd'.

(b) Design Steps for Shaft Considering Bending Moment

When bending moment due to impeller blades, pulleys, gears, etc. cannot be neglected, equivalent torsion moment (T_e) is calculated.

Step 1

Calculation of T_e

(a) For ductile material, using Guest's formula (Pandeya and Shah, 2006)

$$T_e = \sqrt{B^2 + T^2} = \frac{\pi}{16} d^3 \tau_w \tag{12.16}$$

where τ_w is working torsional shear stress.

(b) For brittle material, using Rankine's formula (Pandeya and Shah, 2006)

$$B_e = \text{Bending moment.}$$

$$B_e = \frac{1}{2}\left\{B + \sqrt{B^2 + T^2}\right\} \tag{12.17}$$

$$= \frac{\pi}{32} d^2 \sigma_w$$

T is calculated as per the previous section. The designed torque (= T_d) is calculated for a given shock factor.

Therefore,

$$T_d = T \times \text{shock factor.}$$

If key is with the shaft,

$$T_d = \frac{\text{T} \times \text{shock factor}}{\text{key factor}} \tag{12.18}$$

Calculation of bending moment

Forces cause bending moment to the shaft in the same place. One 'B' is calculated which is used in the above calculation. When forces are acting in two different planes, equivalent bending moment is calculated as per Equation (12.19).

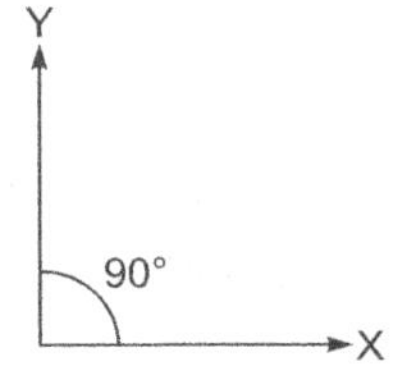

$$B = \sqrt{B_x^2 + B_y^2} \tag{12.19}$$

B_x is the bending moment in X-direction. B_Y is the bending moment in Y-direction. The general calculation of bending moment is described in the appendix to this chapter.

For any angle,

$$B = \sqrt{B_x^2 + B_y^2 + 2B_x B_y \cos\theta} \tag{12.20}$$

For the designed bending moment (B_d), due to reversal of stress, is defined by Equation (12.20).

$$B_d = B \times \text{fatigue factor } (ff) \tag{12.21}$$

For the presence of key in the shaft,

$$B_d = \frac{B \times ff}{KF} \tag{12.22}$$

where KF is the key factor.

The 'ff' and 'SF' can be considered from Table 12.4 (Pandeya and Shah, 2006).

KF is calculated from the following relation.

$$KF = 1 - 0.2w - 1.1h \tag{12.23}$$

where

w is the ratio of width of key way to the diameter of the shaft and

h is the ratio of height or depth of key way to the diameter of the shaft.

12.4.7 Design of Pin Key/Sunk Key

Sunk key is made of shaft material or of slightly softer material (Fig. 12.7).

Table 12.4 *Guidelines for shock factor and fatigue factor*

	SF	*ff*
Stationary shaft		
Gradually applied load	1	1
Suddenly applied load	1.5 – 2	1.5 – 2
Rotating shaft		
Gradually applied load	1	1
Suddenly applied load		
Minor shock	1 – 1.5	1.5 – 2
Heavy shock	1.5 – 3	1.5 – 3

Design steps are described here.

Step 1

As
$$KF = 1 - 0.2w - 1.1h$$
where

$$h = \frac{\text{height or depth of key way}}{\text{diameter of the shaft}}$$

$$w = \frac{\text{width of key way}}{\text{diameter of the shaft}}$$

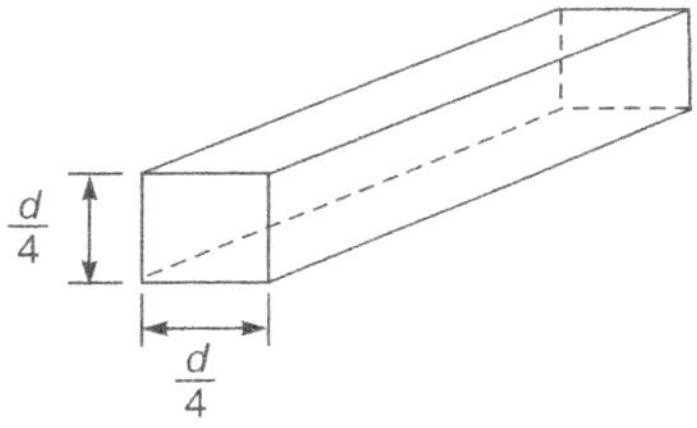

Figure 12.7 Sketch of a key.

Step 2

From
$$MHP = \frac{2\pi TN}{4500}, \text{ calculate } T.$$

Step 3

T' is the actual torque.

$\therefore$
$$T' = T(100 + n')$$
$$n' = \% \text{ of overload}$$

Step 4

$\therefore$
$$T' = p \times \frac{d}{2} \text{ and then calculate } p$$
$$p = \frac{T}{r}$$

'r' is the radius of the shaft.

Calculation of length

Two failures, viz., crushing and shear failures, may be considered for the purpose of the design.

Considering crushing failure

$$\frac{d}{8} \times l_1 \times \sigma_{cwk} = P \tag{12.24}$$

where σ_{cwk} = working crushing stress for key material

Considering shear failure

$$\frac{d}{4} \times l_2 \times \tau_{wk} = P \tag{12.25}$$

l_1 and l_2 may be calculated from the above relations. Desired length is the greater value between l_1 and l_2. If the greater value of 'l' is found to be less than 1.5d, then 'l' = 1.5d may be considered for the actual purpose.

12.5 AGITATOR ASSEMBLY

Satisfactory sealing of stirrer shaft is considered to be one of the most difficult problems in construction of a satisfactory bioreactor.

12.5.1 Drive Configuration

There are two configurations for the entry of the drive assembly to the bioreactor, viz, top entry and bottom entry. Bottom entry is advantageous for the following reasons:

- more space in head plate for the insert of different ports
- easy removal of baffles and impellers
- shorter stirrer shaft length

12.5.2 Types of Stirrer Assembly

The development of stirrer assembly is for the following reasons:

- possible leak
- jamming of shaft
- proper maintenance
- risk of contamination

They are of the following types described by Solomon (1969).

Packed Gland

Shaft is sealed by several layers of packing rings made from PTFE threads, asbestos or cotton. At high agitator speed, packing suffers quick wear.

Bush Type Seal

This is made from olite or Teflon with or without oil seals. Sometimes oil seals and felt washers are used to provide aseptic condition.

Mechanical Seal

Carbon/ceramic seals are employed for more flexible agitator assembly used in modern bioreactor.

Magnetically Driven Agitators

This avoids the needs to introduce a shaft through the vessel. PTFE coated magnetic rods are used, which are either located in bearings on the bottom.

12.5.3 Types of Agitators

Agitator serves following tasks:

- Dispersion of air and the higher transfer rates at gas/liquid interface and to extract CO_2.
- Creation of constant living conditions by homogeneous distribution of reactant, pH, temperature, etc.
- Improvement of heat transfer at the heat transfer surfaces for dissipating heat generated by the biological and mechanical work.

Important impellers are multiblade disc impellers, propeller, turbine, INTERMIG impellers, etc.

Multiblade Disc Impellers

They produce radial flow and a high energy dissipiation density in the proximity of the agitator (Kipke, 1984).

Propeller Impellers

They create an axial flow and are used for low shear forces (Narendranathan, 1986).

Turbine Impellers

They are based on the propeller impeller design.

INTERMIG Impellers

They create both radial and axial flows (Rehm, 1980).

Impellers are selected for aerobic systems for supplying sufficient oxygen. To maximize their efficiency, Taguchi and Kimura (1970) reported an empirical relation for optimized spacing of the impeller on the agitator shaft (Solomons, 1972).

$$S_m = \left[\frac{\log m}{2\{(m-1)\log m - \log(m-1)!\}}\right] \times \left[H - \left\{0.9 + \frac{(m-1)\log m - \log(m-1)!}{\log m}\right\} D_i\right] \tag{12.26}$$

where

S_m is impeller spacing
H is liquid depth
D_i is impeller diameter
m is number of impellers.

Example 12.1

Design a reactor for the enzymatic conversion of starch to glucose by the action of α-amylase, β-amylase and glucoamylase. The kinetics of glucose synthesis is given by the following equation.

$$\frac{dC_G}{dt} = \frac{V_{m_1} C_{S_0} \exp(kt)}{K_{m_1} + C_{S_0} \exp(kt)}$$

where

$$k = \frac{V_{m_1}}{K_{m_1}}$$

C_G is the concentration of glucose

Enzyme preparation is in powder form.

C_{S_0} is the initial concentration of reactant

$= 50$ % (w/v) starch solution

$V_{m1} = 5.14 \times 10^{-6}$ (mol/(ml)(min))

$K_{m1} = 9.08 \times 10^{-6}$ (mol/ml)

Maximum concentration of glucose is achieved in 4 hours. Assume other data reasonably.

Solution

Basis: 1000 lb of glucose per day

$$\text{From } \frac{dC_G}{dt} = \frac{5.14 \times 10^{-6} \times 0.5 \exp\left(\frac{5.14}{9.08} \times 4 \times 60\right)}{9.08 \times 10^{-6} + 0.5 \exp\left(\frac{5.14}{9.08} \times 4 \times 60\right)}$$

$$= 5.14 \times 10^{-6} \text{ mol/(ml)(min)}$$

Production of glucose = 1000 lb/day

i.e., 0.69444 lb/min

Calculation of volume of reactor

Production rate is achieved from the kinetic equation = 5.14×10^{-6} mol/(ml)(min)

Molecular weight of glucose = 180

Production rate for glucose = 2.027×10^{-6} lb/(ml)(min)

$$\text{Volume of reacting medium required} = \frac{0.69444}{2.027 \times 10^{-6}} \text{ ml} = 342.6\ l$$

Assuming 90% conversion of reactant to glucose, starch (reactant) entering in the reactor in 4 hours

$$= 0.5 \times \frac{342.6}{0.9} \text{ kg}$$

$$= 190.3 \text{ kg}$$

Starch solution entering in the reactor = 380.667 l in 4 h

Therefore, feed rate = 95 l/h

Weight of starch entering into the reactor = 47.5 kg/h

Catalyst (enzyme) requirement

(a) Glucoamylase is 1% of the reacting mass = 0.01 × **380.667 kg (weight/volume)**

= 3.80667 kg for one reactor volume

∴ Rate $= \frac{3.8067}{4}\frac{\text{kg}}{\text{h}}$

= 0.9516 kg/h

(b) β-amylase requirement is 0.4 % of the reacting mass.

∴ Amount of β-amylase = 0.004 × 380.667 kg for one reactor volume.

∴ Rate of β-**amylase requirement = 0.380661 kg/h.**

Assuming the ratio of α-amylase to β-amylase = 1 : 1 in the reaction mixture.

∴ Total amount of catalyst per reactor volume of reactant = (1.5226 × 2 + 3.8067) kg

= 6.8519 kg

As the density of the catalyst is not known, it is assumed that the volume of 6.8519 kg of catalyst at soluble state is 7 l.

∴ Total volume of reaction mass = (342.6 + 7) l

= 349.6 l

An allowance for extra head space of 25% of the reacting mass is assumed in this case. The total volume of the reactor = 349.6 l + (349.6 × 0.25) l

= 437 l

Input to the reactor

Flow rate of starch (reactant) $= \frac{342.6}{4}$ l/h

= 85.65 l/h

Addition of glucoamylase = 0.9516 kg/h

Addition of α-**amylase = 0.38066 kg/h**

Addition of β-**amylase = 0.38066 kg/h**

Reactor configuration

Assuming height to diameter ratio of 1 : 1,

∴ The volume of the reactor $= \frac{\pi D^3}{4}$

$= 437 \times 10^3\ \text{cm}^3$

∴ Diameter of the reactor = 82.25 cm

Height of the reactor = 82.25 cm

The reactor is an externally-jacketed vessel. Steam is passed through the jacket to maintain a constant temperature in the reaction fluid.

Calculation of shell thickness

Following data are considered for this purpose.

Design temperature = $90^\circ \pm 2\ ^\circ$C

Steam pressure = 15 psi

Since it is a jacketed vessel with 15 psi steam pressure, there is no other internal or external pressure acting on the shell. So 15 psi is considered as the design pressure.

The design pressure $= 1.086 \times 10^5$ N/m^2

Let us consider the material of construction as IS 1570 (1961). Related data useful for the design are mentioned here.

Value of elastic modulus (E) = 190,000 MN/m^2

Allowable stress (f) = 139 MN/m^2

Specific weight = 77,000 N/m^3

In the beginning, a thickness of 5 mm is assumed.

Now, an elastic and plastic deformation will determine the feasibility of the assumption.

Check for elastic failure

Using the simplified equation

$$P_e = kE\left(\frac{t}{D_0}\right)^m$$

where, P_e is the maximum allowable pressure at thickness t

k and m are constants depending on $\left(\frac{D_0}{L}\right)$,

D_i is the inside diameter of the reactor vessel

D_o is the outside diameter of the reactor vessel

In this case,

$$D_i = 82.25 \text{ cm}$$

$$D_o = (822.5 + 5) \text{ mm} = 827.5 \text{ mm}$$

$$L = 822.5 \text{ mm}$$

$$\therefore \quad \left(\frac{D_0}{L}\right) = \frac{827.5}{822.5} = 1.006$$

From the standard chart, $\frac{D_o}{L} = 1$, $k = 0.87$ and m = 2.49.

Substituting these values, we can calculate P_e from the above formula

$$P_e = 0.87 \times 190000 \times \left(\frac{0.5}{82.75}\right)^{2.49} \text{ MN/m}^2$$

$$= 4.988 \times 10^5 \text{ N/m}^2$$

P_e is more than the design pressure (= 1.085×10^5 N/m^2). Therefore, the design is safe.

Check for plastic failure

The safe design pressure (P) for a thickness 't' is given by

$$P = 2f\left(\frac{t}{D_o}\right)\frac{1}{1 + \dfrac{1.5U\left(1 - 0.2\dfrac{D_o}{L}\right)}{100\left(\dfrac{t}{D_o}\right)}}$$

where, U, the roundness factor = ± 5 % for class I vessel below 1000 mm diameter (IS 4503).

Therefore,

$$P = 2\times 139\left(\frac{0.5}{827.5}\right)\frac{1}{1+\dfrac{1.5\times 0.05\left(1-0.2\left(\dfrac{827.5}{822.5}\right)\right)}{100\left(\dfrac{0.5}{827.5}\right)}}$$

$= 0.084\ \text{MN/m}^2$

$= 8.4 \times 10^4\ \text{N/m}^2$

Since this pressure is less than the design pressure, the design is unsafe with the assumed thickness of 5 mm.

So we assume a thickness of 6 mm and check again for plastic deformation which gives a pressure of $3.042 \times 10^5\ \text{N/m}^2$.

After repeating the above calculation the thickness of 6 mm is found to be safe.

Adding 2 mm as corrosion and other allowances, the total thickness is (6 + 2) mm = 8 mm.

Hence, the reactor dimensions are

Selected shell thickness = 8 mm
Minimum inside diameter = 822.5 mm
Minimum outside diameter = 838.5 mm
Height = 840 mm

Jacket design

Jacket inside diameter should be equal to the outside diameter of the reactor plus 30 mm thickness so that a gap of 15 mm is maintained between the reactor vessel and the jacket for the passage for steam.

A thickness of 5 mm is taken without checking, as the pressure is 15 psi and the gap between the jacket and shell is small.

So the ultimate jacket dimensions are

Thickness of the jacket plate = 5 mm
Inside diameter of the jacket = 878 mm
Outside diameter of the jacket = 883 mm

Selection of the head of the reactor

A standard hemispherical dished-end head is chosen in the bottom and top enclosures. The dimensions are

Inner crown radius = 840 mm
Inner knuckle radius = 0.06 × 840 mm
= 34.4 mm

Straight flange is of 50 mm.

EXERCISES

12.1 A vertical drive of short centre type is required from an electric motor to a fermenter. The drive is transmitted to the agitator shaft by a pulley fly wheel and keyed to the shaft of the agitator. The belt is to be stretched cotton combined with rubber. It is required to design the drive using the data mentioned below:

Weight of the belt = 0.01 kg/cm^3
hp of the drive = 0.07 KW
Centre to centre distance between pulley = 200 cm.
Motor pulley diameter = 20 cm.
Rpm of the agitator = 300
Working tensile strength for the belt material = σ_{+wb} = 25 kg/cm belt width
% slip in this tension = 3%
Friction factor between belt and pulley = 0.35
Material of construction for both of the pulley = Cast iron
Belt thickness = 4 mm

Pulley on the motor side has 4 arms of elliptical section for which an assumption can be made $b = 2.5$ h. b and h are section parameters for arms.

Design the pulley on the agitator side and motor side, allowing the effect of the key way in the shaft.

12.2 A fermenter is to operate at 5 psig working pressure during the fermentation. Maximum allowable pressure is rated at 35 psig. Calculate the thickness of the end plate for the fermenter. Material of construction is SS and the reactor volume is 50 l. Do you prefer similar end plates for the top and bottom ends?

REFERENCES

Kipke KD (1984) Improvement of the fermentation parameter by directed selection of the stirring system. *Chem. Tech.* (Leipzig) **13** No. 8, 46-51.

Klapp E (Ed) (1980) *Apparate and Anlagen Technik*, Springer, Heidelberg.

Lyderson BK, D'Elia NA and Nelson KL (Eds) (1994) *Bioprocess Engineering: Systems, Equipment and facilities*, John Wiley & Sons Inc., New York.

Narendranathan TJ (1986) Designing fermentation equipment-fermenter design requirements. *Chemical Engineering*, **425**, 23–31.

Pandya NC and Shah CS (Eds) (2006) *Machine Design*, 17 th edn., Charotar Publishing House Pvt Ltd, Anand, India.

Rehm HJ (Ed) (1980) *Industrielle Mikrobiologie, 2. Aufl.*, Springer Verlag, Berlin, Heidelberg, New York.

Ruge J (Ed) (1974) *Handbuch der Schureißtechnik*, Berlin.

Solomons GL (Ed) (1969) *Materials and Methods in Fermentation*, I&L Press, Oxford.

Solomons GL (1972) Improvements in the design and operation of the chemostat. *Journal of Applied Chemistry and Biotechnology* **22**: 217-228.

Taguchi H and Kimura T (1970) Studies on geometric parameters in fermentor design. I. Effects of impeller spacing on power consumption and volumetric oxygen transfer coefficient. *Journal of Fermentation Technology*, **48**, 117.

Transparency, Form and Function: Fermenter Manufacturing – Art, Bioengineering, Ale, Dreipunut Verlag, Wald, 2006.

Vogel HC (Eds) (1983) *Fermentation and Biochemical Engineering Handbook: Principles, Process design and equipment*, Noyes Publications, New Jersey.

APPENDIX 12

General Calculation of Bending Moment

Let us take one example of a simple supported beam DE with forces applied at A, B, and C points of Figure 12.8).

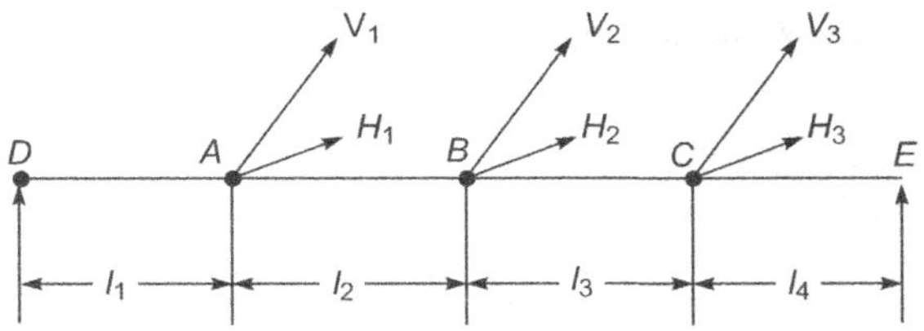

Figure 12.8 Forces acting on simple supported beam.

Two components of forces applied at A, B, and C points are shown as vertical (V) and horizontal (H forces.

$$A \rightarrow (V_1, H_1)$$
$$B \rightarrow (V_2, H_2)$$
$$C \rightarrow (V_3, H_3)$$

We consider resultant force (R_{DH}) in the horizontal direction.

$\therefore$ Bending moment at

A $\quad R_{DH} \times l_1$

B $\quad R_{DH} \times (l_1 + l_2) - H_1 l_2$

C $\quad R_{DH} \times (l_1 + l_2 + l_3) - H_1(l_2 + l_3) - H_2 l_3$

Same procedure is adapted for vertical forces.

Therefore, the resultant bending moment at each point is described here.

at A $\quad \sqrt{B_{H_A^2} + B_{V_A^2}}$

at B $\quad \sqrt{B_{H_B^2} + B_{V_B^2}}$

and at C $\quad \sqrt{B_{H_C^2} + B_{V_C^2}}$

Index

AUTHOR'S PROFILE

Professor Tapobrata Panda is currently with IIT Madras. He was with the IIT Kharagpur before moving to the IIT Madras in 1987. He received his Ph.D. in Biochemical Engineering at the Indian Institute of Technology Delhi, M. Tech. and B. Tech. in Food Technology and Biochemical Engineering from Jadavpur University, Kolkata, and B.Sc. in Chemistry from University of Calcutta. Prof. Panda carried out his advanced research in protoplast system in Technical University of Vienna in 1985. As a Visiting Scientist in the Department of Chemical Engineering, Iowa State University, Ames, in 1994, Dr. Panda studied on site-directed mutagenesis under Indo-US program. He received first All India Biotech Association Award (AIBA, Delhi) in 1998, besides other distinctions in the academic programs. He was Visiting Faculty to the Asian Institute of Technology, Bangkok under deputation from IIT Madras.

From the research team of Prof. Panda at IIT Madras, 20 Ph.Ds have graduated by contributing in the areas of enzyme systems, kinetics, process optimization, and development of microbial products. The h-index(web of science) of his publication is 19. Also, his research group has been extensively involved in research on the BioMEMS, biological synthesis of nanoparticles, and the design of therapeutic molecules. His collaborative research with the Korea Research Institute of Biosciences and Biotechnology has contributed an important avenue on esterase. Prof. Panda is a member of the Editorial Board of The Open Biotechnology Journal (Bentham Science Publisher) and Open Enzyme Inhibition Journal. He was a member of the TAPPI, USA.

Made in the USA
Monee, IL
07 March 2025

13638541R00282